Driving Agribusiness With Technology Innovations

Theodore Tarnanidis
University of Macedonia, Greece

Maro Vlachopoulou
University of Macedonia, Greece

Jason Papathanasiou
University of Macedonia, Greece

A volume in the Advances in Business Strategy and Competitive Advantage (ABSCA) Book Series

IGI GLOBAL
DISSEMINATOR OF KNOWLEDGE

www.igi-global.com

Published in the United States of America by
IGI Global
Business Science Reference (an imprint of IGI Global)
701 E. Chocolate Avenue
Hershey PA, USA 17033
Tel: 717-533-8845
Fax: 717-533-8661
E-mail: cust@igi-global.com
Web site: http://www.igi-global.com

Library of Congress Cataloging-in-Publication Data

Names: Tarnanidis, Theodore, 1977- editor. | Vlachopoulou, Maro, 1956- editor. | Papathanasiou, Jason, editor.
Title: Driving agribusiness with technology innovations / Theodore Tarnanidis, Maro Vlachopoulou, and Jason Papathanasiou, editors.
Description: Hershey : Business Science Reference, [2017]
Identifiers: LCCN 2016054475| ISBN 9781522521075 (hardcover) | ISBN 9781522521082 (ebook)
Subjects: LCSH: Agricultural industries. | Technological innovations.
Classification: LCC HD9000.5 .D68 2017 | DDC 338.1--dc23 LC record available at https://lccn.loc.gov/2016054475

This book is published in the IGI Global book series Advances in Business Strategy and Competitive Advantage (ABSCA) (ISSN: 2327-3429; eISSN: 2327-3437)

British Cataloguing in Publication Data
A Cataloguing in Publication record for this book is available from the British Library.

Advances in Business Strategy and Competitive Advantage (ABSCA) Book Series

Patricia Ordóñez de Pablos
Universidad de Oviedo, Spain

ISSN:2327-3429
EISSN:2327-3437

Mission

Business entities are constantly seeking new ways through which to gain advantage over their competitors and strengthen their position within the business environment. With competition at an all-time high due to technological advancements allowing for competition on a global scale, firms continue to seek new ways through which to improve and strengthen their business processes, procedures, and profitability.

The **Advances in Business Strategy and Competitive Advantage (ABSCA) Book Series** is a timely series responding to the high demand for state-of-the-art research on how business strategies are created, implemented and re-designed to meet the demands of globalized competitive markets. With a focus on local and global challenges, business opportunities and the needs of society, the **ABSCA** encourages scientific discourse on doing business and managing information technologies for the creation of sustainable competitive advantage.

Coverage

- Strategic alliances
- Co-operative Strategies
- Small and Medium Enterprises
- Adaptive Enterprise
- Resource-Based Competition
- Core Competencies
- Differentiation Strategy
- Value Chain
- Ethics and Business Strategy
- Entrepreneurship & Innovation

IGI Global is currently accepting manuscripts for publication within this series. To submit a proposal for a volume in this series, please contact our Acquisition Editors at Acquisitions@igi-global.com or visit: http://www.igi-global.com/publish/.

The Advances in Business Strategy and Competitive Advantage (ABSCA) Book Series (ISSN 2327-3429) is published by IGI Global, 701 E. Chocolate Avenue, Hershey, PA 17033-1240, USA, www.igi-global.com. This series is composed of titles available for purchase individually; each title is edited to be contextually exclusive from any other title within the series. For pricing and ordering information please visit http://www.igi-global.com/book-series/advances-business-strategy-competitive-advantage/73672. Postmaster: Send all address changes to above address.

Titles in this Series

For a list of additional titles in this series, please visit: www.igi-global.com/book-series

Handbook of Research on Small and Medium Enterprises in Developing Countries
Noor Hazlina Ahmad (Universiti Sains Malaysia, Malaysia) T. Ramayah (Universiti Sains Malaysia, Malaysia) Hasliza Abdul Halim (Universiti Sains Malaysia, Malaysia) and Syed Abidur Rahman (Universiti Utara Malaysia, Malaysia)
Business Science Reference • copyright 2017 • 479pp • H/C (ISBN: 9781522521655) • US $255.00 (our price)

Technology-Driven Productivity Improvements and the Future of Work Emerging Research and Opportunities
Göran Roos (University of Adelaide, Australia)
Business Science Reference • copyright 2017 • 255pp • H/C (ISBN: 9781522521792) • US $175.00 (our price)

Entrepreneurship and Business Innovation in the Middle East
Philippe W. Zgheib (Lebanese American University, Lebanon)
Business Science Reference • copyright 2017 • 357pp • H/C (ISBN: 9781522520665) • US $190.00 (our price)

Comprehensive Problem-Solving and Skill Development for Next-Generation Leaders
Ronald A. Styron, Jr. (University of South Alabama, USA) and Jennifer L. Styron (University of South Alabama, USA)
Business Science Reference • copyright 2017 • 389pp • H/C (ISBN: 9781522519683) • US $205.00 (our price)

Diasporas and Transnational Entrepreneurship in Global Contexts
Sanya Ojo (University of East London, UK)
Business Science Reference • copyright 2017 • 334pp • H/C (ISBN: 9781522519911) • US $200.00 (our price)

Integrating Art and Creativity into Business Practice
Giovanni Schiuma (University of Basilicata, Italy & University of the Arts London, UK) and Antonio Lerro (University of Basilicata, Italy)
Business Science Reference • copyright 2017 • 260pp • H/C (ISBN: 9781522520504) • US $170.00 (our price)

Sustainable Entrepreneurship and Investments in the Green Economy
Andrei Jean Vasile (Petroleum-Gas University of Ploiesti, Romania) and Domenico Nicolò (University Mediterranea of Reggio Calabria, Italy)
Business Science Reference • copyright 2017 • 339pp • H/C (ISBN: 9781522520757) • US $185.00 (our price)

IGI GLOBAL
DISSEMINATOR OF KNOWLEDGE
www.igi-global.com

701 East Chocolate Avenue, Hershey, PA 17033, USA
Tel: 717-533-8845 x100 • Fax: 717-533-8661
E-Mail: cust@igi-global.com • www.igi-global.com

Table of Contents

Detailed Table of Contents

Efficiency assessment in agriculture is a research field were quite important methodologies have been implemented. Data Envelopment Analysis (DEA) in one of the most recognized approaches due to the considerable advantages of it. In this paper the implementation of DEA Window analysis assesses efficiency scores of the primary sectors of EU member states on both operational and environmental level, verifying considerable efficiency differences among them and a continuous improvement after the application of the latest Common Agricultural Policy (CAP) reform. Regarding prognostication of crop and animal output, as well as Green House Gas (GHG) emissions, the application of Artificial Neural Networks (ANNs) is being proposed, succeeding satisfactory quality characteristics for the models being proposed for operational and environmental predictions in EU agriculture.

Especially in the food sector, fraud and counterfeiting are affecting the trust of consumers, who are more and more oriented to chose products basing on quality and traceability attributes rather than the price. Recently, the Electronic Product Code Information Services (EPCIS) standard was introduced to provide specifications for the representation of product traceability information. The collection and analysis of such information allows supply chains to be monitored and controlled through virtualization. Several applications of EPCIS were presented in literature, even if most of them are mainly focused on enabling technologies, with less emphasis on assessing how the available information can be used for a control at a higher level. This chapter review the relevant literature available on this topic, and present an architecture allowing the traceability of information about products throughout the entire supply chain by exploiting both the EPCIS standard and a NoSQL database. An application showing the potentiality of the proposed system in a case study is also reported.

Consumers purchase food from different sources, mainly via traditional/long chains where hypermarkets are the final link between producer and consumer. However, consumers are seeking direct relationships with producers. This, together with the increase of social media usage offer producers the potential to build short chains for promoting/selling their products. The aim of this work is to summarize the role that online short food supply chains could play as an opportunity for SMEs in the agri-food sector. Moreover, it highlights a new perspective based on social media as potential short supply chains. To this end, a thorough review of the literature has been carried out, together with an online survey where social networks as food marketing channels have being studied. The chapter concludes pondering about the different food products/sectors that could take advantage of the creation of short supply chains and of the wider use of social networks as marketing tools.

Local food production is becoming increasingly popular in developed post-modern economies. Attention has been directed to developing such forms of food supply by adapting information connectivity. A case study of a local food network in Norway indicates that local food supply paradoxically attempts to mimic the dominant industrialised modes of food production. It is suggested that the fact that local food supply is "personal" and associated with close proximity makes it more closely resemble service supply chains. Applying contingency theory, a conceptual model is developed that indicates how the local food supply must take into consideration the degree to which customer value is associated with tailoring food supply. The high need for tailored local food production implies that information connectivity should support mutual adaptation while, in cases of less need for tailoring information, connectivity should seek automation. Local food production is always a hybrid of these approaches.

The shift in agricultural production and agribusiness may be a solution in reducing unemployment and particularly that of young people which is dramatically high in several European countries that are experiencing the negative consequences of the recent global financial crisis that led to a dramatic decline in their GDP per capita and has affected all sectors of economic activity, including agriculture. The overall scope of this chapter is to present an Agricultural Entrepreneurship and Social Innovation Framework that can lead to a new business model with social aspects, contribute to the economic growth and sustainability and hence combat the phenomenon of unemployment and poverty in rural areas that have been seriously affected by the recent financial crisis.

However, these oscillations can be smoothed out if the adjustment time of the order placement is increased. Under an extreme event that reduces the demand for milk, behavior changes and the system remains in dis-equilibrium for the entire simulation. Once again, adjustment times remain the leverages that can influence and mitigate those negative effects. Finally, a more robust and collaborative decision-making process among the actors of the chain could be beneficial for all not only under normal circumstances, but also in the presence of extreme uncertainty.

Preface

Information and Communications Technologies (ICT) deal with the advancement of pioneering technologies. Specifically, in the context of business sector, the term refers to the automation of business processes, by manipulating efficiently all the necessary information (i.e. to acquire, to store and to analyze). E-commerce developments and access of information via the World Wide Web helped firms to restructure their traditional mechanisms. Thus, more and more businesses, around the globe seek to implement innovative mechanisms through the use of digital environments.

Initially the evolution of ICT has been used to industrialized economies and to specific sectors (i.e. information and technology producing industries). Latest ICT was expanded to less developed countries and across all sectors of the economy. While there are many books available on digital environment, we found that most of those try to address the necessary steps for Agri-entrepreneurs in order to take advantage of the electronic and internet-based mechanisms. However, to date, to the best of our knowledge, in the agriculture-business economies, and especially in less developed countries, no contributions were found in the economic literature on ICT that address this issue.

We argue that agricultural growth depends on rural and social development and it remains central to poverty reduction. Moreover, agriculture-based economies face huge problems, stemming from land and water constraints, pure rural infrastructure, and pure innovative mechanisms, lack of public and private investments and volatile climate and weather conditions. Hence, technology innovations in agribusiness provide a plethora of benefits, such as improved data/information/knowledge sharing, cost reduction, upgraded exchanges and coordination management. Similarly, the agriculture sector is facing continuous changes at a local, national and international level. Firms are mainly interested in producing and selling goods to the consumer global markets based on the creation of mutual beneficial partnerships with several intermediaries (i.e. suppliers and other funding bodies). The smooth integration of new technologies transforms the existing and planned ICT platforms (i.e. production, logistics, and other marketing and management processes) on the basis of creating long-term gains in the new digital environment. Hence, the main objective of the book is to check whether companies in the agribusiness sector take advantage of the open networks and advances in information and communication technologies. In today's digital economy web-based applications are pertinent for businesses to energize their core competencies and capabilities.

We have organized this book into two main parts. The first group of chapters concerns with the use of information communications technologies in the Agribusiness sector. Whereas, the second group of chapters develops effective techniques and tools for emerging case studies in the specific study domain.

The target audience of this book consists of professionals and researchers working in the field of information and knowledge management in various disciplines, e.g. information and communication sciences, administrative sciences and management, sociology, computer science, and information technology. And also, we hope readers will find this book essential for upgrading their existing work and education. Finally, we like to express our special appreciation to all contributors of the accepted chapters.

Section 1

Agribusiness Productivity–Innovation–Sustainability–Environment

Chapter 1
Assess and Prognosticate Operational and Environmental Efficiency of Primary Sectors of EU Countries:
Implementation of DEA Window Analysis and ANNs

George Vlontzos
University of Thessaly, Greece

Panos M. Pardalos
University of Florida, USA

ABSTRACT

Efficiency assessment in agriculture is a research field were quite important methodologies have been implemented. Data Envelopment Analysis (DEA) in one of the most recognized approaches due to the considerable advantages of it. In this paper the implementation of DEA Window analysis assesses efficiency scores of the primary sectors of EU member states on both operational and environmental level, verifying considerable efficiency differences among them and a continuous improvement after the application of the latest Common Agricultural Policy (CAP) reform. Regarding prognostication of crop and animal output, as well as Green House Gas (GHG) emissions, the application of Artificial Neural Networks (ANNs) is being proposed, succeeding satisfactory quality characteristics for the models being proposed for operational and environmental predictions in EU agriculture.

DOI: 10.4018/978-1-5225-2107-5.ch001

INTRODUCTION

It is a continuous goal of the European Union (EU) Common Agricultural Policy (CAP) to improve both operational and environmental efficiency of agricultural holdings, aiming by this way to increase the competitiveness of EU primary sectors as a whole in a globalised production and trading framework. The quantification of this approach is being expressed by the 20-20-20 strategy which focuses on increasing the energy efficiency by 20%, reducing the CO_2 emissions by 20% and produce 20% of overall energy consumed by renewable energy resources (European Commission, 2011). One of the most important policy reforms for the EU agriculture was the implementation of the Agenda 2000, with the establishment of a totally new framework for subsidies management, decoupled from both crop and animal production. Since the year 2005 the new subsidy scheme has come into force, providing by this way the ability to the EU to fully comply with the last World Trade Organization (WTO) agreement of the Uruguay Round (European Commission, 2013). Under this new framework, the subsidy scheme has a pure supportive role on the producers' income, increasing by this way the impact of their managerial decisions on the improvement of efficiency of their holdings.

Up to now, the implementation of Data Envelopment Analysis (DEA) has contributed substantially towards this goal.The non-parametric approach of this methodology, in accordance with the absence of a priori assumptions formulates a framework where it is easily applicable. Another quite important advantage of it is the ability to use multiple inputs and outputs for efficiency assessment, increasing by this way the objectivity of the results being obtained when handling real life tasks. Therefore, it has been used for efficiency assessment of very important sectors of the economy, like banking and health care. Nevertheless, it has also been used in various cases in agriculture, as well as in the food industry, trying to evaluate the efficiency rates of inputs used, as well as the outputs achieved. For almost every sector of the economy there is a variation of efficiency goals, beyond the typical economic ones. Consumers demand for the adoption of environmentally friendly production methods, in addition with the need for a continuous increase of energy efficiency, outlined new perspectives for the use of DEA with fairly successive research results. The European primary sector, especially after the implementation of the Agenda 2000, has put a lot of effort to meet these goals, by applying a series of policies focused on improving the environmentally friendly profile of production methods, as well as increasing their energy efficiency. Due to the fact that it is important for policy makers to detect efficiency trends overtime, the DEA Window analysis is a suitable tool for such assessments. It has been used for rather diversified economic activities, verifying by this way the validity of the method and its acceptance as a research tool by researchers (Yang and Chang, 2009; Pjevcevic *et al*, 2012). In this paper it is used to assess efficiency and identify efficiency change of EU countries primary sectors in operational, and environmental terms, as the outcome of significantly diversified agricultural policies, quantifying by this way their positive or negative impact, providing at the same time hints for counteractive actions.

Quite important is also the ability of policy makers to assess the level of success of policies being planned, before their implementation. Throughout the years it has been proven that this is not an easy task, due to the fact that this level of success is heavily depended on the assumptions being made as well as the suitability of the models being used for such estimations. Regarding agriculture, there are models focusing on the impact of policies on agricultural trade and development, as well as other tasks like biophysical and environmental ones. All these widely recognised models are based on various mathematical methodologies, providing useful information for significant issues of agriculture, like land management, agricultural trade and agricultural income. There are though other prognostication methodologies, being already used in many scientific fields with significant success, like the Artificial Neural Networks (ANNs). In this paper ANNs are being used as a tool to estimate future performance of EU countries primary sectors in both operational and environmental terms.

BACKGROUND

DEA has been used for many decades, when Farrell (1957) stated the problem of measurement of productive efficiency. Based on these ideas Charnes et al (1978) introduced DEA as an adequate methodology for quantifying relative deficiencies of multi-input and multi-output production units. The great acceptance and usefulness of DEA is proved by the use of it for efficiency assessment of very important production sectors of the economy, even nowadays (Cook and Seiford, 2008). The most important utilities of DEA are the use of peer groups, the identification of efficient operating practices, the setting of targets, the identification of efficient strategies, the monitoring of efficiency changes over time, and resource allocation (Boussofiane *et al*, 1991). One of the first implementations of this was for the banking sector (Charnes *et al*, 1978; Thanassoulis, 1999). Similar approaches there were for efficiency evaluation for schools (Smith and Mayston, 1987; Thanassoulis and Dunstan, 1994) with satisfactory results. Another very important sector for economies is the energy production one. Special research focus has been given on the electric power plants efficiency on both operational and environmental terms (Sozen *et al*, 2010; Arabi *et al*, 2014). Additionally, DEA has been used for evaluation of ports, prevailing by this way best management practices in a highly competitive sector of international economy (Cullinane *et al*, 2006).

Agriculture and DEA have quite close bonds, because up to now it has been used in various ways, presenting by this way best management practices. Combinations of inputs and outputs being used focus on physical, economic and environmental aspects of production, trying to identify the best mixture of them leading to efficiency measurements, as well as the impact of these categories on efficiency scores. In one of the most competitive agricultural sector, the dairy one, two different DEA models focusing on physical and economic inputs and outputs. Comparison of results demonstrated that it is more important to combine in efficient way both physical and economic resources than focusing on physical output maximization (Stokes, 2007). A similar study identified efficiency scores of different combinations of management practices and feeding (Heinrichs *et al*, 2013; Hansson and Ohlmer, 2008). In the same sector when external operational parameters combined with internal operational characteristics and micro-social issues used to assess efficiency, farm size and management can be either a constraint or a driving force (Hansson, 2007).

The increasing awareness for environmental protection has driven research towards assessing the impact of inputs being used in agriculture on eco-efficiency. Several studies on specific crops assessed eco-efficiency and presented the linkages between eco-inefficiency and input management. Spanish olive growers were proven to be quite eco-efficient with inefficiencies to be closely related with technical inefficiencies. Eco-efficiency was boosted via implementation of agri-environmental projects like university education (Picazo-Tadeo *et al,* 2010; Picazo-Tadeo *et al,* 2012). Eco-efficiency is closely related with land use management too (Kuosmanen and Kortelainen, 2005). The use of DEA for olive trees cultivation provided the ability to measure inefficiencies related with resources management like land and water, in Andalusia were especially water availability is a crucial issue for both inhabitants and cultivations (Gomez-Limon *et al*, 2011). Quite vital issue for farming is labour management too. Application of DEA on citrus cultivation lead to specific alternatives focusing on efficiency improvement in areas where small size of agricultural holdings is a major issue, which is the case in many Mediterranean countries (Martinez and Picazo-Tadeo, 2003). On the same trend for assessing environmental efficiency, a combination of Life Cycle Assessment (LCA) and DEA has been used by researchers regarding agricultural production. LCA is a tool for estimating the environmental impacts of a process or a product. DEA implementation by using LCA results can lead to super efficiency analysis to simplify the selection process of reference performers, which is essential in a benchmarking process. Additionally, there is the ability for inter- and intra- assessment of multiple data sets, because for every process the energy impact has been calculated and used. Furthermore, sustainability issues can be correlated with the economic dimension of every activity being incorporated in these models and the application of DEA Window analysis provide the ability to quantify environmental efficiency variation for a specific time period (Iribarren *et al*, 2010). Applications of this methodology can be found for mussel production, where the DEA targets being obtained can be utilised as virtual cultivation sites with less input use and more output production achieved simultaneously (Lozano *et al*, 2008). Interesting findings were attained when this methodology was used to assess both operational and environmental efficiency for fisheries. This combination is suitable in such cases were multiple input/output data should be used, providing at the same time the ability of not using standard deviations which is usually the case when working with average inventories (Vazquez- Rowe *et al*, 2010). One of the most competitive sectors in animal production is the cow milk one (Silva and Stefanou, 2003). The application of LCA and DEA provided very useful and applicable results, focusing on reducing the operational cost of dairy farms, as well as improving their environmental footprint (Iribarren *et al*, 2011). There are though successful applications in crop production too. LCA and DEA of grape production and vinification presented quantified inefficiencies on both operational and environmental level. In NW Spain a necessity for 30% on average on inputs reduction was assessed, leading to an increase of 28%-39% of environmental gains depending on the impact category (Vazquez- Rowe *et al*, 2012). The applicability of this methodology was verified for arable crops cultivation too. Regarding soybean farming in Iran 46% farms of the sample were found as efficient. The most important contributors to global warming were irrigation and fertilization by 63% and 34% respectively, providing a road map for both efficiency improvement and minimisation of environmental footprint (Mohammadi *et al*, 2013). It has been proved though that DEA methodology autonomously implemented to assess environmental efficiency is a widely accepted approach. Significant advantages of it are the accuracy of results for small data sets and the ability to include undesirable outputs and inputs (Song *et al*, 2012). The continuous focus of both agricultural policies and consumers on environmental issues and the impact of production procedures on them motivate all the participating parties to assess this impact and put serious effort on improving environmental efficiency. Following this rational DEA was used to assess energy efficiency of wheat farms, by separating efficient from inefficient farmers and calculate quantities of inputs being used in a wasteful way. The most important findings originated that only 18% of growers were technically efficient, with the overall technical efficiency to be 0.82. It has been observed also that by implementing energy optimisation the total GHG emissions can be reduced substantially (Khoshnevisan *et al*, 2013). A similar case is the alfalfa production. In this, 46% of growers were found technically efficient, with the average technical efficiency to be 0.84. Optimisation of energy use improved the energy use efficiency by 10.6% (Mobtaker *et al*, 2012). The energy cost of greenhouse vegetable production is the most important one, affecting directly feasibility and competitiveness of agricultural holdings. An integrated input-output analysis measured the energy efficiency of greenhouses producing vegetables like tomatoes and cucumbers. Inputs substantially affecting energy costs are diesel fuel and fertilizers. Quite important is also the energy ratio for the two cultivations, which is 0.69 and 1.48 respectively. In pure economic terms it is indicated that tomato cultivation is more profitable, compared with the cucumber one (Heidari and Omid, 2010).

In the same area, implementation of DEA for the determination of energy efficiency in greenhouse cucumber production, calculated technical efficiency, with 27% of the sample being efficient. In this study CO_2 emissions were included as the major GHG undesirable output (Khoshnevisan *et al*, 2013). Energy use efficiency in greenhouse tomato production is different, compared with the cucumber one, with technical efficiency scores to be on average 0.94, signifying the increased competitiveness of the sector. Regarding energy efficiency, about 25.15% of the total input energy could be saved without reducing tomato yield (Pahlavan *et al*, 2011). The most intensive cultivation in greenhouses is floriculture. Efficiency assessment in rose production is vital for agricultural holding specialised on this, because possible inefficiencies have a direct impact on competitiveness. Such an assessment demonstrated average technical efficiency up to 0.83 and input energy savings of about 43.59% on average can be achieved without reducing rose yield. This percentage can be considered as very significant (Pahlavan *et al*, 2012). The same methodological approach was implemented for assessment of energy efficiency of grape production. The main differences between efficient and inefficient farms were focus on the use of chemicals, diesel fuel and water for irrigation. Education level is positively related with high efficiency scores (Khoshroo *et al*, 2013)

The efficiency issue is not only important on a managerial level, but it is a main issue for policy assessments too. Policy makers seek to plan and implement tools aiming in many cases to improve economic and environmental performance. Therefore, the problem of emission permits reallocation was reached by the implementation of DEA. The most important advantages of, the non-parametric approach, avoids the necessity of gathering information about input and output prices. The first approach was applied for the paper industry in Sweden (Lozano *et al*, 2009). The same methodology was used for reallocation of emission permits for the 15 EU member states regarding agricultural GHGs. The results verified that the reduction and reallocation mechanism applied was fair, benefiting by this way countries operating up or very close to the efficient frontier (Wu *et al*, 2013). The impact of CAP on farming efficiency is a continuous issue for both farmers and EU policy makers. DEA use to olive-growing farms proposed an allocation system for subsidies, having in mind the Agenda 2000 framework. Farm efficiencies were calculated by decomposing DEA scores by means of internalising both positive and negative externalities of agricultural activity (Amores and Contreras, 2009). The EU 2003 CAP reform was one of the most important reforms of CAP since it was first established. One of the major goals of this was the enhancement of the environmental consequences of farming in Europe. Simulation results indicated reductions of production costs and improvement of environmental conditions in rural areas regarding soil, water and GHGs when compared to a scenario without this reform. These indications though have to be verified (Schmid *et al*, 2007; Schmid and Sinabell, 2007). After the last large enlargement of the EU one of the most significant challenges of CAP was to unify farmers' attitudes towards management practices. The 2003 reform was a new policy environment for every member state. More specifically, farmers from new member states was found to be more opposed to policy liberalisation practices, compared with the old ones, declaring at the same time their preference to more agricultural oriented policy tools (Gordon *et al*, 2008). Therefore, it is important to assess and quantify possible differences between old and new member states efficiency, as the outcome of the implementation of the new liberalised policy framework. For many decades CAP was the driven factor for land use management due to the coupled subsidy management system. The decoupling of payments increases the degrees of freedom of farmers' decision making process for their farm activities. This major change was expected to have an impact on land use, but it was found that only 30% of farmers on average in Germany, Portugal and the UK had decided to differentiate their farm activities mix (Tranter *et al*, 2007). The use of DEA for policy efficiency measurement has proved to be a quite appropriate modelling tool, being applied up to now in various cases. Assessment of regional inefficiencies for industry sectors in an important case study, calculating efficiency scores of lead sectors as an evaluation perspective of their future competiveness (Dinc and Haynes, 1999). Regarding development policies,it is accepted that public investments,mainly in infrastructure, aim to attract private investments. Efficiency assessment of such public policy was calculated by the use of DEA identifying investment mixtures attracting successfully private investments (Karkazis and Thanassoulis, 1998; Abello *et al,* 2002; Papajorgji and Pardalos, 2005; Zopounidis and Pardalos, 2010). Focusing on rural development policies, DEA application quantified the impact of them on employment generation in rural areas, being at the same time a useful tool for reallocation of resources among different areas to maximise by this way policy efficiency (Vennesland, 2005). The same method when applied for the evaluation of local actions for LEADER+ project in Greece identified inefficiencies regarding inputs use and proposed corrective alternatives aiming to increase the total efficiency of this project (Vlontzos *et al*, 2014).

Another very important and promising methodology for both performance assessment and prognostication is the Artificial Neural Network (ANN). Regarding performance, a comparison between DEA and ANNs using a data set from the banking sector, demonstrated that although there are among them significant differences, both of them offer quite useful information (Athanassopoulos and Curram, 1996). ANNs have been used for predicting purposes for various economic activities. In agriculture quite important is to establish models for predicting yields, turnovers, and recently undesirable outputs like GHG emissions. A recent study presented an ANN model for predicting wheat yield and GHG emissions having a 11-3-2 structure, with R^2 0.99 and 0.998 for yield and GHG emissions respectively (Khoshevisan *et al*, 2013). On the same trend, ANNs were used to build models for prognostication of environmental parameters in potato production. ANN model having 11-10-6 structure achieved the best performance for this purpose (Khoshevisan *et al*, 2013). Another case study of ANNs for predicting greenhouse basil production determined satisfactory results, having a 7-20-20-1 structure and R^2 of 0.976 (Pahlavan *et al*, 2012). A more policy oriented use of ANNs was applied for cropland change in Romania. This application allows land-change scientists to identify the spatial determinants based on the observed changes and to manage complex factional relationships coexisting in agricultural production process (Lakes *et al*, 2009).

Especially for CAP, prognostication of the impact of various reforms being planned was and is a continuous goal. For this reason the Global Trade Analysis Project (GTAP) applied general equilibrium model has been used taking into consideration the GTAP global data base (Hertel, 1997; McDougall, 1998). In the case of the EU enlargement towards central and eastern European countries the model implementation projected increased agricultural production for these countries and significant financial transfers from EU taxpayers to the central and eastern European farmers. The macroeconomic costs for the EU were found to be limited (Bach *et al*, 2000). Focusing on agricultural land management issues, the combined implementation of GTAP and the biophysical (IMAGE) model for the EU after the enforcement of the 2003 CAP reform showed that there will be no drastic decrease on agricultural land use will occur in the next 30 years due to increased demand for food globally. On the contrary, significant changes on land use are expected in developing areas like Africa (Meijl *et al*, 2006). Another modelling approach is the International Model for Policy Analysis of Agricultural Commodities and Trade (IMPACT) model. It has been developed by the international food policy research institute focusing on connecting food supply chain and water supply and demand (Rosegrant *et al*, 2008). More recent attempts to simulate farm operation in regards of policy recommendations lead to the Farm System Simulator (FSSIM). This is a bio-economic farm model linking micro and macro analysis of farming systems in specific regions (Louhichi *et al*, 2010).

Taking into consideration these methodologies for both assessment and prediction, it is evident that DEA is widely recognised as a powerful and appropriate tool. Therefore, it will be used for CAP assessment treating EU member states as DMUs. Due to the fact that CAP is implemented continuously, DEA Window Analysis will be used, in order to assess and quantify not only the operational and environmental efficiency, but efficiency change on a yearly basis too. Additionally, a new forecasting methodology is being proposed to predict outputs on both operational and environmental level, by the implementation of ANNs. It will be applied separately on an operational and environmental basis, attempting by this way to identify which methodological approach is simpler, regarding ANN structure, and which model performs better for prognostication purposes.

METHODOLOGICAL APPROACH

DEA is a non-parametric model where there is no requirement for a priori specification of inputs and outputs in production function. Every production unit is called Decision Making Unit (DMU) which uses m inputs to produce s outputs. The data set consists of n DMUs. The efficiency of every DMU is measured by using the following model:

$$\max \varphi + \varepsilon\left(\sum_{i=1}^{m} s_i^- + \sum_{r=1}^{s} s_r^+\right)$$

Subject to

$$\sum_{j=1}^{n} \lambda_j x_{ij} + s_i^- = x_{io} \quad i=1,2,\ldots,m$$

$$\sum_{j=1}^{n} \lambda_j y_{rj} - s_r^+ = \varphi y_{ro} \quad r=1,2,\ldots,s;$$

$$\lambda_j \geq 0 \quad j=1,2,\ldots,n$$

The efficient target is

$$\hat{x}_{ij} = x_{io} - s_i^{-*} \quad i=1,2,\ldots,m$$

$\hat{y}_{ij} = \varphi^* y_{io} + s_r^{+*}$ r=1,2,…,s, φ unrestricted in sign

The Technically Efficient Capacity Utilization (TECU), based on observed output (u) is:

$TECU = u/u^* = u/\varphi u = 1/\varphi$

with 0<TECU<1

The DEA CCR model assumes constant returns to scale, meaning that observed input-output combinations can scale up or down proportionally. If the objective is to identify efficiency variations for a specific time period, the implementation of DEA Window Analysis is an appropriate approach. Under this, each DMU is considered as different for each period and the data used in this is panel data. Every EU country is being considered as DMU. The time period examined is from 1998 to 2012, covering seven years before the implementation of the decoupled scheme of subsidies management and eight years after. Therefore, the final number of DMUs for this DEA Window analysis is 28 EU member states X 15 years=420. The window length is set to five (5) years, with 28*5=140 efficiencies to be evaluated for each window. According to the literature the most important inputs in the primary sector are the production factors used which are agricultural land, labor, capital, and the energy inputs like agrochemicals, fertilizers and fuels in addition with electrical power. Agricultural production has both desirable and undesirable outputs. The most important ones are crop and animal production as well as Green House Gas (GHG) emissions. The following tables present the basic statistics of all inputs and outputs being used. The data sets used were extracted from EUROSTAT, followed by own calculations, in order to be appropriately formulated for this analysis.

Table 1. Basic statistics of inputs

	Agricultural Land	Chemicals	Energy	Fertilizers	Fixed Capital Consumption	Labour
Medium	6,660.1	358.8	770.2	516.0	1,791.3	1,199.8
Standard Deviation	11,236.78	596.26	1049.89	1180.28	1800.68	1859.82
Max	35,177.8	3,021.5	4,502.7	4,604.5	12,377.4	7,307.4
Min	9.7	0.5	5.4	1.0	3.8	3.2

Source: Eurostat

Table 2. Basic statistics of outputs

	Animal Output	Crop Output	GHG Emissions
Medium	5,032.9	6,624.2	17.8
Standard Deviation	8,658.58	5,695.11	28.65
Max	25,987.7	44,407.2	100.5
Min	67.3	43.2	0.1

Source: Eurostat

It is obvious that there is quite significant variation for every input and output being used. Such differences in such cases are expectable due to the considerably different sizes of primary sectors of EU countries. It is quite important though to examine which treatment of inputs and outputs, as well as undesirable outputs is more appropriateto evaluate efficiency of EU primary sectors. There are three major approaches for this. The first one ignores possible undesirable inputs or outputs, by using as inputs main production indicators and as outputs the revenues from both crop and animal production. The second approach transformsthe undesirable output of GHGs by multiplying the data set by (-1) and then use a translation vector w to convert all negative undesirable outputs to positive ones. In this paper w= (200). The third one treats the undesirable output as input, although this approach does not fully reflect the production process (Seiford and Zhu, 2002; Jahanshahloo *et al*, 2005; Wang *et al*, 2012). The final model assess efficiency taking into consideration the energy inputs and outputs being involved in the production process as they have already mentioned above. The implementation of these models aims to examine if there are possible differences in efficiency scores among them, and to scrutinize the most suitable one for appraising the impact of CAP on efficiency improvement of primary sectors of EU countries.

For everyone who participates in a direct or indirect way in agricultural production process, it is very important to have the ability to foresee the impact of the implementation of specific policies and interventions in general in the near future. As it was presented in the literature review section, up to now there have developed models for this purpose, taking into consideration parameters affected by the CAP. The recent radical changes though towards a liberalised and more market oriented policy approach, provide the framework for the implementation of reliable models for prognostication purposes, being used for a long period of time for both economic and engineering activities (Chinchuluum *et al*, 2008; Zopounidis and Pardalos, 2010). Such models are the ANNs. The implementation of ANNs is being used to predict crop and animal production, as well as GHGs emissions, by using available data sets, in order to examine the suitability of both of them for prediction purposes. ANNs time series problem definition requires the arrangement of input vectors and target vectors as well. The type series problem being used aims to predict future values of a time series $y(t)$ based on past values of that time series and from past values of a second time series $x(t)$. This prediction form is called Nonlinear Autoregressive with Exogenous input (NARX), with the formula describing it to be the following:

$$y(t)=f(y(t-1),\ldots,y(t-d),x(t-1),\ldots,(t-d))$$

The input and target vectors are randomly divided into three sets, the training, validation and generalization ones. The ration among them is 70%, 15% and 15% respectively. NARX is a two-layer feedforward network consisted of a sigmoid function in the hidden layer and a linear transfer function in the output layer. The output is fed back to the input of the network through delays. The Levenberg-Marquardt algorithm is used for training the network. The comparison of the two networks will verify which approach, the operational or the environmental one, is the most appropriate for prediction purposes of the impact of CAP on efficiency improvement of agriculture of EU countries.

SOLUTIONS AND RECOMMENDATIONS

DEA Window Analysis

As it has already mentioned, three DEA Window models have been applied. The first, and perhaps the most important, finding is that there are no considerable differences in efficiency scores for each country for every treatment of GHGs has been applied. The significance of it becomes more important for the time period after the implementation of the decoupled payments of EU subsidies, where the inclusion or the exclusion of GHGs from the DEA Window models do not differentiate the efficiency scores of EU countries. There are though significant efficiency differences among EU countries, signifying considerable differences among them even though the CAP is implemented for several decades in the EU. For the majority of EU countries the efficiency scores can be considered as satisfactory. There is though significant difference between EU 'old' countries and 'new' ones, differentiating by this way the EU countries before the year 2000 and the countries accessed the EU after the last big enlargement occurred after this specific year. The old EU countries achieve higher efficiency scores, varying from 0.8 to 1, compared with the new ones. This difference exists due to substantial differences of the technology used for agricultural production. This tendency does not apply for the two Scandinavian countries Finland and Sweden, where their efficiency scores are relatively low, being on average 0.5 and 0.6 respectively. The new countries efficiency scores are 0.4 to 0.8, with Romania to be an exception to this propensity, following the efficiency scores of the old EU countries. The following figure presents the average efficiency scores of every country for the 1998-2012 time period.

Figure 1. Average efficiency scores

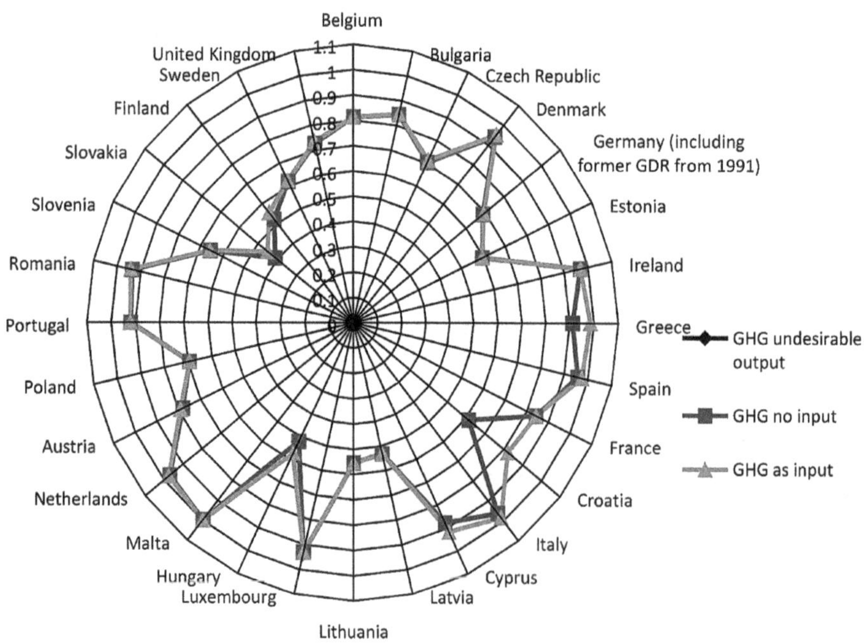

Therefore, if a pure environmental factor does not differentiate efficiency scores, as it happens in other production activities like the energy sector as already mentioned, which are the inputs and outputs diversifying these scores? The answer to this is being provided by the implementation of a different DEA Window model focusing on the pure operational mixture of inputs and outputs, and compares the findings of it with efficiency scores already obtained from the previous models. The criterion used for this selection was the energy dependence of them or not. The non-energy inputs and outputs being used and obtained respectively in the primary sector are the production factors, Land, Labour, and Capital for inputs and Crop and Animal production for outputs. The DEA Window model with this mixture provides the operational efficiency. When the energy dependent ones which are Chemicals, Fertilizers, and Energy for inputs and GHGs as undesirable output are being added in the DEA Window model, the overall efficiency scores are being provided.

The first important finding is that efficiency scores are substantially different, with the operational efficiency to be lower than the overall efficiency for the vast majority of member states. The second important finding is that the efficiency scores in both cases follow the same trend regarding their fluctuations. The third finding is the significant difference in operational efficiency scores between old and new countries, following the same trend with the overall efficiency scores. Finally, the forth finding is that after the year 2005 in most of EU countries the operational efficiency scores are being improved. The following figures present the average efficiency scores being obtained by the two DEA Window models.

The above findings and their comparison between them signifies the direct impact of CAP on agricultural efficiency. Before the implementation of the Agenda 2000 the coupled subsidy management

Figure 2. Average overall efficiency scores

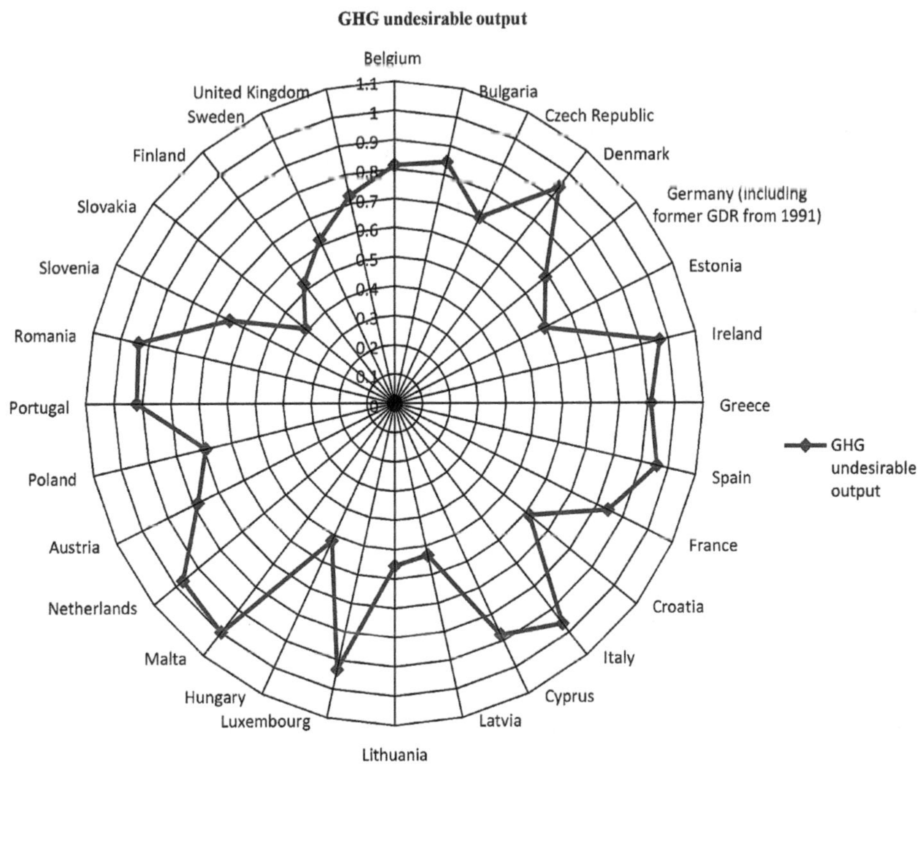

Figure 3. Average operational efficiency scores

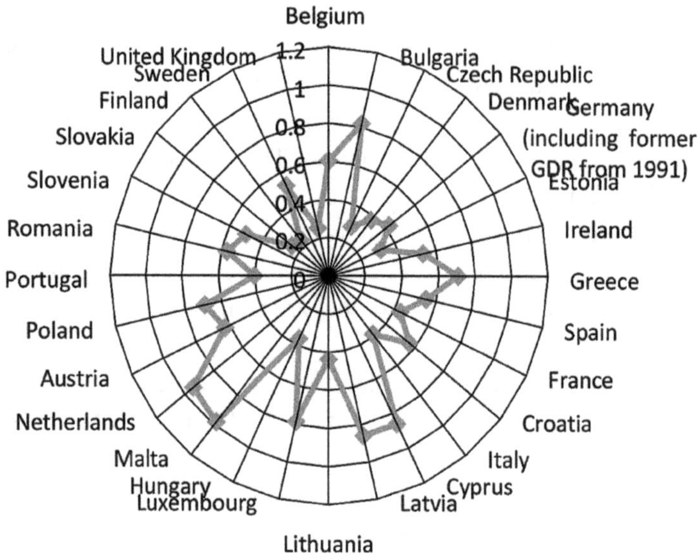

had a direct impact especially on land management, by motivating farmers cultivating specific crops, focusing mainly on arable ones. Such directives had both direct and indirect impacts on the other two production factors, because arable crops are characterised as labour and capital non intensive ones. On the contrary though, the EU did not subsidise the use of energy dependent inputs, like fertilizers or fuels, providing total freedom to farmers and the market in general, to shape supply and demand. The elimination of this intervention, in 2005, offered them increased degrees of freedom on formulating, in a more efficient way, the management of their agricultural holdings. It has to be mentioned though that this subsidy framework, due to the historical approach, did not provide a substantial income support to all farmers, which is the case in the ongoing programming period. Therefore, it is proven that when CAP does not intervene on agricultural holdings' management issues, or farmers have the ability to decide about the quantities of inputs they use, based on market forces, they have the ability to gradually improve their efficiency, as well as maintaining it at satisfactory levels.

Artificial Neural Networks

For the implementation of ANNs two data sets were used, following the same approach with the DEA Window models. All models were trained, tested and validated by using the MATLAB$_{®}$ 2015$_{b}$ software. The first ANN aims to prognosticate crop and animal output, by using as inputs only the non-energy dependent ones, which are agricultural land, labour and capital. The second ANN aims to prognosticate not only the operational outputs, based on the operational inputs, but the GHGs too, by adding as inputs all the relevant energy depended ones, like fertilizers, agrochemicals and fuels.

The best performance of the first model was achieved by applying 12 hidden neurons and 3 delays. For this structure the Mean Square Error (MSE) is 0.052629 at the 5th epoch, with 11 epochs being tried. This score is significantly low and can be considered as acceptable. The network was created and trained in an open loop form, in order to have the ability to get correct past outputs during the training period

Figure 4. MSE scores operational ANN

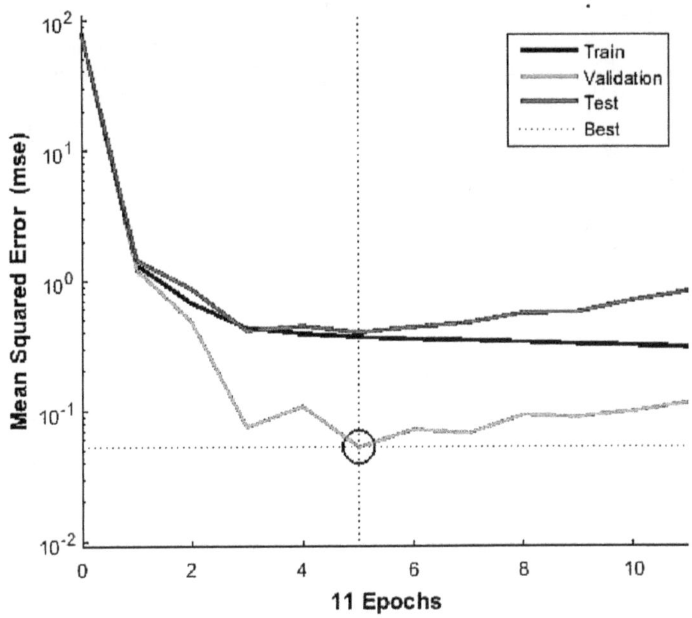

and produce the correct current outputs. The R^2 for validation was 0.9744 which is acceptable too, with R^2 for all three stages of the model to be 0.93524.

The second ANN succeeds the best performance is a different structure. It consists of 11 hidden neurons and 4 delays. The following figure presents the MSE for training, test and validation of the model, achieving the best MSE for validation 0.11325 at epoch 5 after implementing 11 epochs. The R^2 for validation is 0.97365 and the R^2 for all three stages of the model to be 0.98344. Comparing the two models it is obvious that although the qualitative characteristics of both of them are quite satisfactory, the ANN using the energy dependent inputs and undesirable output performs better, because it requires a simpler structure and the R^2 overall score is higher too. These findings provide considerable hints that using energy dependent data for efficiency estimations in agriculture is more safe, compared with the use of pure operational one, signifying at the same time that using market oriented data sets lead to more reliable forecasting results. It remains to be seen in the near future though, when there will be available data from non-energy pure operational inputs not affected by policy interventions, if this qualitative difference between ANNs will still remain or not.

CONCLUSION

Efficiency in agriculture, especially after the recent and radical reforms of CAP towards more liberalised subsidy management practices, is a top priority issue for farmers, policy makers and taxpayers. Implementation of DEA Window methodology quantified operational and environmental efficiency on an EU member state level, as well as efficiency changes before and after the implementation of the new CAP reform. Data availability provided the ability to assess the impact of the most important inputs in agricultural production process on efficiency scores. It is proven that when farming managerial practices

Figure 5. MSE scores environmental ANN

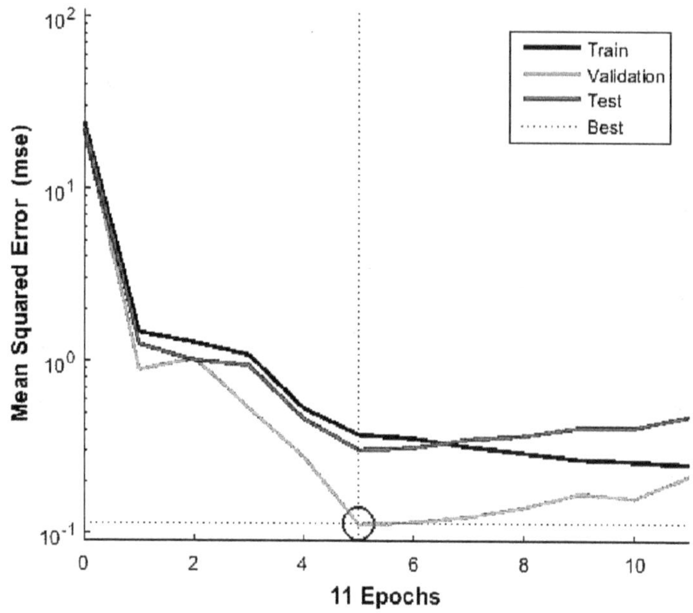

are driven by market forces, there is an improved efficiency outcome, verifying that CAP reforms are heading towards the right direction, having as precondition a globalised trading environment for agricultural products. This is justified by the gradually increased efficiency of EU countries primary sectors, as well as the continuous increase of agricultural income based on market forces and not on subsidies. Implementation of ANNs propose a new methodological approach for *ex ante* policy evaluation, utilising knowhow from other activities, like engineering and economics, which are more market oriented, compared with the majority of agricultural products being produced in the EU.

The widely accepted advantages of this methodology are expected to provide safer prognoses regarding operational activities and environmental safety, increasing by this way the level of success of CAP, improving at the same the utility of financial transfers from taxpayers to farmers.

REFERENCES

Abello, J., Pardalos, P. M., & Resende, M. (2002). *Handbook of massive data sets*. Kluwer Academic Publishers Norwell. doi:10.1007/978-1-4615-0005-6

Amores, A., & Contreras, I. (2009). New approach for the assignment of new European agricultural subsidies using scores from data envelopment analysis: Application to olive-growing farms in Andalusia (Spain). *European Journal of Operational Research, 193*(3), 718–729. doi:10.1016/j.ejor.2007.06.059

Arabi, B., Munisamy, S., Emrouznejad, A., & Shadman, F. (2014). Powerindustryrestructuringandeco-efficiency changes:A new slacks-basedmodelinMalmquist–LuenbergerIndexmeasurement. *Energy Policy, 68*, 132–145. doi:10.1016/j.enpol.2014.01.016

Athanassopoulos, A., & Curram, S. (1996). A Comparison of Data Envelopment Analysis and Artificial Neural Networks as Tools for Assessing the Efficiency of Decision Making Units. *The Journal of the Operational Research Society*, *47*(8), 1000–1016. doi:10.1057/jors.1996.127

Bach, C. F., Frandsen, S. E., & Jensen, H. G. (2000). Agricultural and Economy-Wide Effects of European Enlargement: Modelling the Common Agricultural Policy. *Journal of Agricultural Economics*, *51*(2), 162–180. doi:10.1111/j.1477-9552.2000.tb01221.x

Boussofiane, A., Dyson, R. G., & Thanassoulis, E. (1991). Applied data envelopment analysis. *European Journal of Operational Research*, *52*(1), 1–15. doi:10.1016/0377-2217(91)90331-O

Charnes, A., Cooper, W. W., & Rhodes, E. L. (1978). Measuring the efficiency of decision making units. *European Journal of Operational Research*, *2*(6), 429–444. doi:10.1016/0377-2217(78)90138-8

Chinchuluun, R., Lee, W. S., Bhorania, J., & Pardalos, P. M. (2009). Clustering and Classification Algorithms in Food and Agricultural Applications:A Survey. In Advances in Modeling Agricultural Systems. Springer Science + Business Media, LLC.

Cook, W., & Seiford, L. (2009). Data envelopment analysis (DEA) – Thirty years on. *European Journal of Operational Research*, *192*(1), 1–17. doi:10.1016/j.ejor.2008.01.032

Cullinane, K., Wang, T.-F., Song, D.-W., & Ji, P. (2006). The technical efficiency of container ports: Comparingdata envelopment analysis and stochastic frontier analysis. *Transportation Research Part A, Policy and Practice*, *40*(4), 354–374. doi:10.1016/j.tra.2005.07.003

Dine, M., & Haynes, K. E. (1999). Sources of regional inefficiency. An integrated shift-share, data envelopment analysis and input-output approach. *The Annals of Regional Science*, (33): 469–489.

European Commission. (2013). *Overview of CAP reform 2014-2020.* Retrieved May 19, 2016, fromhttp://ec.europa.eu/agriculture/policy-perspectives/index_en.htm

Farrell, M.J., (1957). The measurement of productive efficiency. *Journal of the Royal Statistical Society, Series A, General, 120*(3), 253–281.

Gorton, M., Duarin, E., Davidova, S., & Latruffe, L. (2008). Attitudes to agricultural policy and farming futures in the contextof the 2003 CAP reform: A comparison of farmers in selected established and new Member States. *Journal of Rural Studies*, *24*(3), 322–336. doi:10.1016/j.jrurstud.2007.10.001

Hansson, H. (2007). Strategy factors as drivers and restraints on dairy farmperformance: Evidence from Sweden. *Agricultural Systems*, *94*(3), 726–737. doi:10.1016/j.agsy.2007.03.002

Hansson, H., & Ohlmer, B. (2008). The effect of operational managerial practices on economic, technical and allocative efficiency at Swedish dairy farms. *Livestock Science*, *118*(1-2), 34–43. doi:10.1016/j.livsci.2008.01.013

Heidari, M. D., & Omid, M. (2011). Energy use patterns and econometric models of major greenhouse vegetable productions in Iran. *Energy*, *36*(1), 220–225. doi:10.1016/j.energy.2010.10.048

Heinrichs, A. J., Jones, C. M., Gray, S. M., Heinrichs, P. A., Cornelisse, S. A., & Goodling, R. C. (2013). Identifying efficient dairy heifer producersusingproduction costs and data envelopment analysis. *Journal of Dairy Science, 90*(11), 7355–7362. doi:10.3168/jds.2012-6488 PMID:24054291

Hertel, T. (1997). *Global Trade Analysis: Modeling and Applications*. Published by Cambridge University Press.

Iribarren, D., Hospido, A., Moreira, M. T., & Feijoo, G. (2011). Benchmarking environmental and operational parameters through eco-efficiency criteria for dairy farms. *The Science of the Total Environment, 409*(10), 1786–1798. doi:10.1016/j.scitotenv.2011.02.013 PMID:21371738

Iribarren, D., Vasquez-Rowe, I., Moreira, M. T., & Feijoo, G. (2010). Further potentials in the joint implementation of life cycle assessment and data envelopment analysis. *The Science of the Total Environment, 408*(22), 5265–5272. doi:10.1016/j.scitotenv.2010.07.078 PMID:20797768

Jahanshahloo, G. R., Lotfi, F. H., Shoja, N., Tohidi, G., & Razavyan, S. (2005). Undesirable inputs and outputs in DEA models. *Applied Mathematics and Computation, 169*(2), 917–925. doi:10.1016/j.amc.2004.09.069

Karkazis, J., & Thanassoulis, E. (1998). Assessing the Effectiveness of Regional Development Policies in Northern Greece Using Data Envelopment Analysis. *Socio-Economic Planning Sciences, 32*(2), 123–137. doi:10.1016/S0038-0121(97)00023-2

Khoshnevisan, B., Rafiee, S., Omid, M., & Mousazadeh, H. (2013). Applying data envelopment analysis approach to improve energy efficiency and reduce GHG (greenhouse gas) emission of wheat production. *Energy, 58*, 588–593. doi:10.1016/j.energy.2013.06.030

Khoshnevisan, B., Rafiee, S., Omid, M., & Mousazadeh, H. (2013). Reduction of CO_2 emission by improving energy use efficiency of greenhouse cucumber production using DEA approach. *Energy, 55*, 676–682. doi:10.1016/j.energy.2013.04.021

Khoshnevisan, B., Rafiee, S., Omid, M., Mousazadeh, H., & Sefeedpari, P. (2013). Prognostication of environmental indices in potato production using artificial neural networks. *Journal of Cleaner Production, 52*, 402–409. doi:10.1016/j.jclepro.2013.03.028

Khoshnevisan, B., Rafiee, S., Omid, M., Yousefi, M., & Movahedi, M. (2013). Modeling of energy consumption and GHG (greenhouse gas) emissions in wheat production in Esfahan province of Iran using artificial neural networks. *Energy, 52*, 333–338. doi:10.1016/j.energy.2013.01.028

Khoshroo, A., Mulwa, R., Emrouznejad, A., & Arabi, B. (2013). A non-parametric Data Envelopment Analysis approach for improving energy efficiency of grape production. *Energy, 63*, 189–194. doi:10.1016/j.energy.2013.09.021

Kuosmanen, T., & Kortelainen, M. (2005). Measuring Eco-efficiency of Production with DataEnvelopment Analysis. *Journal of Industrial Ecology, 9*(4), 59–72. doi:10.1162/108819805775247846

Lakes, T., Muller, D., & Kruger, C. (2009). Cropland change in southern Romania: A comparison of logistic regressions and artificial neural networks. *Landscape Ecology, 24*(9), 1195–1206. doi:10.1007/s10980-009-9404-2

Lozano, S., Iribarren, D., Moreira, M. T., & Feijoo, G. (2009). The link between operational efficiency environmental impacts. A joint application of Life Cycle Assessment and Data Envelopment Analysis. *The Science of the Total Environment*, *407*(5), 1744–1754. doi:10.1016/j.scitotenv.2008.10.062 PMID:19056104

Lozano, S., Villa, G., & Brannlund, R. (2009). Centralized reallocation of emission permits using DEA. *European Journal of Operational Research*, *193*(3), 752–760. doi:10.1016/j.ejor.2007.07.029

McDougall, R. A., Elbehri, A., & Truong, T. P. (1998). Global Trade Assistance and Protection: The GTAP 4 Data Base. Center for Global Trade Analysis, Purdue University.

Meijl, H. V., Rheenen, T. V., Tabeau, A., & Eickhout, B. (2006). The impact of different policy environments on agricultural land use in Europe. *Agriculture, Ecosystems & Environment*, *114*(1), 21–38. doi:10.1016/j.agee.2005.11.006

Mobtaker, H. G., Akram, A., Keyhani, A., & Mohammadi, A. (2012). Optimization of energy required for alfalfa production using data envelopment analysis approach. *Energy for Sustainable Development*, *16*(2), 242–248. doi:10.1016/j.esd.2012.02.001

Mohammadi, A., Shahin, R., Jafari, A., Dalgaard, T., Trydeman Knudsen, M., Keyhani, A., & Hermansen, J. E. etal. (2013). Potential greenhouse gas emission reductions in soybean farming: A combined use of Life Cycle Assessment and Data Envelopment Analysis. *Journal of Cleaner Production*, *54*, 89–100. doi:10.1016/j.jclepro.2013.05.019

Pahlavan, R., Omid, M., & Akram, A. (2011). Energy use efficiency in greenhouse tomato production in Iran. *Energy*, *36*(12), 6714–6719. doi:10.1016/j.energy.2011.10.038

Pahlavan, R., Omid, M., & Akram, A. (2012). Energy input-output analysis and application of artificial neural networks for predicting greenhouse basil production. *Energy*, *37*(1), 171–176. doi:10.1016/j.energy.2011.11.055

Pahlavan, R., Omid, M., Rafiee, S., & Mousavi-Avval, S. H. (2012). Optimization of energy consumption for rose production in Iran. *Energy for Sustainable Development*, *16*(2), 236–241. doi:10.1016/j.esd.2011.12.001

Papajorgji, P. J., & Pardalos, P. M. (2005). *Software Engineering Techniques Applied to Agricultural Systems: An Object-Oriented and UML Approach (Applied Optimization)*. Secaucus, NJ: Springer-Verlag New York, Inc.

Picazo-Tadeo, A., Beltran-Esteve, M., & Gomez-Limon, J. A. (2012). Assessing eco-efficiency with directional distance functions. *European Journal of Operational Research*, *220*(3), 298–309. doi:10.1016/j.ejor.2012.02.025

Picazo-Tadeo, A., Gomez-Limon, J. A., & Martinez, E. R. (2011). Assessing farming eco-efficiency: A Data Envelopment Analysis approach. *Journal of Environmental Management*, *92*(4), 1154–1164. doi:10.1016/j.jenvman.2010.11.025 PMID:21193265

Pjevčević, D., Radonjić, A., Hrle, Z., Čolić, V., (2012). DEA Window Analysis for Measuring Port Efficiencies in Serbia. *Traffic Management Review, 24*(1), 63-72.

Schmid, E., & Sinabell, F. (2007). On the choice of farm management practices after the reform of the Common Agricultural Policy in 2003. *Journal of Environmental Management*, *82*(3), 332–340. doi:10.1016/j.jenvman.2005.12.027 PMID:17095138

Schmid, E., Sinabell, F., & Hofreither, M. F. (2007). Phasing out of environmentally harmful subsidies:Consequences of the 2003 CAP reform. *Ecological Economics*, *60*(3), 596–604. doi:10.1016/j.ecolecon.2005.12.017

Seiford, L. M., & Zhu, J. (2002). Modeling undesirable factors in efficiency evaluation. *European Journal of Operational Research*, *142*(1), 16–20. doi:10.1016/S0377-2217(01)00293-4

Silva, E., & Stefanou, S. E. (2003). Nonparametric Dynamic Production Analysis and the Theory of Cost. *Journal of Productivity Analysis*, *19*(1), 5–32. doi:10.1023/A:1021865018717

Smith, P., & Mayston, D. J. (1987). Measuringefficiency in the public sector. *Omega*, *15*(3), 181–189. doi:10.1016/0305-0483(87)90068-5

Song, M., An, Q., Zhang, W., Wang, Z., & Wu, J. (2012). Environmental efficiency evaluation based on data envelopment analysis: A review. *Renewable & Sustainable Energy Reviews*, *16*(7), 4465–4469. doi:10.1016/j.rser.2012.04.052

Sozen, A., Alp, I., & Ozdemir, A. (2010). Assessment of operational and environmental performance of the thermal power plants in Turkey by using data envelopment analysis. *Energy Policy*, *38*, 6194–6203.

Stokes, J. R., Tozer, P. R., & Hyde, J. (2007). Identifying Efficient Dairy Producers Using Data Envelopment Analysis. *Journal of Dairy Science*, *90*(5), 2555–2562. doi:10.3168/jds.2006-596 PMID:17430960

Thanassoulis, E. (1999). Data Envelopment Analysis and Its Use in Banking. *Interfaces*, *29*(3), 1–13. doi:10.1287/inte.29.3.1

Thanassoulis, E., & Dunstan, P. (1994). Guidingschools to improved performance usingdata envelopment analysis: An illustrationwith data from a local education authority. *The Journal of the Operational Research Society*, *45*(11), 1247–1262. doi:10.1057/jors.1994.198

Tranter, R. B., Swinbank, A., Wooldridge, M. J., Costa, L., Knapp, T., Little, G. P. J., & Sottomayor, M. L. (2007). Implications for food production, land use and rural development of the European Unions Single FarmPayment: Indications from a survey of farmersintentions in Germany, Portugal and the UK. *Food Policy*, *32*(5-6), 656–671. doi:10.1016/j.foodpol.2007.04.001

Vasquez-Rowe, I., Iribarren, D., Moreira, M. T., & Feijoo, G. (2010). Combined application of life cycle assessment and data envelopment analysis as a methodological approach for the assessment of fisheries. *International Journal of Life Cycle Assessment*, *15*(3), 272–283. doi:10.1007/s11367-010-0154-9

Vasquez-Rowe, I., Villanueva-Rey, P., Iribarren, D., Moreira, M. T., & Feijoo, G. (2012). Joint life cycle assessment and data envelopment analysis of grape production for vinification in the RíasBaixas appellation (NW Spain). *Journal of Cleaner Production*, *27*, 92–102. doi:10.1016/j.jclepro.2011.12.039

Vennesland, B. (2005). Measuring rural economic development in Norway using data envelopment analysis. *Forest Policy and Economics*, *7*(1), 109–119. doi:10.1016/S1389-9341(03)00025-X

Vlontzos, G., Arabatzis, G., & Manos, B. (2014). Investigation of the relative efficiency of LEADER+ in rural areas of Northern Greece. *International Journal of Green Economics*, *8*(1), 37–48. doi:10.1504/IJGE.2014.064470

Wang, K., Wei, Y. M., & Zhang, X. (2012). A comparative analysis of Chinas regional energy and emission performance: Which is the better way to deal with undesirable outputs? *Energy Policy*, *46*, 574–584. doi:10.1016/j.enpol.2012.04.038

Wu, H., Du, S., Liang, L., & Zhou, Y. (2013). DEA-based approach for fair reduction and reallocation of emission permits. *Mathematical and Computer Modelling*, *58*(5-6), 1095–1101. doi:10.1016/j.mcm.2012.03.008

Yang, H.-H., & Chang, Ch. (2009). Using DEAwindowanalysistomeasureefficienciesofTaiwansIntegrated-telecommunicationfirms. *Telecommunications Policy*, *33*(1-2), 98–108. doi:10.1016/j.telpol.2008.11.001

Zopounidis, C., & Pardalos, P. M. (2010). *Handbook of Multicriteria Analysis*. Applied Optimization, Springer-Verlag Berlin Heidelberg. doi:10.1007/978-3-540-92828-7

Chapter 2
A Virtual Supply Chain Architecture to Grant Product Transparency in Agribusiness

Giulia Bruno
Politecnico di Torino, Italy

ABSTRACT

Especially in the food sector, fraud and counterfeiting are affecting the trust of consumers, who are more and more oriented to chose products basing on quality and traceability attributes rather than the price. Recently, the Electronic Product Code Information Services (EPCIS) standard was introduced to provide specifications for the representation of product traceability information. The collection and analysis of such information allows supply chains to be monitored and controlled through virtualization. Several applications of EPCIS were presented in literature, even if most of them are mainly focused on enabling technologies, with less emphasis on assessing how the available information can be used for a control at a higher level. This chapter review the relevant literature available on this topic, and present an architecture allowing the traceability of information about products throughout the entire supply chain by exploiting both the EPCIS standard and a NoSQL database. An application showing the potentiality of the proposed system in a case study is also reported.

INTRODUCTION

Today, consumers pay more and more attention to product quality and transparency of food ingredients, origins and production processes. Also, due to the rise of foodborne diseases, consumers in the food sector are even more interested in having ensured food quality and safety (D'Angelo et al. 2014). This requires knowing the origin of the product and its pathway from the producer to the final seller. Consumers are oriented to chose products basing on the credence attributes rather than the price, and fraud and counterfeiting cause the lost of consumers' trust. For this reason, the problem of integrating data through the supply chains is becoming an important research topic (Badia-Melis et al. 2015). Food traceability systems allow for all supply-chain actors and the National regulatory authorities to identify

DOI: 10.4018/978-1-5225-2107-5.ch002

the source of a food quality problem and initiate procedures to remedy it. To guarantee the safety and trust of consumers, a strong collaboration along the supply chain is needed (Rota et al. 2013).

Even if traceability of products has been introduced since the 1990s (Cheng & Simmons, 1994), only large enterprises, characterized by a considerable use of information and communication technology, employ efficient and automated systems (Gandino et al. 2009). Small enterprises only rarely implement supply chain management systems, since for them, adding traceability to their normal operation has caused the decrease of efficiency and the increase of costs (Cimino & Marcelloni, 2012). For these reasons, a considerable challenge is to develop systems specifically suited for small-scale enterprises.

By recording product transitions, barcodes and Radio Frequency Identification (RFID) have provided partial solutions to this issue (Kelepouris et al. 2007, Solanki & Brewster 2014). Traceability of food product in the supply chain has gained considerable importance, particularly following a number of food safety cases during delivery (Abdul Kadir et al. 2015). However, companies and organizations today are not able to manage an extended network of suppliers and distributors. Thus, fraud and counterfeiting are rising and are difficult to discover.

The Electronic Product Code Information Service (EPCIS) is a standard that provides specifications to keep track of products (Främling et al. 2013). Food supply chains handle a large variety of objects, depending on the type of food product and the stage of the supply chain. At the farm, main objects are seeds, feed, fertilizers, and farm resources. After processing, they become discrete objects when they are packaged, shipped and distributed to retailers. The collection and analysis of the product information from different stages allows supply chains to be virtualized and remotely monitored, optimized and controlled (Verdouw et al. 2016). By utilizing expressiveness of the EPCIS standard, all the processes are visible to the users so that they can control the safe of foods, e.g., by avoiding dangerous place-of-origins (Byun & Kim 2015). To reach this goal, all the roles involved have to incrementally provide additional information regarding the product stages. EPCIS proposes a mechanism to exchange and share data, but the semantics of data is informally defined and its interpretation is left up to the individual implementing engines (Solanki & Brewster 2015).Hence, the aim of this chapter is to present a collaborative architecture to allow the traceability of information throughout the entire supply chain.

To this aim, the chapter revises the relevant and most recent literature on the topic, and it describes the available standards to track products and the events they generate inside the supply chain. The need of storing a huge amount of heterogeneous data suggests the adoption of a non relational database to better manage the data generated by the system. Thus, also a section recalling the advantages of NoSQL databases in comparison with relational ones is present. Finally, the procedure to map EPCIS events to non relational documents in a case study is reported. The last section illustrates conclusions and states future works.

LITERATURE REVIEW

Despite technical and cost challenges, the application of RFID in supply chain management is growing around the world (Musa & Abba Dabo 2016). Particularly, RFID was introduced in the agrifood industry (Gandino et al. 2009), also together wireless sensor networks for monitoring the environment conditions (Exposito et al. 2013). D'Angelo et al. (2014) highlighted that another key issue to add values is to integrate the traceability system with the supply-chain management, and thus use traceability data to manage the business process and improve its performance. Abate et al. (2014) proposed to exploit a semi-active RFID module to remotely monitor the storage and transport conditions. Their system is able to analyze both the route travelled by each package and its environmental conditions (e.g., temperature, humidity and brightness variations).

Examples of application of EPCIS to trace products in food supply chian are described in several works for different kinds of products. Cimino & Marcelloni (2012) presented a system realized in Italy for the wine supply chain, following the model defined by the Wine Traceability Working Group of GS1 (GS1, 2008). Works related to wine traceability are also present in many different countries, e.g., Albania (Vukatana et al. 2016), Czech Republic and Moldova (Anikina & Djordjevic 2015), and France (Touzard & Maffezzoli 2015). Mainetti et al. (2013) proposed a system for the fresh vegetable supply chain, to trace products from cultivation to transformation. A low cost and an easy access for the consumers is ensured by the use of DataMatrix instead of RFID tags to tag the final products.

In the fish market, traceability systems based on EPCIS to store and manage fish information from farmer to retailer were studied (Hsu et al. 2009, Gunnlaugsson et al. 2011, Yan et al. 2012). Parreño-Marchante et al. (2014) described the adoption of an EPCIS based system in two business cases, by highlighting the advantages and obstacles in its exploitation. In the meat supply chain, due to recent meat crises and scandals, reference architectures based on EPCIS standard to grant transparency were presented (Kassahun et al. 2013, Thakur & Forås 2015, Kassahun et al. 2016). The works of Kassahun addressed the design of a traceability architecture by defining the actors, the components and the interactions of the system. They explained the steps to design a concrete architecture and they implemented such a system for a beef supply chain in Germany. Particularly, they used the concept of event defined by EPCIS, and mapped events such as the birth of an animal, the splitting of carcass into sides, etc. in different kinds of EPCIS events. Then, they realize a application to track and trace products given their code or their location. In the context of meet supply chain, another the key focus is to maintain the right temperature during handling and processing, because changes in the meat quality resulting from temperature variations are irreversible. Thus, temperature recording and monitoring is a pre-requisite, even if most temperature monitoring systems are only strip chart recorders or data loggers, which are not automated and require manual inspection. The aim of Thakur & Forås (2015) is to integrate such time-temperature data in the traceability system, as an EPCIS event. They have shown that event-based data capture gives the ability to highlight anomalies or food incidents.

Also the provision of dairy products is critical, especially because once a safety accident happens, the source of the accident could be found and the defective products could be recalled immediately, which could maximally protect consumers' benefit and rebuild an enterprise's reputation. Tian (2016) addressed this issue by proposing a control system based on the GS1 international standards. Jakkhupan et al. (2015) designed a traceability system for the rice supply chain in Thailand. Their system is composed of a centralized data service, operated by a trusted party, to link data coming from a set of distributed data services, each of which operated within each company to manage the internal data. In this way, producer companies are able to identify the raw materials and the sources of the ingredients of the product. Furthermore, since every rice bag is linked to the lot number, the lot number is linked to the paddy, and the paddy is linked to the farmer, if a defect is found in the paddy (e.g., insecticide contamination), the traceability system would be able to reveal the current locations of the rice bags that were produced in the same lot of manufacturing, to be recalled before arriving to customers.

A ginseng traceability system based on EPCIS was proposed by Hwang et al. (2015) to provide information to both to the consumer and the local competent authorities. Previously, farmers used hand-written logs to record physical phenomena, such as luminance, relative humidity, temperature, and CO2 emission in a time-consuming and inefficient process. Then, the processors manually entered the product-related information into the information system. However, due to the fact that the growing period of ginseng is very long (4 to 6 years), such information was frequently lost, and the inspection process for all phases was manually checked and labor-intensive. To overcome this problem, the authors proposed an RFID-based system on an EPC wireless sensor network.

The EPCIS standard is also adopted for other kinds of products not in the agribusiness sector. Particularly, in the pharmaceutical domain, it is usedto grant the security of the drug supply chain (NamGung et al. 2012, Byun & Kim 2015). Other studies exploit EPCIS for sustainability reasons to overcome purchase barriers of eco-friendly products and increasing consumers' product trust, for example of wood products (Appelhanz et al. 2016). The wide adoption of systems to manage digitized information and events related to products allows the virtualization of the supply chain, thus enabling all the supply chain actors to monitor, control, plan and optimize business processes remotely and in real-time through the Internet, based on virtual objects instead of observation on-site (Verdouw et al. 2016). Furthermore, advanced food traceability systems help not only to minimize unsafe or poor quality products, but also improve the value-added business in the food supply chain (Chen 2017).

GS1 STANDARDS

The GS1 system originated in the United States in 1973 and it provides for the use of unambiguous codes to identify goods, services, assets, and locations worldwide to overcome the limitations of using intra-company specific coding systems and to make trading much more efficient and responsive to customers (GS1 2016). These codes can be represented in barcodes or RFID to enable their electronic reading. In addition, the GS1 system also provides for supplementary information, such as best before dates, serial numbers, and batch numbers, which can appear in barcode form. By following the principles and design of the GS1 system, users can design applications to process GS1 system data automatically. The system logic guarantees that data captured from GS1 endorsed barcodes produces unambiguous electronic messages, and processing can be fully pre-programmed. The GS1 system of standard is graphically represented in Figure 1. It is composed of three main elements: (i) GS1 identification keys, used to identify objects, things or locations, (ii) GS1 data carriers, i.e., media that can hold GS1 identification keys and are used for automatic data capture, and (iii) GS1 communication standards, for the electronic data interchange.

Identification Keys

The main GS1 identification keys are the Global Trade Item Number (GTIN), the Global Location Number (GLN) and the Serial Shipping Container Code (SSCC).

- GTIN is used to uniquely identify trade items, which are products and services that are priced, ordered or invoiced at any point in the supply chain. Each trade item that is different from another has its own separate GTIN. Their main function is to provide a way to uniquely identify any item so it can be looked up in a database at any point during the supply chain and from any place in the world.
- GLN is used to identify locations and legal entities. Locations can be a physical place such as a warehouse, a storage cabinet or a specific shelf within a store, while legal entities are companies,

divisions of a company, or functions that takes place within a legal entity. Using a GLN rather than a proprietary internal numbering system for locations gives a company significant advantages, because it provides a standardized way to uniquely identify locations important to the supply chain.
- SSCC is used to identify individual logistic units, i.e., any combination of items put together in a carton, case, pallet or truck, that needs to be managed through the supply chain. The SSCC enables a unit to be tracked individually, providing benefits for order and delivery tracking and automated goods-receiving.

Another important key is the Global Individual Asset Identifier (GIAI), which is used to identify fixed assets of any value within a company that need to be identified uniquely, such as a vehicle, a computer or a desk. Having a unique identifier for their assets allows a business to identify, track and manage them across their entire life. They provide a quick way to be able to look up an asset in a database so its use, location or state can record. The usage of such codes to identify elements in a food supply chain is shown in Figure 2.

Data Carriers

GS1 provides a large set of data carriers, i.e., media that can hold GS1 identification keys and other attribute data. In fact, the same content can be encoded onto different kinds of carriers, depending on what use will be made of it. The most used data carriers are reported in the following.

- The EAN/UPC barcode is the longest-established and most widely used GS1 data carrier, the one used on most products since it is read by laser scanning devices.
- The GS1-128 barcode has become the gold standard in the logistics field, since it can carry all GS1 identification keys, as well as variable information like serial numbers, expiration dates, and

Figure 1. GS1 system of standards for identifying objects, data capture and data interchange

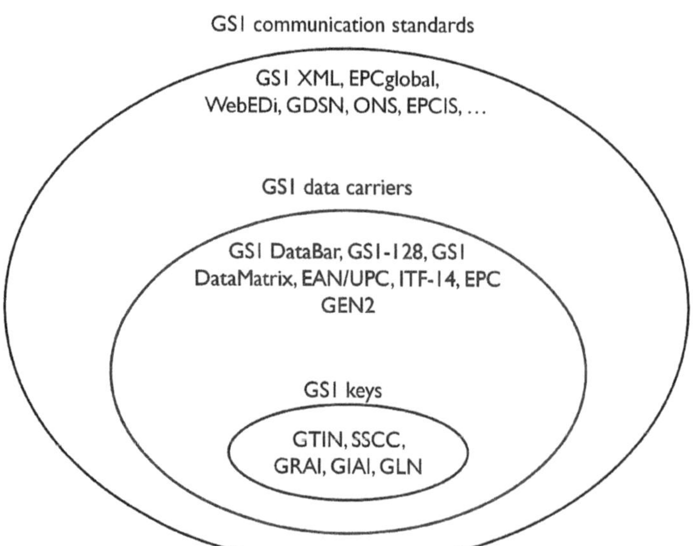

Figure 2. GS1 identification keys in the food supply chain

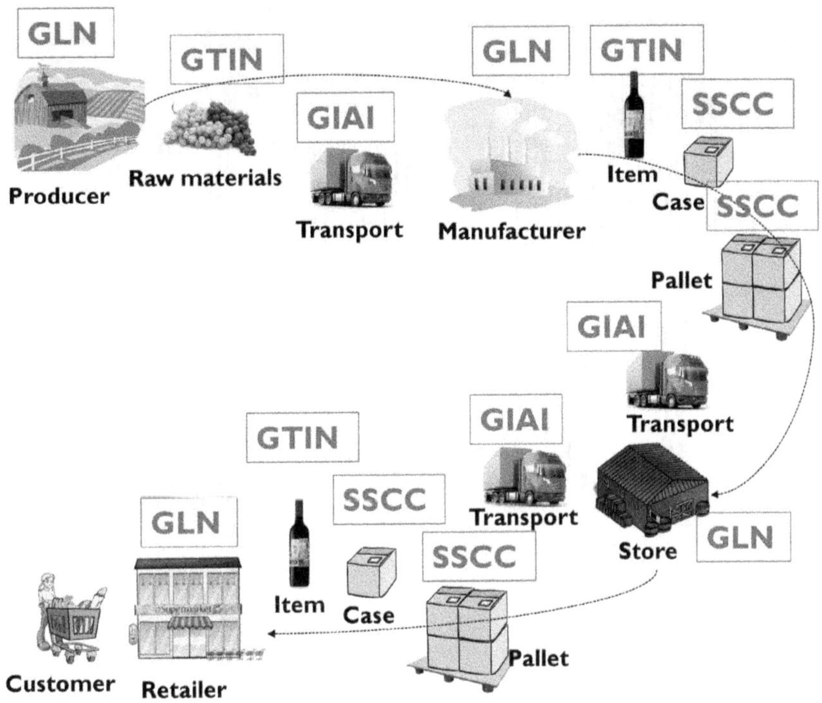

measures. It is a linear symbol, and it has taken greater importance in recent years been very flexible, configurable and quite adaptable to a wide variety of needs and use cases. It can also be read by a variety of commercially available laser scanners.

- The DataBar symbol can carry more information and identify smaller items than EAN/UPC barcode and can be also scanned at retail point of sale. As a result, GS1 DataBar enables GTIN identification for fresh variable measure and hard-to-mark products like loose fruit and vegetables, jewelry and cosmetics.
- The DataMatrix is a two-dimensional symbol that allows a wealth of information to be encoded in a very compact space. However, it is not intended to be used at high volume "omnidirectional" retail point-of-sale environments like supermarkets, because it is only compatible with applications whose reading systems use camera-based scanners.
- The EPC/RFID tag uses Radio-Frequency Identification technology to encode GS1 keys in the GS1 Electronic Product Code (EPC). RFID works via a microchip, which stores the relevant data and reflects the data to a reader antenna by means of electromagnetic waves. Since these waves can pass through solid materials, the chips may be shielded by adhesive film or integrated directly inside the packaging or product. They can offer a significant time-saving advantage over other data carriers: the reader and transponder do not need to be in each other's line of sight and the transmission/reading process is ultra-rapid.

The graphical representations of the described data carriers are provided in Figure 3.

Figure 3. GS1data carriers: (a) EAN/UPC barcode, (b) GS1-128 barcode, (c) DataBar, (d) DataMatrix, (e) RFID

EPCglobal Network

The EPC global Network is a network used to share product data between partners. The basis of the information flow in the network is the Electronic Product Code (EPC) which is stored for each product by the RFID tag. EPC is the emerging way to globally identify entities. It includes all the GS1 identification keys, ensuring full interoperability with existing systems. Today, all EPC-identified objects are serialized, which means they carry a unique serial number: an EPC assigned to one object is thus different than an EPC assigned to another. This allows unique, accurate and specific identification of individual objects. The network manages dynamic information on individual products, including data regarding the movement of an object throughout the product life cycle. The EPCglobal Network consists of several components, one of them is the EPC Information Services (EPCIS).

EPCIS

EPCIS provides a data model serialized as an XML schema for capturing information artifacts that encapsulate the geographical progress and status of an item or set of items during the different phases of the supply chain. This is achieved by capturing data generated through the scanning of a barcode or RFID tag and encapsulating it as an event. Accordingly, to EPCIS 1.1 (EPCIS 2014), four core event types are defined:

- ObjectEvent, used to register a general event occurring to a physical or digital object;
- AggregationEvent, used to indicate a physically aggregation objects (i.e., objects that are physically constrained to be in the same place at the same time, such as items on a pallet);

- TransformationEvent, used when a set of input objects is consumed and transformed into output objects;
- TransactionEvent, used when one or more objects are linked or delinked with business transactions.

The field of each event represents the four possible dimensions of an EPCIS event (i.e., what, when, where and why). Particularly, they contain information regarding (1) the physical or digital objects participating in the event, (2) the period of time in which the event occurred, (3) the location in which the event was registered, and (4) the business context of the event. The meaning of the dimensions varies depending on the event type. Figure 4 shows how EPCIS events occur in a food supply chain.

In addition to the basic EPCIS types, the standard can be extended by specific organizations to make it more suitable to their business. In fact, every event has an extension point, which can be used to attach additional data. Even if EPCIS specification proposes a mechanism to exchange and share data, the XML schema defines only the structure of the data to be recorded, while the semantics of data is informally defined and its interpretation is left up to the individual implementing engines (Solanki & Brewster 2015). Therefore, additional information regarding the four dimensions of the event (e.g., temperature, humidity, etc.) can be collected and stored.

Core Business Vocabulary

Together with EPCIS, GS1 defined the Core Business Vocabulary (CBV), which defines specific data values to populate part of the EPCIS data model. CBV includes two different vocabularies, Standard Vocabulary and User Vocabulary.

Figure 4. GS1EPCIS events along the food supply chain

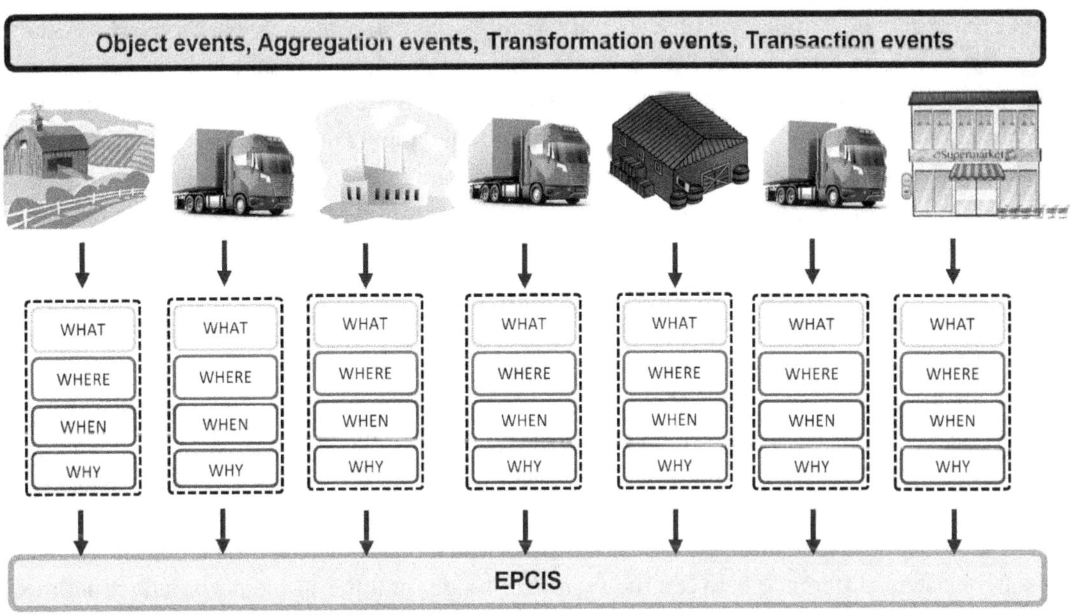

The definition and meaning of the elements of the Standard Vocabulary are agreed in advance by the partners compliant with the standard. For example, the EPCIS standard defines the field "business step," whose possible values (e.g., shipping, receiving) are defined in the Standard Vocabulary. Standard Vocabularies are defined in the following fields of EPCIS: business steps, dispositions, business transaction types, and source and destination types.

On the contrary, the vocabularies of the User Vocabulary are controlled by a single organization, which has not constraints in deciding their meaning. While the CBV does not specify particular user vocabulary elements, it recommends syntax templates for the end users in defining their own elements. User Vocabularies can be defined in the following fields of the EPCIS standard: physical or digital objects, locations including both read points and business locations, business transaction identifiers, source/destination identifiers, and transformation identifiers.

NO-SQL DATABASES

In the 1990s, many people believed that relational databases will be replaced with databases that replicate the in-memory data structures to disk, due to the growth of object-oriented programming languages. However, even if the object-oriented languages became the major force in programming, object-oriented databases faded into obscurity. Relational databases prevailed due to their standard language of data manipulation (SQL) and a growing professional divide between application developers and database administrators. They continued to dominate the enterprise computing world in the 2000s, but during that decade cracks began to open in their dominance (Sadalage & Fowler 2013).

Despite the advantages that relational databases provide, due to an explosion of data and thus the need on the one hand to scale data and on the other hand to manage heterogeneous formats, in the last decades the exploitation of NoSQL databases is on the rise. They will not replace relational ones, which are more mature and widely installed, but for some specific purposes such as handling unstructured massive data even requiring a high level of scalability, NoSQL databases are a better choice (Leavitt 2010).

The primary reason for moving away from the relational model is to make scaling out easier. Data set sizes for applications are growing at an incredible pace. Increases in available bandwidth, and cheap storage have created an environment where even small-scale applications need to store more data than many databases were meant to handle. Scaling up, i.e., getting a bigger machine, has drawbacks: large machines are often very expensive, and eventually a physical limit is reached where a more powerful machine cannot be purchased at any cost. The alternative is to scale out, i.e., to buy other machines and create a cluster. This is both cheaper and more scalable; however, it is more difficult to administer a thousand machines than it is to care for one.

The second reason for adopting a non relational database is the limit of a fixed schema. Relational databases, organize data into a structure of tables and rows, and all operations in SQL consume and return relations, which leads to the mathematically elegant relational algebra. This foundation provides a certain elegance and simplicity, but it also introduces the limitation that values in a relational tuple have to be simple and cannot contain any structure, such as a nested record or list. This limitation isn't true for in-memory data structures, which can take on a much richer structure than relations. As a result, there is a consistent difference between the relational model and the in-memory data structures, and this cause frustration in application developers.vTo address these issues, several kinds of non-relational databases emerged. Today, a wide variety of NoSQL database products on the market, which are built to

Figure 5. Relational and non relational database popularity trends
(image source: http://db-engines.com/en/ranking_trend)

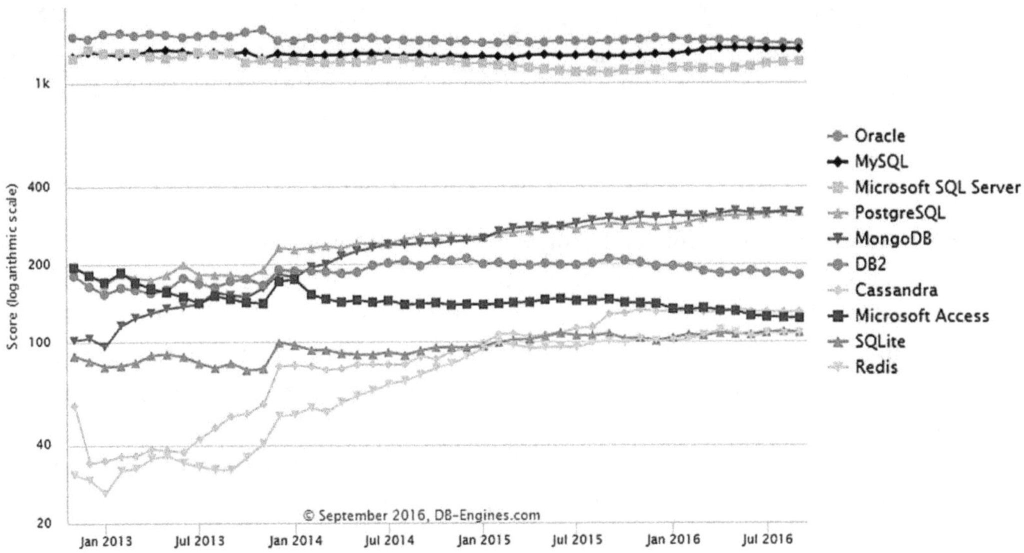

fit specific purposes, and MongoDB (www.mongodb.com) is the most popular among them (see Figure 5). It is considered the right balance between features and complexity, having the features that really matter to the vast majority of today's web applications (Chodorow 2013).

MongoDB

MongoDB is a document-oriented database. Thus, it replaces the concept of a "row" with a more flexible model, the "document". A document is represented in the JavaScript Object Notation (JSON), an open and text-based data exchange format (Crockford 2006). Like XML, it is human-readable and platform independent, and enjoys a wide availability of implementations, being usable in virtually any scenario where applications need to exchange or store structured information as text.

An example of JSON document is the following, which stores the data regarding John Backus, the creator of the computer programming language FORTRAN, who was the winner of the McDowell Award and the Draper Prize. The document is characterized by an identification code (id), and contains a set of other key-values items, where the value can be an array of item or a subset of other key-value items.

```
{
    '_id': 1,
    'name': { 'first': 'John', 'last': 'Backus' },
    'contribs': [ 'Fortran', 'ALGOL', 'Backus-Naur Form', 'FP' ],
    'awards': [
        {
            'award': 'W.W. McDowell Award',
            'year': 1967,
            'by': 'IEEE Computer Society'
```

```
        }, {
            'award': 'Draper Prize',
            'year': 1993,
            'by': 'National Academy of Engineering'
        }
    ]
}
```

By allowing embedded documents and arrays, the document-oriented approach makes it possible to represent complex hierarchical relationships with a single record. This fits naturally into the way developers in modern object-oriented languages think about their data. There are also no predefined schemas: a document's key and values are not of fixed types or sizes. Without a fixed schema, adding or removing fields as needed becomes easier. Generally, this makes development faster as developers can quickly iterate.

Then, MongoDB is designed to scale out. Its document-oriented data model makes it easier for it to split up data across multiple servers. MongoDB automatically takes care of balancing data and load across a cluster, redistributing documents automatically and routing user requests to the correct machines. When a cluster need more capacity, new machines can be added and MongoDB will figure out how the existing data should be spread to them.

Other interesting features provided by MongoDB are the following (Chodorow 2013).

- **Indexing***:* It supports generic secondary indexes, allowing a variety of fast queries, and provides unique, compound, geospatial, and full-text indexing capabilities as well.
- **Aggregation***:* It supports an "aggregation pipeline" that allows building complex aggregations from simple pieces and allow the database to optimize it.
- **Special Collection Types***:* It supports time-to-live collections of data that should expire at a certain time, such as sessions. It also supports fixed-size collections, which are useful for holding recent data, such as logs.
- **File Storage**: It supports an easy-to-use protocol for storing large files and file metadata.

Due to the fact that EPCIS events can be easily translated into a JSON document format (Byun & Kim 2015) and that the efficiency of non-relational versus relational databases in managing EPCIS data was recently proven (Kang et al. 2016), MongoDB can be considered an optimal choice for the data repository of the transparency system.

VIRTUAL ARCHITECTURE FOR PRODUCT TRANSPARENCY

Architecture

The collaborative architecture proposed to grant transparency across the supply chain is shown in Figure 6. In addition to the solid lines showing the item's flow, the information flows are shown as dotted lines. Red thin dotted lines represent an exchange of EPCIS events, while green large dotted lines represent an exchange of transparency information. Each actor involved in the supply chain uses the EPCIS standard to capture events related to products. The captured EPCIS events are sent to the Product tracking platform, where they are stored and analyzed.

The product tracking platform consists of three main elements: the event capturing interface, the EPCIS data repository and the query interface. The capture interface receives the EPCIS events captured by the companies in the supply chain. Procedures of event captures are custom logic specific for each company. An EPCIS event, while containing the information regarding the product identifier, also includes information regarding the business context in which data is obtained.

The captured EPCIS events are stored in a MongoDB database. Through the query interface, the applications interact with the data repository in which all the EPCIS events are stored and organized. Through a mobile App the consumer is able to read the EPC of the product of interest and consult the transparency information stored related to the product. Some information are valid for each product, such as the times and places the product crossed during its lifetime Other information vary from one product to another. For example, for a frozen food, the temperature range at which the product was kept can be retrieved.

Not only the customer can access to the stored information, but also all the other companies in the supply chain can control the events occurred to the products during the previous or following stages. In this way the transparency of the supply chain is granted not only for the consumer, but also to all the companies involved.

Figure 6. EPCIS-based architecture for product transparency in food supply chain

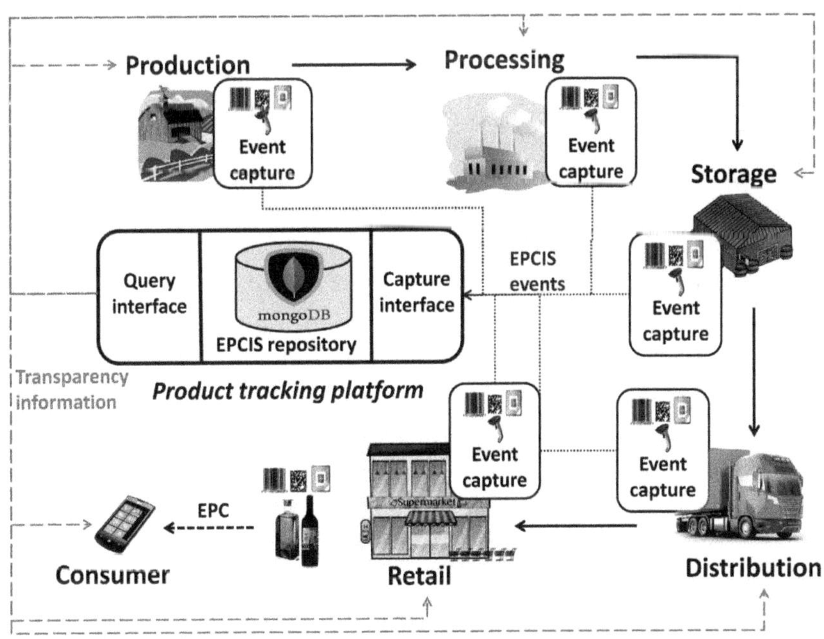

Application

An example of the XML document used to capture an EPCIS event related to the creation of a product is shown in the following.

```
<?xml version="1.0" encoding="UTF-8" standalone="yes"?>
 <epcis:EPCISDocument xmlns:epcis="urn:epcglobal:epcis:xsd:1"
xmlns:xsi="http://www.w3.org/2001/XMLSchema-instance" schemaVersion="1">
  <EPCISBody>
    <EventList>
      <ObjectEvent>
        <eventTime>2016-08-14T18:34:20Z</eventTime>
        <recordTime>2016-08-14T18:34:20Z </recordTime>
        <eventTimeZoneOffset>-02:00</eventTimeZoneOffset>
        <epcList>
          <epc>urn:epc:id:sgtin:0814141.181335.234</epc>
        </epcList>
        <action>ADD</action>
        <bizStep>urn:epcglobal:cbv:bizstep:commissioning
        </bizStep>
        <disposition>urn:epcglobal:cbv:disp:active</disposition>
        <readPoint>
          <id>urn:epc:id:sgln:0814141.00300.1</id>
        </readPoint>
        <bizLocation>
          <id>urn:epc:id:sgln:0814141.00300.0</id>
        </bizLocation>
        <bizTransactionList>
          <bizTransaction type="urn:epcglobal:cbv:btt:po"> urn:
    epc:id:gdti:0814141.05432.2345 </bizTransaction>
          </bizTransactionList>
      <extensions>
        <temperature> 25 </temperature>
      </extensions>
      </ObjectEvent>
    </EventList>
  </EPCISBody>
</epcis:EPCISDocument>
```

In the body of the document, a list of EPCIS event can be registered. In this case a single EPCIS event is represented. In an event, a list of product can be involved, but in this case a single product is considered. The product is identified by its EPC, which is codified through the Serialized Global Trade Item Number (SGTIN) standard. For each event the kind of action (add, delete or observe) is specified.

Then, information regarding the business stem (bizStep) and the disposition is given, accordingly to the CBV definitions. The identification of the EPC reader (readPoint) and of the location at which the event is registered are also reported. For each event, a set of transaction can be associated, identified by their Global Document Type Identifier (GDTI). Finally, the user-defined extensions can be set. In this case, the additional information coming from a temperature sensor is included.

The EPCIS event is translated into a JSON document to be stored in MongoDB. The JSON document corresponding to the previous XML document is the following.


```
{
"_id": ObjectId(""),
"eventType": "ObjectEvent",
"eventTime": ISODate("2016-08-14T18:34:20Z"),
"recordTime": ISODate("2016-08-14T18:34:20Z"),
"eventTimeZoneOffset": -02:00,
"epcList": [
        "urn:epc:id:sgtin:0814141.181335.234"
 ],
"action": "ADD",
"bizStep": "urn:epcglobal:cbv:bizstep:commissioning",
"disposition": "urn:epcglobal:cbv:disp:active",
"readPoint": "urn:epc:id:sgln:0814141.00300.1",
"bizLocation": "urn:epc:id:sgln:0814141.00300.0"",
"bizTransactionList": [
                {
                "type": "urn:epcglobal:cbv:btt:po",
        "value": urn:epc:id:gdti:0814141.05432.2345                }
 ],
"extensions": [
                {
                "name": "temperature",
        "value": 25
                }
 ]
}
```


By having all the events stored in MongoDB in a collection named events, the query to retrieve all the events related to the product of interest of EPC code X in which the wine was observed at a temperature higher that 30 degrees is the following.

```
db.events.find (
  {
    "epcList": "X",
    "extensions":
```

```
      {$elemMatch: {name: "temperature", value {$gt: 30}}}
   }
)
```

DISCUSSION AND CONCLUSION

Different kinds of companies are involved in a food supply chain, each of them operating in different markets and selling different products. This often causes the fact that data collected in each stage are not accessible by companies involved in different stages. As a result, the consumer is not aware of the process followed by the product during the previous stages, but also that production or processing companies do not have information regarding storage, distribution and retail. The lack of communication across the stages clearly affect the trust of the chain, since each actor is not aware of the history of the product during the other stages. In fact, the managers of the companies in the production or processing stages are interested in knowing if the products are correctly stored and distributed along the supply chain. Thus, they need to know when their products arrive to the storages, how long remain there, how long is the journey to the retails, etc. Also the managers of retail companies are interested in information collected during other stages, such as the origin of the products they sell in order to trust their quality. Furthermore, all these information are of interest also for the final customer.

This chapter provides a reference to the development of a virtual architecture to allow the traceability of products along a whole supply chain to assure transparency to all the actors involved, including customers. The trust of actors belonging to the supply chain is granted by the adoption of the EPCIS standard, maintained by the GS1 organization. All the captured EPCIS events are stored in a NoSQL data repository, particularly MongoDB. The storage of such data allows the execution of specific queries to retrieve tracking information regarding products.

The companies involved in all the different stages of the supply chains will benefit from the proposed architecture, in terms of operational efficiencies, market access and risk mitigation. In fact, the proposed architecture will allow supply chains to prove that their foods are safe and unrelated from food safety incident. Furthermore, it also will help eliminating illegal activities, since it gives suppliers and customers the confidence that what they are buying is legal and fairly traded. Finally, it will maintain customer trust.

Future works will address the investigation of useful extensions to enrich the EPCIS standard and the related queries to retrieve information of interest for different aspects of the product. Currently, only a few extensions to the standard attributes were proposed, mainly connected to environmental variables such as temperature. Depending on the kind of products, more structured set of extensions can be designed, thus allowing a differentiation of information coming from different phases of the supply chain. For example, a set of organoleptic properties observed in the products can be stored in the production phase, while the set of machining parameters can be stored in the processing phase. Furthermore, new fields can be added to the events, such as an identification of the person that is responsible for an event, and also new kinds of events can be defined, in addition to the standard ones.

REFERENCES

Abate, F., Di Leo, G., Paolillo, A., & Paciello, V. (2014) An advanced traceability system based on semi-active RFID devices. *20th IMEKO TC4 International Symposium and 18th International Workshop on ADC Modelling and Testing Research on Electric and Electronic Measurement for the Economic Upturn*, 1106-1109.

Abdul Kadir, E., Shamsuddin, S. M., Supriyanto, E., Sutopo, W., & Rosa, S. L. (2015). Food Traceability in Supply Chain Based on EPCIS Standard and RFID Technology. *TELKOMNIKA Indonesian Journal of Electrical Engineering*, *13*(1), 187–194. doi:10.11591/telkomnika.v13i1.6919

Anikina, E. V., & Djordjevic, M. (2015). International Experience of winemaking production traceability system use. *Journal of Hygienic Engineering and Design*, *12*, 85–89.

Appelhanz, S., Osburg, V. S., Toporowski, W., & Schumann, M. (2016). Traceability system for capturing, processing and providing consumer relevant information about wood products: System solution and its economic feasibility. *Journal of Cleaner Production*, *110*, 132–148. doi:10.1016/j.jclepro.2015.02.034

Badia-Melis, R., Mishra, P., & Ruiz-García, L. (2015). Food traceability: New trends and recent advances. A review. *Food Control*, *57*, 393–401. doi:10.1016/j.foodcont.2015.05.005

Bukeviciute, L., Dierx, A., & Ilzkovitz, F. (2009) The functioning of the food supply chain and its effect on food prices in the European Union. *European Economy, Occasional Papers*, 47.

Byun, J., & Kim, D. (2015). Enhanced Object Traceability on Unified and Linked EPCIS Events. *Lecture Notes in Computer Science*, *9418*, 232–246. doi:10.1007/978-3-319-26190-4_16

Chen, R.-Y. (2017). An intelligent value stream-based approach to collaboration of food traceability cyber physical system by fog computing. *Food Control*, *71*, 124–136. doi:10.1016/j.foodcont.2016.06.042

Cheng, M. L., & Simmons, J. E. L. (1994). Traceability in manufacturing systems. *International Journal of Operations & Production Management*, *14*(10), 4–16. doi:10.1108/01443579410067199

Chodorow, K. (2013). *MongoDB - The definitive guide*. O'Reilly.

Choi, J. B., Rogers, J., & Jones, E. C. (2015) The impact of a shared pharmaceutical supply chain model on counterfeit drugs, diverted drugs, and drug shortages. *Portland International Conference on Management of Engineering and Technology*, 1879-1889. doi:10.1109/PICMET.2015.7273165

Cimino, M. G. C. A., & Marcelloni, F. (2012). Enabling Traceability in the Wine Supply Chain. *Lecture Notes in Computer Science*, *7200*, 397–412. doi:10.1007/978-3-642-31739-2_20

Crockford, D. (2006). *RFC 4627, Network Working Group*. Available online: http://www.ietf.org/rfc/rfc4627.txt

D'Angelo, G., De Luca, G., & Rampone, S. (2014). Food safety and traceability system based on RFID technologies and services. *First International Workshop on Global Sustainability Inside and Outside the Territory*.

EPCIS. (2014). *EPC Information Services (EPCIS) Version 1.1 Specification.* GS1 Standard Version 1.1, May 2014.

Exposito, I. J. A., Gay-Fernandez, I., & Cuinas, A. (2013). Complete Traceability System for a Wine Supply Chain Using Radio-Frequency Identification and Wireless Sensor Network. *IEEE Antennas and Propagation Magazine*, *55*(2), 255–267. doi:10.1109/MAP.2013.6529365

Främling, K., Parmar, S., Hinkka, V., Tätilä, J., & Rodgers, D. (2013). Assessment of EPCIS Standard for Interoperable Tracking in the Supply Chain. *Studies in Computational Intelligence*, *472*, 119–134.

GS1. (2008). *GS1 Working Group, Wine Supply Chain Traceability.* GS1 Application Guideline, Brussels, Belgium.

GS1. (2016). *GS1 General Specifications.* Available online at http://www.gs1.org/docs/barcodes/GS1_General_Specifications.pdf

Gandino, F., Rebaudengo, M., & Sanchez, E. R. (2009). On Improving Automation by Integrating RFID in the Traceability Management of the Agri-Food Sector. *IEEE Transactions on Industrial Electronics*, *56*(7), 2357–2365. doi:10.1109/TIE.2009.2019569

Gunnlaugsson, V., Thakur, M., Forås, E., Ringsberg, H., Gran-Larsen, Ö., & Margeirsson, S. (2011) EPCIS standard used for improved traceability in the redfish value chain. *International Conference on Modern Information Technology in the Innovation Processes of the Industrial Enterprises*, 182-192.

Hsu, Y.-C., Chen, A. P., & Wang, C.-H. (2008) A RFID-Enabled Traceability System for the Supply Chain of Live Fish. *IEEE International Conference on Automation and Logistics*, 81-86.

Hwang, Y.-M., Moon, J., & Yoo, S. (2015). Developing A RFID-based Food Traceability System in Korea Ginseng Industry: Focused on the Business Process Reengineering. *International Journal of Control and Automation*, *8*(4), 397–406. doi:10.14257/ijca.2015.8.4.36

Jakkhupan, W., Arch-int, S., & Li, Y. (2015). An RFID-based traceability system A case study of rice supply chain. *Telecommunication Systems*, *58*(3), 243–258. doi:10.1007/s11235-014-9866-7

Kang, Y.-S., Park, I.-H., Rhee, J., & Lee, Y.-H. (2016). MongoDB-Based Repository Design for IoT-Generated RFID/Sensor Big Data. *IEEE Sensors Journal*, *16*(2), 485–497. doi:10.1109/JSEN.2015.2483499

Kassahun, A., Hartog, R. J. M., Sadowski, T., Scholten, H., Bartram, T., Wolfert, S., & Beulens, A. J. M. (2014). Enabling chain-wide transparency in meat supply chains based on the EPCIS global standard and cloud-based services. *Computers and Electronics in Agriculture*, *109*, 179–190. doi:10.1016/j.compag.2014.10.002

Kassahun, A., Hartog, R. J. M., & Tekinerdogana, B. (2016). Realizing chain-wide transparency in meat supply chains based on global standards and a reference architecture. *Computers and Electronics in Agriculture*, *123*, 275–291. doi:10.1016/j.compag.2016.03.004

Kelepouris, T., Pramatari, K., & Doukidis, G. (2007). RFID-enabled traceability in the food supply chain. *Industrial Management & Data Systems*, *107*(2), 183–200. doi:10.1108/02635570710723804

Leavitt, N. (2010). Will NoSQL databases live up to their promise? *Computer*, *43*(2), 12–14. doi:10.1109/MC.2010.58

Mainetti, L., Mele, F., Patrono, L., Simone, F., Stefanizzi, M. L., & Vergallo, R. (2013). An RFID-Based Tracing and Tracking System for the Fresh Vegetables Supply Chain. *International Journal of Antennas and Propagation. Article ID*, *531364*, 1–15.

Musa, A., & Abba Dabo, A.-A. (2016). A Review of RFID in Supply Chain Management: 2000–2015. *Global Journal of Flexible Systems Management*, *17*(2), 189–228. doi:10.1007/s40171-016-0136-2

NamGung, K., Choi, Y., Park, S., & Jun, C. (2012). The Development of e-Pedigree Model for Securing Transparent Pharmaceutical Distribution Channel in Korea. *Convergence and Hybrid Information Technology*, 226-234.

Parreño-Marchante, A., Alvarez-Melcon, A., Trebar, M., & Filippin, P. (2014). Advanced traceability system in aquaculture supply chain. *Journal of Food Engineering*, *122*, 99–109. doi:10.1016/j.jfoodeng.2013.09.007

Rota, C., Reynolds, N., & Zanasi, C. (2013). Sustainable Food Supply Chains: The Role of Collaboration and Sustainable Relationships. *International Journal of Business and Social Science*, *4*(4), 45–53.

Sadalage, P. J., & Fowler, M. (2013). *NoSQL Distilled*. Addison-Wesley.

Solanki, M., & Brewster, C. (2014). Modelling and Linking Transformations in EPCIS Governing Supply Chain Business Processes. *Lecture Notes in Business Information Processing*, *188*, 46–57. doi:10.1007/978-3-319-10491-1_5

Solanki, M., Brewster, C. (2015). OntoPedigree: Modelling pedigrees for traceability in supply chains. *Semantic Web*.

Thakur, M., & Forås, E. (2015). EPCIS based online temperature monitoring and traceability in a cold meat chain. *Computers and Electronics in Agriculture*, *117*, 22–30. doi:10.1016/j.compag.2015.07.006

Tian, F. (2016). A Quality and Safety Control System for China's Dairy Supply Chain Based on RACCP & GS1. *13th International Conference on Service Systems and Service Management*, 1-6.

Touzard, J.-M., & Maffezzoli, C. (2015). *Glamur project multidimensional comparison of a local and global wine supply chain*. INRA.

Verdouw, C. N., Wolfert, J., Beulens, A. J. M., & Rialland, A. (2016). Virtualization of food supply chains with the internet of things. *Journal of Food Engineering*, *176*, 128–136. doi:10.1016/j.jfoodeng.2015.11.009

Vukatana, K., Sevrani, K., & Hoxha, E. (2016). Wine Traceability: A Data Model and Prototype in Albanian Context. *Foods*, *5*(11), 1–11.

Yan, B., Hu, D., & Shi, P. (2012). A traceable platform of aquatic foods supply chain based on RFID and EPC Internet of Things. *International Journal of RF Technologies*, *4*(1), 55–70.

KEY TERMS AND DEFINITIONS

EPCIS: A GS1 standard that enables trading partners to share information about the physical movement and status of products as they travel throughout the supply chain.

GS1: An international non-profit organization with the aim of defining standards for product traceability.

MongoDB: A NoSQL database that stores data in the form of "document" and automatically manages replication and sharding of data.

NoSQL: database: A kind of database that provides a mechanism for storage and retrieval of data not modeled with the tabular relations used in relational databases.

RFID: Radio-frequency identification, which contain electronically stored information and it is attached to objects to make them automatically identifiable and traceable.

Supply Chain: A system of organizations, people, activities, information, and resources involved in the transformation of natural resources, raw materials, and components into a finished product from the supplier to the customer.

Traceability: The ability to verify the history and location of an item by means of documented recorded identification.

Transparency: The availability of information about how the product is produced, such as the presence of GMO ingredients in food products.

Chapter 3
Factors Influencing Acceptance and Use of ICT Innovations by Agribusinesses:
A Conceptual Framework

Adamkolo Mohammed Ibrahim
Universiti Putra Malaysia, Malaysia & University of Maiduguri, Nigeria

Md. Salleh Hj. Hassan
Universiti Putra Malaysia, Malaysia

Sarina Yusuf
Universiti Putra Malaysia, Malaysia

ABSTRACT

This chapter proposes a modified conceptual framework for investigating the influence of cognitive, management characteristics and organizational size factors on information and communication technology (ICT) adoption by agribusinesses. Agro-based small and medium-scale enterprises (SMEs) often deal in commodities that have shorter shelf life. Given that, researchers often face challenges determining the appropriate conceptual framework to adopt, which yields results that proffer both practical and theoretical solutions to business problems, hence, it is imperative for agri-preneurs to harness technology for maximum profit and food security. The unified theory of use and acceptance of technology (UTAUT) model, which has four key predictors, was adopted with the integration of two external variables: SME Managerial Characteristics and SME Organizational Size. Factor analysis shows that five out of the six predictors loaded strongly. The study concludes that researchers in technology adoption should consider integrating organization and management quality variables into their research frameworks.

DOI: 10.4018/978-1-5225-2107-5.ch003

INTRODUCTION

Background

The literature has identified Information and Communication Technology (ICT) as consisting of a wide range of technological applications: telecommunications technologies, digital broadcast technologies and electronic information facilities (Chitura, Mupemhi, Dube & Bolongkikit 2008). ICT, therefore, encompasses a range of technologies, information and resources. Furthermore, ICTs have been identified as having the potential to boost small and medium-scale enterprises (SMEs) (Ismail, Jeffery & Van Belle 2011). However, SMEs are often reluctant to relinquish the traditional approaches and means of conducting business and firm operation and embrace technological change. For SMEs to be able to compete favorably in the modern market that is driven by technological innovations the numerous benefits of ICT should be taken into consideration. There is also the need for more awareness creation on the benefits of ICT to SMEs (Chitura, et al. 2008).

This study was conducted to satisfy the urge to determine the agro-based firms' organizational expectancy of ICT use in business based on the UTATUT model perspective. The data was collected from employees of small and medium agribusinesses that are based in Selangor using a survey questionnaire. The theoretical perspective of the Unified Theory of Acceptance and Use of Technology (UTAUT) model formulated by Venkatesh, Morris, Davis and Davis (2003) was adopted. Even though the model seeks to understand organizational employees' technology adoption behavior, it focuses on the cognitive perspective and downplays or ignores possible influence organizational size and management characteristics could have on the organizations ICT adoption drive. This study seeks to understand the correlation between organizational and managerial characteristics alongside the four key predictors of the model of ICT adoption.

ICT use being the criterion variable, the four key predictors of the model (performance expectancy, effort expectancy, social influence and facilitating conditions in) in addition to aforementioned two integrated predictors that were derived from the SME size and SME management characteristics literature were applied to predict ICT use. The model has been a comprehensive theme in ICT adoption studies for over one decade due to its parsimonious power to explain a range of organizational and individual ICT adoption expectations and use behavior (Ahmad, Tarmidi, Ridzwan, Abdul Hamid & Abdul Roni 2014; Taiwo & Downe 2013).

Although the literature has widely documented that the UTAUT model explains many organizational and individual ICT adoption issues, its application in the study of Malaysian agro-based SMEs that are based in Selangor is limited. Hence, this study was conducted to reveal the critical factors that influence ICT use behavior among Malaysian agro-based firms. Moreover, this study is expected to be of immediate importance to the developing SME sub-sector in Malaysia's economy and to contribute to the current literature of ICT and (agro-based) SMEs in the country.

Meanwhile, SMEs have been identified as a major business sector across the world, covering a wide range of industries, such as agro-based or farm-based, metal-based, cosmetic-based, garment-based, etc. In most countries, the number of SMEs tremendously exceeds the number of large enterprises and companies. One of the advantages of SMEs is that they contribute strongly to the gross domestic products (GDP) and provide ample employment opportunities in most countries (Malhan 2015; Do, Mazzarol, Volery, Geoff & Reboud 2015). The literature has given numerous instances where ICT contributes to the economic development of nations in a couple of ways: as an important channel to convert innovative ideas into economic opportunities; as the basis for competitiveness through the revitalization of social and productive networks as a source of new employment; and as a way to increase productivity (Sonawane 2014).

Moreover, since this chapter focuses on a research, conceptual framework, findings of the research study are not presented and discussed.This chapter was structured into two broad parts, namely literature review and methodology. In the literature review section, relevant literature was reviewed under these sub-sections: UTAUT model, Malaysian agro-based SMEs, SME organizational size as a factor influencing ICT use and SME management characteristics as factors influencing ICT use among Malaysian SMEs.

STATEMENT OF RESEARCH PROBLEM

This conceptual framework was designed with the primary aim of bridging or closing the practical lacuna observed in the original UTAUT model variables because the four key variables (PE, EE, SI and FC) focus on the cognitive dimensions rather than environmental (organizational) and managerial nuances. The literature is quite replete with studies focusing on SMEs ICT adoption/use behavior/intention, often dominated by theoretical solutions to organizational problems rather than practical or both. Observations have revealed that business organizations, governments and non-governmental organizations (NGOs) often sponsor or conduct research studies with the aim of understanding the factors that affect both the theoretical and practical performance of their business. However, often, they end up discovering more of theoretical implications. This poses a great challenge to researchers and businesses.

The ICT adoption literature proposes quite a number of models. Among these models, eight have been very popular cited, namely, the Theory of Reasoned Action (TRA), the Theory of Planned Behavior (TPB), the Technology Acceptance Model (TAM), the Motivational Model (MM) and the model Combining the Technology Acceptance Model and the Theory of Planned Behavior (C-TAM-TPB). The rest is the Model of PC Utilization (MPCU), the Innovation Diffusion Theory (ID) and the Social Cognitive Theory (SCT) (AlAwadhi & Morris 2008; Taiwo & Downe 2013; Taiwo, Downe & Mahmood 2012). The UTAUT model is a unified model that was developed by Vankatesh et al. (2003) based on social cognitive theory with a combination of the eight models (Venkatesh, et al. 2003; Venkatesh, Thong & Xu 2012) mentioned above.

Furthermore, Dholakia and Bayo-Moriones, et al. (2013) and Kshetri (2004) argue that the influence of organizational and managerial factors in organizational technology (innovation) adoption is inevitable. Therefore, this study was further prompted by the urge to understand the influence of managerial characteristics and organizational size on SMEs ICT use. Hence, a modified UTAUT model that comprises a hybrid of three types of predictor variables, namely, cognitive variables (PE, EE, SI and FC), an organizational size variable (OS) and a managerial characteristics variable (C&I). Venkatesh, Morris and Ackerman (2000) posit that such modification is permissible in ICT use studies, especially with the UTAUT model because of its flexibility.

LITERATURE REVIEW

The UTAUT Model in Research

The UTAUT model is a unified model that was formulated by Venkatesh, et al. (2003) based on social cognitive theory with a combination of eight key ICT adoption research models. The eight theories are the Theory of Reasoned Action (TRA), the Theory of Planned Behavior (TPB) and the Technology Acceptance Model (TAM). The remaining theories are the Motivational Model (MM), a model Combining the Technology Acceptance Model and the Theory of Planned Behavior (C-TAM-TPB), the Model of PC Utilization (MPCU), the Innovation Diffusion Theory (IDT) and Social Cognitive Theory (SCT) (Taiwo & Downe 2013; Taiwo, Downe & Mahmood 2012). Therefore, it has been widely accepted that the UTAUT model surpassed the eight individual models, with an adjusted variance (R^2) of 70%.

The model uses four key determinants of ICT use and intention: performance expectancy, which is "the degree to which an individual believes that using the system will help him or her to attain gains in job performance" (Venkatesh, et al. 2003, p.447). The second determinant is effort expectancy, which is "the degree of ease associated with the use of the system" (Venkatesh, et al. 2003, p. 450). The third determinant is social influence, which is "the degree to which an individual perceives that important others believe he or she should use the new system" (Venkatesh, et al. 2003, p.451). The fourth determinant is facilitating conditions, which is "the degree to which an individual believes that an organizational and technical infrastructure exists to support the use of the system" (Venkatesh, et al. 2003, p.453). The behavioral intention construct, whose core concept is "the subjective probability of a person that he or she will perform the behavior in question" (Fishbein & Ajzen 1985, p. 288) is the dependent variable. The UTAUT model also considers moderators influencing the four predictors namely, gender, age, experience and [1]*voluntariness* (automaticity) of (ICT) use (AlAwadhi & Morris 2008).

AlAwadhi and Morris (2008) performed a survey of 880 students in the adoption of e-government services using the UTAUT model. The study revealed that performance expectancy, effort expectancy and peer influence determine behavioral intention. Similarly, facilitating conditions and behavioral intentions determine use of e-government services. Cheng, Liu and Qian (2008) found that performance expectancy and social influence determine users' behavioral intention toward Internet banking. Similarly, an empirical study in the use of Web 2.0 performed by Fang, Li and Liu (2008) suggest that performance expectancy, social influence and effort expectancy significantly predict firm managers' intention to engage in knowledge sharing using (Kaynak, Tatoglu & Kula 2005; Taiwo & Downe 2013; Taiwo, Downe & Mahmood 2012).

Result of an empirical investigation conducted by Maldonado, Khan, Moon and Rho (2009) on 240 secondary school students in Peru in the acceptance of an e-learning technology suggests that social influence significantly predicts behavioral intention. Wu, Tao and Yang (2007) investigated the acceptance of 3G services in Taiwan and found performance expectancy and social influence as predictors of behavioral intention. They also found that performance expectancy, effort expectation, social influence and facilitating conditions as predictors of user behavior.

ICT adoption in the business by SMEs could provide many benefits to both retailers and consumers. Such benefits may come in the form of implementing and using online transaction applications such as e-commerce, e-shopping, and even e-banking. In addition, entrepreneurs can access narrow market segments that are widely distributed while buyers benefit by accessing global markets with larger product availability from a variety of sellers at reduced costs (Mohapatra 2013); and product quality and the creation of new methods of selling existing products could improve.

The use of ICT has become almost inevitable in businesses (Bayo-Mariones & Lera-Lopez 2007). However, globally, SMEs lag behind in terms of applications of ICT (Mohapatra 2013; Burke, FitzRoy & Nolan 2002). However, in spite of the many potential advantages of the use of ICT in SMEs, its adoption remains limited, particularly in developing countries. For example, a survey conducted by Verizon found that 36% of small businesses established web sites primarily to advertise and promote their business, compared to 9% that established web site to sell or market their goods and services online. Similarly, in a survey of 444 SMEs in 2002, a study found that many SMEs were reluctant to conduct transactions online, more than 80% were only using the Internet to communicate (via e-mail) and gather business information (Mohapatra 2013). Some of the unique characteristics of SMEs are identified as lack of business and ICT strategy, limited access to capital resources, greater emphasis on using ICT and IS to automate rather than 'informate' (what do y mean;) the influence of major customers and limited information skills (Burke, et al. 2002; Mohapatra 2013).

SME Organisational Size as Factor Influencing ICT Use

The literature focusing on the determinants of ICT adoption suggests a grouping of variables into three categories, namely characteristics of the SME, management factors, which in general refer to entrepreneurial, management characteristics, and a group of variables related to the firm's environment (Bayo-Mariones & Lera-Lopez 2007). This study focused on the first two variables. Moreover, SME characteristics include those variables related to structural factors of the SMEs. The relationship between ICT adoption and SME size has been extensively studied (Bayo-Mariones & Lera-Lopez 2007). According to the Schumpeterian arguments large firms are better placed to develop and exploit new ICTs, because of economy of scale, better working conditions and greater ability to benefit from scale economies. Rogers (2003) argues that SMEs may be alternatively adoptive to new technology as they respond to changes in customers' needs and socioeconomic conditions. However, the debate on the role of SME size is still ongoing, as empirical studies have reached mixed conclusions.

SME Management Characteristics as Factors Influencing ICT Use

The unique nature of SMEs has proven to be an important consideration for any research interest (Dholakia & Kshetri 2004). A past study has found that firm managers play a significant role in the establishment, development and advancement of the organization (Abdullah, Shamsuddin, Wahab & Abdul Hamid 2012), which significantly influences the firms' policies and operations. Therefore, any major decisions regarding technology adoption would reside on the SMEs' managers (Taalika 2004).

Several scholars have claimed that technology adoption is based on three stages of cognitive, affective and behavioral behaviour (Bayo-Moriones & Lera-Lopez 2007; Bayo-Moriones, Billo´n & Lera-Lopez 2013; Higon 2011; Petroni & Rizzi 2001; Rogers 2003). They explained that at the cognitive stage, SMEs' managers become aware of the technology and through analysis of benefits and feasibility, they develop feelings towards it. If the feeling is favorable, the SME will move to behavioral stage in terms of actual use of ICT, which is translated into organizational willingness. Hence, it is suggested that the success of any technology adoption will depend on various factors such as technology characteristics, external characteristics and organizational characteristics (Abdullah et al. 2012).

Management quality (creativity and innovativeness) and firm size factors (population of employees and capital-base) have also been found to influence ICT adoption behavior among entrepreneurs and workers (Idota, Bunno & Tsuji 2011; Idota, Ueki, Bunno, Shinohara & Tsuji 2014). Firms and individuals' distinctive expectations of ICT use may either permit or limit change, innovation and performance (Fink & Disterer 2006). Therefore, it is important to investigate whether these perceptive expectations affect adoption and use of technology in business.

Malaysian Small and Medium Agribusinesses

Agribusiness is a term that is often used to denote an aggregate view of agriculture and business-related activities that cover the myriad of functions and processes that are involved in modern food production and distribution (Food and Agriculture organization [FAO] 2013). Agro-based SMEs have been given various definitions, though the key theme of the concept has been retained. Cited in FAO (2013) the United States Agency for International Development (USAID) (2008) defined agro-based SMEs as any business related to agriculture, including farming, processing, exporting, input suppliers, trading and retailing. Also cited in *Ibid*, Encyclopaedia Britannica (2011) broadly defined agro-based SMEs as agriculture operated by the business; specifically, a part of a modern economy devoted to the production, processing and distribution of food, fibre products and by-products including the financial institutions that fund these activities. Similarly, Sharma (2013) saw agro-based SMEs as those entrepreneurial firms that are engaged in the sourcing, production, processing (manufacturing) and distribution (marketing) of farm produces. Those three definitions were given from functionalist perspective.

However, agro-based SMEs are often defined from a size perspective, i.e., capital-base (annual turnover) and number of employees (Bernaert, Poels, Snoeck & De Barker 2013; SMECORP 2014). Agro-based SMEs in Malaysia are grouped under manufacturing industry. The Small and Medium Enterprises Corporation (SMECORP) of the country has provided a working definition for SMEs based on their size and capital base. Small enterprises are business firms with an annual turnover ranging from RM340,000 to RM17 million, or comprising of between 5 and 75 employees, or both. Medium enterprises are business firms with an annual turnover ranging from RM17 million to RM56 million or comprising between 75 and 200 employees, or both (SMECORP 2015).

Furthermore, recent statistics indicate that SMEs constitute 99.2% of total business establishments in Malaysia (645,136) and employs over 4,854,142 people, which is 56.4% of the total SMEs employment (8,460,971) size in 2012 and contributes 31% to the GDP. Agricultural sector with 34,188 SMEs (6.2%), contributes 6.6% to GDP and grows at a 3.3% rate per annum (DOSM 2014a,b; SMECORP 2014), which is low compared to services and manufacturing sectors that recorded 7.2% and 5.7% annual growth rates respectively.

Hypotheses of the Study

This study tested 12 hypotheses. Six hypotheses focused on the correlation between the predicting and criterion variables while the remaining six hypotheses focused on the four moderating variables (gender, age, experience and voluntariness (automaticity) of (ICT) use), which moderate the relationships between the predicting and criterion variables mentioned above (see Table 1).

Table 1. Hypotheses of the study with attributes

Hypothesis		Variable(s)	Statistic
H_1	There was a significant relationship between performance expectancy and ICT use.	PE & IU	Pearson Product Correlation/ Factor Analysis
H_2	There was a significant relationship between effort expectancy and ICT use.	EE & IU	"
H_3	There was a significant relationship between social influence and ICT use.	SI & IU	"
H_4	There was a significant relationship between facilitating conditions and ICT use.	FC & IU	"
H_5	There was a significant relationship between SME (organizational) size and ICT use.	SS & IU	"
H_6	There was a significant relationship between SME managerial creativity and innovativeness and ICT use.	C&I & IU	"
H_7	The influence of performance expectancy on ICT use would be moderated by gender, age and experience.	PE, IU, GDR, Age & EXP	Moderating Effect
H_8	The influence of effort expectancy of on ICT use would be moderated by gender, age and experience.	EE, IU, GDR, Age & EXP	"
H_9	The influence of social influence on ICT use would be moderated by voluntariness (automaticity) of (ICT) use.	SI, IU, & VOU	"
H_{10}	The influence of facilitating conditions would be moderated by gender, age and experience.	FC, IU, GDR, Age & EXP	"
H_{11}	Gender, age and experience moderated the relationship between SME (organizational) size and ICT use.	SS, IU, GDR, Age & EXP	"
H_{12}	Gender, age and experience moderate the relationship between SME (organizational) size and ICT use.	SS, IU, GDR, Age & EXP	"

Note: PE = Performance Expectancy
EE = Effort Expectancy
FC = Facilitating Conditions
SI = Social Influence
SS = SME (organizational) Size
C&I = SME Creativity and Innovativeness
IU = ICT Use
GDR = Gender
EXP = Experience
VOU = Voluntariness of Use

METHODOLOGY

Conceptual Framework of the Study

The conceptual framework was designed based on the UTAUT model (Venkatesh, et al. 2003) perspective, with some modifications. The UTAUT model is robust, comprehensive and is one of the most used models in ICT adoption studies (AlAwadhi & Morris 2008). Because this study focused on organisational ICT use by both employees and managers (workers) of SMEs, the study deemed it necessary to incorporate external predictor variables into the framework.

The adjustments that were performed on the model, which yielded this conceptual framework, were few. The original four predictors in the model were left unmodified. However, adopting the modification approach adopted by AlAwadhi and Morris (2008) in the study they conducted on the use of e-Government where the researchers adopted the UTAUT model with some adjustments, this study integrated two additional predictors to the model. The resulting research framework therefore consisted of six key predictors (independent variables) namely, performance expectancy, effort expectancy, social influence, facilitating conditions, SME Managerial Creativity and Innovativeness and SME Organisational Size (see Figure 1).

The UTAUT predictors are moderated by four variables, namely gender, age, experience and voluntariness of use. However, as part of the modification only the moderating influence of gender and age was investigated for the two externally incorporated predictors (see Bayo-Moriones & Lera-Lopez 2007; Dholakia & Kshetri 2004; Rahman & Ramos 2014). There is a large number of studies on the influence of firm management characteristics and firm size on ICT adoption (see Bayo-Moriones & Lera-Lopez 2007; Bayo-Moriones, Billo´n & Lera-Lopez 2013; Dholakia & Kshetri 2004; Higon 2011).

Many other similar studies on firm leadership characteristics and size provided the researcher a strong, empirical underpinning to merge the four original predicting variables in the UTAUT model with the two aforementioned derived predictors and investigate their influence on the ICT use behavior of the agro-based SMEs. However, prior to that, after performing an extensive and critical review of literature in firm characteristics studies the researcher perceived the urge to determine whether the SME organizational size and SME management qualities influence the use of ICT by Malaysian agro-based SMEs. Therefore, this study modelled the conceptual framework into a six-predictor research model

Figure 1. The conceptual framework of the study
Note: IVs = Independent variables; DV = Dependent variable
Source: Adopted with modifications from Venkatesh, et al. (2003)

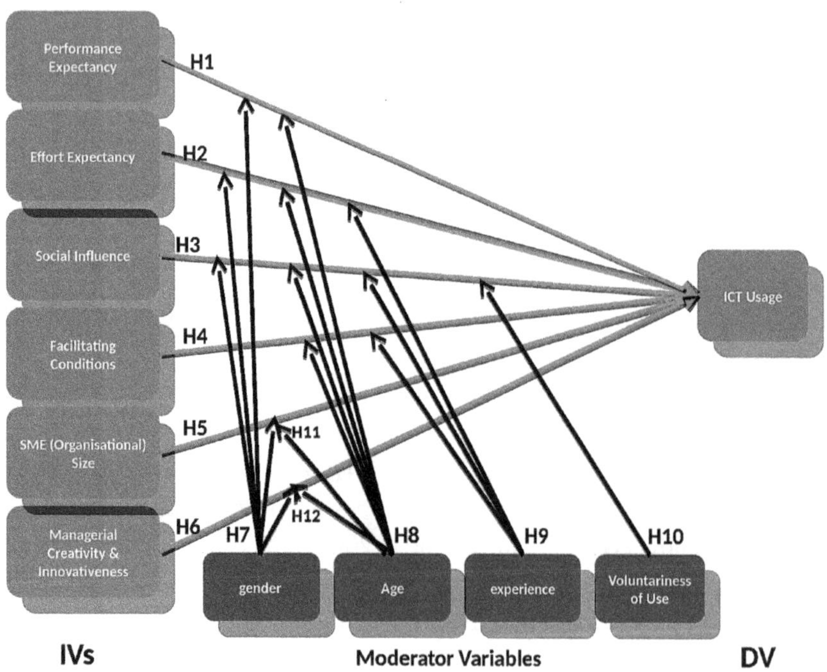

with ICT use as the independent variable. However, this study does not claim that the UTAUT Model could not provide an appropriate scale to measure the concept. Rather, this study wanted to explore (innovate in) another approach that combines concept-specific and concept-inclusive constructs to solve the research problem.

The SME managerial creativity and SME organizational size constructs that were integrated are concept-specific constructs, directly focusing on a particular concept (Rahman & Ramos 2014). The UTAUT Model is largely a concept-inclusive model, focusing on measuring a wide range of ICT adoption problems (AlAwadhi & Morris 2008; Venkatesh, et al. 2003). This chapter focused on determining ICT use behavior rather than ICT use behavioral intention as suggested by Venkatesh et al. (2003) in the model. This was because the ICT adoption literature has been already 'replete' with studies focusing on use intention (AlAwadhi & Morris 2008). In addition, it has been nearly two decades since the new technologies have evolved (Njoh 2012); hence, adoption may have substantially improved from mere intention to actual use. This study aimed to contribute toward closing that literature gap, which Ramayah, Ling, Taghizadeh & Rahman (2016) also suggest exists in the Malaysian ICT adoption literature. The conceptual framework provided a cogent research model from which a valid and reliable instrument was designed that was employed to collect data, which yielded interesting results. Table 2 shows the degree of the reliability of the constructs (both for pre-test and actual data collection) and the cumulative standard deviation of the various variables.

The preliminary/summary of the descriptive findings presented in Table 2 was for actual study only. Interestingly, as one of the two integrated predictors, SME managerial creativity and innovativeness scored the highest Mean value (M = 4.25, SD = 0.020) the other incorporated predictor, SME organizational size recorded the lowest mean value (M = 3.49, SD = 0.997). Both constructs, however, recorded a high degree of reliability, α = 0.888, 0.847 and α = 0.857, 0.757 for both pre-test and actual study respectively. In addition, both variables were measured with eight items in the scale respectively. Performance expectancy scored higher mean value of M = 4.14, SD = 0.035, with a high Chronbach alpha coefficient of α = 0.899, 0.796 for both pre-test and actual study respectively.

Although the dependent variable, ICT use scored a moderate mean value of M = 3.95, SD = 0.054, measured with 17 items, the construct recorded the highest degree of reliability α = 0.965, 0.951 for both pre-test and actual study respectively. With the exception of the independent variable, which was

Table 2. Cronbach alpha of the scale with a summary of descriptive statistics of result yielded by the conceptual framework

Variable	No. of items	Cronbach alpha (α)		SD	Mean
		Pre-test study (n = 40)	Actual study (n = 382)	(Actual study)	
SME Managerial Creativity and Innovativeness	8	0.888	0.847	4.25	0.020
Performance Expectancy	10	0.899	0.796	4.14	0.035
Facilitating Conditions	10	0.848	0.849	4.07	.039
Social Influence	10	0.868	0.889	4.03	0.018
ICT Use	17	0.965	0.951	3.95	0.54
SME Organisational Size	8	0.822	0.808	3.85	0.040
Effort Expectancy	10	0.857	0.757	3.49	0.997

measured with 17 items and the each of the two imported variables, which were measured with eight items each, the rest of the variables were measured with 10 items each; and each of the variables recorded either a very high or just high range of degree of reliability.

CONCLUSION, IMPLICATIONS, AND RECOMMENDATIONS FOR FUTURE RESEARCH

Conclusion

This chapter determined the critical factors that influence the use of ICT by Malaysian agro-based SMEs. The perspective of the UTAUT model was adopted. After the review of relevant literature, which gave strong theoretical and conceptual underpinning for an integrated conceptual framework to be designed, the six-predictor research model that was yielded explained 31% of the variance in ICT use after running confirmatory factor analysis CFA.

The results of this study were quite unpredicted, in that one of the two external variables that were incorporated into the UTAUT Model i.e., SME Managerial Creativity and Innovativeness (C&I) scored the highest mean value, followed by Performance Expectancy (PE), which is one of the four original variables of the model. Initially, the expectations were that the original variables of the UTAUT model would yield more important results than the externally incorporated variables. Furthermore, running confirmatory factor analysis eliminated the Social Influence (SI) variable due to poor loading (see Hair, et al. 2010), consequently dislodging two of the 12 hypotheses that were derived from the (SI) variable. The (analytical) fit of the structural research model was not affected negatively, as it recorded a good RMSEA value of 0.079 (see Kenny 2014).

The modified conceptual framework and the resultant structural model have validly and reliably explained almost one-thirds (31%) of the variance in the ICT use behavior of Malaysian agro-based SMEs. Moreover, given the size of the sample (400 agro-based SMEs, workers) and the population size (834 agro-based SMEs in Selangor state, Malaysia with 78,855 workers) (see SMECORP 2014), the results of the study can be generalized among a population of one million SME workers (see Sekaran 2003).

Implications of the Study

Theoretical Implications

This chapter reveals that adopting a hybridization approach by integrating external constructs from managerial and organizational characteristics into the UTAUT model yields comprehensive outcomes. Because the model was formulated to explain ICT adoption variations from organization employees' perspective that was why this approach was adopted. Hence, this study argues that in any organizational ICT adoption study the issue bordering the correlational influences between managerial characteristics and ICT adoption behavior on one hand and between organizational characteristics and ICT adoption behavior on the other hand cannot be ruled out (see Bayo-Moriones, et al. 2013; Dholakia & Kshetri, 2004; Venkatesh, et al. 2003). Furthermore, the discovery of this study that SME managerial and organizational size characteristics have a significant and positive influence on ICT use has lent credence to the claim (argument) posed (above).

Practical Implications

Arguably, many entrepreneurial studies surrounding ICT adoption are primarily conducted in order to solve business problems from practical points of view (see Dholakia & Kshetri 2004; Martin & Matlay 2001). However, quite a number of such studies end up proffering theoretical solutions. In order to avoid this challenge, this study adopted the hybridised perspective and discovered that innovative researchers could proffer more effective and comprehensive solutions regarding technology adoption in small and medium-scale firms. These solutions do not only focus on the cognitive variables inherent in the model but also on investigating the influence of related external variables such as management characteristics and organisational size alongside the inherent variables and integrating them into the model. Chief executive officers of SMEs, governments and other stakeholders in the innovative development of SMEs should identify organizational management, quality and organizational size as key factors that affect technological innovation in SMEs. In addition, they need to determine the degree and direction of such influence on the ICT adoption behavior of the firms with a view to finding lasting proactive and result-oriented solutions for both industrial and socio-economic development.

Recommendations for Future Research

This chapter recommends that future research should adopt the modified conceptual model of this study and conduct similar study in a different context (setting/milieu). Presumably, it might yield different results or support the current study. Another perspective to this particular recommendation is that future research should retain the methodology of this conceptual model and adopting the original UTAUT model without any modification. Given that the conceptual model suggested by this chapter was modified by incorporating two constructs to the original four key predictors, thus making the model have six predictors (refer to Figure 1). The two additional constructs were derived from business firms organisational and management characteristics literature. Each of the two imported constructs was conceptualised to determine the influence of organisational size (see Bayo-Moriones & Lera-Lopez 2007; Dholakia & Kshetri 2004) and SME management creativity and innovativeness (see Higon 2011; Martin & Matlay 2001) on technology (ICT) use respectively. [2]This recommendation was prompted by the urge to close a construct gap created by the elimination of Social Influence construct (which is one of the four original key constructs of the UTAUT model) due to poor component loading in factor analysis.

This chapter recommends that future research should retain the methodology and adopt the model without modification. Future research should identify additional variables, which might provide explanation of ICT use by SMEs more comprehensively. Retaining the current research model, this chapter recommends that future research should determine ICT use by large enterprises, thus focusing on either employees or management personnel only given that the present research combined both categories of SME workers.

Since this study adopted a quantitative approach, this chapter recommends that future research should adopt qualitative approach in order to conduct an in-depth observation of the causal relationships between the variables, especially from experts' points of view to elicit rich information that quantitative approach does not support. In addition, it will be interesting to understand the ICT usage behavior of non-agro-based SMEs. Hence, this study recommends that further research should focus on non-agro-based businesses. Generally, however, the effect of ICT use behavior on overall business operations and strategic directions would be of interest for future researchers.

REFERENCES

Abdullah, N. H., Shamsuddin, A., Wahab, E., & Abdul Hamid, N. (2012). Preliminary qualitative findings on technology adoption of Malaysian SMEs. *2012 IEEE Colloquium on Humanities, Science and Engineering Research (CHUSER 2012).*

Ahmad, N. N., Tarmidi, M., Ridzwan, I. U., Masdiah Abdul Hamid, M., & Rusli Abdul Roni, R. (2014). The application of the Unified Theory of Acceptance and Use of Technology (UTAUT) in for predicting the usage of e-Zakat online system. *International Journal of Science and Research, 3*(4), 63–70.

Alawadhi, A., & Morris, A. (2008). The use of the UTAUT model in the adoption of *e*-government services in Kuwait. *Proceedings of the 41st Hawaii International Conference on System Sciences.* doi:10.1109/HICSS.2008.452

Bayo-Moriones, A., Billon, M., & Lera-Lopez, F. (2013). Perceived performance effects of ICT in manufacturing SMEs. *Industrial Management & Data Systems, 113*(1), 117–135. doi:10.1108/02635571311289700

Bayo-Moriones, A., & Lera-Lopez, F. (2007). A firm-level analysis of determinants of ICT adoption in Spain. *Technovation, 27*(6-7), 352–366. doi:10.1016/j.technovation.2007.01.003

Bernaert, M., Poels, G., Snoeck, M., & De Barker, M. (2014). Enterprise architecture for small and medium-sized enterprises: A starting point for bringing EA to SMs, based on adoption models. In *Information Systems for Small and Medium-Sized Enterprises* (pp. 67–96). Berlin: Springer. doi:10.1007/978-3-642-38244-4_4

Burke, A. E., FitzRoy, R. L., & Nolan, M. A. (2002). Self-employment, wealth and job creation: The roles of gender, non-pecuniary motivation and entrepreneurial ability. *Small Business Economics, 19*(3), 255–270. doi:10.1023/A:1019698607772

Cheng, D., Liu, G. & Qian, C. (2008). On determinants of user acceptance of Internet banking: A theoretical framework and empirical study. *Advanced Management of Information for Globalized Enterprises.*

Chitura, T., Mupemhi, S., Dube, T., & Bolongkikit, J. (2008b). Barriers to electronic commerce adoption in small and medium enterprises: A critical literature review. *Journal of Internet Banking and Commerce, 13*(2), 1–13.

Cudec, R. (2000). Exploratory factor analysis. In Handbook of Applied Multivariate Statistics and Mathematical Modeling. New York: Elsevier Inc. doi:10.1016/B978-012691360-6/50011-2

Dholakia, R. R., & Kshetri, N. (2004). Factors impacting the adoption of the internet among SMEs. *Small Business Economics, 23*(4), 311–322. doi:10.1023/B:SBEJ.0000032036.90353.1f

Do, H., Mazzarol, T., Volery, T., Geoff, S., & Reboud, S. (2015). SME owner-manager's anticipated returns to commercialisation: The influence of organisational factors. *60th International Council for Small Business (ICSB) World Conference 2015: Entrepreneurship at a Global Crossroads.*

DOSM. (2014a). *SMEs Census 2011.* Available at www.statistics.my.gov/

DOSM. (2014b). *Statistics Yearbook Malaysia 2012.* Retrieved from http://www.statistics.gov.my/portal/download_Buku_Tahunan/files/BKKP/2012/pdf/TAB10.pdf

Fang, W. C., Li, M. W., & Liu, C. W. (2008). Measurement of the knowledge-sharing efficacy of Web2.0 site constructed on the basis of knowledge-based systems by applying the model of UTAUT: Evidence of the early adopters. *Proceeds of the 3rd International Conference on Innovative Computing Information and Control (ICICIC'08).*

FAO. (2013). *Enabling environments for agribusiness and agro-industries development: Regional and country perspectives. Agribusiness and Food Industries Series 1* (pp. 1–61). Rome: FAO, United Nations.

Fink, D., & Disterer, G. (2006). International case studies: To what extent is ICT infused into the operations of SMEs? *Journal of Enterprise Information Management, 19*(6), 608–624. doi:10.1108/17410390610708490

Fishbein, M., & Ajzen, I. (1985). Belief, attitudes, intention and behaviour. An introduction to theory and research. Reading, MA: Addison-Wesley.

Hair, J. F. J., Black, W. C., Babin, B. J., Anderson, R. E., & Tatham, R. L. (2010). *Multivariate data analysis a global perspective. Pearson Education Inc.*

Higon, D. A. (2011). The impact of ICT on innovation activities: Evidence for UK SMEs. *International Small Business Journal, 30*(6), 684–699. doi:10.1177/0266242610374484

Idota, H., Bunno, T., & Tsuji, M. (2011). Empirical analysis of internal social media and product innovation: Focusing on SNS and social capital. *Proceedings of 22nd European Regional ITS Conference in Budapest.*

Idota, H., Ueki, Y., Bunno, T., Shinohara, S., & Tsuji, M. (2014). Role of ICT in the innovation process based on firm-level evidence from four ASEAN economies: An SEM approach. *25th European Regional Conference on the International Telecommunications Society (ITS).*

Ismail, R., Jeffery, R., & Van Belle, J.-P. (2011). Using ICT as a value adding tool in South African SMEs. *Journal of African Research in Business & Technology, 2011*, 1–12. doi:10.5171/2011.470652

Kaynak, E., Tatoglu, E., & Kula, V. (2005). An analysis of the factors affecting the adoption of electronic commerce by SMEs: Evidence from an emerging market. *International Marketing Review, 22*(6), 623–640. doi:10.1108/02651330510630258

Kenny, D. A. (2014). *Measuring model fit.* Available online at http://davidakenny.net/cm/fit.htm#RMSEA

Kenny, D. A., Kaniskan, B., & McCoach, D. B. (2014). The performance of RMSEA in models with small degrees of freedom. *Sociological Methods & Research.*

MacCallum, R. C., Browne, M. W., & Sugawara, H. M. (1996). Power analysis and determination of sample size for covariance structure modelling. *Psychological Methods, 1*(2), 130–149. doi:10.1037/1082-989X.1.2.130

Maldonado, U. P. T., Khan, G. F., Moon, J., & Rho, J. J. (2009). E-learning motivation, students' acceptance/use of educational portal in developing countries. *Proceeds of the 4th International Conference on Computer Sciences and Convergence Information Technology.*

Malhan, S. (2015). Impact of demographic and entrepreneur's profile on level of supply management chain practices in selected small scale and medium enterprises (SMEs) in retail sector. *International Journal of Business Management and Scientific Research, 2*.

Malhotra, N. K. (2004). *Marketing research: An applied orientation*. Upper Saddle River, NJ: Prentice-Hall.

Martin, L. M., & Matlay, H. (2001). Blanket approach to promoting ICT in small firms: Some lessons from the DTI ladder adoption model in the UK. *Internet Research, 11*(5), 399–410. doi:10.1108/EUM0000000006118

Mohapatra, S. (2013). Sustainability in e-commerce adoption in small and medium enterprises (SMEs): A case study in Odisha, India. *International Journal of Green Computing, 4*(2), 12–23. doi:10.4018/jgc.2013070102

Njoh, A. J. (2012). Implications of spatial and physical structures for ICT as a tool of urban management and development in Cameroon. *Habitat International, 36*(3), 343–351. doi:10.1016/j.habitatint.2011.06.006

Nunnally, J. C. (1978). *Psychometric theory*. New York, NY: McGraw-Hill.

Petroni, A., & Rizzi, A. (2001). Antecedents of MRP Adoption in small and medium-sized firms. *Benchmarking: An Integrated Journal, 8*(2), 144–156. doi:10.1108/14635770110389852

Rahman, H., & Ramos, I. (2014). Open innovation in SMEs: Prospects and Challenges. In *Information Systems for Small and Medium-sized Enterprises* (pp. 313–335). Berlin, Germany: Springer. doi:10.1007/978-3-642-38244-4_16

Ramayah, T., Ling, N. S., Taghizadeh, S. K., & Rahman, S. A. (2016). Factors influencing SMEs website continuance intention in Malaysia. *Telematics and Informatics, 33*(1), 150–164. doi:10.1016/j.tele.2015.06.007

Rogers, E. M. (2003). *Diffusion of Innovations* (5th ed.). New York: Free Press.

Sekaran, U. (2003). *Research methods for business: A skill building approach*. New York, NY: Wiley.

Sharma, P. (2013). Rural entrepreneurship, science and technology and innovations in farm-based entrepreneurship venture: A case study of a farmer inventing accustomed machine for tillage farming. *International Journal of Innovative Research and Development, 2*(4), 6-14. Available at www.ijird.com

SMECORP. (2014). *Definition of SMEs (Appendix 4)*. Available at www.smecorp.gov.my

SMECORP. (2015). SME development framework: The Malaysian case. *Proceedings of the Cairo AMC-CBE-WBG Conference*. Available at http://www.cbe.org.eg/NR/rdonlyres/95EA10FA-62EE-4469-A40E-A952FDE2B434/2737/CairoSMEConferenceJan2015Session3DatoHafsahMalaysi.pdf Retrieved 6 March 2016.

Sonawane, M. B. (2014). Entrepreneurship (entrepreneurial mind) in emerging economies. *Journal of Business Management and Social Sciences Researches, 3*(2), 24-26. Available at www.borjounals.com

Taalika, S. (2004). Assessing the managerial influence as a facilitating factor in innovation adoption process. *Proceedings of the IAMOT Conference*.

Taiwo, A. A., Downe, A. G., & Mahmood, A. K. (2012). User acceptance of E-government: Integrating risk and trust dimensions with UTAUT Model. *Proceedings at the International Conference on Computer and Information sciences.*

Taiwo, A. D., & Downe, A. G. (2013). The Theory of User Acceptance and Use of Technology (UTAUT): A meta-analytic review of empirical findings. *Journal of Theoretical and Applied Information Technology, 49*(1).

Teo, T., Su Luan, W., & Sing, C. C. (2008). A cross-cultural examination of the intention to use technology between Singaporean and Malaysian pre-service teachers: An application of the Technology Acceptance Model (TAM). *Journal of Educational Technology & Society, 11*(4), 265–280.

Venkatesh, V., Morris, M. G., & Ackerman, P. L. (2000). A longitudinal field investigation of gender difference in individual technology adoption decision-making processes. *Organizational Behavior and Human Decision Processes, 83*(1), 36–60. doi:10.1006/obhd.2000.2896 PMID:10973782

Venkatesh, V., Morris, M. G., Davis, G. B., & Davis, F. D. (2003). Use acceptance of information technology: Toward a unified view. *Management Information Systems Quarterly, 27*(3), 425–478.

Wu, Y. L., Tao, Y. H., & Yang, P. (2007). Using UTAUT to explore the behaviour of 3G mobile communication users. In *Industrial Engineering and Engineering management, 2007 IEEE International Conference on* (pp. 199-203). IEEE.

ADDITIONAL READING

Abd Rahman, A., Kamarulzaman, A., Hirawaty, N. & Sambasivan, M. (2013). A study on organizational culture, performance and technological adoption: Behaviors of Malaysian food processing SMEs. *Pertanika Journal of Social Sciences and Humanities, 21*(S), 231-256.

Adamkolo, M. I., Hassan, M. S., & Yusuf, S. (2016 Dec.). Determinants of information technology adoption among Malaysian farm-based enterprises. *Pertanika Journal of Social Sciences and Humanities, 24*(4), ISSN: 0128-7702.

Al-Qeisi, K., Dennis, C., Hegazy, A., & Abbad, M. (2015). How viable is the UTAUT model in a non-Western context? *International Business Research, 8*(2), 204–219. doi:10.5539/ibr.v8n2p204

Aleke, B., Ojiako, U., & Wainwright, D. W. (2011). ICT adoption in developing countries: Perspectives from small-scale agribusinesses. *Journal of Enterprise Information Management, 24*(1), 68–84. doi:10.1108/17410391111097438

Bhuiyan, A. B., Said, J., Ismail, M. D., Jani, M. F. M., Yong, D., & Fic, G. (2016). The innovation drivers, strategies and performance of food processing SMEs in Malaysia. *Geografia Malaysian Journal of Society and Space, 12*(2), 154–166.

Hilmi, M. F., Ramayah, T., Mustapha, Y., & Pawanchik, S. (2010). Product and process innovativeness: Evidence from Malaysian SMEs. *European Journal of Soil Science, 16*(4), 556–565.

Läpple, D., Renwick, A., & Thorne, F. (2015). Measuring and understanding the drivers of agricultural innovation: Evidence from Ireland. *Food Policy*, *51*, 1–8. doi:10.1016/j.foodpol.2014.11.003

Rahman, H., & Ramos, I. (2010), Open innovation in SMEs: From closed boundaries to networked paradigm. In *Proceedings of the Issues in Informing Science and Information Technology Education Conference,* Cassino, Italy June 19–24, (pp. 471–487).

Spinelli, R. (2016). The Determinants of IT Adoption by SMEs: An Agenda for Research. In *Organizational Innovation and Change* (pp. 41–52). Switzerland: Springer International Publishing. doi:10.1007/978-3-319-22921-8_4

Taylor, P. (2015). The importance of information and communication technologies (ICTs): An integration of the extant literature on ICT adoption in small and medium enterprises. *International Journal of Economics. Commerce and Management*, *3*(5), 1–22.

Zaremohzzabieh, Z., Samah, B. A., Muhammad, M., Omar, S. Z., Bolong, J., & Shaffril, H. A. M. (2016). An investigation into factors influencing rural youth entrepreneurs intentions to use ICT: A case of Malaysia. *International Journal of Entrepreneurship and Small Business*, *27*(4), 480–504. doi:10.1504/IJESB.2016.075010

KEY TERMS AND DEFINITIONS

Agribusinesses (Agro-Based Enterprises): These entrepreneurships deal with unprocessed and semi-processed agricultural and related products and products respectively for the purposes of making profit.

Creativity: As used in this chapter, this term is related to a firm's management function that refers to the art of improving product quality and designing product packaging in such a way that they are marketable and promotional unique and original.

ICT Adoption: This term refers to a willingness to accept and use a particular technology for the purposes of boosting a business entrepreneurship.

ICT (Also Concisely Referred to as the Technology): ICT is a popular acronym that refers to information and communication technology. There are numerous definitions of ICT. However, as used in this chapter, the term simply refers to any electronic and digital device or system that is employed to retrieve, process, archive and disseminate information (including data) between and or among various users that may be linked online or otherwise. It also refers to any such device or system that can be used to process raw materials or add value to semi-finished products and yield superbly finished products.

ICT Use: This term refers to the deliberate and actual (as opposed to intention or willingness to employ/use) employment of a technology, device or system by an individual, a group or an organization in order to satisfy various needs.

Innovation: This term refers to the strategic adoption and application (use) of ICT in a business entrepreneurship aimed at improving the business. It also refers to the acceptance of ICT-related strategic business interventions by an enterprise management with a view to maximize profit.

Organisational Size: This term refers to a business organization's population of permanent workers, number of branches and monetary value of its annual turnover.

SMEs (Firms): SMEs or SME is an acronym that fully means small and medium-scale enterprises. The term refers to two types of business enterprises that are literally categorized based on their size (see definition of organizational size above). Collectively, SMEs are also referred to as firms.

ENDNOTES

1. Voluntariness of use refers to the use of a technology voluntarily (because of internalization and habituation due to long-time use and experience), as opposed to the use of a technology under compulsion (AlAwadhi & Morris 2008; Venkatesh, et al. 2003).
2. This little explanation is an allusion to one of the findings (which are not in this chapter) of the research study conducted with this conceptual framework. The allusion becomes necessary in order to justify some assertions.

Chapter 4
The Ecovillage:
Concept and Applications

Azizul Hassan
Cardiff Metropolitan University, UK

Geoffrey Wall
University of Waterloo, Canada

ABSTRACT

This paper examines ecovillages in the context of tourism and research development. Four ecovillages form Bangladesh are selected as cases of this study. Data were mainly collected by focus group discussions harnessing a cross validity check of the given statements and arguments. A critical explanatory type analysis illustrated and evaluated the ecovillage concept application. Ecovillages are validated as an idea, a useful concept and as a practice resulting in tourism, research and sustainable livelihood practices. The concept also found as providing substantial and supplementary economic opportunities for its residents.

INTRODUCTION

The purpose of this paper is to examine the 'ecovillage' concept in the context of tourism and research practices. Conceptually, ecovillages are purposefully built settlements recognizing the existence of positive relationships between the environment and the society while in most cases, underpinned by the moral imperatives of their proponents (Trainer, 1998). Ecovillages are diverse in contents, including permaculture, renewable energy production or environment friendly community buildings. This type village concept practically affects host communities by increased employment opportunities, income generation and well-being of residents through improved living standards. In principle, ecovillages are built within a participatory process relying on indigenous knowledge and local practice capacities. In cases, ecovillages attract innovative technologies supported by foreign funding mechanisms, and promote 'clean and green' environments in ways that are culturally appropriate. Ecovillages as such create ap-

DOI: 10.4018/978-1-5225-2107-5.ch004

peal to a range of tourists either for leisure or having research interests in underpinning the creation, the operational activities, the environments and the resident lifestyles that result from this defined concept.

The development of ecotourism concept is attached to the community based movements that dated back as early as in the 1960s and 1970s. At a later stage, such movement literally was promulgated in many development initiatives including tourism. Eventually, in 1996, the Global Ecovillage Network (GEN) was established as a network of ecovillages and 'intentional communities' to share ideas and promote the ecotourism concept. In custom, GEN initiated the diffusion of this concept. Since then, this concept has been applied in more than seventy countries. The building of ecovillages tends to follow an established set of guidelines addressing the environment and surroundings in designing, implementing and operating stages. The ecovillage concept evidently gets support from the non-government organizations (NGOs).

Very often, the ecovillage concept is aimed for ensuring the betterment of its residents' livelihood within its outside neighborhoods. Ecovillage residents are described as the 'Intentional communities' that proponent the common success of this concept. Such communities in their historic origin perspective can be traced back to the Utopian communities of the industrial revolution or even farther before (Holloway, 1951; Hardy, 2000). In an ecovillage, the resident members are normally like-minded, having common social, religious, spiritual, geographical or ethnic visions (Gilman, 1991). Typically, residents of an ecovillage adopt responsible behaviors towards the environment. In return, Ecovillage offers them a safer, better and sustainable living environment with adequate resources to maintain their dependents as self-reliant.

Characteristically, the ecovillage concept varies enormously in the orientation and degree of success. General tourists in ecovillages are often driven by the common motive of curiosity. Still, its divergent features attract tourism researchers having diverse motives and research goals (Christian, 2003). However, the extents to which the ecovillage concept can achieve goals are debatable. The ecovillage concept is often preceded by discussions of sustainability and, in a tourism context, the benefits that this concept application could bring to destination communities, resident-tourist relationships and tourist satisfaction. The ecovillage concept, though not widely discussed in the tourism literature, has the potential to enrich discussions of appropriate development and associated policy initiatives, as well as the sustainable use of resources. However, given the imprecise nature of the ecovillage concept, there is a need to explore the meanings of this concept in greater detail, as well as to investigate its success in stimulating the creation of sustainable communities in reality. Accordingly, this paper examines the nature of the ecovillage concept. It then explores implementation aspects of this concept in Bangladesh leading to a critical explanation of the notion as a potential construct to inform tourism research and development.

CONTEXT

The exact meaning of the term 'ecovillage' is obscure and a wide agreement on its meaning does not exist. A group of academics and practitioners suggested that the concept was first used by Diane and Robert Gilman, founders of the Context Institute based in Seattle, Washington, in the United States. The concept appeared in 1991 in the Gaia Trust's commissioned report on sustainability (Gilman, 1991). However, others believe that Corrine McLaughlin and Gordon Davidson coined the term as early as in 1985 (McLaughlin and Davidson, 1985). A simple basic and generally acceptable definition of the ecovillage concept was offered by Gilman (1991) as the human settlements aimed to generate less damaging impacts for nature and the environment. Bang (2005) argued that ecovillages are human settlements, typically having fifty to five hundred members who engage in leisure activities, food production supported by creating opportunities for social development, manufacturing and commercial activities. The key goal of ecovillages is meant as to harmlessly integrate the environment with human activities to support developments for an unforeseeable future.

Jackson (2004) viewed the ecovillage concept as an element of a relatively unexplored perspective having huge potential. Jackson (2004) argued that this notion can outline the formulation of a clear relationship existing within a society, responsible natural resources use and sustainable development. According to Litfin (2014), ecovillages can provide an ultimate safeguard against human inflicted negative consequences to favor the nature and the habitats. On the other side Jackson and Svensson (2002) believed that, ecovillages can be an escaping option from the everyday modern industrial life hustle and bustle, particularly for those who search for a more communal and green lifestyle. From this perspective, human living patterns in industrial societies are seen as monotonous. Exhausted energy resources and global economic meltdowns evidence the inability of the existing systems to provide even the most basic needs of mankind. Theoretically, the ecovillage concept is proposed as a practical means of replacing the irresponsible use of natural resources by a better, more efficient and reliable means to support both present and future generations. Thus, the concept is imbued with an ethical and moral position that supports self-reliance and sustainable ways of providing livelihoods (Ross and Wall, 1999).

From the concept application perspective, ecovillages can be defined as purposefully settled communities which aim to become ecologically, socio-economically more self-reliant on a sustainable basis. Also, a social network exists within residents sharing similar cultural, spiritual, ecological and socioeconomic values (Fellowship for Intentional Community, 2005). According to GEN (2014), notions of the 'ecovillage' and the 'intentional community' are very similar. Ecovillage is a traditional and intentional community using local participatory processes to integrate ecological, economic, social and cultural dimensions of sustainability in order to regenerate social and natural environments. Features, sizes and numbers of such villages vary from place to place based on beneficiary demands. Such villages are usually built with integrated systems for sewage disposal, and water and electricity supply as basic human necessities (Van Schyndel Kasper, 2008). This definition ascertains that the common literature on ecovillages is suffused by the notion of sustainability. Sustainability in this context is seen as incorporating with like minded people and community groups (Lim and McAleer, 2005). Sustainability after attaching to development turns as more focused towards the economy. The contribution of the publication of Dasmann, Milton and Freeman's (1973) as 'Ecological Principles for Economic Development' is relevant in this perspective. Dasmann et al. (1973) advocate to establish a dignified living for the present and future generations by maintaining healthy relationships between fragile ecosystems, human settlements and thriving economies.

The involvement of donor agencies in implementing the ecovillage concept is evident. Sustainability appeared as the basic factor propelling the implementation process. This idea of sustainability has been of especial interest to national and international organizations while, their roles vary in countries relying on circumstances (Blewitt, 2012). Contributions of both public and private sector bodies for policy formulations depend on local circumstances backed by cultural differences. Sustainability was given explicit attention in the 'World Conservation Strategy' of the International Union for Conservation of Nature (IUCN). Sustainability was seen as a development process as an active route of changes meeting both existing and future human wants and ambitions (IUCN, 1980). This was later re-stated in the Brundtland Commission Report as 'development that meets the needs of the present without compromising the ability of future generations to meet their own needs' (WCED, 1987: 43). Still, sustainability aspects in the ecovillage concept application remain blurred while, the application, validity and usability at multiple scales are difficult to manage (Honey, 1999).

Environmental quality requires the attention of stakeholders and beneficiaries attached with ecovillages (Jurin, 2012). Community groups and business establishments must be responsive to the possible effects of tourism, as well as their own activities on the environment (Akama and Kieti, 2003). However, tourists like to visit places with high quality environments, helping to reinforce consciousness about nature and habitats. Citing the example of the Mt. Coot-the Botanic Gardens, Ballantyne, et al. (2009) suggested that tourists in such place demanded more awareness and commitments to address conservation issues.

From a tourism perspective of the ecovillage application, sustainability affirms the need for development approaches to both addressing and resolving issues and tensions that result from the interactions between host communities, environment and tourism (Butler, 1999). In a tourism context, the management of the economic, physical and social-cultural impacts are discussed by Mathieson and Wall (1982) while, the pleas for the integration of environment, society and economy are explained by Farrell (1999). These are congruent with the widely advocated three pillars of sustainability.

Yet, within tourism development, the objectives of sustainability proponents can be multifold: for example, to meet demands of the host communities through improving both short and long term living standards; to satisfy the demands of increased number of tourists; and to safeguard the natural environment (Cater, 1993). With respect to the ecovillage concept application, such goals are desired to be achieved within the compass of one small settlement. In order to explore the practical application aspects of the ecovillage concept, selected cases from Bangladesh were considered as presented in the next section.

METHODOLOGY

The four cases that have been selected are in Bangladesh. The basic reason for this selection was to bring results from diverse contexts. The cases of ecovillages in Bangladesh ensured the representation of tourism, research, sustainability and technology adoption. A combined pattern of four non-symmetrical features have been unique to research the ecovillage concept.

The Cases

Four cases from Bangladesh were selected for investigation, reflecting the researchers' access and availability of data and information. These sites were purposefully selected. The selection criteria aimed to reveal varied situations indicating the diversities in the ecovillage concept application. However, given the interest in tourism, the selection criteria was more subjective that concentrated on both tourism and research activities supported by a congenial interaction between the residents, tourists and researchers (Miller and Spoolman, 2012). All four ecovillages were purposefully built and partially funded by donors. This proved that in Bangladesh, the ecovillage concept is financially and strategically supported by international organisations. However, the role of individual entrepreneurs and the Bangladesh government agencies were unclear (Hassan and Forhad, 2013). Details of each of the case study sites are given as below:

- **The Panigram Resort**: This ecovillage is situated on a river 70 kilometers from the Sundarbans, the world's largest mangrove forest. It has been serving the demands of both tourism and researchers. This ecovillage was designed as a boutique resort having facilities as a wellness and spa center to attract general tourists. In reality, it promotes social and environmental well being. The Panigram Resort has been striving to protect the cultural and natural heritage of Bangladesh, while offering distinctive and authentic experiences to tourists. It has easy transportation by air, land and water, and is built from materials such as bamboo and mud to interpret traditional Bangladeshi architectural patterns. This ecovillage used eco-friendly building materials and has organic farming capacities for healthy crop production. This ecovillage strives to develop responsible and harmonious relationships between the communities with nature (The Panigram Resort, 2014).

- **The Dacope Upazila Ecovillage**: The Dacope Upazila Ecovillage is located in Khulna. This ecovillage was built to accommodate the homeless, those seriously suffered by the cyclone Aila in 2009. This ecovillage was built in one of the worst-hit areas above the flood level. This ecovillage provisions sustainable activities and livelihood for the victims. One ecovillage in the Sutrakhali Union to accommodate seventy two families was built by the Bangladesh Government as the largest. Two other small ecovillages having financially and technically supported by the European Commission of Humanitarian Aid and Civil Protection (ECHO) was also built within close distance. These ecovillages are having considerable impacts mostly positive and negative, on local livelihoods and sustainable development practices while, promoting tourism and research activities (The Daily Star, 2014).
- **The Rangpur Ecovillage**: This ecovillage expands over the villages of Charani and Betgari of Gogachara Upazila, Rangpur. This ecovillage is featured with information technology access. It disseminates the Internet technology as an unusual characteristic of conventional rural areas. The Rangpur Ecovillage endeavors for the 'development of women through capacity building and introducing IT for empowerment'. This ecovillage promotes its residents' livelihoods while, positively affecting the area that was characterized by poverty, male domination, illiteracy and a lower awareness of human rights. According to the Eco-Social Development Organization (ESDO) a non government organization dedicated to the conservation of bio-diversity, intervened to empower women in this ecovillage through education and access to information and communication technology to make them self-reliant by working to achieve environmental and social justice. The ESDO claims to achieve a rapid progress in female literacy increase and their empowerment (Eco Connexion, 2014).
- **The Babui Ecoresort***:* The Babui Ecoresort as an ecovillage is situated over twenty acres of land in the Moheshkhali Island in Cox's Bazaar, the longest unbroken sea beach in the world. It was built for tourism having sustainability principles. This ecovillage was built with both financial and policy guidelines from the International Finance Corporation (IFC), the Global Ecotourism Network (GEN) and the International Ecotourism Society (TIES). The Babui Ecoresort is designed to cater to tourist demands, while encouraging sustainable practices for resource usage. The ecovillage is situated in a fragile area. This was a reason to ensure careful environmental considerations at all stages of development (i.e. site evaluation, implementation, resource use and management (The Babui Ecoresort, 2014).

The above four case study sites were built as ecovillages. However, the perspectives, purposes and situations were different. The selection as such clarifies a generalized interpretation of the basic ecovillage concept (Butler, 1999). Tourists visit these ecovillages, for diverse purposes ranging from leisure, experience gathering and research. In general, the first ecovillage is partly concerned to promote scientific or agricultural research, the second places a strong emphasis on promoting research activities, the third strives to support sustainable livelihoods and the fourth focuses on the promotion of ecotourism. However, all ecovillages promote tourism researches.

METHODS

This research relied on both online and offline information sources, gathered from 2013 to 2015. For the secondary data, an intensive literature review was carried out from both published and unpublished sources (newspapers, reports, journals, books and tourism industry reports). This study employed personal observation as an effective tool for field data collection (Hennik, 2013). In order to collect in-depth data in a participative manner, Focus Group Discussions (FGD) were organized in each study site. Data collection ensured both data validity and reliability by cross checks while, the FGDs were being conducted. FGD in this research perspective was found as usable and effective (Shakya, 2009). Each respondent group included three people categories (i.e. planners, entrepreneurs/interest groups and ecovillage members/residents). Participants were selected on the basis of purposive sampling technique emphasizing their direct engagement with ecovillages. The respondent group was selected by random sampling technique. All respondent groups had mixed male and female representation, while the participation invitation was made by person, telephone and e-mail. However, the lack of secondary data was acute in this particular research context.

Each FGD started with welcome speeches by the researchers outlining the aim, purposes and details of the research and the discussions. Participants were requested to provide as much information about their ecovillage as possible during the open discussion sessions. Any formal interview taking were avoided. Average duration of the discussions was less than 30 minutes. The discussions were recorded, transcribed, and cross-checked with other available information to ensure data validity and reliability (Rios & Campo, 2013). Information from primary and secondary sources was then finally analyzed as research findings.

FINDINGS

Findings explored the key themes as attached to this research. The Definition Dilemma: Findings outlined the unclear meaning of the ecovillage concept to its residents. Mostly, this definition ambiguity reflected unfamiliarity of the residents with the ecovillage concept. To make the residents aware about the meaning of the ecovillage concept was albeit challenging when, the literacy rate was relatively

Table 1. Respondent types

Types of Respondents	Panigram Resort	Dacope Upazila Ecovillage	Rangpur Ecovillage	Babui Ecoresort
Planners	P1	D1	R1	B1
Entrepreneurs / Interest Groups	P2	D2	R2	B2
Ecovillage members / residents	P3	D3	R3	B3

very low. Complexities in meaning were clearly present in the words of a beneficiary of the Dacope Upazila Ecovillage, 'Ecovillage... what is that?...the Geram (village) we are living in now is different from traditional ones. But, I really do not know what the ecovillage is or never heard anything about this. Setting up this 'Geram' helped us greatly to support our livelihood and we now have better ways to earn and live' (D3).

The Nature of Ecovillages: Literally, findings reveal a gap between the concept of the ecovillage and its application. The theoretical meaning of this concept is slightly altered when implemented. This alteration became particularly acute when, tourism research and get involved with ecovillages. In its actual feature, the ecovillage concept did not necessarily meant to emphasis on tourism and research rather, resided by intentional community. However, case ecovillages of this study explored differences in terms of purposes, they were built and have been serving. This difference denotes that tourism and research practices cannot be seen as the ultimate reasons to build ecovillages. Although the ecovillages all relied on substantial government and NGO support, they were built initially for different reasons.

Funding from a donor and international agencies has been found as an inseparable element for building ecovillages in a country like Bangladesh with relatively weaker economical structures. In the developing world, funding is usually coming from the international donor agencies as project support. In a few cases, ecovillages are fully funded by the Government of Bangladesh agencies. Rather, these are mainly built either by both financial and technical support from the international agencies. In other cases, ecovillage building relies on both local and international NGOs.

The Dacope Upazila ecovillage spotlighted on accommodating the Aila cyclone victims, while addressing the interests of curious visitors was a secondary and subsequent opportunity. In an underdeveloped country as Bangladesh, natural calamities also act as a facilitator for ecovillages. Still, residents, tourists or researchers in all ecovillage had to the set guidelines. A resident of the Rangpur Ecovillag reported that this ecovillage was built as an example for the rest of the world. A member of the Rangpur Ecovillage and another pointed out, 'We have full awareness about this. The concerned authorities prepare outlines and we had to read the terms and conditions to follow' (R1). Basic findings stressed on some specific areas as related to the ecovillage concept (e.g.: sustainability, tourism and research).

Sustainability: Sustainability as the most common theme permeated in all four case ecovillages. However, the meaning of this term varied in terms of interpretation and contexts of each ecovillge. For example, P2 stated, 'the Panigram Resort is planned in a way that will help to learn sustainability in almost every building stages'. Following a bit similarity, D1 mentioned, 'Visibly the Dacope Upazila less likely emphasises on sustainability but, the residents' livelihood generation practices are truly sustainable in their use of resources'. In contrast, the Rangpur Ecovillage had an unusual view on sustainability as espoused by R3, 'Rangpur Ecovillage is more focused towards technology-based sustainability while, its residents are very much alerted to ensure a better livelihood by technology adoption'. According to B2, 'The Babui Ecoresort is more content to endorse the green tourism product and service development to ensure sustainability'. One entrepreneur of the Babui Ecoresort argued, 'We have freedom to express our opinions and that, in turn, supports sustainability practices within our ecovillage' (B1). On the other side, a contractor of the Babui Ecoresort suggested, 'The activities or outcomes of ecovillages are harmonious to the well-being of the humans with a possibility to continue in the coming decades' (B2).

Tourism Aspects: Although building objectives differ, each ecovillage generally accepts tourists and researchers. Development of the Babui Ecoresort was more firmly embedded in tourism. This ecovillage in its actual sense is the application of entrepreneurial principles to serve tourists. For example, a resident commented that, 'I was less aware about the ecoresort concept, but, after moving to this ecoresort,

I have managed to understand the ideas, features and relevant information of an ecoresort. My visit is mainly for tourism purposes as influenced by the experiences shared information by the earlier tourists' (B3). For example, one entrepreneur of the Panigram Resort stated that, 'We are aware about the opportunities that an ecovillage offers. We are supporting tourism promotion and research interests while, looking at our investment return. To us, an ecovillage is the mean to support tourism and research to ensure business profitability' (P1).

Research Practices: The Panigram Resort focused more on tourism, particularly visitors interested in conducting research. A Panigram informant indicated that 'Very often we see researchers coming here from the United States of America and many other countries to conduct research in areas like horticulture, permaculture and so on (P3)'. This ecovillage is constantly trying to create collaborations with academic institutions to encourage especially academics' and researchers' visitation. It had successes in attracting both domestic and overseas researchers by meeting their demands. According to D2, 'The Dacope ecovillage is also playing roles to attract scientists and thus to promote research based tourism on areas like disaster management and control, waste disposal, and developing technologies for calamity prone areas'. On the other side, the Dacope Upazila ecovillage was interested in presenting the living views and resource use patterns of the resettled flood victims. However, D3 opined that, 'We do not necessarily invite any tourists or researchers to our ecovillage. They come and go on their own'. Among all ecovillages, the Panigram Resort was more oriented and capable of welcoming tourists and researchers.

The core goals for implementing the ecovillage concept were concerned with tourism, and sustainability. Research and technology adoption appeared at the later stages to facilitate the two earlier goals. This study identifies the existence of tensions regarding control and local political issues leading to power practices. All these can possibly subvert the basic goals of the ecovillage concept. For example, the Bangladesh cases of ecovillages showed that local authorities may become overly concerned about exploiting the potentials of ecovillages. For such reasons, the ecovillage concept in few cases missed its set aims and goals. Community aspects of the ecovillage concept are relevant offering the residents in a settlement with communal feelings. This turned as a relevant feature of the ecovillage concept. Stakeholders, planners and beneficiaries mostly addressed the sustainable and human activity issues in an ecovillage. Even having dissimilar beliefs, ecovillages are meant to be inhabited by like-minded peoples in order to establish a cohesive and self-reliant community (Community, 2005).

PERSONAL STORIES

The circumstances that led the building of the case ecovillages were dissimilar. These were challenging relating peoples' harder life experiences. However, building of these ecovillages certainly eased living struggles of some residents as revealed below:

A male informant of the Dacope Upazila Ecovillage named Mr. Ali was a flood victim. He and his family members lost all of their possessions becoming a hardcore landless before compelled to move to this ecovillage. They were forced to take shelter on the railway embankment and had to stay there for weeks. This was the only place in the area staying above the flood level. Mr. Ali's only 5 years old daughter died from contaminated water drinking. Mr. Ali was completely helpless, being unable to do anything for surviving. The time when the ecovillage was built, he became a member and started living there. The loss of his daughter was a big shock and he was still trying to cope with it. Mr. Ali was a strong supporter of activities that support tourism in the ecovillage. The water is frequently sold to tourists coming to the ecovillage while, the ingenious taste of the water encouraging the tourists to promote it. He was confident to find out a solution to water purification through interactions and knowledge exchanges with tourists. Relying mainly on indigenous knowledge he developed an uncommon technology for drinking water purification. This is an example that the ecovillage concept can offer to bring livelihood, tourism, research and sustainability under a common umbrella.

Sadika was a resident of the Rangpur Ecovillage. Getting forced for marriage at a very early age, she was frequently beaten mercilessly by her husband. The only reason was her inability to bring money from her parents to the husband as a dowry. The frequency of mistreatment increased after she gave birth to a daughter as her husband deserved a baby boy. After each incident, she dreamed of freedom from these trials. As an under aged girl and a victim of forced marriage, she visibly became incapable to cope with the stress level. After moving to this ecovillage, she was able to get a decent life with her only daughter, a college student. Before, Sadika had no knowledge about technology especially, using a computer. She participated a training course organized by the ESDO, a local non-governmental organization. Now, she and her daughter are very familiar with computer operations and using the information technology. She works freelance and manages her livelihood by designing leaflets and other items used for tourism marketing purposes. She felt a sense of accomplishment with changes that the Rangpur Ecovillage brought in her life. The ecovillage ensured her well-being, '*The gadget that I saw in a dream is in my hand.... Am I still dreaming?*' This example thus narrates the role of an ecovillage in guiding people to get engaged in tourism activities for livelihood generation.

The two above personal stories are examples of the ecovillage residents to get involved in tourism and research. These instances can be hardly as featured as random. Livelihood generation activities supported by tourism and research in ecovillages are quite common. Such improvements also lead to the betterment of livelihood options and capacities of the ecovillage concept.

DISCUSSION

Participant resident beneficiaries may be motivated by the need of a dwelling and may or may not understand the ecovillage concept well. In all of the Bangladesh cases, local people had very little knowledge about the conceptual underpinnings of ecovillages. Thus, there may be a gap in understanding and knowledge sharing between local residents and distant supporters who may have elaborate conceptualizations of what an ecovillage should be like (GEN, 2014). Sometimes this can give rise to bitter disagreements. The cases from Bangladesh represent sustainable livelihood development based upon green production, the adoption of advanced technology, the development of tourism and promotion of research. The study thus confirms that, ecovillages building in practice may be or may not be designed with tourism in mind. This is mainly because these ecovillages are primarily resided by distinctive communities that attract tourists having curiosity in education and research. However, more or less in practice, the ecovillage concept follows theoretical soundness. As evidenced, the Rangpur Ecovillage and the Babui Ecoresort were featured as an outcome the conceptual framework of Bang (2005), Gilman (1991) and Jackson (2004). In principle, the ecovillage concept adores the following aspects:

Figure 1. The basic aims to bring the ecovillage concept in practice

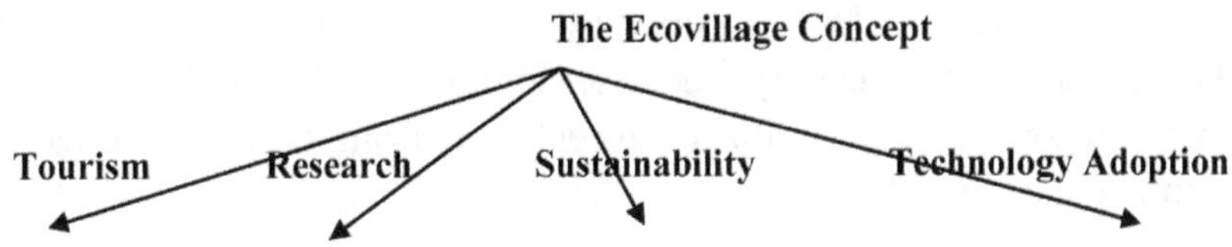

The Ecovillage Concept

Tourism practices in these ecovillages offered an added source of income and acted as a mean of diversifying local livelihoods. Tourism in ecovillages indirectly supported self-supporting initiatives of the ecovillage residents. Tourism also created awareness to controlled resource uses, improved water supply, sanitation and alternative sources of food production and collection.

In terms of research, the Bangladesh cases of ecovillages clearly revealed that the ecovillage concept can promote and nurture research interests, thereby attracting both domestic and international 'scientific' tourists. For example, the Panigram Resort emphasized on agriculture, environment and sustainability. This ecovillage attracted researchers having interests in eco-friendly building materials and organic farming for healthy crop production. With limited capacities and resources, this ecovillage also encouraged tourism research. By creating opportunities and serving research activities, private sector entrepreneurs managed to ensure substantial financial support. In contrast, the Dacope Upazila ecovillage was established to support destitute people suffering from natural calamities. As a result, sustainability was in the center. Still, this ecovillage got tourists resulting in positive impacts on local livelihoods through employment generation and the sale of local products.

In case ecovillages of this study, sustainability was found as an important feature of the ecovillage concept. In all examples, sustainability expected to enhance the residents' living standards. Sustainability was given attention both for the natural and the livelihood well-being. In this perspective, sustainability meant a better way of livelihood generation. Ecovillages focus on creating human settlements that will not harm the earth's biodiversity and ecological balances. On the other side, beneficiaries of such villages are claimed to be more self-reliant by reducing fossil fuel consumption, by using renewable energy sources, by growing food for their own consumptions, and by using efficient technologies to meet demands for electricity, heating, water and sewage disposal. These ecovillages may also promote the use of shared vehicles for transportation, as well as common stoves, cooking utensils, and central heating systems to minimize environmental impacts and reduce costs.

The provision of technology adoption aiming to empower the ecovillage residents and the woman in particular were somehow new to the conceptual features of an ecovillage. The Rangpur Ecovillage with the Eco-Social Development Organization (ESDO) offered information technology adoption services mainly for the women residents. This adoption helped for ensuring self-sufficiency and women empowerment to those residents, in a great way.

However, processes of designing, describing and evaluating the contributions of the ecovillage concept required further and clear explanation to the residents. However, tourism and research practices in the case ecovillages were more visible comparing to sustainable and technology adoption. One of the key reasons can possibly be the financial outcomes from tourism and research practices. The ecotourism concept was not originated from Bangladesh. Still, the country added some exclusive features to it as research and technology adoption as well as tourism. In Bangladesh, the ecovillage concept still remains at its infancy stage, meaning as experimental and less supported by the effective policies and planning approaches. Arguably, there are no universal mandatory guidelines for both building and operating the activities in ecovillages. Rather, the goals of the ecovillage concept sway from tourism to research, sustainability or technology adoption due to the diversity of the global geographical and socio-political approaches.

CONCLUSION

Concept of the ecovillage and its implementation does not necessarily guided by any set standard requirement when, the ecovillages are aimed to address basic human needs. This is evidenced in Bangladesh where, supportive planning policies and measures were mostly lacking. Relying on findings, this study summarises a set of generic features of the ecovillage concept as: ecovillages are mostly built to serve specific purposes, the residents are like-minded peoples or even turns into a community after settling, residents practice sustainable resource use, residents ideological underpinnings are collaborative, often resident women's have strong development desires, the livelihood generation systems are rooted in establishing harmonious relationships between the humans and their environment, tourism, research and technology adoption uncommonly or infrequently accompany the ecovillage concept. Somewhat paradoxically, the implementation of the ecovillage concept very often requires the acquisition of external funding. The geo-environmental settings also play crucial roles in building the ecovillages as in Bangladesh, the flood-prone regions mostly have ecovillages built on over the flood level regions.

Residents of the ecovillages are a bit different from mainstream societies. These ecovillages commonly attract tourists and researchers wishing to witness the novel lifestyles of the residents while, research activities normally cover agricultural, horticultural or organic. By conforming the very generic principles of the ecovillage concept, the Bangladesh cases of ecovillages validate its generalised implementation across the world. The concept supports the achievement of sustainable livelihood generation based on shared knowledge and experiences. Nevertheless, the appearances and characteristics of ecovillages differ greatly. Also, these ecovillages are subject to the environmental and cultural contexts within which they are built. Even the sizes of ecovillages vary substantially. However, common attributes of ecovillages are collective values, facilities and resources aiming to harness a congenial human-environment relationship.

The findings of this study from Bangladesh concerning ecovillages can constitute interesting spaces in elsewhere in the world, in terms of tourism practices and research activities. This study suggests that the ecovillage concept is implementable in locations having geographical and socio-economical varieties. The concept can also constitute a distinctive form of tourism destination. Ecovillages can attract visitors for educational and scientific purposes having interests in learning from experiences. At the same time as tourism destinations, ecovillages can possibly offer an interesting setting to undertake research. The manageable sizes make them better selection for creating a research setting. The ecological and social attributes of ecovillages are linked to diverse livelihood strategies. This research is expected to diversify the existing knowledge of the ecotourism concept. Contributions of this study surely expand the extent of present researchers on ecovillages. As yet, they have attracted limited attention in Bangladesh both from the public and private sectors. Reduced awareness of the residents about the ecotourism concept was a major restriction for data generation.

A comprehensive research, including different interest groups can possibly bring out a wide range of information about the ecovillage concept.

REFERENCES

Akama, J. S., & Kieti, D. M. (2003). Measuring Tourist Satisfaction with Kenyas Wildlife Safari: A Case Study of Tasavo West National Park. *Tourism Management, 24*(1), 73–81. doi:10.1016/S0261-5177(02)00044-4

Ballantyne, B., Packer, J., & Hughes, K. (2008). Environmental Awareness, Interests, and Motives of Botanic Gardens Visitors: Implications for Interpretive Practice. *Tourism Management, 29*(3), 439–444. doi:10.1016/j.tourman.2007.05.006

Bang, J. M. (2005). *Ecovillages: A Practical Guide to Sustainable Communities.* New Society Publishers.

Blewitt, J. (2012). *Understanding Sustainable Development.* London: Earthscan.

Butler, R. W. (1999). Sustainable Tourism: A state-of-the Art Review. *Tourism Geographies, 1*(1), 7–25. doi:10.1080/14616689908721291

Cater, E. (1993). Ecotourism in the Third World: Problems for Sustainable Tourism Development. *Tourism Management, 14*(2), 85–90. doi:10.1016/0261-5177(93)90040-R

Christian, D. L. (2003). *Creating a Life Together: Practical Tools to Grow Ecovillages and Intentional Communities.* New Society Publishers.

Dasmann, R. F., Milton, J. P., & Freeman, P. H. (1973). *Ecological Principles for Economic Development.* London: John Wiley.

Dawson, J. (2006). *Ecovillages: New Frontiers for Sustainability.* Totnes: Green Books.

Eco Connexion. (2014). *Rangpur Ecovillage.* Retrieved from: bit.ly/1nbUsgl

Farrell, B. H. (1999). Conventional or Sustainable Tourism? No Room for Choice. *Tourism Management, 20*(2), 189–191.

Fellowship for Intentional Community. (2005). *Communities Directory: A Comprehensive Guide to Intentional Communities and Cooperative Living.* Rutledge, MO: Fellowship for Intentional Community.

Franklin, S. (1995). Science as Culture, Cultures of Science. *Annual Review of Anthropology, 24*(1), 163–184. doi:10.1146/annurev.an.24.100195.001115

Gilman, R. (1991). The Ecovillage Challenge - The Challenge of Developing a Community Living in Balanced Harmony - With Itself as Well as Nature - is Tough, but Attainable. *In Context, IC#29.* Retrieved from: http://bit.ly/19AH38j

Global Ecovillage Network. (2014). *What is an Ecovillage?* Retrieved from: http://bit.ly/1b4GgTC

Goodwin, H. (1995). Tourism and the Environment. *Biologist (Columbus, Ohio), 42*(3), 129–133.

Hardy, D. (2000). *Utopian England: Community Experiments 1900-1945.* London: E. and F.W. Spon.

Hassan, A., & Forhad, A. (2013). The Role of NGOs in the Sustainable Development of Bangladesh. *Present Environment and Sustainable Development, 7*(2), 59–72.

Hawkins, D. E. (1993). Global Assessment of Tourism Policy: A Process Model. In D. G. Pearce & R. W. Butler (Eds.), *Tourism Research: Critiques and Challenges* (pp. 175–200). London: Routledge.

Hennik, M. M. (2013). *International Focus Group Research*. Cambridge, UK: Cambridge University Press.

Holloway, M. (1951). *Heavens on Earth: Utopian Communities in America*. New York: Library Publishers.

Honey, M. (1999). *Who Owns Paradise? Ecotourism and Sustainable Development*. Washington, DC: Island Press.

Jackson, H., & Svensson, K. (2002). *Ecovillage Living: Restoring the Earth and Her People*. Totnes: Green Books.

Jackson, R. (2004). The Ecovillage Movement. *Permaculture Magazine*, *40*, 25–30.

Jurin, R. (2012). *Principles of Sustainable Living: A New Vision for Health, Happiness, and Prosperity*. Champaign, IL: Human Kinetics.

Laarman, J. C., & Perdue, R. R. (1989). Tropical Science and Tourism: The Case of OTS in Costa Rica. *Annals of Tourism Management*, *10*(1), 29–38. doi:10.1016/0261-5177(89)90032-0

Lim, C., & McAleer, M. (2005). Ecologically Sustainable Tourism Management. *Environmental Modelling & Software*, *20*(11), 1431–1438. doi:10.1016/j.envsoft.2004.09.023

Litfin, K. T. (2014). *Ecovillages: Lessons for Sustainable Community*. Cambridge, UK: Polity Press.

Marhieson, A., & Wall, G. (1982). *Tourism: Economic, Physical and Social Impacts*. Harlow: Pearson.

McLaughlin, C., & Davidson, G. (1984). *Leadership and Power in Intentional Communities. In Context*, *7*, 23.

Miller, G., & Spoolman, S. (2012). *Living in the Environment: Principles, Connections, and Solutions*. Belmont: CENGAGE Learning.

Rios, V. R., & Campo, E. P. (2013). *Business Research Methods: Theory and Practice*. Madrid: ESIC Editorial.

Ross, S., & Wall, G. (1999). Ecotourism: Towards Congruence between Theory and Practice. *Tourism Management*, *20*(1), 123–132. doi:10.1016/S0261-5177(98)00098-3

Shakya, M. (2009). *Risk, vulnerability and tourism in developing countries: The case of Nepal*. Berlin: Logos Verlag Berlin GMBH.

The Babui Ecoresort. (2014). *About Us*. Retrieved from: http://bit.ly/1kySE15

The Daily Star. (2014). *Ecovillage Raises Hope for Aila Victims*. Retrieved from: http://bit.ly/199aimH

The International Union for Conservation of Nature (IUCN). (1980). *World Conservation Strategy*. Gland, Switzerland: International Union for the Conservation of Nature and Natural Resources. Retrieved from: http://bit.ly/1bWsFdJ

The Panigram Resort. (2014). *Home*. Retrieved from: http://www.panigram.com

Trainer, F. E. (1998). Towards a Checklist for Ecovillage Development. *Local Government, 3*(1), 79–84.

Van Schyndel Kasper, D. (2008). Redefining Community in the Ecovillage. *Human Ecology Review, 15*, 12–24.

West, P. (2008). Tourism as Science and Science as Tourism Environment, Society, Self, and Other in Papua New Guinea. *Current Anthropology, 49*(4), 597–626. PMID:19230266

West, P., & Daniel, B. (2006). Some Unexpected Consequences of Protected Areas: An Anthropological Perspective. *Conservation Biology, 20*, 609–616. doi:10.1111/j.1523-1739.2006.00432.x PMID:16909541

World Commission on Environment and Development (WCED). (1987). *Our Common Future*. Oxford, UK: Oxford University Press.

ADDITIONAL READING

Hassan, A. (2012a). Package Eco-tour' as Special Interest Tourism Product-Bangladesh Perspective. *Developing Country Studies, 2*(1), 1–8.

Hassan, A. (2012b). Key Components for an Effective Marketing Planning: A Conceptual Analysis. *International Journal of Management and Development Studies, 2*(1), 68–70.

Hassan, A. (2012c). Rationalization of Business Planning Through the Current Dynamics of Tourism. *International Journal of Management and Development Studies, 2*(1), 61–63.

Hassan, A., & Ahmed, F. (2013). The Role of NGOs in Preventing Unemployment in Bangladesh. *Kuwait Chapter of Arabian Journal of Business and Management Review, 2*(9), 1–12. doi:10.12816/0001227

Hassan, A., & Burns, P. (2014). Tourism Policies of Bangladesh – A Contextual Analysis. *Tourism Planning and Development, 11*(4), 463–466. doi:10.1080/21568316.2013.874366

KEY TERMS AND DEFINITIONS

Community: A community as a social unit is the group of people having shared common values.

Ecovillage: These are purposefully built settlements that recognize the existence of positive relationships between the environment and the society.

Environmental Sustainability: Environmental sustainability relates to the rates of indefinite continuation of non-renewable resource depletion, pollution control and renewable resource harvest.

Tourism: In the simplest meaning, tourism is travelling for pleasure that as a business also involves the attraction, accommodation and entertainment of tourists.

Chapter 5
Eco-Certification and Transparency in Global Food Supply Chains

Philippos Karipidis
Alexander Technological Educational Institute of Thessaloniki, Greece

Dimitrios Tselempis
Alexander Technological Educational Institute of Thessaloniki, Greece

Loukas Tsironis
University of Macedonia, Greece

ABSTRACT

This chapter portrays the information flow for sustainability issues along the globalized food supply chain and explores the eco-certification decisions of farm businesses, viewing them as the first upstream chain participant. This examination is based on the literature to connect eco-certification with transparency and to portray traceability schemes for sustainability issues in food supply chains, followed by information technology-based systems and applications supporting traceability. The third section presents the eco-certification decisions at the supply chain level in four subsections. It first builds a theoretical framework regarding the downstream firms' sustainability-related decisions by offering conceptual definitions. Next the farm business decision logic is given, followed by the discrete choice model. The specialization of the model is presented in the third subsection, followed by the results, discussions, and implications for practitioners. Some conclusions and implications for future research are offered in the last section.

INTRODUCTION

The food supply is implicated for its environmental impact at all stages of the food life cycle, while the production stage contributes the highest share of the average household footprint for food consumption (Garnett, 2011; Weber & Matthews, 2008). Thus, it is expected that the agro-food sector can become substantially less harmful for the environment if sustainability-related changes in the structure and the

DOI: 10.4018/978-1-5225-2107-5.ch005

quality of food production take place. For example, if the conventional food production is transformed to organic, then the use of synthetic agrochemicals can be minimized, the water consumption can be reduced, and the energy gain can reach even 41.5% of the total energy requirements in food production and supply from the farm gate to the consumer's table (Schwarz, Schuster, Annaert, Maertens, & Mathijs, 2016; Tubiello et al., 2014; Ziesemer, 2007). Such a shift to a more environmentally responsible agriculture can be directed by public authorities' interventions and nongovernmental organizations' pressures and can be driven by the market mechanisms (Mylan, Geels, Gee, McMeekin, & Foster, 2014; Tselempis, Karipidis, Pavloudi, & Semos, 2015; Weber & Matthews, 2008).

In the case that such a change is driven by market forces, environmentally responsible agriculture could be the result of the responses of the food supply chain actors to environmentally sensitive demand. For example, by 2008, the major United Kingdom supermarkets had announced targets to improve the sustainability performance of their supply chains both as a response to postpone regulation and to defend against nongovernmental organization criticism and as a reaction to a perception in the market regarding a cultural shift toward greater sensitivity to sustainability problems and the rise of "green consumers" (Mylan et al., 2014). This provides a new competitive context for supermarkets. In other words, consumer choices can generate incentives for producers and marketers by indicating preferences for eco-friendly products or by rewarding them with price premiums, thus leading the food supply chain actors to adopt proactive environmental management strategies in order to become more attractive to customers (Delmas & Grant, 2014; Hartmann & Moeller, 2014; Karipidis & Sartzetakis, 2013; Karipidis, Tsakiridou, Aggelopoulos, & Belidis, 2010; Weber & Matthews, 2008).

Viewing environmentally friendly food production and supply through the lens of a supply chain, we adopt Ahi and Searcy's (2013) suggestion that coordination in sustainable supply chains can be a good starting point. They defined sustainability as the creation of coordinated supply chains with key interorganizational business systems designed to efficiently and effectively manage the material, information, and capital flows associated with the procurement, production, and distribution of products or services in order to meet stakeholder requirements. Combining this with Kottila and Rönni's (2010) findings that consumers are requesting better placement and assortment of information available at stores, food supply chain actors must create coordinated sustainable supply chains through the integration of environmental considerations as well as communicate messages to consumers (using signals) to elevate their own positions against market competitors. For example, they can implement programs that certify that they use inputs or sell food grown in sustainable cultivations or animal production units avoiding negative effects on the environment, from the reduction of carbon footprints to the elimination of eco-toxic substances.

The main objective of this chapter is twofold. It first aims to portray the flow of sustainability–related information to downstream food supply chain partners by reviewing the literature. Taking into consideration that the overall supply chain performance can be improved if all supply chain partners embed sustainable practices into their activities (Ahi & Searcy, 2013; Hassini, Surti, & Searcy, 2012) and that a substantial reduction of environmental consequences can be achieved by improvements in the food production stage, then the study's second aim is to analyze the eco-certification decisions of farm businesses and identify the factors accelerating certification. Thus, the chapter's main contribution is that it analyzes the farm business eco-certification decisions within the framework of information sharing and the relationships between supply chain members in order to achieve and trace the sustainability in food supply chains.

The remainder of the chapter is structured as follows. The next section is composed of four subsections based on a literature review. It connects the eco-certification decision with transparency and portrays traceability schemes for sustainability issues at the food supply chain, followed by information technology (IT) based systems and applications that facilitate traceability. Some definitions regarding the eco-certification schemes are given at the end of the section. The third section presents the eco-certification decisions at the supply chain level in four subsections. More specifically, it first builds a theoretical framework regarding the downstream firm's sustainability-related decisions by offering selected conceptual definitions. Next the logic of farm business decisions is given, followed by the discrete choice model. The specialization of the model is presented in the third subsection, followed by the results, discussions, and implications for practitioners. Some conclusions and implications for future research are offered in the last section.

BACKGROUND

Transparency by Eco-Certification

With a focus on supplier-buyer relationships, transparency is defined as an individual's subjective perception of being informed about the relevant actions and properties of the other party in the interaction (Eggert & Helm, 2003), and process transparency is achieved if everyone can see and understand the necessary aspects and status of an operation at all times (Womack & Jones, 2003). For example, the environmental issues of the agricultural production process should be properly seen and understood by all downstream supply actors. Taking into consideration the place of each food supplier in the supply chain, it is apparent that they are claimants of transparency while they also deliver information to other actors, consumers, and public authorities. Their motivation for transparency is to (1) respond to the differentiated consumers' demand, (2) optimize business processes because product and process attributes can be coupled with process performance by improving information exchange through integrated information systems, (3) add value by labeling products according to food product attributes, (4) be able to quickly recall products from markets when incidents occur or link to the downstream supply chain to limit the incident and minimize costs, and (5) comply with legislative regulations (Trienekens, Wognum, Beulens, & van der Vorst, 2012).

The need for sustainability-related transparency in agricultural and food markets is ascertained by Kottila and Rönni's (2010) findings that consumers are requesting better placement and assortment of information available at stores and that the scarcity of communication between consumers and the other actors in a supply chain jeopardizes consumers' access to organic products, leading to a gap between sales figures and potential demand. This is especially recognized in a globalized food context. Regarding the validity of information provided, Darnall and Vázquez-Brust (2016) highlighted that consumers' trust in information providers varies. Consumers' trust of government and environmental nongovernmental organizations to provide credible environmental information encourages them to use eco-labels sponsored by these entities. However, consumers' distrust of private business to provide credible environmental information discourages their use of business association-sponsored eco-labels. This consumer distrust can be eliminated if eco-label sponsors proceed in certification by using third-party auditors.

It is apparent that each food supply chain actor must deliver a set of information along the supply chain concerning environmentally harmful activities, such as the use of synthetic fertilizers, pesticides and antibodies, soil degradation, greenhouse gas emissions, food waste and toxic emissions, animal production waste and animal welfare, and water and energy use. Although information systems and technology can support information exchange and transparency in food chains, these can never guarantee the integrity of a complex set of information and transparency that is wished for. The transmission of such information becomes possible by working according to sustainability standards and having the right arrangements with all supply chain partners (Darnall & Vázquez-Brust, 2016). This implies that flexible information system solutions for specific supply chain relationships should always be accompanied by matching governance mechanisms and by adopting the proper standards (Trienekens et al., 2012; Wognum, Bremmers, Trienekens, van der Vorst, & Bloemhof, 2011). Figure 1 portrays a typical food supply chain, including reflections on environmental consequences and information sharing by supply chain actors.

Because food supply chains, especially at the upstream portion, mainly consist of small- and medium-sized enterprises causing diseconomies of scale, the sustainability-related standards can also limit the administrative burdens. These standards, called eco-certification schemes, can include environmental management schemes or quality management schemes that incorporate environmental issues. These emerged as important market-based mechanisms aimed at internalizing externalities by aligning private incentives with social values attached to natural resources and the environment (Basu, Chau, & Grote, 2003; Bougherara, Grolleau, & Thiébaut, 2005). As Hatanaka, Bain, and Busch (2005) noted years before, a wide range of third-party certifications are becoming an alternative for conventional producers, building a mechanism that also signals to consumers the nature of food production practices through the use of eco-labels. Thus, it becomes possible to exchange sustainability-related information between supply chain actors, as well as among supply chain actors and society and public authorities, thus enhancing transparency.

Figure 1. Environmental consequences of activities of food supply chain actors

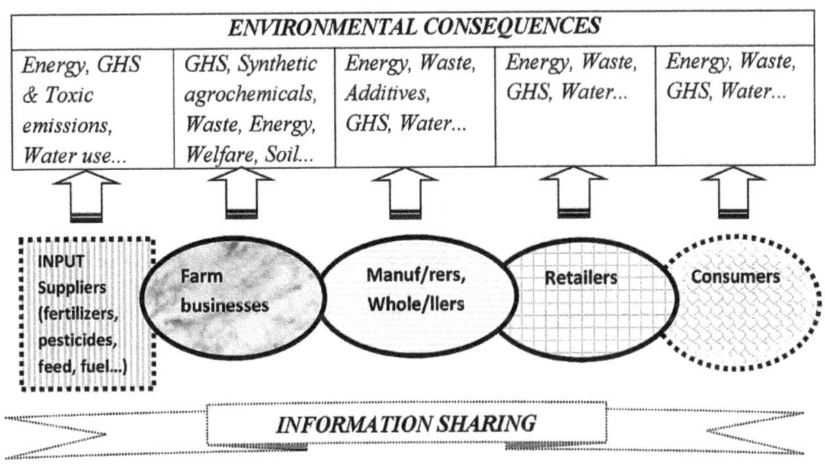

Tracing Sustainability Issues

The transition to sustainable supply chain management and practices can assist organizations to reduce their total carbon footprints and optimize their end-to-end operations and thus achieve greater profitability (Ahi & Searcy, 2013). A crucial presupposition for this achievement is that consumers are sufficiently informed about the sustainability issues in food production and marketing processes so that they include eco-friendly products in their buying choices (Karipidis et al., 2010; Karipidis & Sartzetakis, 2013). Thus, supply chain participants endeavor to ensure the availability of sustainability-related information, thus meeting the requirements for traceability in the supply chain and enhancing market transparency. The set of information each food supply chain must share with customers, consumers, and society depends on its place in the supply chain and its production and supply activities (see Figure 1).

Regarding traceability in sustainable food supply chains, the United Nations Global Compact and Business for Social Responsibility (2014) defined traceability as "the ability to identify and trace the history, distribution, location and application of products, parts and materials, to ensure the reliability of sustainability claims, in the areas of human rights, labour (including health and safety), the environment and anti-corruption" (p. 6). In the context of sustainability, according to these same organizations, "traceability is a tool to assure and verify sustainability claims associated with commodities and products, ensuring good practice and respect for people and the environment all along the supply chain" (p. 6).

In an attempt to describe the schemes that enable the tracing of sustainability claims, three main models can be identified following the definitions provided by the United Nations Global Compact and Business for Social Responsibility, which take into consideration the approach followed in tracking a claim: product segregation, mass balance, and book and claim (see Figure 2). In the product segregation model, the eco-certified products are physically separated from the noncertified products at each stage along the supply chain. This ensures that customers (and consumers) know that 100% of a product, such as organic cheese, consists of eco-certified or noncertified materials.

This can be viewed in two product segregation model forms: bulk commodity and identity preserved. The bulk commodity form separates eco-certified from noncertified products but allows mixing of eco-certified products from different producers who adopt a certain certification scheme. The identity preserved form doesn't allow mixing of eco-certified products throughout the supply chain and provides

Figure 2. Traceability schemes, model forms, and claims

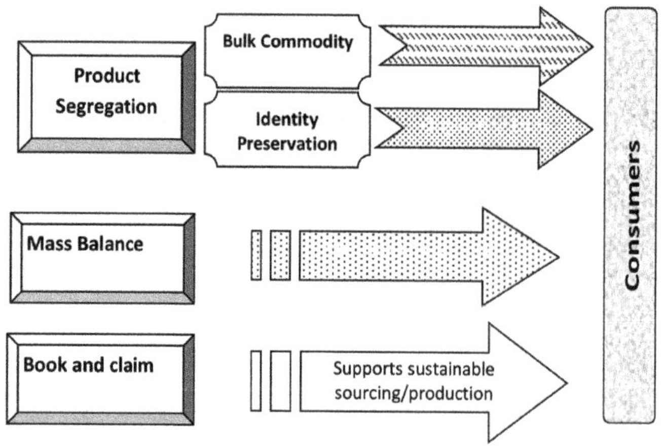

traceability from a specific farm business or manufacturer to the final customer, thus enabling the traceability of products back to the originating farm or production site. It is criticized for being costly and requiring advanced technology because all material sources must be strictly separated, controlled, and monitored at each stage of the supply chain, as it requires supply chain actors to know all their suppliers and collect and verify data at all levels throughout the supply chain.

In the mass balance model, eco-certified and noncertified products can be mixed. The exact volume of eco-certified products entering the food chain is controlled, and equivalent volumes of the eco-certified products that leave the supply chain are sold as certified. Customers may not know whether their specific share of the product contains eco-certified or noncertified products, but they face claims such as "product contains X percent of certified ingredients," which can be used by supply chain actors.

The book and claim model does not seek to have traceability at each stage in the food supply chain. It relies on the link between the volume of the eco-certified product at the beginning of the supply chain and the amount of eco-certified product offered at the end of the chain. Supply chain participants can obtain sustainability certificates for the volume of eco-certified products entering into the supply chain. Certified and noncertified products flow freely throughout the supply chain. It is not certain that the products a supply chain participant sells contain a certified proportion, but its production has supported sustainable sourcing, enabling the chain participant to use claims such as "the product supports sustainable sourcing" or "the product supports production of essential commodities."

It is apparent from this discussion that food supply chain participants can choose between different sustainability claims and different traceability schemes so that sustainability can be properly traced along the food supply chain.

Information Technology-Based Traceability Systems

Because some traceability systems can be costly and require advanced technology, the new developments in IT-based traceability systems can provide supply chain participants with technological tools that enable them to mitigate such problems. More specifically, the provision of advanced information capabilities to the consumer, who is the last participant in the food supply chain, can become a source of competitive advantage for companies that achieve a smooth information flow along the supply chain. Examining the technological and nontechnological aspects related to IT deployment and supply chain collaboration supports this (Pramatari, 2016).

The implementation of traceability systems based on IT in food supply chains has grown quickly, as upstream supply chain participants must respond to the increasing customer, consumer, society, and public authority interests to trace a set of information regarding quality attributes and characteristics, production conditions, environmental consequences, region of origin, animal welfare, agrochemical residues, genetic modification, and the like (Folinas, Manikas, & Manos, 2006; Wognum et al., 2011). The information-sharing needs of supply chain actors attracted the interest of many researchers to build and propose several IT-based systems and applications that facilitate and support food traceability, such as Feng, Fu, Wang, Xu, and Zhang (2013); Kim and Woo (2016); and Yan, Hu, and Shi (2012). In an extensive literature review, Badia-Melis, Mishra, and Ruiz-García (2015) presented the latest technological advancements enabling or supporting food traceability. Selected examples are given next.

The most recent advances in food traceability IT systems that contribute to the enhancement of effectiveness in information sharing along supply chains, as well as in transaction cost reduction, include a set of innovative implementations of radio frequency identification. This technology uses radio waves to automatically identify objects that have been used to improve product and order tracking and real-time location tracking. Near field communication is a promising wireless communication technology that facilitates mobile phone usage, offering a range of services from payment and loyalty applications to access keys for offices and houses. Eventually it integrates all such services into one single mobile phone. It is typically for use over a very short distance for making payments when the product arrives to the end consumer. This service, combined with the wireless monitoring devices, offers control in situ or in the moment information while the goods are on the go. The combination of radio frequency identification and near field communication allows the end consumer to know the complete history of the purchased product.

The quick response (QR) code can embed text, video, advertisements, personal information, and so on, and can be integrated into users' smartphone applications, enabling them to scan and decode information and messages about food products. In the same vein, the critical tracking event focuses on the events that manipulate the products in the supply chain. New trends in traceability offer a whole new point of view by focusing on the parts implicated in the food supply chain, such as the actors involved, the processes, and the elements added to the food and the product itself. Some years ago, Feng et al. (2013) developed a cattle/beef traceability system that integrates radio frequency identification technology with a personal digital assistant and barcode printer. They obtained real-time and accurate data acquisition and transmission, achieving high efficiency of information tracking and tracing across the supply chain.

Recently, Pramatari (2016) reported that interesting findings may be gained by going beyond the technological aspects relating to IT deployment, such as the governance rules and relations between supply chain partners, the organizational issues associated with existing and new practices, the motives for adoption, and the pertinent costs and benefits for the various partners. Taking into account the suggestion of Herzfeld Drescher and Grebitus (2011) that the global spread of a particular certification standard can be modeled as any other organizational innovation, and that third-party certification reflects a broader shift from public to private governance in the food sector (Hatanaka et al., 2005; Henson & Reardon, 2005), we can view eco-certification as an organizational innovation enabling effective and efficient governance of the food supply in a globalized food market.

The eco-certification that each supply chain participant gains enables him or her to share with stakeholders a set of information regarding a large number of environmental consequences that he or she mitigates. The advanced IT traceability systems support such a sharing, enabling consumers, customers, society, and public authorities to concisely trace information about the environmental responsibility of the supply chain participants. It allows for the elimination of the information asymmetry problem between suppliers and customers, reduction of the transaction cost in the food supply chain, and more effective sustainable supply chain management. Through trace codes, consumers can acquire information on suppliers and products, which can strengthen their intention to purchase products and improve their faith in their origins (Piramuthu, Farahani, & Grunow, 2013). Thus, upstream food supply chain participants are motivated to become environmentally responsible.

Eco-Certification Schemes

Apart from the competitive enhancement of food companies choosing to implement an environmental management scheme (EMS), some risks can arise for those companies that do not implement one. More specifically, the supply chain liability effect creates strong risks for the focal firm because higher responsibility attributions increase consumers' anger and propensity to boycott. Because consumers hold a focal firm responsible for behaviors and incidents upstream, an increasing number of supply chain participants are voluntarily committing to corporate sustainability. To ensure compliance, firms increasingly realize the relevance of their supply chains and their dependence on suppliers' environmental or social practices (Grimm, Hofstetter, & Sarkis, 2016; Hartmann & Moeller, 2014). Supplier management strategies to ensure compliance with corporate sustainability standards in the supply chain focus on suppliers' eco-certification.

As is apparent from the earlier sections of this chapter, the set of sustainability-related information that is shared by supply chain participants to build competitive advantage and brand integrity/awareness depends on the food they supply, their marketing strategy, and the activities they carry out. For example, a global food chain participant with activities that build its marketing mix faces opportunities in certain target markets to provide signals and messages about a set of sustainability issues connected with its activities and the food it offers. Thus, it can ask upstream participants to implement certain EMSs, meeting the requirements that are taken by the systems it implements.

Over the past two decades, the basic issues of sustainability have been internalized into food quality assurance and food quality management schemes worldwide (Manikas, Hamann, & Sentic, 2016). Thus, a large number of eco-certification schemes are available for firms intent on meeting sustainability requirements ranging between EMSs incorporating quality-related issues and quality management systems incorporating environmental issues. Thus, a set of 465 eco-certification schemes and eco-labels is recorded worldwide today, meeting the differentiated needs and requirements of customers, consumers, society, and public authorities. Examples of sustainability standards are the international ISO 14001, the European Eco-Management and Audit Scheme (EMAS), the industry association's Global Good Agricultural Practice (GlobalG.A.P.), and specific corporations such as The Nature Conservancy (TNC), the Forest Stewardship Council (FSC), and the Marine Stewardship Council (MSC). Some national standards are also available in many countries, such as AGRO 2.1–2.2 in Greece, which combines selected requirements of ISO 14001 and ISO 9001.

ECO-CERTIFICATION DECISIONS

Theoretical Framework

In previous years, some researchers confirmed the necessity for a collaborative view of sustainable food supply chain management. Pérez Mesa and Galdeano-Gómez (2015) provided evidence of the benefits of active cooperation strategies for perishable produce suppliers, implying that the stable relationship between suppliers and retailers renders firms more competitive by achieving synergies. Hassini, Surti, and Searcy (2012) proposed a framework for sustainable supply chain metrics that involved each supply chain participant from the supplier to the consumer and noted that all supply chain partners should embed sustainable practices into their activities in order to improve the overall supply chain performance. In accordance with this, Wong (2013) reported that sustainability management in the supply chain has shifted from an individual business responsibility to a supply chain partner responsibility, and Hartmann and Moeller (2014) highlighted that firms should work to ensure sustainable behavior throughout the supply chain because consumers hold a focal firm responsible for behaviors and incidents upstream.

Supply chain management that ensures compliance with sustainability standards usually focuses on suppliers' and sub-suppliers' responses to sustainability requirements. Research findings suggest that firms can improve suppliers' and sub-suppliers' compliance with sustainability standards by actively managing them through assessment and collaboration (Grimm et al., 2016). Laari, Töyli, Solakivi, and Ojala (2016) confirmed that manufacturers with strong green supply chain management practices, combined with arm's length environmental monitoring of suppliers, are likely to perform well in environmental issues. If food supply chain participants (manufacturers, wholesalers, and retailers) are seeking to improve performance and achieve environmental goals, then they must form more collaborative relationships with suppliers. A necessary presupposition for supply chain participants to manage their relationships with suppliers is to better understand their sustainability-related behaviors.

In the framework proposed by Hassini et al. (2012) for sustainable supply chain metrics, performance is measured in three dimensions of sustainability: economic, social, and environmental. Combined with Ahi and Searcy's (2013) definition concerning the sustainable supply chain coordination, and the highlights of Massaroni, Cozzolino, and Wankowicz's (2015) findings regarding the dimensions of supply chain sustainability, it can be assumed for eco-certification that (1) the economic dimension refers to eco-certification-related costs, revenues, profitability, and return on investment for each supply chain participant; (2) the social dimension is connected with the working conditions, such as safe working environments, the quality of life for community members, human training, sourcing practices, and the impact of these practices on communities; and (3) the environmental dimension embraces environmentally friendly technologies and practices connected with the use of synthetic agrochemicals, irrigation/water consumption, soil protection, packaging, material sources, carbon emission, machines, transportation, transport practices, energy sources, energy consumption, waste management, and pollution. Thus, we shape the framework presented in Figure 3, which portrays the links between the supply chain participants (downstream and upstream) and the consumers/customers, society, and environment in accordance with the three sustainability dimensions.

Figure 3. Value creation for customers, society, and environment

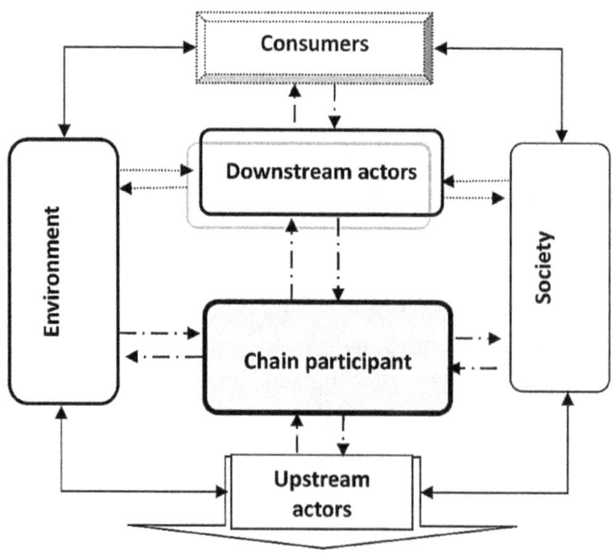

We assume that a food production or marketing firm faces the challenge for ensuring compliance with sustainability standards in a collaborative, sustainable supply chain. Its supplier management strategies can be effectively and efficiently implemented by focusing on eco-certification of its suppliers and sub-suppliers in order that they be controlled and monitored by third-party auditors (eco-certifiers). For example, a chain participant will seek suppliers certified according to the GlobalG.A.P. or organic standard if it intends to achieve a certification according to the ISO 14001 standard.

Following Porter and Kramer's (2011) suggestion for shared value creation, it can be assumed that the food marketing firm is seeking to create shared value in order to be more effective and far more sustainable than the majority of current business efforts in the social arena. Its management has to balance stakeholders' claims, which means that it attempts to achieve a balance among economic, social, and environmental sustainability if it takes into consideration the value created for the customers and suppliers, workers and community, and environment, respectively. In such an attempt, it cooperates with upstream participants by asking them to implement the proper EMSs and to be certified by a third party in order to communicate its eco-friendliness to downstream supply chain players, society, and public authorities.

As is apparent from the previous discussion, each upstream supply chain participant can choose to adopt an eco-certification scheme, depending on the product it produces and its goals and taking into consideration the downstream participant's requirements. Because the food production stage contributes the highest share on the average household footprint for food consumption, reaching 83%, the most crucial sustainability enhancement is expected to be achieved from the primary production stage. For example, transforming the conventional food production to organic, the synthetic agrochemicals are eliminated, the water consumption and the energy requirements are substantially minimized, the soil is protected, and the ecological degradation is minimized.

Downstream participants of a sustainable food supply chain, especially a long global chain attempting to formulate and implement a supplier strategy, must successfully segment their suppliers in order to determine which of them to engage, the type of engagement, and the level of resources required to manage the supply. Focusing on the primary production stage—not all suppliers or farm businesses are intent on being certified, and there are differences in the timing of certification (Tselempis et al., 2015)—downstream participants could benefit if they choose farm businesses that substantially contribute to the competitiveness of the food supply chain to be enhanced, such those being eco-certified. They could also seek to encourage selected suppliers who are not willing to be certified to accelerate the eco-certification (Handschuch, Wollni, & Villalobos, 2013; Hattam, Lacombe, & Holloway, 2012). Thus, they must understand the eco-certification decisions and factors impacting them, such as the characteristics of farm businesses and the relationships with buyers/customers. In addition, it could be useful for downstream supply chain participants to know the factors discouraging their suppliers to decide or accelerate the certification (barriers) in order to determine the type of relationship with them and the resources required to manage it.

Farm Business Eco-Certification Decisions

Downstream participants of the sustainable food supply chain require farm businesses to be certified under certain eco-certification schemes, which aim to demonstrate their environmental friendliness. Because the certification is a costly and long-term effort for farm businesses, a high percentage of them do not intend to be eco-certified. These farm businesses are mostly small and require a great deal of encouragement to become certified (Handschuch et al., 2013; Hattam et al., 2012). In addition, farm businesses decide on eco-certification at different stages, and some accelerate the eco-certification decision in order to attain certification at an early stage while others are certified at a later stage.

Analyzing the farm business eco-certification behavior can provide food marketers with information to help them properly build relationships with those producers who are more willing to be eco-certified and to predict the acceleration of certification decisions. They can also identify the factors that encourage or discourage farm businesses to be certified in order to properly support them to accelerate the eco-certification decision. Because eco-certification accompanied with the proper eco-labels contributes to environmental protection, enhanced quality of life, and mitigation of the information asymmetry problem, the present analysis could be also useful for public authorities. Thus, the main objective of this section is to analyze the eco-certification decisions of farm businesses in the framework of the globalized sustainable food supply chain and identify the factors that impact these decisions.

We assume an environmentally responsible farm business attempts to integrate eco-friendliness into its strategy by implementing an EMS and by communicating its eco-friendly efforts to customers, consumers, society, and public authorities. These farm business activities imply specific costs, including investments in green technologies and practices, purchasing green inputs, investing in soil protection and improvement, investing in water and energy use reduction, research and development costs, training and information acquisition, implementation of an eco-management system, and eco-certification fees. Going beyond the Karipidis and Sartzetakis (2013) proposition, we assume that the farm business plans to properly transform its production process in order to provide eco-certified products to consumers and its customers who are wholesalers, manufacturers, or retailers. Thus, it shares value with consumers, customers, society, and public authorities, as portrayed in Figure 4.

Because eco-friendly actions imply certain costs, a farm business's eco-certified output will be offered at a higher cost relative to the noncertified product. Although economic, social, and environmental value is created, we focus on the economic value the farm business creates by providing customers with an eco-certified product, assuming that the eco-certification is indirectly connected with social and environmental value. The farm business profit function is

Figure 4. Farm business decision in conjunction with the selected factors impacting it

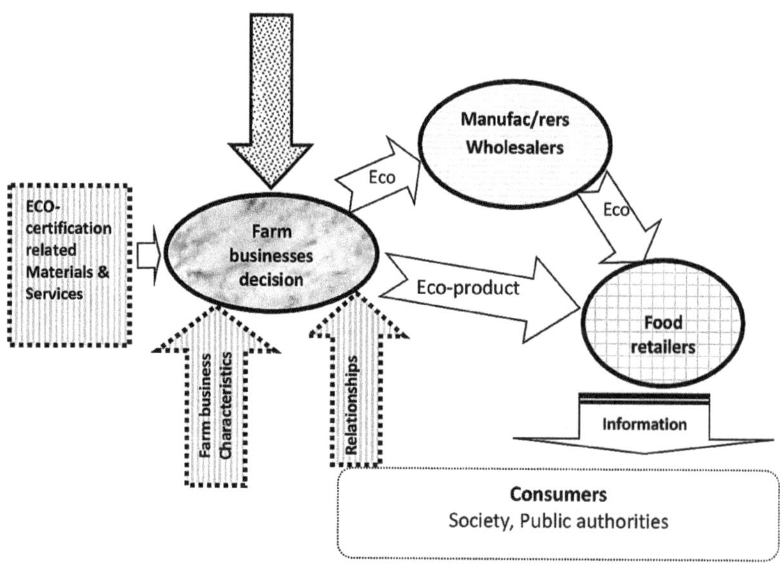

$$\pi = qp - c\,, \tag{1}$$

where q is the quantity of products, p is the price, and c is the cost before certification. After deciding on eco-certification, the farm business chooses its inputs to meet eco-certification requirements, and it incurs costs c+Δc. We assume that it provides customers with certified products, thus expecting its cost to be covered by the customers' preferences or/and willingness to pay premium prices. If the profit increase is negative, then the farm business's owner will not approve the eco-friendliness unless it either expects indirect benefits, such as improved market position and brand image, or attempts to avoid penalties. If, however, the profit increase is positive, eco-certification creates economic value and therefore a direct incentive for approving eco-friendliness. Finally, if the profit increase is zero, then the farm business owner is indifferent between adopting the environmentally responsible process and continuing with the status quo, potentially deciding to seek long-term benefits from improved branding and future penalties avoidance. Thus, the farm business's profit function helps us to connect the quantity of certified product and its price and eco-certification-related costs to the profit, thus enabling us to foresee the factors impacting eco-certification decisions.

As eco-certification is interwoven with a multitude of organizational, process, and technological changes and innovations, we resort to methodological tools used in the case of innovation adoption. The discrete choice models have been used to examine how decisions are influenced by the conditions under which they are realized (Herzfeld et al., 2011; Karipidis & Tselempis, 2014). In the present research, the farm business's certification decision is based on aspirations to increase profits by boosting its competitiveness as well as the competiveness of the supply chain.

Implementing EMS by adopting an eco-certification scheme can be measured against whether the farm business implements the EMS at an early stage before others or at a later stage by using the proper scale of measurement or by including the farmer in one of three to five categories, between innovators and laggards. The focus of the present analysis is the acceleration of the certification decision, which reflects the timing of the eco-certification decision measured by the time (t) the farm business is under eco-certification and offers eco-certified products. In the discrete choice model, the dependent variable (y) is a rough categorization of a continuous but unobserved variable y*. The acceleration of certification model can take the following form:

$$y_i = \mathrm{a}_k^{fb} X_k^{fb} + \mathrm{b}_j^{rc} X_j^{rc} + \mathrm{d}_l^{eb} X_l^{eb} + u_i \tag{2}$$

y_i is used as a proxy for y_i*. According to this model, the farm business chooses alternative *i* (timing). The explanatory variables reflect the economic, social, and behavioral characteristics (X^{fb}_k) of the farm business, which can impact the acceleration of eco-certification. In an attempt to capture more factors that can impact the acceleration of certification, such as the farm business's relationships with customers and certain factors of the outer business environment that can discourage farm businesses from being eco-certified (barriers), we introduce in the model some variables (X^{rc}_j) and (X^{eb}_i) reflect them respectively.

The present analysis uses a three-point scale to measure the time: If the farm business implements EMS at an early stage and stays under the certification for the maximum time, then t takes the maximum value (t = 3). If the farm business implements EMS at the latest stage, then it stays under the certification for the minimum time (t = 1).

Specialization of the Model

Many studies investigate the factors of the internal farm business environment that have an impact on farmers' certification-related decisions and identify some market and public policy factors impacting them (Asfaw, Mithöfer, & Waibel, 2010a, 2010b; Karipidis & Tselempis, 2014; Kersting & Wollni, 2012; Masakure, Cranfield, & Henson, 2011; Muriithi, Mburu, & Ngigi, 2011; Soltani, Azadi, Mahmoudi, & Witlox, 2014; Veldstra, Alexander, & Marshall, 2014). Based on previous studies conducted worldwide, we select fifteen factors connected with the economic, social, and behavioral characteristics of farm businesses that can impact the certification decision, six factors connected to the relationship of farm businesses with their customers as the buyers of primary agricultural products (wholesalers, manufacturers, or retailers), and three factors that can be viewed as barriers to certification, such as no sufficient provision of information, no technical support provision to farm businesses, and the requirements of the eco-certification scheme. These factors are presented in Table 1 and grouped into the three

Table 1. Factors that can impact eco-certification decisions

GROUP OF FACTORS	FACTORS	SIGNS
Economic, Social and Behavioural characteristics of farm business	Family size	+
	Special education	+
	Education level	+
	Frequency of computer use	+
	Participation in associations	+
	Prospect years	+
	Income from agriculture	+
	Successor existence	+
	Years in agriculture	±
	Implementation of GAP before certification	+
	Profession before the farming	±
	Insufficient knowledge	-
	Low economic capability	-
	Insufficient substructure	-
	Insufficient workforce	-
Relationships with customers	Written contracts	+
	Verbal agreements	+
	Quality control by buyers	+
	Penalties by buyers	+
	No buying preference for certified	-
	No price premium for certified	-
Barriers	Barrier -information provision insufficient	-
	Barrier -no technological support	-
	Barrier –certification scheme requirements	-

categories. Because the profit function presented earlier enables us to calculate the possible positive or negative impact of each factor on the profit, and thus the impact on the farm business's eco-certification decision, the expected signs of the factors are given in the same table.

We estimated different forms of the ordered choice model by selecting combinations of variables that reflect the factors given in Table 1. The selection is based on the covariance matrix of independent variables with the dependent variable, taking into consideration the relevant correlation coefficients. The estimations were conducted by using the data described next in model estimation and results section. After the elimination of the variables that do not substantially contribute to the variability of the dependent variable, the ordered model becomes

$$\begin{aligned} ACCELERATION &= \mathrm{a}_1\mathrm{COMPUTERUSE} + \mathrm{a}_2\mathrm{INCOMEAGRICULTURE} \\ &+\mathrm{a}_3\mathrm{SUCCESSOREXIST} + \mathrm{a}_4\mathrm{GAPBEFORE} + \mathrm{b}_1\mathrm{WRITCONTRACT} \\ &+\mathrm{d}_2\mathrm{NOPREFER} + \mathrm{d}_1\mathrm{BARTECHSUP} + u_i \end{aligned} \quad . \tag{3}$$

where the COMPUTERUSE variable is measured with a five-point scale and reflects the frequency of using a personal computer—that is, how often the farm business owner uses a personal computer; the INCOMEAGRICULTURE variable reflects the share (% percentage) of farm business income from agricultural activities; the SUCCESSOREXIST variable, which is a binary variable, reflects the existence of the successor who will continue and maintain the eco-certification; the GAPBEFORE variable, which is also a binary variable, reflects the implementation of a code of good agricultural practices in farm business fields in the past; the WRITCONTRACT variable, which is measured with a five-point scale, reflects the frequency the farm business signs written contracts with the buyers/purchasers of the products it produces; the NOPREFER variable, which is measured with a five-point scale, reflects the frequency the farm business perceives that the buyers/purchasers do not exhibit a preference for eco-certified products; and the BARTECHSUP variable, which is measured with a five-point scale, reflects the frequency the farm business isn't provided with agricultural technical support.

Model Estimation, Results, and Discussion

Based on the previous studies focusing on quality certification, including environmental issues, a survey was developed in the summer of 2011 and a small-scale pretest with some in-depth interviews was conducted in the winter of the same year. All the necessary modifications were made to take into account the comments and suggestions received, which primarily concerned the difficulty in answering the questionnaire (including its size), the clarity of the questions, and their order. The final questionnaire was delivered to participants in the spring and summer of 2012. The questionnaires were completed by conducting interviews with 250 fruit and vegetable producers, who are dispersed all over the region of Central Macedonia–North Greece, which is the main fruit-and-vegetable producing region in Greece. The respondents were randomly selected from a database held by the regional services of the Ministry of Agricultural Development and Food, which includes farm businesses implementing quality management systems incorporating environmental issues and EMSs. After discarding several problematic questionnaires, we were able to use a total of 231 in our analysis.

Estimation regarding the acceleration of the eco-certification model (3) was conducted using the Eviews program by choosing the ordered choice model form (extreme value of error distribution); we used the Huber/White option to compute robust (quasi-maximum likelihood) estimators, in the sense that consistent estimates of parameters are produced, even if the distribution is incorrectly specified. The results are presented in Table 2, including coefficient estimates, Z-statistics, and p-values. It was determined that the null hypothesis was rejected and that all of the independent variables affected the variability of the dependent variable because the log likelihood value of –169.5453 was highly significant (prob. 0.000). The average score of the time respondents are under eco-certification is 2.42 (>2), meaning that most of the farm businesses participating in the study implement eco-certification schemes at an early stage rather than at a late stage. Because the model estimation is based on aggregated data, including the adoption of four eco-certification schemes with differences among them, and because it doesn't include some factors of the external business environment previously studied (Karipidis, Tselempis, Karypidou, & Aggelopoulos, in press), we do not expect the interpretation capacity of the model to be high. Although there are no previous studies similar to the present, the interpretative ability (pseudo R-squared 0.221335) can be considered adequate when compared with the interpretative ability of analogous studies (Tselempis et al., 2015) or studies that estimate adoption models in the case of one or two certain certification schemes (Handschuch et al., 2013).

Regarding the fifteen economic, social, and behavioral characteristics that can shift farm business decisions, the results indicate that the acceleration of eco-certification is affected by four variables. More specifically, the COMPUTERUSE variable positively impacts the acceleration, implying that it is more probable that the farm business accelerates eco-certification if its owner frequently uses a personal computer. Taking into consideration those reported in the chapter's "Information Technology-Based Traceability Systems" subsection, it can be explained by the fact that IT facilitates the communication of the sustainable food supply chain participants with customers, consumers, society, and public authorities, making the management of environmental issues more effective and efficient at both the individual level and at the supply chain level. The INCOMEAGRICULTURE variable positively impacts the acceleration of certification, which means that when the share of the farm business income from agricultural activities increases, it becomes more probable that the farmer accelerates the eco-certification. This is explained because the high share of income from agriculture enables the farm business to achieve economies of scale, thus making eco-certification more attractive to it.

Table 2. Estimation results of the ordered model

VARIABLE	Coefficient	Std. Error	z-Statistic	Prob.
COMPUTERUSE	0.237578	0.087300	2.721388	0.0065
INCOMEAGRICULTURE	0.012137	0.004392	2.763414	0.0057
SUCCESSOREXIST	0.471863	0.202113	2.334646	0.0196
GAPBEFORE	0.829647	0.270865	3.062955	0.0022
WRITCONTRACT	0.764281	0.310116	2.464500	0.0137
NOPREFER	-0.292036	0.122606	-2.381903	0.0172
BARTECHSUP	-0.226153	0.123930	-1.824850	0.0680
Pseudo R-squared	0.221335	LR statistic	96.38616	
Prob(LR statistic)	0.000000			

The third variable for which the results indicate that it substantially contributes to the variability of the dependent variable—that is, the acceleration of certification—is the SUCCESSOREXIST variable. It implies that in the case when there is a successor to the farm business owner, the probability that the farm business will accelerate the eco-certification increases because the eco-certification becomes a good start for the successor who can continue and maintain the eco-certification and gain economies of scale. There is no point at which any of these results conflict with the results of previous studies examining farmers' certification decisions (Asfaw et al., 2010b; Kersting & Wollni, 2012; Masakure et al., 2011). These extend the findings of Karipidis and Tselempis (2014) concerning the maintenance of quality certification by incorporating environmental issues as well as the results of Tselempis et al. (2015) who studied certification as a response of farm businesses to demand.

The acceleration of eco-certification is also positively affected by the GAPBEFORE variable, which reflects a behavioral characteristic of the farm business. It means that in the case when the farm business implemented a good agricultural practices program in the past, the probability that the farm business will accelerate the eco-certification decision increases. It implies that experience makes the farm owner more willing to decide on eco-certification at an early stage because the implementation of good agricultural practices can be viewed as a preparatory step for farm businesses, which helps them to proceed in eco-certification, leading to a reduction in the cost of eco-certification.

Following the findings of Grimm et al. (2016) and Laari et al. (2016) regarding the relationships of sustainable supply chain participants with the suppliers and sub-suppliers and the assessment and monitoring of suppliers, the marketers, who are downstream participants, should use these findings to segment farm businesses, which are their suppliers, in order to determine which of them to engage and the resources required to manage the supply. More specifically, the first step would be to create three segments of certified suppliers according to the three groups of acceleration, linked with the characteristics impacting eco-certification decisions. In the second step, marketers would proceed in a segmentation of the non-eco-certified suppliers, based on the four characteristics that make more probable the acceleration of certification, such as the frequency of computer use, the income from agriculture, the existence of a successor, and the implementation of good agricultural practices in the past. It would enable downstream participants to (1) choose from the noncertified farm businesses those for which it is more probable to proceed in eco-certification by somewhat encouraging them, (2) define the type of engagement with them, and (3) identify and count the resources they have to use in order to encourage some of them to accelerate the eco-certification decision.

Regarding the relationship between farm businesses and the buyers of agricultural products (manufacturers, wholesalers, retailers), it is observed that two of the six factors presented in Table 1 play a crucial role in eco-certification decisions. More specifically, the WRITCONTRACT variable contributes substantially to the variability of the dependent variable by positively impacting the acceleration. It implies that if the farm business owner frequently signs written contracts with buyers/customers, then that makes him or her more willing to accelerate the eco-certification decision. This confirms the report by Karipidis, Chrysochou, and Karypidou (2016) that exporting firms face the certified product as a strength for their exporting marketing mix, and in some cases, they partially or wholly undertake the cost of eco-certification. The second variable, which substantially contributes to the variability of the dependent variable, is the NOPREFER variable, which reflects the frequency of no preference of buyers for eco-certified products. More specifically, it is found that, as expected, it negatively impacts the acceleration of eco-certification by implying that if the farm business owner perceives that the buyers/customers do not express a stable preference for eco-certified products, then he or she does not accelerate the certification decision.

Regarding the variables reflecting those factors that act as barriers to farm businesses in proceeding with eco-certification, it is found that the BARTECHSUPPORT variable substantially contributes to the dependent variable's variability. It implies that if agricultural technical support is not provided to the farm business owner, then it makes him or her less eager to accelerate the eco-certification decision because it becomes more difficult for the farmer to assess whether the implementation and certification expenses will be counterbalanced by corresponding benefits and because the eco-certification effort can become more costly. Taking into consideration the conclusions of Hartmann and Moeller (2014), Hassini et al. (2012), and Wong (2013) that the overall supply chain performance can be improved if all supply chain partners embed sustainable practices into their activities, the marketers could lead their suppliers to accelerate eco-certification by the proper supplier strategies. For example, they would incorporate a stable preference for the products of eco-certified farm businesses into their supply strategies in order to encourage them to accelerate eco-certification. They would also incorporate written contracts with farm businesses into their strategies and provide them with the proper agricultural technical support to accelerate eco-certification.

CONCLUSION

The transformation of food production to a more sustainable process in which food supply chain participants properly respond to the environmentally responsible demand can be supported by providing consumers with sustainability-related information that helps them to include eco-friendly products in their buying choices. In such an attempt, supply chain participants must choose between a broad set of traceability systems and eco-certification schemes in combination with the advanced IT-based systems and applications to facilitate information sharing and to ensure that the requirements for transparency and food traceability can be met. As an ultimate result, by sharing sustainability-related information along the globalized food supply chain, a competitive advantage and brand integrity/awareness can be achieved.

To ensure compliance with sustainability requirements, food supply chain participants increasingly realize the relevance of their supply chains and their dependence on suppliers' environmental practices. Because the farm business is the upstream chain participant with a major contribution in environmental consequences of the food supply, downstream participants (manufacturers, wholesalers, retailers) require it to implement a form of EMS certified by a third party (eco-certification) to provide consumers with signals and messages about sustainability issues. In order for the downstream participants to implement an effective and efficient suppliers' strategy, they must understand farm business eco-certification decisions and determine the drivers of these decisions.

By combining a shared value framework for chain participant decisions with a farm business decision model, the acceleration of eco-certification was analyzed in conjunction with selected factors that enable downstream supply chain participants to segment their suppliers. It was concluded that there are some factors connected with economic, social, and behavioral characteristics of farm businesses accelerating the eco-certification decision and thus enabling farm buyers of agricultural products to segment their suppliers by defining with which of them to engage, the type of engagement, and the resources needed to manage the relationship with them. It was also found that, regarding the relationship between downstream supply chain participants and farm businesses, certain factors, such as the assignment of written contracts with farm businesses, the indication of a stable buying preference for eco-certified products, and the provision of agro-technical support, accelerate the eco-certification decision. This implies that these factors can be viewed as tools to enable marketers/buyers of agricultural products to encourage farm businesses to accelerate the eco-certification decision.

These findings are especially useful for downstream global chain participants in food supply chains with small food producers. A weakness of this study is that the empirical investigation of the eco-certification model is based on data collected by farm businesses as participants in fruit and vegetable supply chains. It would be useful for eco-certification decisions to be examined in different supply chains, such as cotton, cereals, cheese and milk, and meat chains. Furthermore, because downstream supply chain participants choose different environmental signals and IT-based traceability systems, it would be useful to examine their sustainability-related decisions in conjunction with the relationships to upstream supply chain participants. This would be especially interesting for those cases with the highest environmental consequences.

REFERENCES

Ahi, P., & Searcy, C. (2013). A comparative literature analysis of definitions for green and sustainable supply chain management. *Journal of Cleaner Production*, *52*, 329–341. doi:10.1016/j.jclepro.2013.02.018

Asfaw, S., Mithöfer, D., & Waibel, H. (2010a). Agrifood supply chain, private-sector standards, and farmers health: Evidence from Kenya. *Agricultural Economics*, *41*(3–4), 251–263. doi:10.1111/j.1574-0862.2010.00443.x

Asfaw, S., Mithöfer, D., & Waibel, H. (2010b). What impact are EU supermarket standards having on developing countries export of high-value horticultural products? Evidence from Kenya. *Journal of International Food & Agribusiness Marketing*, *22*(3–4), 252–276. doi:10.1080/08974431003641398

Badia-Melis, R., Mishra, P., & Ruiz-García, L. (2015). Food traceability: New trends and recent advances. *Food Control*, *57*, 393–401. doi:10.1016/j.foodcont.2015.05.005

Basu, A. K., Chau, N. H., & Grote, U. (2003). Eco-labeling and stages of development. *Review of Development Economics*, *7*(2), 228–247. doi:10.1111/1467-9361.00188

Bougherara, D., Grolleau, G., & Thiébaut, L. (2005). Can labelling policies do more harm than good? An analysis applied to environmental labelling schemes. *European Journal of Law and Economics*, *19*(1), 5–16. doi:10.1007/s10657-005-5273-6

Darnall, N., Ji, H., & Vázquez-Brust, D. A. (2016). Third-party certification, sponsorship, and consumers' ecolabel use. *Journal of Business Ethics*, 1–17. doi:10.1007/s10551-016-3138-2

Delmas, M. A., & Grant, L. E. (2014). Eco-labeling strategies and price-premium: The wine industry puzzle. *Business & Society*, *53*(1), 6–44. doi:10.1177/0007650310362254

Eggert, A., & Helm, S. (2003). Exploring the impact of relationship transparency on business relationships: A cross-sectional study among purchasing managers in Germany. *Industrial Marketing Management*, *32*(2), 101–108. doi:10.1016/S0019-8501(02)00224-9

Feng, J., Fu, Z., Wang, Z., Xu, M., & Zhang, X. (2013). Development and evaluation on a RFID-based traceability system for cattle/beef quality safety in China. *Food Control*, *31*(2), 314–325. doi:10.1016/j.foodcont.2012.10.016

Folinas, D., Manikas, I., & Manos, B. (2006). Traceability data management for food chains. *British Food Journal, 108*(8), 622–633. doi:10.1108/00070700610682319

Garnett, T. (2011). Where are the best opportunities for reducing greenhouse gas emissions in the food system (including the food chain)? *Food Policy, 36*(1), S23–S32. doi:10.1016/j.foodpol.2010.10.010

Grimm, J. H., Hofstetter, J. S., & Sarkis, J. (2016). Exploring sub-suppliers compliance with corporate sustainability standards. *Journal of Cleaner Production, 112*, 1971–1984. doi:10.1016/j.jclepro.2014.11.036

Handschuch, C., Wollni, M., & Villalobos, P. (2013). Adoption of food safety and quality standards among Chilean raspberry producers – Do smallholders benefit? *Food Policy, 40*, 64–73. doi:10.1016/j.foodpol.2013.02.002

Hartmann, J., & Moeller, S. (2014). Chain liability in multitier supply chains? Responsibility attributions for unsustainable supplier behaviour. *Journal of Operations Management, 32*(5), 281–294. doi:10.1016/j.jom.2014.01.005

Hassini, E., Surti, C., & Searcy, C. (2012). A literature review and a case study of sustainable supply chains with a focus on metrics. *International Journal of Production Economics, 140*(1), 69–82. doi:10.1016/j.ijpe.2012.01.042

Hatanaka, M., Bain, C., & Busch, L. (2005). Third-party certification in the global agri-food system. *Food Policy, 30*(3), 354–369. doi:10.1016/j.foodpol.2005.05.006

Hattam, C. E., Lacombe, D. J., & Holloway, G. J. (2012). Organic certification, export market access and the impacts of policy: Bayesian estimation of avocado smallholder times-to-organic certification in Michoacán Mexico. *Agricultural Economics, 43*(4), 441–457. doi:10.1111/j.1574-0862.2012.00595.x

Henson, S., & Reardon, T. (2005). Private agri-food standards: Implications for food policy and the agri-food system. *Food Policy, 30*(3), 241–253. doi:10.1016/j.foodpol.2005.05.002

Herzfeld, T., Drescher, L. S., & Grebitus, C. (2011). Cross-national adoption of private food quality standards. *Food Policy, 36*(3), 401–411. doi:10.1016/j.foodpol.2011.03.006

Karipidis, P., Chrysochou, P., & Karypidou, I. (2016, June). *Does the Greek food supply respond to the world challenges?* Paper presented at the second International Conference of Development and Economy, Thessaloniki, Greece.

Karipidis, P., & Sartzetakis, E. (2013). Firm's decisions based on consumers' choices in eco-certified food markets. *Economics Research International*, 1–10. doi:10.1155/2013/920164

Karipidis, P., Tsakiridou, E., Aggelopoulos, S., & Belidis, A. (2010). Consumers' purchasing and store switching intentions in respect of eco-marked products. *International Journal of E-Business Research*, 2(6), 511–524. doi:10.1504/IJEBR.2010.035702

Karipidis, P., Tselempis, D., Karypidou, I., & Aggelopoulos, S. (in press). Market-driven or policy-directed quality certification? *Sinergie: Italian Journal of Management*.

Karipidis, P., & Tselepmis, D. (2014). Farmers intention to maintain quality certification. *EuroMed Journal of Business, 9*(1), 93–110. doi:10.1108/EMJB-04-2013-0015

Kersting, S., & Wollni, M. (2012). New institutional arrangements and standard adoption: Evidence from small-scale fruit and vegetable farmers in Thailand. *Food Policy*, *37*(4), 452–462. doi:10.1016/j.foodpol.2012.04.005

Kim, Y. G., & Woo, E. (2016). Consumer acceptance of a quick response (QR) code for the food traceability system: Application of an extended technology acceptance model (TAM). *Food Research International*, *85*, 266–272. doi:10.1016/j.foodres.2016.05.002

Kottila, M., & Rönni, P. (2010). Communication between actors of food chains: Case studies of two organic food chains in Finland. In A. Lindgreen, M. K. Hingley, D. Harness, & P. Custance (Eds.), *Market orientation: Transforming food and agribusiness around the customer* (pp. 171–186). Surrey, UK: Gower.

Laari, S., Töyli, J., Solakivi, T., & Ojala, L. (2016). Firm performance and customer-driven green supply chain management. *Journal of Cleaner Production*, *112*, 1960–1970. doi:10.1016/j.jclepro.2015.06.150

Manikas, I., Hamann, K., & Sentic, A. (2016). Quality management schemes and connections to the concept of sustainability in the food chain. In E. Iakovou, D. Bochtis, D. Vlachos, & D. Aidonis (Eds.), *Supply chain management for sustainable food networks* (pp. 233–254). Chichester, UK: Wiley. doi:10.1002/9781118937495.ch9

Masakure, O., Cranfield, J., & Henson, S. (2011). Factors affecting the incidence and intensity of standards certification evidence from exporting firms in Pakistan. *Applied Economics*, *43*(8), 901–915. doi:10.1080/00036840802600103

Massaroni, E., Cozzolino, A., & Wankowicz, E. (2015). Sustainability in supply chain management – A literature review. *Sinergie: Italian Journal of Management*, *33*(98), 331–355. doi:10.7433/1115

Muriithi, B. W., Mburu, J., & Ngigi, M. (2011). Constraints and determinants of compliance with EurepGap standards: A case of smallholder French bean exporters in Kirinyaga district, Kenya. *Agribusiness*, *27*(2), 193–204. doi:10.1002/agr.20261

Mylan, J., Geels, F. W., Gee, S., McMeekin, A., & Foster, C. (2014). Eco-innovation and retailers in milk, beef and bread chains: Enriching environmental supply chain management with insights from innovation studies. *Journal of Cleaner Production*, *107*, 20–30. doi:10.1016/j.jclepro.2014.09.065

Pérez Mesa, J. C., & Galdeano-Gómez, E. (2015). Collaborative firms managing perishable products in a complex supply network: An empirical analysis of performance. *Supply Chain Management: An International Journal*, *20*(2), 128–138. doi:10.1108/SCM-06-2014-0185

Piramuthu, S., Farahani, P., & Grunow, M. (2013). RFID-generated traceability for contaminated product recall in perishable food supply networks. *European Journal of Operational Research*, *225*(2), 253–262. doi:org/10.1016/j.ijpe.2013.04.035

Porter, M. E., & Kramer, M. R. (2011, January–February). Creating shared value: How to reinvent capitalism—and unleash a wave of innovation and growth. *Harvard Business Review*, 1–17.

Pramatari, K. (2016). Information technology for food supply chains. In E. Iakovou, D. Bochtis, D. Vlachos, & D. Aidonis (Eds.), *Supply chain management for sustainable food networks* (pp. 183–204). Chichester, UK: Wiley. doi:10.1002/9781118937495.ch7

Schwarz, J., Schuster, M., Annaert, B., Maertens, M., & Mathijs, E. (2016). Sustainability of global and local food value chains: An empirical comparison of Peruvian and Belgian asparagus. *Sustainability*, *8*(4), 344. doi:10.3390/su8040344

Soltani, S., Azadi, H., Mahmoudi, H., & Witlox, F. (2014). Organic agriculture in Iran: Farmers barriers to and factors influencing adoption. *Renewable Agriculture and Food Systems*, *29*(2), 126–134. doi:10.1017/S1742170513000069

Trienekens, J. H., Wognum, P. M., Beulens, A. J., & van der Vorst, J. G. (2012). Transparency in complex dynamic food supply chains. *Advanced Engineering Informatics*, *26*(1), 55–65. doi:10.1016/j.aei.2011.07.007

Tselempis, D., Karipidis, P., Pavloudi, A., & Semos, A. (2015). Is the quality certification in crop and vegetable production a market driven choice in Greece? *Agricultural and Food Economics*, *3*(13), 1–12. doi:10.1186/s40100-015-0032-7

Tubiello, F. N., Salvatore, M., Cóndor Golec, R. D., Ferrara, A., Rossi, S., Biancalani, R., & Flammini, A. (2014). *Agriculture, forestry and other land use emissions by sources and removals by sinks*. Rome, Italy: FAO.

United Nations Global Compact and Business for Social Responsibility. (2014). *A guide to traceability: A practical approach to advance sustainability in global supply chains*. New York, NY: United Nations Global Compact Office.

Veldstra, M. D., Alexander, C. E., & Marshall, M. I. (2014). To certify or not to certify? Separating the organic production and certification decisions. *Food Policy*, *49*(2), 429–436. doi:10.1016/j.foodpol.2014.05.010

Weber, C. L., & Matthews, H. S. (2008). Food-miles and the relative climate impacts of food choices in the United States. *Environmental Science & Technology*, *42*(10), 3508–3513. doi:10.1021/es702969f PMID:18546681

Wognum, P. N., Bremmers, H., Trienekens, J. H., van der Vorst, J. G., & Bloemhof, J. M. (2011). Systems for sustainability and transparency of food supply chains – Current status and challenges. *Advanced Engineering Informatics*, *25*(1), 65–76. doi:10.1016/j.aei.2010.06.001

Womack, J., & Jones, D. (2003). *Lean thinking: Banish waste and create wealth in your corporation*. New York, NY: Simon & Schuster.

Wong, C. W. (2013). Leveraging environmental information integration to enable environmental management capability and performance. *Journal of Supply Chain Management*, *49*(2), 114–136. doi:10.1111/jscm.12005

Yan, B., Hu, D., & Shi, P. (2012). A traceable platform of aquatic foods supply chain based on RFID and EPC Internet of things. *International Journal of RF Technologies: Research and Applications*, *4*(1), 55–70. doi:10.3233/RFT-2012-0035

Ziesemer, J. (2007). *Energy use in organic food systems*. Rome, Italy: FAO.

Chapter 6
Social Marketing:
A New Marketing Tool for the Food Sector

Ahmed Elghannam
Damanhour University, Egypt

Francisco J. Mesías
University of Extremadura, Spain

ABSTRACT

Consumers purchase food from different sources, mainly via traditional/long chains where hypermarkets are the final link between producer and consumer. However, consumers are seeking direct relationships with producers. This, together with the increase of social media usage offer producers the potential to build short chains for promoting/selling their products. The aim of this work is to summarize the role that online short food supply chains could play as an opportunity for SMEs in the agri-food sector. Moreover, it highlights a new perspective based on social media as potential short supply chains. To this end, a thorough review of the literature has been carried out, together with an online survey where social networks as food marketing channels have being studied. The chapter concludes pondering about the different food products/sectors that could take advantage of the creation of short supply chains and of the wider use of social networks as marketing tools.

INTRODUCTION

Food marketing and distribution sectors have experienced remarkable changes in recent years, mainly in terms of concentration processes that have taken place mostly among large food companies (MAGRAMA, 2008). As a consequence, the configuration of the producer-distributor relationship has undergone serious changes, since it moved from a situation in which manufacturers dominated the conditions of purchasing of their products, to a new context in which distributors have enhanced their bargaining position (Oubiña, 2000) and got the capacity of driving the demand. Consequently, this situation has generated a dissatisfaction state for producers, who cannot find a stable market and revenue for their activities and accordingly they make a little profit or even losses. Since consumers are demanding a variety of

DOI: 10.4018/978-1-5225-2107-5.ch006

products and services with the highest quality, best price, and added value, important changes in the producer-distributor relationship are taking place. Consumers are also looking for more information with regard to origin, safety and healthiness of the purchased food. All these issues have to be considered in the light of the great importance of the agri-food sector both in the European Union and in Spain. Within the European Union, the Spanish food industry is placed fifth in terms of the net sales value of the agri-food sector after Germany, France, Italy and United Kingdom (MAGRAMA, 2014). The food and beverage industry in Spain is considered the first industrial branch, as stated by the latest statistical survey (INE, 2013), representing 20.6% of net sales, 18.2% of employed people and 15.3% of the added value. In 2013, total net sales stranded for 91,450 million Euros, representing an increase of 1.4% over the earlier year. Meat industry represented 22.1% of that figure; animal feed comes behind with 9.7%, followed by fats and oils (9.4%) and dairy (9.3%) (MAGRAMA, 2014).

There are different shopping chains used by Spanish consumers to purchase food and drinks. In this regard, the relative importance of supermarkets (53.6% of market share in 2012) has gradually increased as compared with specialized shops (22.7% of market share) that together with other formats have a less noticeable presence. Nevertheless, the choice of point of purchase displays certain differences in the Spanish market depending on whether consumers are to buy fresh or processed food. In the first case, specialized shops are considered one of the most preferred choices, with 31.1% of the market share of meat and with 40.6% in fresh fruits. Regarding processed food, free service channels have become the preferred choice for households (supermarkets account for 70.7% of the sales of milk or 59.2% of sales of olive oil) (MERCASA, 2013). Furthermore, other formats that weren't common until recently, like delicatessen stores, 24 hours shops and sales on the Internet have been considered to be used by consumers (MAGRAMA, 2008). On the other hand, 70% of purchases are made via traditional (long) chains where hypermarkets or supermarkets are the final link between producer and consumer.

Nonetheless, a sector of the population is still seeking more direct relationships with the producer, claiming their right to choose the products they consume and to be informed about the source and model of production (MAGRAMA, 2013). Thus, it would be of a great value for traditional enterprises to deal directly with consumers by creating short food supply chains (SFSC), reducing the passageway through several links in the food chain, hence facilitating the traceability of food products and a better price transmission between producers and consumers. In this sense the vast increase in the social media usage has offered the potential for producers to build up new short chains for promoting and selling their products in a rapid, inexpensive and direct way, also contributing to reduce market margins by enhancing direct transactions.

Social media could have various uses in the food marketing sector, as it would facilitate the role of marketing managers to identify their customers' profile, their preferences and the way they perceive certain products. Consequently companies may perfectly define their target customers and change their marketing policies accordingly. Social networks are also developing a new system in order to enable the online purchasing process through their platforms, something that will open new opportunities for food manufacturers. Within this context, the main objective of this work is to review the role that online short food supply chains could play as an opportunity for small and medium enterprises (SME) in the agri-food sector. Furthermore, to highlight a new viewpoint relying on social media platforms as possible short supply chains for food SMEs. The specific questions to address are: (1) how can short food supply chains be defined? (2) how could social media applications serve as a short food supply chain? (3) which types of food would consumers buy through social networks?

The chapter is organized as follows. First of all, the concept of short food supply chain and the main related perspectives identified in literature will be described. Subsequently, a new approach of short supply chain will be presented within the context of internet purchases and offline models. The chapter will also indicate the existing situation concerning e-marketing practices in food SMEs, emphasizing the empowering role of information and communication technologies. Finally, the results of a consumer survey regarding willingness to buy food products through social networks will be presented. The survey has been completely developed online, due to the nature of the study, where social networks as a food marketing channel are being studied. Therefore, it was designed in an online format using Google Forms, and including socio-demographic characteristics and frequency habits regarding the use of social media. Accordingly, it was decided to use Facebook, Linkedin and Google+ to distribute the questionnaire. As a whole, this chapter concludes with a reflection about the different food products and food sectors that could take advantage of both the creation of short food supply chains and a wider use of social networks as marketing tool within those chains.

BACKGROUND

A literature review shows different definitions of short supply chains. For instance, the definition adopted by the European Council (Santini & Gomez y Paloma, 2013) is "a supply chain formed by a limited number of economic agents, committed to cooperation, local economic development and socio-economic relations between producers and consumers in a close geographical area".

Also, it is worthy here to differentiate between two main types of short supply chains. On the one hand, we find the direct short chains, where the number of intermediaries is zero and on the other hand the indirect short chains, which only have a single intermediary between producers and consumers (Mundubat, 2012).

The marketing via these short supply chains can be done in relation with the different marketing structures available or used. We can differentiate those structures according to the use of the Internet as online and off-line short supply chains. The online chain allows either the possibility to purchase products directly through the network or just offers online support so that consumers are put into direct contact with sellers. Online chains may include: i) online platforms, such as the ones used for broadcasting purposes, (the user only has access to the information and contact details of the producer) and those that allow purchasing online directly; and ii) on-line sales websites of producer or manufacturer where electronic commerce takes place (MAGRAMA, 2013).Whereas the offline short channels are those who do not offer their products via the web. Within the so-called offline chains, there are various purchasing outlets such as producer markets, direct sale shops, shipping door-to-door and consumer groups.

According to the above, a more inclusive definition of a short supply chain would be "a chain in which the number of intermediaries is equal to or less than one whether the transaction takes place through online or offline platforms" (Figure 1). This concept is widely applicable to the food sector, in which case we would refer to short food supply chains.

A literature review identifies three main types of short food supply chains (Marsden et al., 2000; Renting et al., 2003) that are briefly described in the following paragraphs.

Figure 1. Information flows in short food supply chain models

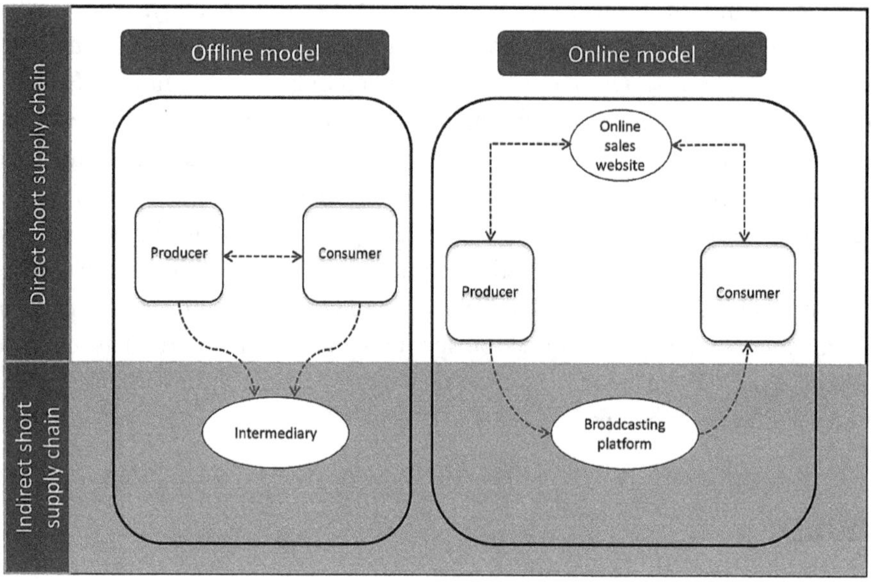

1. **Face-to-Face**: In these short chains producers sell their products directly to consumers on a face-to-face basis. Farm gate sales, pick-your-own and farmers markets are some examples of face-to-face short food supply chains.
2. **Spatial Proximity**: The main characteristic of this model is that products are produced and sold through local market channels in the specific region of production. It includes farm shop groups, food service outlets, local food retailers and consumer cooperatives. Other interesting examples also reported in the literature about spatially proximate short food supply chains are Community supported agriculture (CSA) (Brown & Miller, 2008), Solidarity purchasing groups (GAS) (Migliore, 2014) and Associations for the maintenance of peasant agriculture (AMAP). All these types share the same essential principles whereby subscribers receive a share of the harvest in return for money and labor, although they could vary slightly according to different regions and countries (Santini & Gomez y Paloma, 2013).
3. **Spatially Extended**: In this case the products are sold not only to local consumers but also to consumers in other regions. Therefore, under this model labeling and certification programs could be used to differentiate these products emphasizing the quality dimension (Abatekassa, 2011) such as in the case of Fair trade and Protected Designation of Origin (PDO).

In terms of social and economic impacts, there is evidence that short food supply chains favor interaction and direct connection between farmers and consumers, thus promoting the development of confidence and social capital. This can also result in the development of community sense and co-existence and can even cause changes in eating and shopping habits and enhance social and environmental awareness. Moreover, economic benefits can be noticed regarding rural development and economic regeneration, as local agricultural systems and short chains have a greater effect on the local economies than long ones, with implications also for maintaining local employment especially in rural areas (Santini and Gomez y Paloma, 2013).

In addition to quality issues, social and environmental concerns are the main reasons why consumers are interested in short chains. Thus, it would be of great value for SMEs to take a part in this type of marketing. They can even use short and long food chains combined in order to get the greatest benefit and to reduce the risk. Already in 2013, and based on a study on short commercialization chains in the agri-food sector, the Spanish Ministry of Food and Agriculture started promoting food marketing via short supply chains (MAGRAMA, 2013). To this end, it recommended SMEs to bet on the development of marketing using websites or mobile phone applications, along with building the presence in social networks, which can be considered as affordable and not overly complicated tools for non-experts (MAGRAMA, 2013).

METHODOLOGY

Data Collection and Survey Design

With the aim of identifying types of food that consumers would be interested in buying directly through social networks, an online survey was developed during December 2015 using Google forms – online (www.docs.google.com). A pretest was presented to 15 consumers (not included in the final sample) in order to assess the clarity of the questions. A convenience sampling adjusted to the sex and age of the Spanish population was adopted in this study. Finally, a total of 120 questionnaires have been considered for analysis. Table 1 shows the socio-demographic characteristics of the final sample compared with those of Spain. Although the sample size can be considered insufficient for a market study, due to the introductory nature of this work this number of surveys was deemed sufficient for the purposes of the study.

As a result of the main features of the study, where social networks as a food marketing tool were going to be studied, it was decided to use Facebook, LinkedIn and Google+ to distribute the questionnaire. A link to the questionnaire was sent to respondents together with the following introductory message: "Although at present is not common to purchase directly on social networks, some platforms are developing new tools that would give you the opportunity to buy directly from the social site. All the process, from purchasing to payment would be carried out from the same app". They were then asked to list all the food they would buy on social networks, in what is defined as a free listing task.

Table 1. Socio-demographics characteristics of the sample

Variable	Spain	Sample
Gender		
Men	48.60	48
Women	51.40	52
Age		
18 - 35 years	24.5	28
36 - 50 years	30.2	31
> 50 years	45.3	41

Free listing is a qualitative technique based on asking individuals to list as many items or ideas as possible related to a topic or product. It is a simple but powerful tool that can provide an understanding of consumers' attitudes when they face a completely new setting, as for example food purchasing through social media. Results from a free listing exercise are usually analyzed considering that the most important item is the one which receives the higher number of mentions from the participants.

DATA ANALYSIS

Once the data were collected they were analyzed using content analysis (Stewart & Shamdasani, 1990), a research technique used to make replicable and valid inferences from texts or other meaningful materials (Krippendorff, 2004). In order to develop this task, the answers were categorized using as a basis the food classification found in online pages of major Spanish supermarkets, i.e. Mercadona and El Corte Inglés.

Initially, a search for recurrent terms was developed. Subsequently, those terms with similar meaning were grouped into categories. Products were grouped according to the main categories and subcategories used by supermarkets, but also taking into consideration food concepts mentioned by the participants. Frequencies of each of the products listed by respondents were counted separately. The frequencies in every category were determined by counting the number of consumers that used the same word or an equivalent term. Finally, percentages of each category/concept were calculated by dividing the total of frequencies of each category/subcategory or concept between the total of terms mentioned by respondents.

RESULTS AND DISCUSSION

Agri-Food E-Marketing

E-marketing is a subset of e-business that uses electronic means to perform marketing transactions and accomplish certain marketing goals for an organization (Petrovic, 2010). It therefore implies the application of digital technologies to contribute to the marketing activities of an enterprise so as to strengthen the relationship with customers and create added value for the product. It includes both direct response marketing and indirect marketing elements, and uses a range of technologies to help connect businesses to their customers (Tsekouropoulos et al., 2011).

The internet can be used to facilitate purchasing transactions among all kinds of actors: among consumers, among businesses, between businesses and consumers (Grunert & Ramus, 2005). Studies dealing with consumer behavior in internet shopping concluded that people have a wide range of different motivations and different approaches which trigger their behavior and which include not only the pros of convenience, financial benefits and easy information accessing but also hedonic aspects of e-commerce like enjoyment, normative beliefs and self-efficacy (Mandilas et al., 2013; Shang et al., 2005; Joines et al., 2003)

Over the last decade, online shopping has provided an open window for producers to market their products and has become one of the most rapidly growing forms of shopping (Zhu et al., 2014). Potential uses of e-marketing are interesting for the agri-food sector due to both globalization in markets and fragmentation in supply (Hausen et al., 2006). However, adoption of this approach by businesses is low (Canavari et al., 2016), in particular by small and medium sized enterprises which represent the majority of agricultural production (European Commission, 2005; Fritz & Canavari, 2008; Bewley & Russell, 2010; Canavari et al., 2010; Lehmann et al., 2012).

According to the e-Business Index 2006, calculated by e-Business Watch (European Commission, 2007) for 10 different sectors in 10 EU countries, the food sector is found in the lowest two ranks in this benchmarking which express e-business adoption as "a percentage of firms in a sector with a certain activity", regardless the size of the firms (European Commission, 2007). Given that the European food sector, as it was mentioned before, is dominated by small and medium sized firms, we can interpret that these results mainly reflect the SMEs' relationship with new technologies, as many small companies still face diverse problems to get digitally connected with their suppliers and customers.

In a small enterprise, information management -as a part of an e-business strategy- can be effectively and efficiently achieved with the use of less sophisticated and less expensive systems compared with those used by large companies (European Commission, 2007). Moreover, the conclusion of the European e-business report (European Commission, 2007), showed the great potential of e-business for SMEs, which could be noticed as follows:

1. While SMEs need to cooperate, for example by building networks, information and communication technology (ICT) usage facilitates cooperation in many ways (e.g. through project management tools, or online collaboration tools for design).
2. Also, current technological developments hold opportunities for small companies, for example, Voice-over-IP telephone and mobile e-business solutions. Moreover, ICT companies are increasingly addressing the SME market by developing affordable, small-sized solutions such as Enterprise Resource Planning (ERP) which are computer systems organized for resource administration in an organization, or Customer Relationship Management (CRM) suites which are computer systems to support the management of relations with customers, sales and marketing. Such software can include several features to manage the company's sales and customers.
3. Finally, many SMEs are forced to expand their market area. E-Commerce can be an opportunity -maybe the only way- for them to go global.

Social Networks as an E-Marketing Tool

Recently, the spectacular development of internet use -above all the Web 2.0 and online social networks- has aroused great interest in the marketing sector. Many companies have decided to incorporate social marketing (Marketing via social networks) to support their commercial activities (Mata & Quesada, 2014). From a market perspective, social networks can be considered as collections of individuals which create a marketplace. This situation presents valuable opportunities to do business based on the potential benefits that a company can get from such networks to promote its brands or products (Dooley et al., 2012; Mata & Quesada, 2014). From this perspective, both online social networks and e-commerce may be considered complementary e-marketing tools and not substitutes.

The proliferation of social media applications such as online communities, social networking sites or blogs gives the public new means for receiving, and, more importantly, providing information (Rutsaert et al., 2013). It is worth mentioning that social networks offer their services for free most of the times, relying on advertisement revenues to cover their expenses. This means that marketing aspects are the core factor of success for such type of sites.

The free service together with the revolutionary increase in the use of social media offer producers the potential to build a new short channel for promoting and selling their products in a rapid, low-cost and direct way. It may also contribute to reducing market margins by enhancing direct sales (Business to consumer or B2C). In addition, by operating through social networks companies can create and manage their own pages to communicate directly with their customers (followers or fans, as they are known in social network jargon) thus saving advertising costs.

Businesses could start their free advertising campaigns by sharing pictures, information and even videos about their products. Additionally, a direct flow of information among consumers (consumer to consumer or C2C) is supported. This enables fans to spread the word about the company performance by sharing and commenting with their friends while getting them to leave their experience about the product, its advantages, and disadvantages. The study of Sturiale & Scuderi (2013) provides evidence about the significant impact that purchasing experience has on consumers' intention to spread the word among their peers.

Social Networks as an Opportunity for SME Agri-Food Companies

Social networks can offer interesting uses for enterprises as a chain for food marketing. In this way they can facilitate the role of marketing managers to identify their customers' profile, their preferences and the way they perceive certain products. Thus companies may perfectly define their target segments and change their marketing policies when needed.

Under the request of producers, social networks can also show ads to specific people who might be interested in their message, as these applications are designed to help advertisers to find relevant/suitable customers through the use of new technologies like cookies. Cookies offer a useful way for advertisers to understand if the sale of a product on its website is connected to an ad on the social network. Cookies are also used to learn whether someone who saw an ad later visited the advertiser's site. Moreover, Facebook and other social sites provide reports about the performance of the ads they show, such as how many people viewed or clicked on ads or demographic information about the people who viewed an ad. This information can help advertisers and producers to understand and measure the effectiveness of their ads, which helps them show better and more interesting ads to people.

The dependence of social networks on advertising revenues led them to improve their offer with some new marketing tools. For example, Facebook has recently added the "purchase button" to companies' pages with the idea of linking their Facebook pages with their online shops in order to facilitate the online purchasing process through social platforms. In Spain, firms such as El Corte Ingles and Carrefour (food retailers) or Navidul and Oleoestepa (food producers) are some examples of companies that have already taken the initiative to add this purchasing button to their Facebook pages, a move that most SMEs in the agri-food sector have not yet followed. The peculiarities of food marketing could be the reason for this slow movement towards a wider use of social media as a new and strong marketing tool within the Spanish agri-food sector.

In addition, one of the greatest problems any SME may face is the creation of its brand image. Every company needs to communicate constantly with its -present or future- customers either to introduce new products and services or just to provide them with information about their existing products. An adequate solution here is to invest in online brand positioning strategies. Hence, social networks could play an essential role as cheap and simple short chains through which the company can offer a more social image for its customers. Once consumers browsing a brand page perceive emotional and informational supports that satisfy their needs, it would be natural for them to commit to this page. In turn, consumers themselves would participate in the co-creation of a strong brand for the company (Wang & Hajli, 2014).

Consumer Willingness to Buy Food Through Social Networks

This section shows the results of the survey that was previously described and whose aim was to get an insight into those types of food products that would be liable to be bought through social networks. Therefore, Table 2 shows the categories of food products stated by the consumers in the free listing task, together with their percentage of mention. As can be observed, food cupboard is the most frequently food group mentioned by respondents with about 32% of total answers, followed by beverages and drinks with 14%.

The high willingness of consumers towards such types of food products may be derived from their specific characteristics as dry, canned or bottled food, mainly non-perishable and long-lasting. These products are characterized by a long shelf life and airtight containers that would ensure their good conditions for consumption. So that consumers would have no fears about freshness or refrigeration conditions during purchasing in an online environment and specially, during the subsequent delivery.

Table 2. Products that consumer would buy through social networks

Groups	Products	Percentage
Food cupboard	Jam, honey and sugar	6
	Rice, legumes and Pasta	7
	Coffee and tea	2
	Cereals and flours	4
	Preserved food	5
	Oils	5
	Soups and ready meals	3
	Total	32
Appetizers	Nuts	4
	Appetizers and snacks	2
	Total	6
Drinks/beverages	Water, juices & soft drinks (sodas)	6
	Wine	8
	Total	14
Processed meat products and cheeses	Processed meat products and cheeses	5
	Cured ham	1
	Total	6
Bakeries/ bread	Bread and bakery	6
Fruits and vegetables	Fruits and vegetables	7
Frozen	Frozen	5

Categories with less than 5% of total mentions were excluded

Also noteworthy are the results for the "Processed meat products and cheese" group, which ranked third with almost 10% of answers, a figure which shows that a considerable percentage of consumers are willing to purchase them directly via social networks. Although in this case these products cannot be considered as non-perishable, most of them have long usage periods, and they are part of the traditional Spanish food.

In the case of Bread, fruits and vegetables; although consumers always tend to get them fresh, it was also found a certain group of consumers who declared their intention to buy them online. This finding is in accordance with the interesting results found by Ramus and Asger Nielsen (2005) in relation to online purchasing of perishables like vegetables, fruit, meat and bread. Those authors stated that for consumers a major concern was the helplessness to control freshness of the purchased products. Nevertheless, there were also online grocery shoppers who declared that meat and vegetables ordered via the internet were fresher than supermarket groceries as they believed they were supplied almost directly from the farmers.

Although the participants had been asked to list the foods they would buy via social networks, some of the answers made reference to food concepts, and not really to food. Table 3 shows the food concepts that participants would buy through social networks.

As expected, most of the concepts that appear refer to the non-perishable character of the food, with the most mentioned being "tinned/packaged foods" followed by "non-perishable foods". However, Table 3 also presents two interesting aspects, such as "renowned brands" and "organic foods". These findings could indicate that some consumers, who look for specific products (e.g. certain brands or organic products which perhaps are not always easy to find in their usual marketing channels) may be open to consider buying food via social platforms. So it is worthy here to highlight the potential role that social networks could play as an alternative short supply chain within the Spanish organic market, where organic foods are mainly marketed using direct marketing or via specialised shops (Tarkiainen & Sundqvist, 2005).

SOLUTIONS AND RECOMMENDATIONS

In order to propose solutions and recommendations dealing with the issues discussed in this chapter the first aspect that must be taken into account is the variety of potential uses of social networks in the agri-food sector. For example, in the case of top-range foods (Delicatessen) or those with Protected Designation of Origin, in which the ability to identify consumers with very specific characteristics is nowadays only available to large enterprises with powerful marketing research departments. With the use of social networks, any food producer can identify and interact with clients with high potential interest. Something similar could happen with organic food, where we now find a large number of small

Table 3. Food concepts that participants indicated they would buy through social networks

Concept	Percentage
Tinned/packaged	29.7
non-perishable	28.1
Renowned brands	10.9
Organic	6.2
Gourmet/ security seal	7.8

producers that on many occasions have to sell their products through conventional chains due to the lack of specific chains. Social networks can supplement this deficiency by providing farmers a direct connection and interaction with their customers. In the case of organic food, various researchers have stated that its limited availability in the market is one of the main determinants that drives consumers to buy online (Ramus & Nielsen, 2005).

Obviously, there are products in which the advantages are not so clear, such as those highly perishable, where the problem is more of logistics than of marketing. Even if it can be considered that social networks can help to improve this logistic problem (through the fast and cheap broadcast of information, from y towards the consumer), companies can still benefit from this and other valuable information coming directly from their present and prospective customers.

In addition, one of the greatest problems any SME may face is the creation of its brand image. Every company needs to communicate constantly with its -present or future- customers either to introduce new products and services or just to provide them with information about their actual products. An adequate solution here is to invest in online brand positioning strategies. Hence, social networks could play an essential role as cheap and simple short chains through which the company can offer a more social image for its customers.

FUTURE RESEARCH DIRECTIONS

We are aware that this study has some limitations mainly due to the introductory character of the survey, with the use of a representative -although small- sample of consumers and qualitative methodologies. Another limitation is that the study refers to a single time point (2015). The fast and constant evolution in the digital sectors will force a constant update of this and other studies dealing with food marketing through social media. These limitations open new ways for further studies leading to several directions:

1. Firstly, we consider that it is necessary to increase the sample in order to get representative results at a national or even European level. The increasing globalization and the development of the single European market are opening great opportunities for small and big agri-food companies, which find that their target consumers are not only those on the closest city or region.
2. Further extend in the analysis of the factors that can influence consumers' willingness to buy food through social media is needed. Nowadays there is only limited evidence about how prone are different types of consumers to purchase food online, but almost nothing if the channels being considered are the various social networks currently in operation.
3. Finally, we have the intention to analyze the characteristics of the companies that are already using social networks to promote and sell their products. By comparing them with those other which are reluctant to invest in these new frontiers of food marketing, interesting information regarding subsectors or types of companies with potential can be obtained. Benchmarking strategies or incentives could then be implemented in order to increase the overall competitiveness of the agri-food sector.

CONCLUSION

In the present context of global marketing, small and medium enterprises must look for potential sources of advantage that can help them compensate their flaws. This situation is even more difficult in the agri-food sector, with a fragmented business fabric and great importance of perishable products. Nevertheless, the widespread use of ICT and especially of social networks can open opportunities for these companies, particularly to build up short supply chains.

Small and medium agri-food firms have within their reach a powerful tool that can compensate some of the disadvantages derived from their (lack of) size. In this context, the possibility to interact directly with their customers building short food supply chains must be highlighted as one of the most promising lines of growth for this sector. The development of these tools will allow companies to meet market trends and satisfy consumers' demands, who want to know what they eat and where it comes from. Firms will also be able to develop more accurate and cheaper marketing strategies, thus improving their position in the markets and gaining competitiveness, essential aspects to survive and thrive in today's global food markets. Although consumers may be reluctant to buy food online, some products that do not require cold chain to be delivered or that are not affected by transport delays (preserves, canned/bottled food) are the most likely to be accepted for online purchasing.

Fruits and vegetables are one of the most important productions of the Spanish agri-food sector. Yet due to their perishable nature are among the less likely food products to be sold online. Social networks can play an essential role here as the tool that would allow producers to convey real time information to their customers about valued aspects such as harvesting schedules or ripening state.

ACKNOWLEDGMENT

The authors would like to acknowledge the funding provided by the Egyptian Ministry of Higher Education and the Egyptian Sector of Cultural Affairs and Missions which made this research possible.

REFERENCES

Abatekassa, G., & Peterson, H. C. (2011). Market access for local food through the conventional food supply chain. *International Food and Agribusiness Management Review*, *14*(1), 63–82.

Bewley, J. M., & Russell, R. A. (2010). Reasons for slow adoption rates of precision dairy farming technologies: evidence from a producer survey. *Proceedings* of *the First North American Conference on Precision Dairy Management.*

Brown, C., & Miller, S. (2008). The impacts of local markets: A review of research on farmers markets and community supported agriculture (CSA). *American Journal of Agricultural Economics*, *90*(5), 1298–1302. doi:10.1111/j.1467-8276.2008.01220.x

Canavari, C., Fritz, M., Hofstede, G. J., Matopoulos, A., & Vlachopoulou, M. (2010). The role of trust in the transition from traditional to electronic B2B relationships in agri-food chains. *Computers and Electronics in Agriculture*, *70*(2), 321–327. doi:10.1016/j.compag.2009.08.014

Canavari, M., Deiters, J., Fritz, M., Hofstede, G. J., Sprundel, G. V., Pignatti, E., & Fernández, M. C. et al. (2016). A typology of trust when we introduce information technology. In M. Canavari, M. Fritz, & G. Schiefer (Eds.), *Food supply networks: trust and e-business* (pp. 142–159). CABI Publishing. doi:10.1079/9781845936396.0142

Dooley, J. A., Jones, S. C., & Iverson, D. (2012). Web 2.0: An assessment of social marketing principles. *Journal of Social Marketing*, *2*(3), 207–221. doi:10.1108/20426761211265195

European Commission. (2005). e-Business Watch. *ICT and Electronic Business in the Food and Beverages Industry. Report 2005 edition. ICT adoption and e-business activity in 2005.* Retrieved July, 2016, from http://www.e-business-watch.org/

European Commission. (2007). e-Business Watch. *The European E-Business Report2006/07edition. A Portrait Of E-Business In 10 Sectors Of The EU Economy, 5th Synthesis Report of the e-Business.* Retrieved July, 2016, from http://www.e-business-watch.org/

Fritz, M., & Canavari, M. (2008). Management of Perceived e-Business Risks in Food-Supply Networks: e-Trust as Prerequisite for Supply-Chain System Innovation. *Agribusiness*, *24*(3), 355–368. doi:10.1002/agr.20163

Grunert, K. G., & Ramus, K. (2005). Consumers Willingness To Buy Food Through The Internet: A Review Of The Literature And A Model For Future Research. *British Food Journal*, *107*(6), 381–403. doi:10.1108/00070700510602174

Hausen, T., Fritz, M., & Schiefer, G. (2006). Potential Of Electronic Trading In Complex Supply Chains: An Experimental Study. *International Journal of Production Economics*, *104*(2), 580–597. doi:10.1016/j.ijpe.2005.04.010

INE. (2013). *Industrial Companies Survey*. CNAE 09. Retrieved July, 2016, from http://www.ine.es/jaxiT3/Tabla.htm?t=2539

Joines, J. L., Scherer, C. W., & Scheufele, D. A. (2003). Exploring Motivations For Consumer Web Use And Their Implications For E-Commerce. *Journal of Consumer Marketing*, *20*(2), 90–108. doi:10.1108/07363760310464578

Krippendorff, K. (2004). Reliability in content analysis: Some common misconceptions and recommendations. *Human Communication Research*, *30*(3), 411–433.

Lehmann, R. J., Reiche, R., & Schiefer, G. (2012). Future Internet And The Agri-Food Sector: State-Of-The-Art In Literature And Research. *Computers and Electronics in Agriculture*, *89*, 158–174. doi:10.1016/j.compag.2012.09.005

MAGRAMA. (2008). *La Distribución Agroalimentaria Y Transformaciones Estratégicas En La Cadena De Valor*. Retrieved July, 2016, from https://www.google.es/search?q=La+Distribuci%C3%B3n+Agroalimentaria+Y+Transformaciones+Estrat%C3%A9gicas+En+La+Cadena+De+Valor&ie=utf-8&oe=utf-8&gws_rd=cr&ei=p3YSVvFFxLZ43KeIsA0

MAGRAMA. (2013). *Canales Cortos De Comercialización En El Sector Agroalimentario*. Retrieved July, 2016, from http://www.magrama.gob.es/es/alimentacion/servicios/observatorio-de-precios-de-los-alimentos/estudioccc_tcm7-272187.pdf

MAGRAMA. (2014). *Industria Alimentaria En 2013-2014*. Retrieved July, 2016, from http://www.magrama.gob.es/es/alimentacion/temas/industria-agroalimentaria/_informeindustriaalimentaria2013-2014_tcm7-156934.pdf

Mandilas, A., Karasavvoglou, A., Nikolaidis, M., & Tsourgiannis, L. (2013). Predicting Consumers Perceptions in On-line Shopping. *Procedia Technology*, *8*, 435–444. doi:10.1016/j.protcy.2013.11.056

Marketing4Ecommerce. (2013). El comercio electrónico alcanzacifras record en España. *Marketing4Ecommerce*. Retrieved July, 2016, from http://marketing4ecommerce.net/el-comercio-electronico-alcanza-cifras-record-en-espana/)

Marsden, T., Banks, J., & Bristow, G. (2000). Food supply chain approaches: Exploring their role in rural development. *Sociologia Ruralis*, *40*(4), 424–438. doi:10.1111/1467-9523.00158

Mata, F. J., & Quesada, A. (2014). Web 2.0, Social Networks And E-Commerce As Marketing Tools. *Journal of Theoretical and Applied Electronic Commerce Research*, *9*(1), 56–69. doi:10.4067/S0718-18762014000100006

MERCASA. (2013). *Informe Sobre Producción, Industria, Distribución Y Consumo de Alimentación en España*. Retrieved July, 2016, from http://www.mercasa.es/noticias/post/informe_2013_sobre_produccion__industria__distribucion_y_consumo_de_alimentacion_en_espana

Migliore, G., Forno, F., Guccione, G.D., & Schifani, G. (2014). Food Community Networks as sustainable self-organized collective action: a case study of a solidarity purchasing group. *New medit: Mediterranean journal of economics, agriculture and environment- Revue méditerranéenne d'economie, agriculture et environment*, *13*(4), 54-62.

Mundubat (2012). *Circuitos Cortos De Comercialización En Euskal Herria*. Retrieved July, 2016, from http://www.baserribizia.info/index.php/euskalherria/ekimenak/4134 dossiersobrecircuitosc ortosdecomercializaccioneneuskalherria

Oubiña, J. (2000). *Las Relaciones De Poder Y Conflicto En Los Canales De Distribucion De Productos De Granconsumo: Un Analisis De La Satisfaccion En Las Relaciones Fabricante-Distribuidor Desde La Perspectiva Del Fabricante* (Unpublished doctoral dissertation). Universidad Autonoma de Madrid, Spain.

Petrovic, D. (2010). *What is e-Marketing*. Retrieved July, 2016, from http://analogik. com/articles/231/e-marketing

Ramus, K., & Nielsen, N. A. (2005). Online grocery retailing: What do consumers think? *Internet Research*, *15*(3), 335–352. doi:10.1108/10662240510602726

Renting, H., Marsden, T. K., & Banks, J. (2003). Understanding alternative food networks: Exploring the role of short food supply chains in rural development. *Environment & Planning A*, *35*(3), 393–411. doi:10.1068/a3510

Rutsaert, P., Regan, Á., Pieniak, Z., McConnon, Á., Moss, A., Wall, P., & Verbeke, W. (2013). The use of social media in food risk and benefit communication. *Trends in Food Science & Technology*, *30*(1), 84–91. doi:10.1016/j.tifs.2012.10.006

Santini, F., & Gomez y Paloma, S. G. (Eds.). (2013). *Short Food Supply Chains and Local Food Systems in the EU: a state of play of their socio-economic characteristics*. Luxembourg: Publications Office of the European Union.

Shang, R. A., Chen, Y. C., & Shen, L. (2005). Extrinsic Versus Intrinsic Motivations For Consumers To Shop On-Line. *Information & Management*, *42*(3), 401–413. doi:10.1016/j.im.2004.01.009

Stewart, D. W., & Shamdasani, P. N. (1990). *Focus groups – Theory and practice*. Newbury Park, CA: Sage.

Sturiale, L., & Scuderi, A. (2013). Evaluation Of Social Media Actions For The agrifood System. *Procedia Technology*, *8*, 200–208. doi:10.1016/j.protcy.2013.11.028

Tarkiainen, A., & Sundqvist, S. (2005). Subjective norms, attitudes and intentions of Finnish consumers in buying organic food. *British Food Journal*, *107*(11), 808–822. doi:10.1108/00070700510629760

Tsekouropoulos, G., Andreopoulou, Z. S., Koliouska, C., Stavroula lefa, Koutroumanidis, T., & Batzios, C. (2011). e-Marketing and Internet Functions of Agricultural Products In SME In Greece. *Proceedings of the International Conference on Information and Communication Technologies*, 213-224.

Wang, Y., & Hajli, M. N. (2014). Co-creation in branding through social commerce: The role of social support, relationship quality and privacy concerns. *Proceedings of Twentieth Americas Conference on Information Systems*.

Zhu, X., Zhang, Q., Zhang, L., & Yang, J. (2014). Consumers' Demand And Online Promotion Of The Food Retailing Through The E-Commerce Websites In China. *Advance Journal of Food Science and Technology*, *6*(1), 102–111.

KEY TERMS AND DEFINITIONS

Cookie: Is a small piece of data that sent automatically from a certain website and stored in someone's internet browser while the user is browsing the website. Once the user visit a website, the browser notifies about the user's previous activity by sending the cookie back to the server.

E-Business: It is the fact of conducting business online, that is, the application of information and communication technologies (ICT) to conduct all the activities related to business. This could involve the purchasing and selling of goods and services, together with providing technical support or customer services.

E-Commerce: Electronic commerce is the performing of buying or selling process of products or services throughout electronic networks, such as the Internet or online social networks. The term often used in conjunction with e-business, and although they refer to different concepts, they are often used in an undifferentiated way.

E-Marketing: Also known as Digital marketing, the term refers to the application of digital technologies to contribute to the marketing activities of an enterprise so as to strengthen the relationship with customers and create added value for the product. It is based mainly on the Internet, but also includes mobile phones, display advertising, and any other digital medium.

Protected Designation of Origin (PDO): is one of the Quality Schemes for food in the European Union. It is a kind of geographic indication applied to an agricultural product or foodstuff whose quality or characteristics are fundamental and exclusively to the geographical environment in which it is produced, transforms and develops. It differentiates products created in a given area, against producers from other areas who would like to take advantage of the good name that created the originals, in cultivation or manufacture. In order to get the PDO, the entire product must be traditionally and completely manufactured (prepared, processed and produced) within the specific region and thus acquire unique properties.

Small and Medium Enterprise: is belonging to the category of micro, small and medium-sized enterprises which, in a general definition, consists of enterprises that employ less than 250 people and which achieve an annual return not more than 50 million Euros, and/or an annual balance sheet total not exceeding 43 million Euros. Within the SME category, a small enterprise is defined as an enterprise which employs not more than 50 persons and their annual turnover and/or annual balance sheet does not exceed 10 million Euros, whereas a micro enterprise is an enterprise which employs fewer than 10 persons and their annual turnover and/or annual balance sheet does not exceed 2 million Euros.

Social Media: Social media are computer-mediated tools made up of a set of actors (such as individuals or organizations) that are related according to some criterion (professional relationship, friendship, etc.). They are normally symbolizing the actors as nodes and relationships as lines connecting them. The type of connection representable in a social network is a dyadic relationship or interpersonal tie. Within this concept, people, companies and even organizations can co-create, co-share, or exchange information, interests, views, and other virtual contents like pictures or videos. Till now there is no unanimity among the authors to propose a specific typology for social media however, they have some common features: (1) social media are Web 2.0 Internet-based applications; (2) user-generated content, as users create their own profiles for the website, and website facilitates the development of online networks by connecting a user's profile with those who share the same interest.

Chapter 7
Reasons for Adapting Information Connectivity in the Short Supply Chains of Local Food Producers

Per Engelseth
Molde University College, Norway

ABSTRACT

Local food production is becoming increasingly popular in developed post-modern economies. Attention has been directed to developing such forms of food supply by adapting information connectivity. A case study of a local food network in Norway indicates that local food supply paradoxically attempts to mimic the dominant industrialised modes of food production. It is suggested that the fact that local food supply is "personal" and associated with close proximity makes it more closely resemble service supply chains. Applying contingency theory, a conceptual model is developed that indicates how the local food supply must take into consideration the degree to which customer value is associated with tailoring food supply. The high need for tailored local food production implies that information connectivity should support mutual adaptation while, in cases of less need for tailoring information, connectivity should seek automation. Local food production is always a hybrid of these approaches.

INTRODUCTION

Abatekassa and Peterson (2011) revealed how food markets are becoming increasingly globalized, a trend that is rooted in industrialised large-scale production of both fresh and processed foods. This chapter focuses on a specific issue within the emerging industry; namely, the use of information technology in local food production with the aim of creating value for practitioners in these types of short supply chains (Engelseth & Hogset, 2016). Local food production is a particular form of industry. The research question that we pose is, more specifically, whether it is necessary to adapt the information systems structure and processes for use in such short supply chains typical of local food production.

DOI: 10.4018/978-1-5225-2107-5.ch007

To consider this issue, it is first pertinent to analyze the societal context of local food production; the outer layer of the research query. This is because information use is fundamentally viewed as being contingent on such societal factors, including paradigmatic world views regarding how the food industry, and people in general, perceive food production. Due to the dominance of this form of food production, the supply chain management of foods is commonly associated with modern large-scale production systems found in complex industrial networks.

What then is mordernism and how does this impact on current food supply? According to Giddens (1991, p. 5), modernism implies a state of mind in which self-identity becomes "… a reflexively organized endeavor"; that is, people in the modern age are not bound by a locality. This also implies that consumption of local "product" is ideally not bound by space. This state of mind impacts on technology use through production. Thus, as it emerged, modernism has encompassed the Industrial Revolution, the mechanization of "man" and mass production to achieve economies of scale.

How then do local foods relate to the domince of moderniosm in food production? Accordingly, "local" food may also be mass-produced and distributed globally. This technological change has also meant that the "traditional" ways of producing foods prior to the Industrial Revolution will never return to the same manner as before. Some sort of "going back to the future" is viable, whereby historic modes of food production such as using traditional marketplaces to distribute foods may still inspire but dot completely drive change in food production as a blueprint may.

Apart from nostalgia, consumers seeking foods that remind them of the old days, traditional modes of local food distribution have features that we can still learn from today. This involves also features of information connectivity and use. Food was previously produced close to consumers and sold direct from farmers or fishermen or at local markets. At traditional markets, consumers and farmers as well as fishermen often developed personal bonds, securing value from a customer perspective through institutionalized business relationships. Connectivity in this form of local food supply was manual and sufficient. The Industrial Revolution led to increased scales of raw-material production, processing, distribution and retail.Information connectivity, involving features of the quality of communicating information that binds these supply chain actors together, was adapted to this modernistic logistics system. Integration was simple and personal. Information systems did exist and predominantly involved personal communication.

With the spread of modernistic production, traditional markets faltered in what we currently term "developed countries". The rise of food production involing collaboration between various specialised producers and service providers and large amounts of long-distance transport also meant that information coinnectivity and information use had to support such modernistic food supply. From a supply chain management viewpoint, this connectivity may range in industrialised food supply chains from weak to strong, depending technically on investments in information technology to standardise resources facilitating automating interaction. This automation is associated with the now common modernistic information system use, and is therefore dependent on technical features associated with network complexity, perceptions of supply risk and interdependency.Information connectivity is an expression of supply chain integration, a cornerstone of this modernistic supply chain management thinking; the information flow supporting production as flows of goods and services (Lambert et al., 1998). According to Closs et al. (2005), managers who run large-scale factories that are responsible for product supply regard information connectivity as playing a vital role in achieving successful logistics management programs. However, this view does not account for small-scale production where goods supply is targeted at a local market. This indicates also a research issue associated with adapting information connectivity and use to the short supply chain structure typical of local food production.

How may we then analyse information provision and use in the short supply chain structure? We turn first to considering the dominant modernistic food production structure. In the literature, most of the effort to integrate information systems used by different supply chain actors are concerned with mass-produced manufactured goods, including food, subject to different degrees of processing, applying what Thompson (1967) termed "long-linked technology".With regard to how production is organized, this form of technology is associated with physical distribution. Foods being long-linked may be described metaphorically as being supplied through "flows". This represents a type of supply associated with mainly sequential interdependencies and thus does not account for all types of supply.

This raises a research issue regarding the degree to which local food supply, clearly a form of physical distribution, may still be denoted as "long-linked", especially since local foods involve short supply chains. This research issue is pertinent since the food supply is, to some degree, not long-linked in nature; this means that information use needs to be adapted to particularities regarding organizing local food supply. The research isse is formulated as follows:

- What is the nature of the local food supply chain structure?
- How does this structure impact on local foods production precesses?
- How do these particularities of local food supply impact on information connectivity in such forms of food supply?

Answering this question should reveal reasons for adapting information systems to local foods supply and point to ways to use information in such forms of production. The chapter is orgainsed with first a literature review that starts by considering features of local foods supply particularities. These particularities are then considered in relation to how production is carried out in this form of supply chain from a contingency theory perspective, especially pointing out how local foods supply may be regarded more as a services type of industry than as a physical distribution type. This is followed by methodological considerations associated with the case study.

Findings are then presented, analysed mainly from a contingency theory perspective, focusing in variation in interdependency, and how this impacts on information connectivity. Theory on service supply chains is also applied to consider whether local foods potentially may be described as resembling servces industry supply chains. Finally, the empirical findings, pointing out how local foods as more like a service type of supply chain, demands information sysetms that enhance connectivity taking into consideration the interdependencies typical of this type of production.

LITERATURE REVIEW

There is no universal definition of the term "local food". The term refers broadly to food products produced close to consumers (Martinez et al., 2010). It is therefore a reflexive concept associated with all activities related to the method of food production and distribution constrained by geographical measures and social-cultural emotions (Amilien et al., 2008). The increasing demand for quality local foods in modern supermarket and other more specialized post-modern shops have marked the re-emergence of local food producers in the food chain. The functionality of foods is widened in the affluent world to encompass a higher degree of consumers seeking self-realization, opening up demand for more expensive, quality foods produced on a smaller scale and at locations closer to the consumers. The local food system contains three issues: (1) how and where food is produced, (2) how food will be distributed to consumers, and (3) consumer, food preferences and options (Darby et al., 2008; King, 2010). These three questions represent the epicenter of a local food system. Focusing on the aspect of distribution, producers of goods and services obtain access to the market through marketing channels, implying variation in the use of intermediaries. These differences also imply variation in the organization and in the use of logistics resources. The fact that a local food supply chain is short implies limitation in use of intermediaries – possibly one or even none at all,which would imply direct distribution. This also simplifies information connectivity since there fundamentally are fewer actors to interconnect in local food supply chains.

While logistics is predominantly associated with economies of scale regarding transport, storage and warehouse handling activities, the emergence of increasing amounts of small-scale local food distribution directs attention to *the economies of small scale*, different from economies of scope since local foods not necessarily involves lowering average cost by producing more types of products The fact that there are numerous organizations and technical ways to release information connectivity raises another question regarding what type of "information connectivity" is best suited for these types of short supply chains that predominantly distribute up-market quality local foods in wealthy post-modernistic economies. Previous research on local foods has indicated that interaction in short local food supply chains more closely resembles service supply chains than the modernistic food chains they seemingly compete with (Engelseth and Hogset, 2016). Following contingency theory (Thompson, 1967), studies of local food supply have revealed that they were not only shorter or "smaller", but that they are also highly flexible since they retained the traditional strong role of the actor and personalized information exchange. Whilst long-linked industrialized food supply is predominantly sequentially interdependent, local food logistics systems are small, simple and flexible. This is because they have a simple supply chain structure and produce small volumes of goods that, through negotiations between sellers and buyers are directed, not according to detailed plans, but based on actual and often relatively frequently transmitted orders.

Food supply chains can be somewhat crudely described as being associated either with physical distribution or services supply.Being rather different from the distribution of mass-produced manufactured goods, is it possible that local food supply more closely resembles services supply? Sampson (2000) described from a chain perspective two different types of service supply. The first type is the single-level bidirectional supply chain, which involves bidirectional interaction between the service supplier and its customer. This service supplier is again dependent on goods supplied through long-linked technology from its own suppliers upstream in the supply chain. The other type of service chain is termed a two-level bidirectional supply chain. In this form,supply to the service supplier also involves bidirectional interaction. Both these types of supply chains are short in configuration; the focus is conceptually modelled to encompass a triad with the service supplier modelled in the center. While the service firm purchases goods and services to carry out its production, these purchases are viewed in Sampson's (2000) models as contextual.

Bringing in the concept of "service supply chains" into the discussion suggests considering the degree to which local food supply is similar or different from service supply chains. This implies considering the degree of reciprocal interdependence in the local food supply and discussing sources of interdependencies in these shorter forms of supply chains. Within contingency theory, Thompson (1967) expressed the existence of three forms of interdependencies: (1) pooled, (2) sequential, and (3) reciprocal. Following Thompson (1967), pooled interdependencies are associated with a mediating technology use, development associated with resource standardization.Sequential interdependency is associated with long-linked technology and reciprocal interdependencies are associated with the use of intensive technology. This latter form of technology is dependent on complex dialectical exchange and is accordingly the most costly form of interaction in developed economies with high personnel salaries since it is predominantly manual in nature. Sequential interdependencies are associated with coordination using various forms of planning tools including the Lean form of production levelling. Pooled interdependencies imply resource combining and interconnection through resource standardization. In cases where resources may be pooled, production may be automated. In principle, all forms of interdependencies may be present in an industry. Variation in supply chain type is associated with whichever form of interdependency is predominant. Thompson (1967) used his interdependency framework to reveal what characterizes different forms of industry, thereby also indicating how production can be developed in cases of variation or combination of these interdependencies.The "triadic" perception of service supply,including the element of interdependencies in business relationships,is shown in Figure 1, below.

Figure 1 shows how interaction with the customer is reciprocally interdependent, while supplier interaction is either sequential or reciprocal. This is because some suppliers to service firms are mass-produced manufactured items. Normally, since the food supply is a form of physical distribution, local foods should mainly be characterized as sequentially interdependent flows of goods. Thompson's framework is, as in the dominant modern logistics, mainly concerned with large-scale manufacturing-based supply. The shortness of transport combined with fragmentation and a heightened need to network, entails increased importance of sales and purchasing functions associated with the exchange economy. This, in turn, entails heightened reciprocal interdependencies, as well as increased potential to pool resources with other smaller or more specialized firms to which they have outsourced tasks. This process also involves considering local food suppliers as resembling service providers more than manufacturing firms. As in services, local food chains are short and predominately reciprocally interdependent. The networked local food suppliers are not anonymous actors since there are few producers, but they stand out as food professionals who are respected in the local society, including by their often large retailer customers. Therefore, proximity is postulated as being more vital than whether the supply is a service or goods in terms of revealing what characterizes local food production.

Following Thompson (1967), given a perception that local food production is more similar to services than physical distribution, local food supply should be characterized as being associated either with predominantly mediating or intense technology.Stabell and Fjeldstad (1998) highlighted the potential for two types of services, either predominantly pooled or reciprocally interdependent. This difference in interdependencies may be considered analogous to the service industry. For instance, to differentiate

Figure 1. The service supply chain

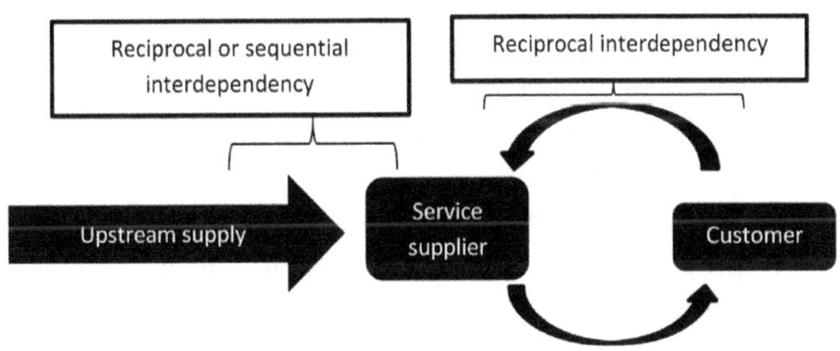

between processes of ordering and preparing a meal at a fast food restaurants from more gourmet-type restaurants (reciprocally interdependent and expensive, "live to eat"). Reciprocal interdependency, as in gourmet restaurants, is also costly to manage and operate. In fast-food chains pooled interdependencies are developed through information standardization; using mediating technology.

This dichotomy indicates that the most obvious path through which local food suppliers can seek to develop efficiencies in information connectivity is through information developed to better support personnel; that is, manual interaction to secure transactions and support the logistics of local food supply. Taking a systems perspective – more precisely, following contingency theory (Thompson, 1967) expanded by theory of loose coupled organizations (Weick, 1976; Orton and Weick, 1990) – implies the development of intensive technology. As loosely coupled systems, rationality and indeterminacy are considered together. This directs attention to the dialectical process between organizations, not seeking generalized solutions, but seeking understandings regarding these inter-organization working in small-scale local food production. This means developing how information supports a predominantly manual form of interaction associated with reciprocal interdependency, improving the dialectics of supplier–customer interaction. This has been suggested as the chosen pathway to developing information connectivity in the local food supply when uncertainty and complexity is high and when these contextual features cannot be changed or handled perceived differently by the local food supplier. There is, therefore, also an alternative potential use of information technology in local food production involving focus on developing the mediating technology. Local foods producers may choose to reduce the predominantly reciprocal interdependency of customer relations,thereby increasing the impact of pooled interdependencies and enabling the use of mediating technology involving standardising business interaction supported by standardised products and packaging as well as standardised information and information systems capturing, processing, storing and sending this information.Simply writing this implies making the gourmet restaurant more like a fast food chain outlet. Reducing interdependency does not mean a weaker or stronger coupling between firms, but rather suggests the need for strategic-driven structural changes.

Developing information connectivity in the local food supply chain involves enabling the traceability of local foods. Traceability needs to be adapted to whether the local food supply chain has an information system that supports predominantly reciprocal or pooled interdependencies. It is clear that capturing data to later facilitate product traceability in supply characterized by pooling interdependencies is simpler than in cases characterized by reciprocal interdependence. In cases of using predominantly mediating technology to integrate the local foods supply chain food, product traceability may be automated to a higher degree. Such automation is not dependent on the scale of production or demands advanced information system competence. This was revealed in Engelseth et al.'s (2014) study of the implementation of electronic product traceability in the upstream portion of a food chain in Thailand. Traceability confirms food's authenticity – a vital product quality component when taking a consumer perspective. Local food quality is a key market differentiation in comparison to modernistic industrialized foods (Engelseth 2015).Finally, the choice to use information technology to support strategy-driven integration and collaboration with either more enhanced pooling to strengthen interaction in cases of upholding reciprocal interdependences. Alternatively, some mix of the two is also associated with market positioning and is in this paper considered a given structural factor.

This study is based on a fundamental assumption that supply chain development is at the verge of a paradigmatic shift driven by both technology and society in general. Technology is assumed to be becoming increasingly modern and society to increase post-modernism. Technology as "tool" means that information systems, functionally, should primarily be adapted to societal demands and technology should play the role of facilitator of this adaptation. Collectively, society is continuously learning,which has an impact on food consumption patterns. Postmodernism indicates that society today encompasses a range of types of food consumption patterns and that food consumption is being fragmented. Both technology and society have at least one thing in common: they both have a past. New technology, merging the fields of logistics and information, together with a capitalistic-rooted competitive market,drives the search for new ways to supply consumers with foods. Local food production is, in a sense, actually a way of going *"back to the future"*.The 1980s films of the same name showed that time travel was a rather messy affair since knowledge, including the values and norms of the future, could not automatically be disseminated in an old-time societal context. This reflects the values of traditional pre-modernistic food supply.

In the present-day context, we will never re-live the food markets of pre-industrial society. However, a range of environmental and ethical concerns, combined with developments in technology such as 3-D printing, may break down the illusion that economies of scale are only associated with mass production. Production should provide customer value. As used in the present study, the concept of "production" encompasses transformation related to achieving customer value; it includes cases of food industry growing, processing and the logistics of food supply. As indicated earlier, the study covers concerns about developing the quality of information-technology-supported proposal negotiations, as well as reduce the need to communicate personally. This implies a need for flexible information connectivity solutions. In addition, since local food suppliers are small businesses with limited economic means, economic information solutions are called for. This implies the use of inexpensive information technology, or co-option or cooperation to invest in more costly information system solutions.

The following case description provides an as-is description of local food supply as a hub-and-spoke system from the perspective of a supermarket in Norway. No detailed considerations regarding types of information resources are discussed. We consider the reasons why information systems in local foods supply need to be adapted, followed by suggestions regarding what should characterize information systems used by local food suppliers from a supply chain perspective. Negatively speaking, this implies managerial concern that information connectivity is not the same as found in predominant modernistic forms of food supply. In addition, local food suppliers are also found to often interact with modernistic food supply chains such as when selling to a retail chain. This implies a question regarding how these types of supply chains react when they interact in relation to purchasing raw materials, logistics and retailing. Different scenarios of resource interdependency include the local food supplier as the submissive actor, or the local food supplier being appreciated and valued by the retailer. This feature of interdependency is also considered in the paper and how this is solved in the studied local food networks.

CASE STUDY RESEARCH STRATEGY, METHOD, AND DATA COLLECTION

This case study was conducted in the vicinity of the small town of Molde, which is located on the north-western coast of Norway. This research strategy was applied, involving first an initial interview with a representative of a supermarket known for its large assortment of local foods. This informant provided a fundamental overview of her local foods supply network as well as fundamental description regarding the importance of local food production from her food retailer perspective. She also provided a list of her local food suppliers.

This main interview was supplemented by five interviews with representatives of different local food producers who were suppliers to this supermarket, and two other key informants. One of these key informants was a local food producer who is active in developing this form of industry. The other key informant was a regional government official who supports the development of local foods through various forms support programs. These key informants provided valuable insight into espåecially the societal, as well as some technical, aspects of local foods production in the studied region of Norway. Since research funds were limited, not all the local producers on the list provided by the retailer were not interviewed. The local food producers were selected using a snowball sampling procedure. The respondents' contact information was gained through the supermarket manager and then a sample of producers was selected from different categories. This method excluded some local food companies, because they did not sell in the supermarket. Snowball sampling is a nonprobability sampling technique. The selection of the sample is not random, which makes it impossible to determine a possible sampling error. Informants were selected mainly by geographical convenience and accessibility. This may influence our results, so we cannot guarantee representativeness of the sample is not guaranteed (Kotz et al., 1999). Personal semi-structered interviews were conducted. We went to the respondents' home or company office directly with a prepared semi-structured interview guide. A general impression from these interviews was that these interviewed local food producers were resourceful and communicative people. Interviews provided a rich amount of qualitative data, often providing unexpected insights. These interviews involved accordingly a high degree of inter-subjectively founded emrgence of data.

We had three types of the interview guide tailored for retailers, producers, and government officials. The interview guide for retailers focused on their perspective, attitude and activity regarding local food. For the producers interview guide, the focus was on the whole supply chain of local food, from harvesting and processing until the customer picks it up. We also included some personal questions about the respondents, which may indicate the future development of local food. The interview guide for the government officials contained questions about policies for helping local farmers and future development of local foods in general in the county in which the supermarket is located.

One limitation of personal interviews is that many pieces of information may be incorrect or exaggerated, because some respondents see the interview as a marketing promotion. Therefore, our case study research may contain some errors. The data analysis used a qualitative method because no accurate numbers were collected during interviews. Most of the collected data were transcribed, including notes were made of observations in the supermarket and of the local foods producers' production facilities.

FINDINGS: THE LOCAL FOODS NETWORK

The case narrative focuses on the interactions and information exchange in the studied local food network. Figure 2 illustrates the studied supply chains leading to the supermarket.

Supermarket Hub

The studied food retailer, located in the Western Norwegian town of Molde, stocks a variety of local foods,such as jam, honey, cheese, fish, cured meat, eggs, lefse (a traditional Norwegian sweet), bread, concentrated juice, sodas, mineral water, carrots and strawberries. This retailer has annual turnover of 135 million NOK, making it a relatively large retailer in the Norwegian market. This retailer strategy

Figure 2. The studied supply network

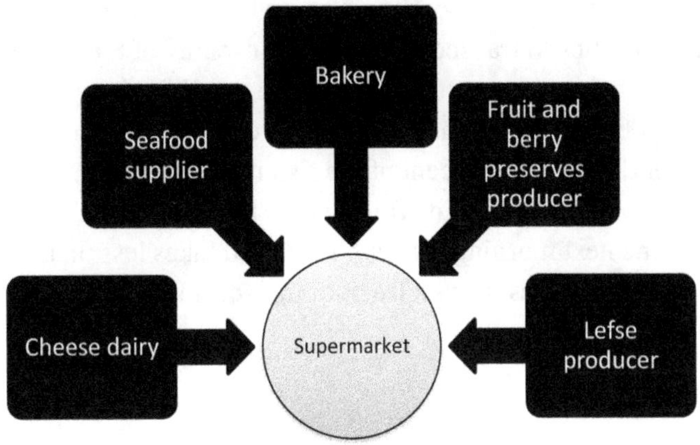

prioritizes local foods and has had a local section in its stores since 2006. The turnover of local foods from this retailer has is tripled since 2006, with local food sales now accounting for 7–8 per cent of the total sales volume. Customers perceive local foods as fresh, having good quality, healthy and tasting good. Local fish and cheese are comparatively more important having a greater annual turnover than other local foods. The retailer applies three different procedures for purchasing local food. One is through a PDA (personal digital assistant), a mobile electronic device that is used to communicate electronically with suppliers. This is the most common way that the supermarket purchases goods and applies to local foods distributed through the distribution center in Trondheim. This mode of purchase is associated to a small degree with local food supplies. Another method involves the retailer ordering goods from local food suppliers by telephone.

The most common purchasing procedure in cases of local food supply to the supermarket is that farmers first contact the supermarket directly by phone to inquire about demand. The quantity of goods ordered is such that local farmers can have a stable profit margin. It is this retailer that enables these small local suppliers to survive. They are very dependent on the supermarket since there is one other supermarket in the same city that has some degree of local foods in their assortment. Food safety regulations in Norway are very strict and are administered by the government through Mattilsynet (http://www.mattilsynet.no/language/english). Like industrial food suppliers, local food suppliers are required to put barcodes and ingredient labels on the packaging of their products. There are no special requirements compared to ordinary food. Furthermore, the supermarket is responsible for checking the quality of the products through visual control and smelling the products. Local farmers usually have a long-term contract with the supermarket. Most contracts are signed or renewed in the central distribution centre used by the supermarket and last for one year. The contract specifies the barcode label in details. The contract may stipulate that local suppliers are not allowed to supply their products to the supermarket's competitors.

Seafood Supplier

The retailer purchases all its fresh seafood from the studied fish retailer, which sells fish at its own store located in the city centre, and also sells to hotels, restaurants, schools, etc. The fish retailer has five staff members in-store: one driver, two salespeople, one main staff, and the owner, who operates the computer and the single landline phone.Most of the fish is sold locally, with only a small quantity going to other places.The company has four main local suppliers, which it calls in the morning to purchase fish. Suppliers use third-party logistics to transport fish. Every evening at 8 p.m. and the next morning at 5 a.m., the fish store receives fish by cars or trucks from the small fishing ports in the vicinity. The fish retailer receives around 1000 kg of different kinds of fish every day, with 700–800 kg of this used for trade. The seafood supplier delivers 80 per cent of the fish it receives to the studied retailer. Every night when the store is closed, the retailer orders its fresh goods by fax. The ordered fish is then delivered to the supermarket at 8 a.m. the next morning. The whole period takes less than 24 hours, to ensure freshness and quality. The fish retailer uses its own truck to transport fish locally.

Bakery

The local bakery is differentiated from other bakery goods suppliers through the quality of its foods, which are hand-made. The bakery is located in the center of town and the location also functions as a popular café. The retailer was one of the bakery's first business customers. The bakery chooses to co-operate with the retailer and spends much time on this relationship to secure its sales. The retailer is the main customer and accounts for 42–45 per cent of the bakery's total sales volume. Both parties report a good personal relationship with each other.

The bakery also supplies a few products to the other supermarket in the town that has taken in some local foods. The bakery's operation is flexible. Canteens and coffee shops daily call in their orders to the bakery. Some business companies also order items for meetings and other types of events. The bakery also supplies six of the canteens of the largest companies in the town. Transport is carried out using the company's own van. It drives directly to the customers at between 7 and 10 a.m. every day. The first customer of the day is a hotel that wants to receive deliveries before 7 a.m. The retailer has a list of different products that require delivery before 8:30 a.m. However, one customer they have in neighboring Ålesund uses its own transportation; drive to Molde to pick up bread every Friday.

Lefse Producer

Lefse is made with potatoes, flour, butter, and milk or cream and is baked on a griddle. It generally resembles a pancake or flat bread and has butter, sugar and cream inside. The lefse producer founded in 2002 when it was awarded government financing. The company mainly supplies products to the studied retailer and some other retailers in the region. Customers can also come to order fresh products and cakes for meetings and parties. The company produces six different types of lefse, which it delivers using its own reefer truck. Products are delivered to the retailers, once a week and twice a week to customers in Trondheim. All of the raw materials, like butter, sugar and cream, are ordinary raw material food supplies; the company does not have local suppliers. The production craft is simple, but the recipes of lefse are old and local and were learned from the owner's ancestors. The main reason why the owners operate the company is to increase income and help local people. The company was established with several owners. Because the two initial owners now are retired, the future development of the company is based on the young people they have hired.

Cheese Dairy

The cheese dairy is located on a farm, where cheese and the production of other dairy products production started in 2003 using the farm's own milk production. The dairy is one several local cheese suppliers of the retailer. The farm owners' motivation for local food production is to work on their own farm, since forming it provides limited revenues. The self-produced cheese adds more culture and therefore tastes richer. The cheese dairy produces nine kinds of cheese and other dairy products, like yoghurts. The dairy has not developed any new products in the past few years. Cheese can be classified as soft cheese, half-hard cheese and hard cheese. Since it is a milk producer, it is also a shareholder of Tine, the largest Norwegian co-operative dairy company. Tine's distribution system covers all of Norway and helps local farms, pick up goods and send directly to their customers. The cheese dairy sells its cheeses to restaurants and hotels at tourist attractions in their region and in some specialty stores in Oslo and Bergen through Tine's transportation system. In the first year, the owners travelled around as salesmen, called at some restaurants and asked if the restaurateurs wanted to taste the products. The yoghurt was delivered there every two weeks based on orders delivered by phone or email. The cheese dairy also sells cheese to various regional supermarkets. The studied retailer is the dairy's fourth-largest customer. It orders, products once a month by text messages. These deliveries are transported using the farmer's own van because they quite often they have something else to do when in town. When transporting to another supermarket in a nearby city, they use Tine's system because they do not go there as often. Both cities are about 30 minutes driving time from the location of the farm.

Fruit and Berry Preserves Producer

The fruit and berry preserves producer is located on a farm in the same area as the cheese producer, and these two farms cooperate in selling each other's products. The owner runs a small farm raising sheep and chicken and growing herbs and berries. The owner started the business because her child was born at that time and she did not want to leave the farm, so she had to create her own job. Twenty years ago, the land on the farm was just a swamp. In the beginning, people just came to the farm for brief visits. Now, it can offer space for small functions, conferences and courses and it has a dining area and accommodation. The farm shop has many kinds of products, mostly seasonal,that are made or processed from the farm. These are a variety of jams, juices, syrups and handmade soap. The owner also sells spices, tea, herbs and eggs in the shop. Some these products come from other producers, as the local farmers' share products with each other.. These goods are differentiated since the production craft is old and traditional; some recipes are ancestral. The farm shop only sells products locally because the transportation cost is high and the glass bottles it sells are heavy. The farm is not profitable and has a very small turnover, but the farmer enjoys running its business. Its main customer is the studied retailer and a few other local shops around their location. Glass bottles and sugar are the two biggest supplied products. The farm receives them once a year. The preserves producer delivers products to its customers once a month using its own truck. On the way back, it also loads some products from other farmers. The farmers also sell these products in their own farm shop. This system helps local farmers build a network to transport together. It can reduce empty transportation costs and share products among local farmers.

ANALYSIS

The reasons for adapting information systems used in local food supply chains are divided into three lines of argumentation: (1) Interactions in local food supply chains resemble service supply chains;(2) initially developing customer and supplier relationships through improving the use of intensive technology; and (3) economizing local food supply through developing the use of mediating technology. This line of argumentation is founded on the presupposition that local food suppliers do not pursue growth and instead aim to remain local food suppliers and thus aim to develop the quality of their information systems and information use in this given short supply chain structure context. The first set of arguments is associated with fundamental reasoning for adapting information system design and information use in the local food supply. The next two lines of argumentations are associated with answering the question of how local food suppliers may develop information connectivity in local food supplies.

Local Food Supply Chain Structure and Interaction Patterns

The local food supply chain may be interpreted as a loosely coupled system (Weick 1976). This means, in practice, that functionality is sought through inter-organizational interaction marketing and logistics between different companies. The case study described here shows little sign of vertical integration and alignment of activities imposed founded on the power of an actor. The studied supply chain is predominantly a negotiated scenario, which implies that information connectivity is a variable. The quality of information connectivity will also impact on the quality of supply,to the degree it supports interaction. Though much less complex than modernistic food supply chains, local food supply chains are definitely adaptive systems.

As indicated in the literature review within the frame of reference, the structures of the studied supply chains flowing local foods to the common supermarket share one decisive characteristic, analytically speaking: these are short supply chains. "Short" is associated with location. The distance between the producer and the supermarket consists of less than 100 km in all cases. In the case of the baker, the distance is only 1 km. This proximity also implies that the supply chains are organizationally simple; the only intermediary is the supermarket. This simplicity includes the fact that the studied local food suppliers are small businesses. This smallness includes the fact that competence focuses on the foods they create and less on managerial processes that support this production. Notions regarding branding, developing customer relationships, and logistics solutions are tried to be kept simple and easy to manage.

In this case study, local foods are also organized differently for industrialized food production. In the same way as the local food producers in the described cases prefer to produce their own product using craft skills, they also prefer to handle marketing and logistics by themselves. This creates an "as-is" distribution system that is simple and personal in nature. It is obviously a case of food production more by people than by machine. Since the number of people in this food production system is low, the complexity of management is also naturally low.

In the case study, distribution and management are exposed as organizational add-ons in the studied local food chains. This reveals a paradox in these local food supply chains. This view of managing and distributing local foods as a peripheral function can be interpreted as the local food suppliers imitating their industrialized food suppliers by applying long-linked technology. It appears that they are subconsciously attempting to squeeze the sequentially interdependent long-linked technology use in the frame of their short supply chains. This drive to manage local food supply based on modern principles is not deliberate, only an expression of the dominance of this management paradigm in food production. In practice, the company finds that marketing and logistics are hassles they deal with their limited spare time, rather than strategically developing their customer and supplier relations.

These local food products appear to more or less "sell themselves" in the context of an affluent post-modernistic societal context. To the degree that there is any strategic thinking about customer relations, it is the retailer that imposes this. However, this retailer views the local food producers as icons; that is, important, but exotic supply chain actors they value mainly because they help differentiate the supermarket's up-market positioning. Given this positioning of the supermarkets, including the high process of the products, there is little initiative to develop information connectivity between the local food supplier. From the local food supplier's perspective, as long as the orders flow in, they focus their attention on producing the goods and delivering them in an as simple a manner as possible. This consideration implies that local food supply chains only superficially resemble service supply chains, where interaction is highlighted to develop customer value. These supply chains, in their "as-is" are, are therefore considered following predominately as long-linked technology in the frame of a shorter distance.

The question arises whether the local food suppliers may or should become more like service supply chains since they are both short in structure,which implies fewer business relationships with the potential for increased intense interaction in them. In the "as-is" state, information connectivity is sought through copying long-linked technology use in modern industrialized supply chains. This is the case since the discourse guiding local food production is mainly the same as it is in industrialized food production. This state is clearly not ideal and developing local food production should aim to better reveal particularities of this mode of food supply.

Intensive Technology to Develop Information Connectivity

By focusing on developing the use of intensive technology, attention is directed to using the existing modernistic paradigm to develop the quality of information connectivity. As discussed in Section 5.1, information connectivity is manual and not viewed as important by both local food supplier and retailer customer. This is a way to economize and also an expression of the fact that these parties perceive the relationship as stable, the level of trust as sufficient, and do not need to be cared for in a costly manner. This section points out that local food suppliers may choose to become more like a service industry suppliers. However, as Stabell and Fjeldstad (1998) showed, interdependencies in the services industry vary. Stabell and Fjeldstad referred to one mode of services supplied as a "value shop", in which the quality of interaction is highlighted. Thompson (1967) pointed out that in cases of reciprocal interdependence, mutual adjustment is a core feature of interaction. The challenge is developing mutual adjustment processes, a form of manual information exchange, and in developed economies with high personnel salaries, a costly form of interaction. This implies that this form of interaction in the local food supply should be kept at a minimum and only used when it is associated with providing value to the customer that substantiates the cost of this form of interaction. In cases where the supply of local foods needs to be "engineered," the use of intensive technology should be developed. In the cases described in this paper, all the producers supply more or less standardized products.

The case descriptions revealed small-scale production that was often based on local resources and traditions. There is a degree of interaction between customers and suppliers, but these interactions are rather simple and not very time-consuming. From a normative viewpoint, unless there is a need to tailor supply to customer demands, the use of the inherently costly intensive technology that involves developing the reciprocal interdependencies is not called for. The main reason founded on the case description to develop the quality of personal interaction is to better show the authenticity of the local food producer as a supplier with a face and history. This is dependent on the importance of this factor in securing the reputation of the local food supplier. To the degree this factor is strategically important, the local food suppliers should strive to become more like a service supply chain and use information technology to support personal interaction with customers. This suggests developing visualizations of the computer–person interface; that is, developing information systems to better support personal interaction.

Mediating Technology to Develop Information Connectivity

An interpretation of local food supply is provided that this form of supply is artificially designed founded on long-linked technology. Reciprocal interdependencies are not well developed. However, there is an alternative mode of developing information connectivity that is the opposite of that indicated in Section 5.2. This suggests that, to the degree that reciprocal interdependencies are encountered, these should be attempted to be reduced. This also implies dismantling seemingly artificially imposed thoughts among local food suppliers that they need to develop their sequential interdependencies through using modernistic production planning and control tools that are commonplace in industrialized food production.

Focusing on mediating technology is associated with services supplied through what Stabell and Fjeldstad (1998) termed as a "value network". In such networks, interdependencies are pooled and developed through activity and resource standardization. It has already been suggested that local foods as physical objects are sufficiently standardized to easily fit into the retailer assortment. In addition, this impacts on how the local foods are informed about. In the case companies, these local foods are painlessly pooled at present since they are sufficiently standardized as supply chain resource objects. From a systems perspective, the quality of pooling resources, including information, is associated with how well they can be coupled together to create synergies. However, this pooling in these personalized local food chains is predominantly manual. This limited use of information technology suggests a potential for development regarding how goods are identified, tracked, traced and informed about in general.

A seamless information flow may be envisioned in the local food supply chain. However, due to the limited economic capabilities of the local food suppliers, this information flow must be economical from the perspective of those suppliers, including limiting investment and information use costs. This point highlights the potential of using third-party information services that are linked to existing information tools of the local food suppliers, such as mobile phones and personal computers.

Conceptual Model

Based on the preceding analyses,we have created the conceptual model shown in Figure 3 below.

The model indicates that the ultimate aim of local food production is to find an appropriate balance of intensity and mediating technology. This balance is dependent on understanding when and where tailoring to customer needs are needed in order to secure customer value. Information connectivity is supported through the use of these technologies. It is vital that information connectivity supports developing customer value in the supply chain. Understanding the nature of customer value is case-specific and requires dedicated case studies. Given the simplicity of local food production, such inquiries should be relatively simple to undertake.

Figure 3. Conceptual model to guide developing information connectivity in local food production

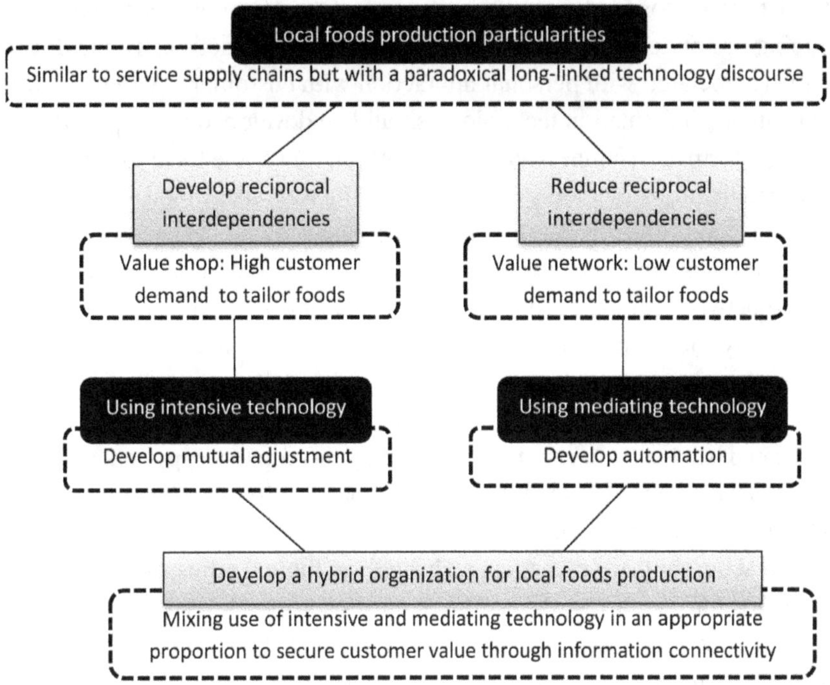

CONCLUSION

Local food production is clearly distinct from the dominant industrialized food production. The local food producers are people that seemingly imitate these modernistic food suppliers by applying long-linked technology in their short supply chains. This artificial use is also associated with focus on the foods themselves, their design and production, rather than developing their supply through enhancing their marketing and logistics. Since this type of food production is characterized by short supply chains, it more closely resembles service supply chains. However, when considering how local food suppliers actually distribute their goods, the local food suppliers are seemingly drowned in the dominant modernistic preconception of what constitutes food supply. Founded on contingency theory (Thompson 1967), the present study has revealed that highlighting interdependencies of local food production show how food supplies are more similar to services supplied than industrialized food production. This perception is predominantly founded on the view that both local food supply and the services industry are associated with short supply chains. This structural shortness implies that both pooled and reciprocal interdependencies need to be highlighted instead of sequential ones.

Following the seminal work of Stabell and Fjeldstad (1998) on the strategical differentiation service offerings, local foods may initially be considered as either value networks or value shops. However, in real business scenarios, the conceptual borderline between these service forms is unclear.Therefore, it is suggested that, like services, local foods are hybrids between these forms. Variation regards the degree to which local foods are other value shops or value networks. This is dependent on the degree to which customer value is dependent on tailoring food supply.

Local food production is highly contingent on the market. This especially true since such supply chains are small, short and local. Accordingly, it is the market context that is suggested to have provided the reason for either developing or not developing the quality of personal interaction. To the degree the quality of the foods offering is associated with personal interaction with customers, with the retailer intermediary or direct with consumers, information technology should be developed to support this interaction. This involves developing information connectivity to support the role of the local food supply as a value shop.

Regarding the practicalities of food production, in all cases, local food suppliers, especially those that do not tailor their food offering to individual consumer preferences, need to develop pooled interdependencies. This implies developing information connectivity to support the role of the local food producer as a value network through reducing the personal factor in the supply chain by increasingly automating it. Currently such connectivity is required to be inexpensive or easy to use. In the case local food producers were found to be both knowledgeable, flexible and communicative. Using IT is however something they do not seem to like nor prioratize. Therefore cheap and simple IT-enabled connectivity solutions are advocated, such as programmes provided by thord-party suppliers to enhance traceability, tracking and trading procedures. These systems should be easy to use, cheap and preferably usable on smartphones. These technologies should enhance pooled interdependencies, simplifying how goods are traded and informed about, but still providing a channel for mutual adjustment through intense interaction when this is called for.

The overall strategic recommendation to the local food industry is to reconsider the current paradoxical use of modernistic food supply planning, implementation and control techniques. Instead, they should develop following a framework of adaptive but not overly complex systems. The notion of interaction and responsiveness are core to this research approach. The supply of local foods is not, at its core, resembles a services industry, but it may be based on similarities in marketing and logistics of the service industry, and thus learn how to develop the supply of local foods, an economically far weaker industrial sector in developed economies. A state of mind needs to be changed among the local foods suppliers, and through this paper the conceptual-level foundations for such a change is provided.

To the degree this framework proves fruitful, this may reveal how to economize local food supply by using state-of-the-art information technology, paving the way to make local food supply an increasingly important source of food supply in the economically developed world. Local foods may, through improved information connectivity in their supply chains that support increasingly economical food supply, be moved from being a post-modernistic curiosity to become a functionally viable mode of geographically constrained mass food supply. If the practitioners and supporting forces understand their particular needs, in association with developing information connectivity, local foods may become the norm rather than the exception in the food industry.

Limitations of this research are associated with this being a case study; generating rich empirical detail rather than generalizability being a main research conern. Understandings created through this research should therefore be tested in other empirical settings. Future research may therefore seek to apply the created analytical framework centered around the model described in figure 3 including using it in different empirical settings including smaller devleoping copuntries and larger countries such as USA, Russia, China or Brazil. Studies may also focus, as development priojects, on practical implications of IT based on thisn framework and report on this, linking in greater detail reasoning for devleoping connectivity put forth through this study with the actual technology (and its suppliers) at hand.

REFERENCES

Abatekassa, G., & Peterson, H. C. (2011). Market Access for Local Food through the Conventional Food Supply Chain. *International Food and Agribusiness Management Review*, *14*(1), 63–82.

Amilien, V., Schjøll, A. &Vramo, L.M. (2008). *Forbrukernes forståelse av lokal mat.* Statens Institutt for Forbruksforskning. Oslo, Fagrapport nr. 1-2008.

Closs, D. J., Swink, M., & Nair, A. (2005). The role of information connectivity in making logistics programs successful. *International Journal of Physical Distribution & Logistics Management*, *35*(4), 258–277. doi:10.1108/09600030510599922

Darby, K., Batte, M. T., Ernst, S., & Roe, B. (2008). Decomposing Local: A Conjoint Analysis of Locally Produced Foods. *American Journal of Agricultural Economics*, *90*(2), 476–486. doi:10.1111/j.1467-8276.2007.01111.x

Engelseth, P. (2015). Customer-Responsive Supply of Local Foods. *Journal of Operations and Supply Chain Management*, *8*(3), 111–119.

Engelseth, P., & Hogset, H. (2016). Adapting Supply Chain Management for Local Foods Logistics. *Proceedings in Food System Dynamics 2016.*

Engelseth, P., Wongthatsanekorn, W., & Charoensiriwath, C. (2014). Food Product Traceability and Customer Value. *Global Business Review*, *15*(4suppl.), 87S–105S. doi:10.1177/0972150914550549

Giddens, A. (1991). *Modernity and Self-Identity.* Cambridge, UK: Polity Press.

King, R. P., Hand, M. S., Di Giacomo, G., Clancy, K., Gomez, M. I., Hardesty, S. D.,... McLaughlin, E. W. (2010). *Comparing the Structure, Size and Performance of Local and Mainstream Food Supply Chains.* USDA ERS Report Number 99.

Kotz, S, Banks, D.L.,& Read, C.B.(1999). *Wiley Series in Probability and Statistics.* Wiley-Interscience.

Lambert, D. M., Stock, J. R., & Pagh, J. D. (1998). Supply Chain Management: Implementation Issues and Research Opportunities. *The International Journal of Logistics Management*, *9*(2), 1–19. doi:10.1108/09574099810805807

Martinez, S., Hand, M., Da Pra, M., Pollack, S., Ralston, K., Smith, T.,... Newman, C. (2010). *Local Food Systems Concepts, Impacts and Issues.* Economic Research Report No. 97, United State Department of Agriculture.

Orton, J. D., & Weick, K. E. (1990). Loosely coupled systems: A reconceptualization. *Academy of Management Review*, *15*(2), 203–223.

Sampson, S. E. (2000). Customer–supplier duality and bidirectional supply chains in service organizations. *International Journal of Service Industry Management*, *11*(4), 348–364. doi:10.1108/09564230010355377

Stabell, C. B., & Fjeldstad, Ø. D. (1998). Configuring value for competitive advantage: On chains, shops, and networks. *Strategic Management Journal*, *19*(5), 413–437. doi:10.1002/(SICI)1097-0266(199805)19:5<413::AID-SMJ946>3.0.CO;2-C

Thompson, J. D. (1967). *Organizations in Action*. New York: McGraw Hill.

Weick, K. E. (1976). Educational organizations as loosely couples systems. *Administrative Science Quarterly*, *21*(1), 1–19. doi:10.2307/2391875

Chapter 8
Stakeholder Agriculture:
Innovation From Farm to Store

Alexandros Antonaras
University of Nicosia, Cyprus

Alexandros Kostopoulos
CSR Hellas, Greece

ABSTRACT

The shift in agricultural production and agribusiness may be a solution in reducing unemployment and particularly that of young people which is dramatically high in several European countries that are experiencing the negative consequences of the recent global financial crisis that led to a dramatic decline in their GDP per capita and has affected all sectors of economic activity, including agriculture. The overall scope of this chapter is to present an Agricultural Entrepreneurship and Social Innovation Framework that can lead to a new business model with social aspects, contribute to the economic growth and sustainability and hence combat the phenomenon of unemployment and poverty in rural areas that have been seriously affected by the recent financial crisis.

INTRODUCTION

According to UN Global Compact (2016), the world's population is expected to grow to nine billion by 2050 and demand on global food systems intensifies every day, while businesses will be critical partners in designing and delivering effective, scalable and practical solutions for food security and sustainable agriculture. Every actor along the agriculture supply chain, including farmers, traders, retailers, investors and consumers have a role to play in advancing food security protecting the environment and ensuring economic opportunity.

Globally recognized organizations like the UN Global Compact have already ranked sustainable agribusiness among their top priority issues. In this context, in 2014, the Food and Agriculture Business (FAB) Principles were launched by UN Global Compact as a voluntary framework to advance the positive impact businesses can have in the food and agriculture space and engage in principle-based collaboration with the UN, governments, civil society and other stakeholders.

DOI: 10.4018/978-1-5225-2107-5.ch008

Focusing on European countries, five decades after the founding of the EU and the implementation of a Common Agricultural Policy, the rural structures of each Member State continue to vary considerably. The main reasons for this variation are the different economic and social-political progress achieved in each member state and the different geographical and climatic data. Thus, for each country the agricultural sector represents a different proportion of their overall economy and contributing differently to the national GDP, employment rates and foreign trade as well as the overall cost of living of the population.

Countries experiencing the negative consequences of the recent economic crisis, of the recent years such as Greece, Cyprus, Spain and Portugal show a dramatic decline in GDP per capita, which has affected all sectors of economic activity including agriculture. In parallel, the agricultural sector in these countries has traditionally been of significant importance to their economies, representing a main economic activity and employment opportunity of a large part of their population, and was relatively higher as a percentage of their GDPs compared to the EU average. It is worth to mention that between 2005 and 2008, the employment rate in the EU, at ages 20-64, rose and reached 70.3%. The trend reversed from 2009 and employment returned to 2006 level at 69%. The following five years employment in EU declined further with an average employment level at 66%, making it very difficult to reach the European target of 75% by 2020 (Eurostat, 2015).

The current lack of sufficient employment opportunities in urban areas, as well as several other obstacles together deepened further youth unemployment in the agricultural sector creating continuous instability guiding to social exclusion and finally poverty. Moreover, farmers have nowadays weak cooperatives leading to the development of individual farming culture that does not support at least a minimum standardization level of their products or farming processes. In addition, the majority of them face overexposure to the financial institutions having difficulties to pay back their loans and as a consequence secure further funding for their future production needs. The limited farming knowledge and absence of formal agricultural training creates further barriers for young people entering the farming sector.,

The shift in agricultural production and agribusiness may be a solution in reducing unemployment and particularly that of young people, which according to the ILO's (2016) World Employment, Social Outlook, in the counties mentioned above it increases continuously and remains dramatically high. To enable such a shift, people, especially young, should be encouraged to remain or return to rural areas and supported to enter the field of agricultural production, which seems to be able to ensure a fair income and provide a chance to young people to build their future with dignity.

There is an urgent need to change the existing and problematic reality. This can be done with the participation of as many as possible players of the production. The creation and support of collective economic rural activities in the form of new, highly-equity cooperatives in the context of social enterprises can create economies of scale and enhance the development of commercial agricultural products by the farmers themselves. These social forms of collective representation bodies and the establishment of regional agricultural chambers can further support rural entrepreneurship, for producers to work together as entrepreneurs. The comprehensive training for farmers arises as a basic need and attending a training course on the "Primary production" can help to de-demonize her.

In addition, the development of an effective network of applied agricultural research to serve farmers in cooperation with the academic community, as well as farming in experimental fields using innovative agricultural applications can contribute to the reversal of the age distribution of farmers. At the same time, the provision of financial support through traditional tools and the development of additional tools of social and solidarity economy, the protection of agricultural means of production, the targeted land use, and the rational use of water may result in the long-term growth of domestic self-sufficiency rate and the development of a National Rural Policy that would be feasible to implement.

Recognizing the need to change the above situation, this chapter presents and analyses the theoretical concept of an innovative approach that has been successfully implemented by a multinational company,(and is briefly presented later in the case study), with the involvement of different stakeholders, including representatives of the primary sector (farmers / producers), secondary (retail, manufacturing) and tertiary sector (banking, transport etc.), in an effort to: [a] secure farmers' income through contractual farming, [b] provide appropriate training to farmers, [c] help securing financial assistance for them provided by financial institutions that undertake the guarantee on behalf of them, [d] support the upgrading of technical means and utilities in order to obtain a higher yield per hectare of production, and [e] develop the appropriate channels to put into the market "unique" products in response to the continuously changing demands of increasingly socially sensitized consumers.

The overall scope of this chapter would be to develop an Agricultural Entrepreneurship and Social Innovation Framework (AgrESIF) that can lead to a new business model with the social aspects and contribute to the economic growth and sustainability, and hence combat the phenomenon of unemployment and poverty in rural areas that have been seriously affected by the financial crisis over the last years. The conceptual Framework links four different forces which shape the potential of an AgrESI project and determine its success and effectiveness, namely the relative global trends, the needs of the rural community, the consumers' behavioral trends and the support available by the responsible corporations, taking into account parameters such as stakeholder expectations, micro-financing needs, international initiatives and the set of entrepreneurial skills required.

GLOBAL TRENDS

Sustainable Farming

Farming sustainably means growing crops and livestock in ways that meet three objectives simultaneously, those of i) economic stability and profit, ii) social benefits to the small farm holders, their families and the local community, and iii) environmental conservation. According to Food and Agriculture Organization of the United Nations (FAO, 2014f) sustainable agriculture depends on a whole-system approach whose overall goal is the continuing health of the land and people, in respect to the Principles of Sustainable Farming. Therefore, sustainable farming concentrates on long-term solutions to problems instead of short-term treatment of symptoms.

More analytically, sustainable farming requires economic sustainability, meaning that the small farm entrepreneurs and/or enterprises are consistently profitable from year to year, the family savings or net worth is consistently going up, while the family debt, where occurs, is consistently going down. Also purchase cost of off-farm feed and fertilizer is decreasing and in some cases reliance on customer payments is decreasing.

On the other hand, social sustainability means that more money circulates within the local economy as the small farmers support other businesses and families in the community. The number of rural families is holding steady or potentially is going up (ATTRA, 2003). Consequently, low skilled young people take over their parents' farms and continue farming as well as highly skilled college graduates with studies in relative fields to agribusiness return to the community after graduation.

Also, environmental sustainability, according to the Millennium Ecosystem Assessment (2005) means that agriculture can be treated as ecosystem management of complex interactions among soil, water, air, plants, animals, climate and people, with the ability to integrate all these factors into a production system that is appropriate for the environment, and as described before, the people, and the economic conditions where the farm is located.

A successful transition from traditional to sustainable farming depends on the farmer's careful monitoring both of progress towards the goals set and of the overall health of the system. Planning and monitoring are particularly important in sustainable agriculture, which relies on natural systems to replace some of the work done by input products like fertilizer and pesticides (ATTRA, 2003). Constant support as the ability to evaluate and re-plan farming procedure according to protocols is vital to the farmer who wishes to farm more sustainably. This transition process should be linked with global trends and international initiatives in order to produce concrete improvements to all stakeholders and mostly to future generations.

The 2030 Agenda for Sustainable Development

Our planet faces multiple and complex challenges in the 21st century. The world's population is expected to grow to nine billion by 2050 and demand on global food systems intensifies every day (UN Global Compact, 2016). 2015 was an important year for the future of agriculture and development. A new set of global Sustainable Development Goals, now on known as "SDGs", will shape the next 15 years of policies, programs and funding (Farming First, 2015). More specifically, on 25 September 2015, the 193 Member States of the United Nations adopted the 2030 Agenda with its Sustainable Development Goals (SDGs), a set of 17 aspirational objectives with 169 targets and many more sub-targets expected to guide the actions of governments, international agencies, civil society and other institutions over the next 15 years (2016-2030).

These ambitious 17 Goals of the 2030 Agenda is a global vision for people, for the planet and for long-term prosperity. They integrate the three dimensions of sustainable development – economic, social and environmental, while at the same time no one goal is separate from the others, and all calls for comprehensive and participatory approaches.

The 2030 Agenda for Sustainable Development is as relevant to developed as it is to developing nations and it charts a plan for the future – shifting the world onto a sustainable and resilient course and leading to transformation. Also, the new 2030 Agenda commits the international community to act together to achieve the Goals and transform our world for today's and future generations.

Agriculture and SDGs

Industrialization drove workers from the fields to the factories with the promise of better opportunities and higher living standards. In employment terms, agriculture accounts for approximately 36% of the global workforce and falling, although the figures mask huge disparities between developed and developing countries (ILO, 2013). For example, according to World Bank (2014) data, in 2010, only around 2% of the US working population was engaged in agriculture, while in India the percentage was approximately 50% and over 80% in Sub-Saharan Africa (including related rural enterprises).

Reaching the SDG targets simply will not be possible without a strong and sustainable agricultural sector (Farming First, 2015). More than any other sector, agriculture is the common thread which holds the 17 SDGs together. Specifically, Sustainable Development Goal #2 calls to: "end hunger, achieve food security and improved nutrition and promote sustainable agriculture." To fulfill this goal, the United Nations has identified a series of specific targets (see 2.1-2.5), as well as the means of implementation (see 2.a-2.c) for reaching them.

Specific target 2.1 aims by 2030, to end hunger and to ensure access by all people, in particular the poor and people in vulnerable situations, including infants, to safe, nutritious and sufficient food all year round. Tackling hunger is not only about boosting food production; it's also about increasing incomes and strengthening markets so that people can access food, even if a crisis prevents them from growing enough themselves (United Nations, 2015).

Specific target 2.2 aims by 2030, to end all forms of malnutrition, including achieving, by 2025, the internationally agreed targets on stunting and wasting in children under 5 years of age, and address the nutritional needs of adolescent girls, pregnant and lactating women and older persons (United Nations, 2015).

According to the Farming First plan, (a coalition of multi-stakeholder organizations developed as a joint call to action in response to the global challenges posed at the 2009 United Nations Commission on Sustainable Development), agriculture can play an important role to address malnutrition in three ways:

1. Biofortified foods, such as vitamin-A enriched rice or sweet potatoes, are bred to have higher amounts of micronutrients and can help provide essential vitamins and minerals;
2. Micronutrient-enriched fertilizers improve soil fertility, helping to support higher yields of more nutritious food and can combat micronutrient deficiencies in humans;
3. Improved agronomic practices can also help, for instance, crop rotation and conservation tillage by encouraging food diversity and preventing nutrient depletion of soils.

Specific target 2.3 aims by 2030, to double the agricultural productivity and incomes of small-scale food producers, in particular women, indigenous peoples, family farmers, pastoralists and fishers, including through secure and equal access to land, other productive resources and inputs, knowledge, financial services, markets and opportunities for value added and non-farm employment (United Nations, 2015).

Agricultural development is inextricably linked to economic growth that benefits the poor. In fact, the World Bank (2016) has estimated that agricultural development is about two to four times more effective in raising incomes among poorer than growth from any sector (and up to 11 times more effective in sub-Saharan Africa). Also, increased productivity, when coupled with better access to markets, can help address hunger directly at the farm level or provide sufficient additional income to buy food at the market. Boosting rural incomes and ensuring ample employment means looking at economic opportunities across the entire rural value chain, from farmers and input suppliers to value-added processing and services, such as transporting and marketing of food (Farming First, 2015).

Specific target 2.4 aims by 2030, to ensure sustainable food production systems and implement resilient agricultural practices that increase productivity and production, that help maintain ecosystems, that strengthen capacity for adaptation to climate change, extreme weather, drought, flooding and other disasters and that progressively improve land and soil quality (United Nations, 2015).

Agriculture is more vulnerable to climate change than any other sector. A warming climate could reduce crop yields by more than 25%, according to the. Agriculture and land use change are also responsible for between 19–29% of global greenhouse gas emissions (United Nations, 2015). But the sector can also help to substantially mitigate against future greenhouse emissions, especially by increasing productivity of land already under cultivation and thus reducing deforestation. A higher price of carbon can help incentivize agriculture's mitigation potential. More productive farms also tend to use less water per unit of crop produced. Innovations such as drip irrigation can also improve agriculture's water use efficiency while still supporting higher productivity.

Reducing food waste is another area which can improve the sustainability and resilience of the agricultural sector. A recent report by WRAP (2015) estimated that one third of all food produced is never consumed, at a total cost as much as $400 billion a year and 3.3 billion metric tons of greenhouse gases being released annually, about 7% of the total emissions. Reducing food waste by 50% globally could save $300 billion a year by 2030 and could feed as many as a billion people. By 2030, this consumer, food waste could cost as much as $600 billion a year, unless we act now to address this problem (Reuters US, 2015).

Specific target 2.5 aims by 2030, to maintain the genetic diversity of seeds, cultivated plants and farmed and domesticated animals and their related wild species, including through soundly managed and diversified seed and plant banks at national, regional and international levels, and promote access to and fair and equitable sharing of benefits arising from the utilization of genetic resources and associated traditional knowledge, as internationally agreed (United Nations, 2015).

Increasing agricultural productivity means that we can produce the food our global population needs while keeping as much other land as natural habitat where biodiversity can flourish (instead of using land more extensively for agricultural and biodiversity simultaneously). Smallholder farmers play a key role, as they hold as much as 75% of the global seed diversity in staple food crops, with the rest being held in gene banks. Urgent action will be needed, since as much as 10% of the biodiversity seen in 2000 may be lost by the year 2030, resulting from land lost to infrastructure as well as from agriculture and climate impacts (EurekAlert, 2015).

More than just its direct impact on hunger and malnutrition, global food system and agriculture are also linked to many global trends and consequently too many of the 17 SDGs.

SDG #1 aims to "end poverty in all its forms everywhere"(UNSDKP, 2015). Growth in agriculture is at least twice more effective in reducing poverty than from any other sector, taking into account that rural people represent the largest segment of the world's extreme poor by far contain more than 70% of the total (Farming First, 2015).

SDG #4 aims to "ensure inclusive and equitable quality education and promote lifelong learning opportunities for all"(UNSDKP, 2015). Agriculture extensions enable farmers to access to the skills, tools, inputs and knowledge they need to thrive. Investment in agricultural extension services yields 80% annual rates of return and can help farmers to double their crop yields(Farming First, 2015).

SDG #5 aims to "achieve gender equality and empower all women and girls" (UNSDKP, 2015). Women produce over half the food globally, so bridging this gap could reduce global hunger by as much as 17% according to Farming First plan. Given equal access to recourses as men, women would achieve the same yield levels, boosting total agricultural output in developing countries by 2½ - 4%. This additional yield could reduce the total number of undernourished people in the world by 100-150m or 12-17%(Farming First, 2015).

SDG #6 aims to "ensure the availability and sustainable management of water and sanitation for all" (UNSDKP, 2015). By 2030, global water demand will increase more than 50%, with agriculture alone requiring more than what can be sustained to feed the world even before domestic and industrial needs are met(Farming First, 2015).

SDG #7 aims to "ensure access to affordable, reliable, sustainable and modern energy for all" (UNSDKP, 2015). By 2030, energy demand is expected to increase as much as 50%, driven mostly by developing world demand. Crops are more likely to be diverted for use as biofuels, doubling or even tripling as a proportion of total use(Farming First, 2015).

SDG #8 aims to "promote sustained, inclusive and sustainable economic growth, full and productive employment and decent work for all" (UNSDKP, 2015). Agriculture is an engine of pro-poor economic growth in rural areas. Entrepreneurship across the rural and food sectors can generate employment and growth. More specifically, according to the data from Farming First plan, 85% of farmers are small holders, who have less than two hectares of land. Also, 43% of the agricultural laborforce in developing countries are women. Moreover, 70% of the youth aged 15-24 in sub-Saharan Africa and south Asia live in rural areas, and they are twice as likely as adults to be unemployed(Farming First, 2015).

SDG #12 aims to "ensure sustainable consumption and production patterns" (UNSDKP, 2015). Nowadays around one third of the food produced is wasted. At the same time average per capita consumption is expected to grow through 2030, despite population increases (Farming First, 2015).

SDG #13 aims to "take urgent action to combat climate change and its impacts" (UNSDKP, 2015). By 2030 agriculture's carbon mitigation potential could reach as much as 7,5% of total global emissions, depending on the price of carbon and adoption of agricultural productivity measures (Farming First, 2015).

SDG #15 aims to "protect, restore and promote sustainable use of terrestrial ecosystems, sustainably manage forests, combat desertification, and halt and reverse land degradation and halt biodiversity loss" (UNSDKP, 2015). Improving the efficiency of farmland can help meet world's growing consumption demand while minimizing the loss of natural habitats and forests for additional cultivation (Farming First, 2015).

As derived from the above, there are significant interactions between SDGs.Staying with the food SDG example, a commonly discussed set of interactions lies in the "nexus" between food, water and energy (Weitz et al. 2014). More specifically, vital for agriculture, water is also required for energy production in cooling thermal power plants and generating hydropower.In addition, energy is required for water pumping and irrigation systems and at the same time water is needed for irrigating agriculture. Finally, according to the ICSU (2016), there are also competing resource requirements, forexample; food production may compete with bioenergy production for the same land or water. All these factors create an active debate in the coming years upon solution space in terms of governance measures or technological options that could transform negative interactions into more positive ones.

LOCAL RURAL COMMUNITY NEEDS

It is widely agreed that in order to avert future food crises, agricultural productivity must be increased in ways that are sustainable, resilient and conducive to rural development and poverty reduction. While the world's cultivated land areas have grown by 12 per cent over the past 50 years, agricultural production has roughly tripled owing to significant increases in the yields of major crops (FAO, 2009).

At the same time, much of the agricultural expansion has been at the expense of forests: between 2000 and 2010, 13 million hectares of forests were lost each year through conversion to other land uses (FAO, 2010). However, forests are crucial for sustainable agriculture as they protect soils, regulate water flows, serve as gene pools and maintain a healthy climate. As a result, agriculture productivity growth is slowing down in many parts of the world.

Family farmers produce at least 56 per cent of all agricultural production worldwide (FAO, 2014a). With supportive, stable policies and greater participation in policy processes, smallholder farmers can respond positively to policy and market opportunities. To realize the full potential of smallholder agriculture, there is a need to remove the constraints that limit its investment capacity. Responsible investments in agriculture are a top priority from the perspective of both donors and recipients (FAO, 2014b).

The first objective is to support investments by smallholders themselves, but their capacity to do so depends on other related investments in collective action, in public goods and supportive rural infrastructure (UNSG, 2014). Secondly, investment in agricultural knowledge, science and technology can contribute to substantial increases in agricultural production over time. Increases in productivity can also contribute to a net increase in global food availability per person and general food security (FAO, 2011).

Also, according to FAO's (Food and Agriculture Organisation of the United Nations) publication under the title "Deep Roots" on the occasion of the appointment of 2014 as International Year of Family Farming (FAO, 2014c), with the expertise of companies and brands that often are the main customers of smallholder farmers and resources of several indirect partners, such as creditors or distributors of their products, smallholder farmers can be supported to achieve better yields, reduce their losses after the harvest, and improve the quality of their staple crops (FAO, 2014d).

Moreover, efforts to gather and share lessons on effective approaches to connect smallholder farmers with markets in a sustainable way and share them widely with stakeholders, as well as investment in smallholder farmers capacity-building are considered necessary for agricultural development in local rural areas (WFP, 2015).

Furthermore, post-harvest handling, storage and transportation of the products, should be reexamined in order to obtain yield sustainable results and boosting local and even national food security over the long term (FAO, 2013).

Generally, investments in infrastructure work better if they support the models of production and markets that are appropriate to smallholder farmers and, furthermore, these investments would need to be bolstered by measures to secure tenure rights (USAID, 2014).

CONSUMERS' BEHAVIORAL TRENDS

During the last decades, a new model of consumption can be seen mostly in wealthy capitalist nations around the world (Lewis & Potter, 2011). Ethical consumption has become an umbrella term covering a wide range of concerns from animal welfare, labor standards and human rights to questions of health and wellbeing and environmental and community sustainability (NRI, 2001).

One of the first polls for the issue by Global Market Insite across 17 countries, including the USA, Australia, Japan, China, India and various European countries, found that 54 per cent of online consumers would be prepared to pay more for organic, environmentally friendly, or Fair Trade products (Horne et all, 2016).

On the one hand, negative modes of campaigning such as boycotts emerged in the nineteenth century continued as global brand-based activism into the twentieth, and on the other hand the shift in the consumer behavior with the development of the general idea of combining ethics and shopping that become a mainstream concept especially in developed countries, guide to the conclusion that if consumers cared about moral issues, then companies and brands that did the right thing would have a larger market share.

Especially for agricultural products ethical consumers purchase those products that minimises social and/or environmental damage, while avoiding products deemed to have a negative impact on society or the environment. But most importantly consumers across the globe are growing increasingly concerned about the origins of the food they eat. Interest in the food safety, local food economy, environmental impacts, and the complexity of the entire food supply chain, has led to the creation of new categories and labels to differentiate food products, such as "bio", "ethical" or "green" (AAFC, 2013)

These developments in consumer behavior can be seen as opportunities for companies, but also farmers as ethical products are expected to see increased demand in the near future as environmental awareness amongst consumers continues to grow. This trend is expected to be seen in nearly all product categories and will have a significant impact on the food sector, as already organic items seeing particularly increased demand (AAFC, 2013).

Within the ethical foods market, manufacturers need to earn customer trust through transparency and wider recognized labels that will promote on a larger scale the overall potential benefits of these products.

Organic foods that were once reserved for niche markets are now becoming more mainstream and with increased exposure and competition. Due to increasing demand and growing availability, unit prices for organic food are expected to decline over the next couple of years, giving the consumer more buying power and subsequently boosting sales(AAFC, 2013).

Fair trade is an alternative approach to conventional trade and is based on a partnership between producers and consumers. Fair trade can make a difference helping producers build sustainable livelihoods, diversify their businesses, and even reinvigorate entire sectors of production (Fairtrade, 2014).

Products that are marked as 'fair-trade' are gaining popularity among today's consumers, although not as quickly as some other ethical food categories, although consumer demand for low prices is seen as a persistent obstacle to market growth by fair trade producers(AAFC, 2013).

At the same time there is a strong need for consumer education regarding ethical products. The lack of international standards, limited product availability in most cases and falsified or inconsistently defined claims happen in the past have created an overall mistrust of ethical labels, so companies and brands will have to invest in earning again consumers' trust by bringing transparency and traceability characteristics of their products. Consistent, clear, and commonly adopted definitions for claims such as "sustainable", "ethical", "fair trade", "carbon neutral" or "social product" would reduce consumer skepticism and increase market potential. According to OECD (2008) study on best Practices in developed countries members of the organization, consumer education as to what these labels represent would also empower consumers to choose products that address the issues of most importance to them.

Last but not least consumer education should also focus on reduction of food losses and waste. Given that many smallholder farmers in developing countries live on the verge of food insecurity, a reduction in food losses in those countries could have an immediate and significant impact on their livelihoods. According to FAO (2014e), if food losses and waste could be halved, the required increase of food available to feed the world population by 2050 would only need to be 25 per cent, and not 60 per cent as currently projected. Considering its nature and causes, halving food losses and waste could be a feasible target based on technical, economic, environmental and social considerations.

RESPONSIBLE CORPORATIONS' INITIATIVES AND SUPPORT

The last few decades of globalisation have generated unprecedented growth but also unprecedented levels of inequality. According to Oxfam's (2014) estimates, one percent (1%) of the world's population will soon own more wealth than the other 99% and the richest 85 people on the planet as much as the poorest half of humanity.

With citizens nowadays demanding responsible business behavior from all kinds of organizations corporations and their leaders (Antonaras et al, 2011) are accountable not only for their specific responsibilities in terms of consultation, impact assessments, benefits and grievance mechanisms but also in relation to wider approach on sustainable development. Galinski argued that having a moral compass leads to more effective business practices — whether in building sales, retaining employees, or reducing litigation and regulation costs (as sited in Antonaras& Memtsa, 2009). It is clear that there is some sort of a relationship between business ethics and business success, or better sustainable excellence (Antonaras& Memtsa, 2009).

In this direction, the 2030 Agenda for SDGs expect both direct and indirect contribution from the business. Direct, through financing and partnering on SDG related projects (e.g. infrastructure) where business will be called upon to invest more in developing countries, or in poor rural areas in developed countries. Indirectly, through an increase of business activity emphasizing on equitable economic growth as a driver for development implies a significant global expansion in the private sector itself. Developing countries as well as rural areas of developed countries need businesses to create more jobs, move people out of poverty and contribute to local problems and needs.

The SDGs clearly prioritize both employment and development. For the agriculture sector this is a long-term balance endeavor with the use of the benefits of agricultural technology and mechanization for greater efficiency on the one hand, while on the other make efforts to avoid potential losses in employment that will affect smallholder farmers and their families.

AGRICULTURAL ENTREPRENEURSHIP AND SOCIAL INNOVATION FRAMEWORK

Taking into consideration all the above forces, a theoretical framework on developing Agricultural Entrepreneurship and Social Innovation (AgrESI) projects (or even businesses) is presented in an effort to provide a simplified model approach and positively influence interested parties in getting involved in such projects. The conceptual framework links the four different forces which shape the potential of an AgrESI project and determine its success and effectiveness, namely the global trends, the needs of the rural community, the consumers' behavioral trends and the support available by the responsible corporations.

It seems that these forces contribute significant input in designing an AgrESI project that will assist people in rural areas to participate in entrepreneurial initiatives, enable organizations to align their CSR related activities to the needs of this vulnerable group taking into account the changing trends in consumer behavior and attempting to respond to the global trends for sustainability as they are expressed, among others, through the 2030 SDGs.

The framework also takes into consideration several other factors such as: the stakeholders expectations and the willingness of responsible corporations to support an AgrESI project, the entrepreneurial skills needed by the people in rural communities in order to ultimately manage the new agri-business, the experiences from similar international initiatives, and the need for micro-financing in order for the new agribusiness to grow and emerge into a sustainable businesses.

Figure 1 below graphically depicts the theoretical framework for agricultural entrepreneurship and social innovation.

The checkerboard arrows indicate that once an AgrESI project is effectively designed and implemented it will soon return value back to the four forces of the framework, in other words it creates shared value. The organizations involved may receive back part of their investment as dividends from their participation in the new agribusiness and benefit from the increased social value that their participation in the AgrESI project will create. Consumers will receive better and high quality products as a result of their preference and demand for products that comply with strict social and environmental criteria. The local rural community will receive the direct value resulting from the new workplaces that will be developed for the local people as a result of the AgrESI projects. Finally, the implementation of AgrESI projects will gradually shape the national trends for sustainable development which in their turn will influence the relative European and international trends.

As explained, the successful initiation of an AgrESI project needs to ensure that stakeholders' expectations are managed, appropriate support, both financial and in kind, is available, and people involved acquiring the necessary entrepreneurial skills and utilize the experiences from similar international initiatives. The significance and importance of these parameters for an AgrESI project are explained below.

Stakeholder Expectations

The adoption of CSR by enterprises is an ongoing process of learning and change. The key issue is to realize their role in the modern socioeconomic environment and to mobilize in adopting CSR initiatives which will assist them in taking a preventative approach to new developments and the new market requirements.

Figure 1. Agricultural Entrepreneurship and Social Innovation Framework (AgrESIF)

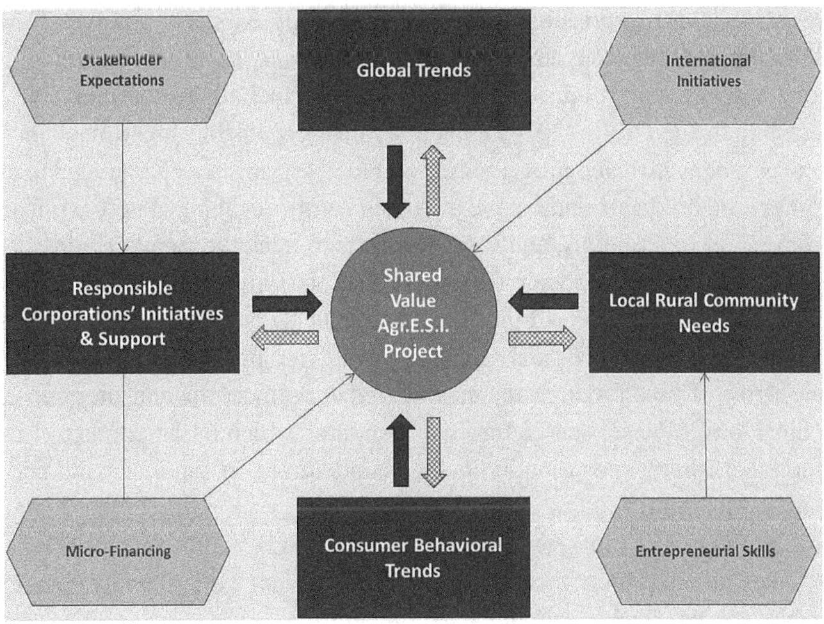

In order to do business sustainably, companies musthave good knowledge and strengthen their reflexes towards all the actors around their sphere of activity. Identifying their stakeholdersis the first step, followed by a second which is to prioritize them, so as to identify therelevancy of their modes of interaction and the relative risks. This practical approach is a core part of CSRpolicies.

Stakeholders whether they act as individuals or groups of individuals they have an impact on the one hand or they are affected on the other by the activities,products or services of a company.Stakeholders cover a wide variety of actors, such as employees and their representatives; customers and consumers; public authorities; national and local communities; Governments, professional organizations, public and international organizations; civil society and NGOs; suppliers and the wider supply chain; media and press; investors and rating agencies; the wider financial community and others.

According to CSR Europe's (2008) Toolbox for Proactive Stakeholder engagement, community involvement and development, promote communication and positive relations between a company and local stakeholders. It facilitates acceptance of corporate activities by the local population. It creates the base and opportunities for synergieswith local decision-makers, who become more positive to support the company in case of difficulty, thus facilitating the long-term sustainability of its activities within the country. In certain delicate contexts, it also tends to reduce incidents and tension, thus preventing provocation and security risks.

A main stakeholder group of a food and agriculture corporation are the small farm holders that act at the same time both as suppliers offering their products, but also at the same time as members of the local community that offers to the corporation the "social license to operate".

When farmers act as a group of individuals and part of the local rural community face sometimes problem to communicate or express their wishes exposingdemands as a shopping list where they ask everything and anything. Businesses have a duty therefore not only to listen, but several times to train local communities in the dialogue process. Also, small farm holders and family farmers, both women and men are the main investors in their own rural community and play a vital role in the sustainability of the community itself but also in the sustainability ofthe food value chain and systems.

Moreover, organizationsrepresenting farmers and their families should strengthen the capacity of those they represent to invest responsibly through provision of assistance in order for farmers to gain improved access to inputs, extension, advisory, specialized financial services, specific education, dedicated training, and access to final consumersthrough retailers including supermarkets.

Equally, according to the CFS's (2014) Principles for Responsible Investment in Agriculture and Food Systems, corporations involved in agriculture and food systems are encouraged to inform and communicate with other stakeholders, conduct due diligence before engaging in new arrangements, conduct equitable and transparent transactions, and support efforts to track the supply chain.

Furthermore, according to the above mentioned Principles processors, retailers, distributors, input suppliers, and marketers are encouraged to inform and educate consumers about the sustainability of products and services and respect national safety and consumer protection regulations. Enterprises involved in the marketing of food products are encouraged to promote the consumption of food which is balanced, safe, nutritious, diverse, and culturally acceptable, which in the context of this document is understood as food that corresponds to individual and collective consumer demand and preferences, in line with national and international law, as applicable.

Last but not least the role of workers in agriculture and food systems is vital.Workers and their organizations play a key role in promoting and implementing decent work, thereby contributing to efforts towards sustainable and inclusive economic development.

Finally, consumer organizations can contribute to the implementation of a new sustainable oriented agreement in the agriculture and food systems by informing and educating consumers about products, farmers, companies and other players about the followed methods with them towards a safer food system that will be able to feed also future generations with dignity.

Entrepreneurial Skills

As mentioned above, the successful implementation of an AgrESI project requires the involved people in rural communities to acquire the appropriate entrepreneurial skills in order to ultimately manage the new agribusiness. People in many rural areas, especially young people, are not only lacking farming and agricultural knowledge and experience, but also lack entrepreneurial and business skills. Potential agribusiness entrepreneurs need to develop a set of skills that will enable them to identify a problem, develop a solution and provide evidence of value for that solution. Therefore, they need to possess appropriate entrepreneurial skills including observation and experimentation skills, creativity and critical thinking skills, communication skills, integration and problem solving skills. In addition, they need to demonstrate commitment against the initiative itself and the other stakeholders involved in the AgrESI project initiative.

According to the Entrepreneurial Learning Initiative (2016)an entrepreneurial mindset can empower ordinary people to accomplish extraordinary things. It can empower people from all walks of life, from every background, culture and discipline. And the implications of entrepreneurial mindset, education reach far beyond enterprise creation.

Entrepreneurship is a mindset that can empower ordinary people to accomplish the extraordinary. Entrepreneurial success does not require revolutionary new ideas (Taulbert& Schoeniger, 2010) but the appropriate skills. After all, entrepreneurs come in all shapes and sizes and their impact extends far beyond the creation of new ventures. Policy makers from the White House to the World Economic Forum have begun to recognize the power of entrepreneurial thinking in all aspects of our increasingly interconnected, globalized societies, including the public, private, academic, and non-profit sectors.

So the challenge for any AGrESI project is to ensure that people to be involved in the new agribusiness initiative should acquire this entrepreneurial mindset. This can be done in cooperation with universities and business incubators that possess the knowledge and the resources to transfer it to the people in the rural areas.

Micro-Financing or In-Kind Support

The majority of the world's population is poor, subsisting on $2-3 per day. Although, according to the UNDP (2009), over 500 million of the world's poor are economically active earning their livelihoods by being self-employed or by working in farms and microenterprises. Poor people constitute the vast majority of the population in most developing countries. Moreover, an overwhelming number of the poor around the world continue to lack access to basic financial services. In the contrary, developed countries despite the global economic crises that increase dramatically financial inequalities and the amount of poor, continues to seemicrofinance as a marginal financial product with high risk. In order to achieve its full potential of reaching a large number of the poor, microfinance should become an integral part of the financial sector. According to the Key Principles of Microfinancing launched by CGAP (the Consultative

Group to Assist the Poor) (2004) financial sustainability is necessary to reach significant numbers of poor people, while sustainablemicrofinancing is about building permanent local financial institutions.

A study of IFAD (2010) describes ways that lack of formal credit affects rural poverty and encourages indigenous financial arrangements to take the place of missing formal finance. Limited access to credit is linked to low technology, agricultural production systems and to people working as an unpaid family labor rather than in self- or wage employment (Meyer, 2011).

Microfinance could be seen as a sustainable mean of poverty alleviation leading to lasting, sustainable development.This development should also be a holistic one covering all aspects of livelihood of small entrepreneurs including small farm holders. From the economic aspect a small amount of financial support in a form of loan helps reduce desperation and anxiety while at the same time allows farmers to make more effective investment choices. Having a limited, butstable income, they can then begin savings with cultivating costs and looking forward to expand their business and eventually becoming more financially secure. From the social aspect microfinancing could help the entire local rural community if the support is given as a program that brings farmers together in groups, enabling them to support each other. Cooperation and trust within the groups strengthens over time and can lead to improved decision making in the wider community. At the same time, for each farmer as an individualto receive a small financial support through a small loan is not just an economic or financial transaction. It is a significant personal moment where each farmer is trusted again and feel valued for his/her family and for the wider community. Also, microfinancing can produce a multiplier effect as in several cases, receivers of small loans become more eager to provide small loans to others especially within their local community.

Given the limited availability of agricultural credit, marketing arrangements that integrate financial services help smallholder farmers participate in markets. Supermarkets, input supply companies and traders often offer inputs on credit as a way to develop preferred supplier relationships with small farmers (Meyer, 2011).Other innovative approaches for supporting individual farmers shouldnot be limited only to financial microloans but also extend to other kinds of support taking into account and assessingeach farmer's needs, character, standard of living, educational level, financial status, repayment capacity, and personal values and vision.

In more detail, individual support could be provided to farmers by other players in the wider value chain by linking them to suppliers with high-quality seeds and fertilizers and negotiating prices on their behalf. Another kind of support could be the provision of formal and informal education in good farming practicesand crop-specific training,provided by experienced and expert agriculturists and relative scientists. Also, providing smallholder farmers with necessary funds to pay the required tuition fees for their children in order to acquire the necessary agricultural education and return back home, could also be seen as an overall investment with high level of long-term return.

Additionally, a way of support could be to link farmers with buyers,assisting them through packaging, labeling and traceabilitytechnology in orderto obtain better market prices. Technology is also a vital success factor. Supportingsmallholder farmers with technological equipment may result to increased efficiency and cost reduction through field-based data collection and proactive management of meteorological changes. Furthermore, empowering women to participate in the farm work through support and on the job training to strengthen their ability to participate in income generation and to better protect their families against loss from unforeseen events as small harvests or bad weather conditions.

Finally, assistance either financial through micro loans either in kind through support can be the solution for smallholder farmers in order to improve their lives of themselves and their families, under the provision that farmers are not bound by unilateral contracts.

International Initiatives

Small scale farmers need support to ensure they are not perpetuating the kinds of practices the SDGs seek to eliminate: as child labor, exploitation of girls and women, or use of forced labour, issues covered by the 10 Principles of the UN Global Compact that is followed in a voluntary basis by thousands of companies lots of them involved with the agriculture sector.

UN Global Compact is the world's largest voluntary corporate sustainability initiative. It was launched in 2000, as a call to companies around the world to align their strategies and operations with universal principles on human rights, labor, environment and anti-corruption, and take actions that advance societal goals. On May 13th, 2014 the UN Global Compact launched the Food and Agriculture Business (FAB) Principles (2014), following a two-year consultation process with over 1,000 businesses, UN agencies and civil society organizations engaged in agriculture, food and nutrition systems. The six FAB Principles aim to help realize sustainable development and empower businesses to contribute to the post-2015 development agenda. The principles were introduced in Rome, Italy, on 13 May 2014.

The Principles respond to calls from the UN Conference on Sustainable Development (UNCSD, or Rio+20) for sustainable development of food production through increasing local investments in a responsible way, reducing waste and the use of water in supply chains and ensuring access to local and global markets. The FAB Principles are the first set of global voluntary business principles ofthe food and agriculture sector, and are designed to serve as umbrella principles that complement existing initiatives on agriculture and food sustainability.

UN Global Compact (2014) business participants in the food and agriculture sector are invited to take an additional, voluntary step to embrace a set of Food and Agriculture Business Principles (see Table 1) and report annually on their progress.

The Principles are aimed at companies in food and agriculture sector wanting to act responsibly and could be followed by all farmers and agribusinesses – regardless of size, crop or location – as a principle-based commitment and as a framework to show their overall orientation toward corporate sustainability.

Table 1. Food and agriculture business principles

Principle	Requirement
Aim for Food Security, Health and Nutrition	Businesses should support food and agriculture systems that optimize production and minimize waste, to provide nutrition and promote health for all people.
Be Environmentally Responsible	Businesses should support sustainable intensification of food systems to meet global needs by managing agriculture, livestock, fisheries and forestry responsibly. They should protect and enhance the environment.
Ensure Economic Viability and Share Value	Businesses should create, deliver and share value across the entire food and agriculture chain from farmers to consumers.
Respect Human Rights, Create Decent Work and Help Communities to Thrive	Businesses should respect the rights of farmers, workers and consumers. They should improve livelihoods, promote and provide equal opportunities.
Encourage Good Governance and Accountability	Businesses should behave legally and responsibly by respecting land and natural resource rights, avoiding corruption, being transparent about activities and recognizing their impacts.
Promote Access and Transfer of Knowledge, Skills and Technology	Businesses should promote access to information, knowledge and skills for more sustainable food and agricultural systems.

Also in 2011, the OECD Declaration on International Investment and Multinational Enterprises, known as OECD Guidelines, that provide an open and transparent international investment environment to encourage the positive contribution of multinational enterprises (MNEs) towards economic and social progress, were revised.The OECD Guidelines are the most comprehensive set of government-backed recommendations on what constitutes responsible business conduct (RBC). They cover nine major areas of RBC: information disclosure, human rights, employment and industrial relations, environment, bribery and corruption, consumer interests, science and technology, competition, and taxation. (OECD, 2015). Large corporations in the area of food and agriculture should take into account these principles when cooperate in the global environment, especially through their supply chains in rural areas of developing countries.

In addition, the Committee on World Food Security (CFS) endorsed on 15 October 2014 at its 41st session the Principles for Responsible Investment in Agriculture and Food Systems, known as CFS-RAI Principles (2014). The principles contain ten core principles related to: food security and nutrition; sustainable and inclusive economic development and poverty eradication; gender equality and women's empowerment; youth; tenure of land, fisheries, and forests and access to water; sustainable management of natural resources; cultural heritage, traditional knowledge, diversity and innovation; safe and healthy agriculture; inclusive and transparent governance structures, processes, and grievance mechanisms; impacts and accountability. An additional section describes the roles and responsibilities of stakeholders.

The objective of the Principles is to promote responsible investment in agriculture and food systems that contribute to food security and nutrition, thus supporting the progressive realization of the right to adequate food in the context of national food security (CFS, 2014).

All the above mentioned volunteer frameworks and principles are addressed to corporations that accept the concept and principles of CSR and recognize the vital role of sustainable development. CSR has a twofold importance. It relates both to their stand-alone / autonomous operation and contribution to the social and natural environment in which they operate, and to their relationship with larger businesses whose suppliers, subcontractors or associates they are. As a result of this twofold importance is that in most cases businesses as part of wider value chains are invited by major clients to operate under specific frameworks and to adopt CSR initiatives to ensure their existing cooperation.

The main aim of businesses operate in the food and agriculture sector is being able to evaluate their priorities and risks and incorporate CSR principles into their own policies, strategies and everyday way of work and thus capitalize on the advantages this approach could bring both in terms of improved competitiveness and maintain sustainable in the future.

THE McCAIN CASE

The Company

Founded by farmers, McCain became a global business, operating under three core values: authenticity, commitment and trust that guide its behavior and business interactions. McCain builds close, long-term relationships with local farmers – some of them are suppliers for three generations – in several markets that operates. McCain focus on Good Food, Good People and Good Business, setting standards for all its operations. Company's involvement is mostly driven, locally and managed by the individual regions, so the support can take many forms such as transfer of knowledge and expertise to growers, donation of products and services, sponsorships that inspire broad participation towards worthwhile goals and, sometimes, direct financial support.

The company supports social business and in 2013 McCain signed an alliance with Nobel Peace Prize Laureate, Muhammad Yunus, to kick-start the 'Campo Vivo' project. This was the first South American joint venture between a multinational company and the Yunus Social Business and the project's aim is to improve the quality of life of Colombian farmers. Campo Vivo benefits underprivileged farming families in Colombia who face many challenges. Some lack the necessary land or resources to produce crops and many farmers have been displaced by the armed conflict in the country. By providing education and support services via the Campo Vivo Cultivation Center of Excellence, the project helps them to successfully develop commercially viable potato, carrot and pea farming businesses in a fair and sustainable way (McCain, 2016).

In addition to local community support, the company funds The McCain Foundation – a philanthropic organization, largely supporting initiatives in the Atlantic provinces of Canada where McCain Foods was founded. The McCain Foundation has five areas of support; arts & culture, community projects, health & wellness, education and environment.

"Karpos Frontidas": An Initiative for Social Product Creation

In 2013, McCain decided to enrich its Corporate Social Responsibility strategy with new innovative approaches and initiated a consultation process with key stakeholders in three European countries (Germany, Greece and Poland) in order to choose the country from which it would start its efforts based on the conclusions of the open dialogue with its stakeholders.

The result of the stakeholder dialogue conducted in the three countries led to the launch of an ambitious intervention in Greece in order to assist local rural farmers to overcome the difficult economic situation in the country, due to the global financial crisis, and in an effort to help them remain at their places, and continue to support their families, local society and country at large.

As a result of that stakeholder engagement process (see: AgrESI - Stakeholder expectations), in 2015, McCain in collaboration with CSR HELLAS –a Greek non-for-profit business driven network for the development of business responsibility and sustainability – created a unique product in the category of fresh vegetables, and especially potatoes. Through this initiative, McCain aims to support Greek small farmers and unemployed young people who returned back to their birthplaces as a result of the high-level of unemployment rates in the cities, in order to get involved with the cultivation of fresh potatoes. "Karpos Frontidas" project is the first pilot step essential for the multiplication of the approach in other products, regions and countries. It also acts as a consumers' awareness approach in order to buy responsibly, yet sustainably.

The project initiates a new relationship between the company and poor farmers living under the poverty line. McCain safeguards the income of farmers through contractual agreements while at the same time is trying to equip them with entrepreneurial skills (see: AgrESI -Entrepreneurial skills)in order to develop the appropriate entrepreneurial mindset.It also provides training, technical means and farming know-how(see: AgrESI - Micro-financing or in-kind support) so as small farmers cultivate potatoes of high quality, as well as gain access to the market under the label of a new social product that has been grown according to the principles of sustainable agriculture (see: AgrESI -International initiatives).

Also, taking into consideration that the new product is grown with special care for the environment and the consumer needs and demands (see: AgrESI -Consumers' behavioral trends), and that it secures a better income for the farmers, this product aims to become the first Social Product in the market of fresh vegetables originated by Greek small poor and underprivileged farmers.

Several other businesses operating within the context of CSR and in order to safeguard their sustainability (see: AgrESI -Responsible corporations' initiatives and support),were invited to cooperate with McCain for the successful implementation of the project. Primarily, as mentioned before McCain (2016b) has the project coordination, while CSR HELLAS acts as consultant on socioeconomic issues. Program integration obtains partially the support of a global ethical investment fund, well known for its support of Greek people and organizations of wider civil society during last years' financial crisis. The provided support from the foundation aimed at strengthening organizations that exhibit strong leadership and sound project management and that can bring considerable, enduring and positive social influences and changes at large.

Furthermore, one of the largest financial institutions in Greece provides financial support (see: AgrESI -Support through Micro-financing or in Kind) to farmers through a new product (credit card) specially designed for small farm holders. With this credit card farmers have the ability to buy their products such as crops, fertilizers, etc. necessary for their farming while McCain safeguards for those micro credits to the bank.

Moreover, McCain through its long-term cooperation with academic institutions offering agriculture education provides training and expertise to farmers, developing special cultivating protocols dedicated to the specific land and farmer's needs. The cooperation with the academic community is continuous through elaboration of research with national or regional scope to determine the investment criteria by region, such as farmer's income, social factors, unemployment rates and analysis of soil ingredients. Also, McCain's business partners offer to small farmers persistence audits of active substances and special discounted rates to the supply of their products, while agronomists and consultants oversee and support farmers (see: Local rural community needs).Furthermore, McCain invests towards renovation and modernization of existing infrastructure for the selection process and packaging, in collaboration with farmers themselves. Last but not least one of the most important services McCain offer to farmers is access to market through its traditional channels of distribution, in order the final product reach consumers. The final product which is a bag of 3 kg of potatoes provides traceability data available to consumers through the use of new technology applications.

In conclusion, the program aimed to tackle poverty and social exclusion through partnership and constructive engagement between businesses and smallholder farmers in disadvantaged rural communities. The programme was designed to meet the four goals below.

Goal 1: Promote awareness, knowledge and uptake of the notion of "Sustainable agriculture", "Farming Sustainability" and "Social Product".

Goal 2: Increase access to formal and informal educational, recreational and business cultural development activities and resources.

Goal 3: Increase smallholder farmers' work readiness, employment prospects and income increase.

Goal 4: Promote active engagement with policy, practice and decision making processes on matters affecting agricultural but also commercial issues.

The main aim of the proposed approach is the gradual withdrawal of McCain as the main supporter of the initiative and the emergence of a self-managed model in which farmers/producers will themselves be involved in the process as the main stakeholders and shareholders of a new agribusiness.

Finally, McCain's initiative is directly aligned with the Sustainable Development Goals (see: Global Trends) and more specifically with Goal#2 that among others "promotes sustainable agriculture"; Goal#8 that "promote sustained, inclusive and sustainable economic growth, full and productive employment and decent work for all"; Goal#12 that "ensures sustainable consumption and production patterns" and Goal#17 that "strengthen the means of implementation and revitalize the global partnership for sustainable development".

FUTURE RESEARCH DIRECTIONS

Although the above case study shows that such projects can be successfully designed and implemented, further work is required in order to analyze the various parameters from the different stakeholders' perspective. In addition, the possibility to design more similar initiatives can be explored by investigating the intentions of corporations that are actively involved in CSR activities and try to identify if these organizations are more likely to support such projects. Socially responsible companies can be identified through CSR Networks operating in many European countries and a quantitative / qualitative survey may be conducted in order to investigate the interest as well as the ability of such organizations to adopt or join respective social innovation programs, which can lead to clear benefits for both the companies themselves and the society at large. At the same time the survey can capture the groups of potential beneficiaries and stakeholders based on the sector of activity of each company concerned.

CONCLUSION

This chapter presented a conceptual framework to enable the development and implementation of Agricultural Entrepreneurship and Social Innovation projects or even enterprises. This new business model which contains various social aspects, can contribute to the economic growth and sustainability and hence combat the phenomenon of unemployment and poverty in rural areas that have been seriously affected by the recent financial crisis in many European countries. An actual case study of such a project was presented describing how a private initiative involving different stakeholders can produce mutual benefits and create social value, and therefore justifying the proposed framework.

Several factors were explained in this chapter, some of which can be seen as innovative approaches to traditional practices, while the development of a clear framework of collaboration among various stakeholders may result in overall improvements in the farming sector. Briefly summarized, farmer's continuous training on modern farming techniques and available technologies should be integrated in their every-day work in order to improve received knowledge and further enrich their skills, keeping focus as well on the development of entrepreneurial perspective. At the same time, communication becomes a high priority for farmers in order to maintain exposure of their products. So, the development of a common communication strategy among farmers from a specific area that cultivate the same product could be seen as an innovative approach for reaching final consumers through appropriate messages that underline the unique characteristics of the product, inform about the adoption of internationally accepted Principles and farming guidelines, and increase overall transparency and traceability.

Moreover, reinforcement of agriculture sector through appropriate financial support could contribute to better promotion and acquaintance of the consumer with the end product, its background and the goals it represents, such as solidarity, fighting youth unemployment, etc. The adoption of regular and systematic communication between all identified stakeholders through effective channels, so that everyone is informed about recent developments, current difficulties and possible opportunities, can be perceived as a prerequisite for the success of the process and as an enhancement of transparency among partners. Social media is a new era for the farming sector that require enrichment regarding farming processes and new technologies that can keep stakeholders, mostly consumers, inform about all dimensions and innovations covered by the agribusiness program, including soil analyses, weather forecasts, safe transportation and storage, etc. In other words, there is a necessity for raising awareness among the consumers regarding branded agricultural products from small farmers, that respect at all stages of the value chain the principles of responsible entrepreneurship and contribute justifiably and transparently to the selling price. Finally, through the appropriate stakeholder engagement and by linking the product to its origin, history, and way of grow, farmers could create the story of the product that accompanies it from farm to store!

REFERENCES

AAFC. (2013). *Agriculture and Agri-Food Canada, Global Analysis Division, Ethical Food Trends in Japan*. Retrieved August 17, 2016, fromhttp://www5.agr.gc.ca/resources/prod/Internet-Internet/MISB-DGSIM/ATS-SEA/PDF/6473-eng.pdf

Antonaras, A., Iacovidou, M., & Memtsa, C. (2011). Measuring social return on investment using the EBEN GR Business Ethics Excellence Model. *Current Issues of Business and Law*, *6*(1), 69–89. doi:10.5200/1822-9530.2011.04

Antonaras, A., & Memtsa, C. (2009). *Business Ethics and Sustainable Excellence*. Presented and published in the proceedings of the 3rd Annual Quality Congress of the Middle East, organised by e-University in partnership with MEQA, Berkley University of California and the International Academy for Quality, Dubai, UAE.

ATTRA. (2003). *Applying the Principles of Sustainable Farming - Fundamentals of Sustainable Agriculture*. Retrieved August 22, 2016, fromhttp://ipm.ifas.ufl.edu/pdfs/Applying_the_Principles_of_Sustainable_Farming.pdf?pub=295%5D

CFS. (2014). *Principles for Responsible Investment in Agriculture and Food Systems, World Food Security (CFS).* Retrieved August 19, 2016, fromhttp://www.fao.org/fileadmin/templates/cfs/Docs1314/rai/CFS_Principles_Oct_2014_EN.pdf

CGAP. (2004). *Key Principles ofMicrofinance - Building financial systems for the poor*. Retrieved August 19, 2016, fromhttps://www.cgap.org/sites/default/files/CGAP-Consensus-Guidelines-Key-Principles-of-Microfinance-Jan-2004.pdf

CSR Europe. (2008). *ProactiveStakeholderEngagement, Practical Guide for companies and stakeholders, CSR Europe Toolbox*. Retrieved August 19, 2016, from http://www.csreurope.org/csr-europes-toolbox-user-manual#.V8VoVyiLSUm

Entrepreneurial Learning Initiative. (2016). *Entrepreneurial Mindset Education and Training Programs*. Retrieved September 1, 2016 from https://elimindset.com/about/

EurekAlert. (2015). *The Global Source for Science News, World crop diversity survives in small farms from peri-urban to remote rural locations, Article by Penn State*. Retrieved August 26, 2016, from http://www.eurekalert.org/pub_releases/2015-02/ps-wcd021115.php

Eurostat. (2015). *Eurostat News Release 81/2015*. Retrieved May 7, 2015, from http://ec.europa.eu/eurostat/documents/2995521/6823708/3-07052015-AP-EN.pdf/7e507ea0-43c7-452f-8e6a-b479c89d2bd6

Fairtrade. (2014). *FairtradeInternational - Sustainable Development Goals and Fairtrade: The case for partnership*. Retrieved August 17, 2016, from http://www.fairtrade.net/fileadmin/user_upload/content/2009/resources/15-10_Sustainable_Development_Report.pdf

FAO. (2009). *How to Feed the World by 2050, High-Level Expert Forum*. Retrieved August 27, 2016, from http://www.fao.org/fileadmin/templates/wsfs/docs/Issues_papers/HLEF2050_Global_Agriculture.pdf

FAO. (2010). *World deforestation decreases, but remains alarming in many countries*. Retrieved August 27, 2016, from http://www.fao.org/news/story/en/item/40893/icode/

FAO. (2011). *The State of Food Insecurity in the World, How does international price volatility affect domestic economies and food security?* Retrieved August 27, 2016, from http://www.fao.org/docrep/014/i2330e/i2330e.pdf

FAO. (2013). *Post-harvest food losses estimation - Development of consistent methodology*. Retrieved August 27, 2016, from http://www.fao.org/fileadmin/templates/ess/documents/meetings_and_workshops/GS_SAC_2013/Improving_methods_for_estimating_post_harvest_losses/Final_PHLs_Estimation_6-13-13.pdf

FAO. (2014a). *Family Farmers: Feeding the world, caring for the earth*. Retrieved August 27, 2016, from http://www.fao.org/docrep/019/mj760e/mj760e.pdf

FAO. (2014b). *CFS-Committee on World Food Security, CFS Chair's Statement for the FAO Regional Conferences 2014*. Retrieved August 27, 2016, from http://www.fao.org/fileadmin/templates/cfs/Docs1314/RC/CFS_Chairs_RC_Statement_EN.pdf

FAO. (2014c). *DEEP ROOTS*. Retrieved August 27, 2016, from http://www.fao.org/3/a-i3976e.pdf

FAO. (2014d). *State of Food and Agriculture in the Africa Region and Comprehensive Africa Agriculture Development Programme Implementation with a Specific Focus on Smallholder Farmers and Family Farming*. Africa Regional Conference, March 2014. ARC/14/3. Retrieved August 27, 2016, from http://www.fao.org/docrep/meeting/030/mj566e.pdf

FAO. (2014e). *Food Losses and Waste in Europe and Central Asia*. Document ERC/14/3. Retrieved August 17, 2016, fromhttp://www.fao.org/docrep/meeting/030/mj621e.pdf)

FAO. (2014f). *Building a common vision for sustainable food and agriculture - Principles and Approaches*. Retrieved August 22, 2016, fromhttp://www.fao.org/3/a-i3940e.pdf

Farming First. (2015). *The story of Agriculture and the Sustainable Development Goals*. Retrieved August 29, 2016, from http://www.farmingfirst.org/sdg-toolkit#home

Horne, , Fien, , Beza, , & Nelson., (Eds.). (2016). *Sustainability, citizenship and cities: theory and practice*. New York: Routledge.

ICSU. (2016). *International Council for Science, A draft framework for understanding SDG interactions by Måns Nilsson, Dave Griggs, Martin Visbeck and Claudia Ringler*. Retrieved August 29, 2016, from http://www.icsu.org/publications/reports-and-reviews/working-paper-framework-for-understanding-sdg-interactions-2016/SDG-interactions-working-paper.pdf

IFAD. (2010). *International Fund for Agricultural Development - Rural Poverty Report 2011: New realities, new challenges: new opportunities for tomorrow's generation, IFAD, Rome*. Retrieved August 20, 2016, from https://www.ifad.org/documents/10180/c47f2607-3fb9-4736-8e6a-a7ccf3dc7c5b

ILO. (2013). *International Labour Organisation. Global Employment Trends 2013*. Retrieved August 29, 2016, from http://www.ilo.org/wcmsp5/groups/public/---dgreports/---dcomm/---publ/documents/publication/wcms_202326.pdf

ILO. (2016). *World Employment Social Outlook - Trends 2016*. Retrieved August 29, 2016, from http://www.ilo.org/wcmsp5/groups/public/---dgreports/---dcomm/---publ/documents/publication/wcms_443480.pdf

McCain. (2016). *KarposFrontidas*. Retrieved September 1, 2016 from http://www.karposfrontidas.gr/

McCain. (2016a). *Responsible Business*. Retrieved September 1, 2016 from http://www.mccain.com/good-business/responsible-business

Meyer, L. R. (2011). *Microcredit and Agriculture: Challenges, Successes, and Prospects*. Retrieved August 20, 2016, from https://www.incofin.com/sites/default/files/attachments/publications/Microfinance%20and%20agriculture.pdf

Millennium Ecosystem Assessment. (2005). *Ecosystems and Human Well-being: Synthesis*. Retrieved August 20, 2016, from http://www.millenniumassessment.org/documents/document.356.aspx.pdf

NRI. (2001). *Natural Resources Institute - University of Greenwich, Ethical Consumers & Ethical Trade: A Review Of Current Literature*. Retrieved August 17, 2016, from http://www.nri.org/projects/publications/policyseries/PolicySeriesNo12.pdf

OECD. (2008). *Promoting Sustainable Consumption, Good Practices In OECD Countries*. Retrieved August 17, 2016, from http://www.oecd.org/greengrowth/40317373.pdf

OECD. (2015). *International Standards Considered In The Oecd-FAO Guidance For Responsible Agricultural Supply Chains*. Retrieved August 19, 2016, from https://mneguidelines.oecd.org/OECD-FAO-Guidance_International-Standards.pdf

Oxfam. (2014). *Even it Up: Time to End Extreme Inequality*. Retrieved August 19, 2016, from https://www.oxfam.org/sites/www.oxfam.org/files/file_attachments/cr-even-it-upextreme-inequality-291014-en.pdf

Reuters, U. S. (2015). *Build better roads in developing world to bolster food supplies: study, Article by Alister Doyle*. Retrieved August 25, 2016, from http://www.reuters.com/article/us-environment-food-idUSKBN0L200520150129

Taulbert, C., & Schoeniger, G. (2010). *Who own the ice house? Eight life lessons from an unlikely entrepreneur*. ELI Press.

UN Global Compact. (2014). *Food and Agriculture Business Principles*. Retrieved August 19, 2016, fromhttps://www.unglobalcompact.org/take-action/action/food

UN Global Compact. (2016). *Food and Agriculture*. Retrieved August 29, 2016, from https://www.unglobalcompact.org/what-is-gc/our-work/environment/food-agriculture

UNDP. (2009). Impact of UNDP microfinance programme on poverty alleviation among farmers in selected local government areas of Kaduna State, Nigeria. *International Journal of Sociology and Anthropology, 1*(6), 99-103. Retrieved August 20, 2016, from http://www.academicjournals.org/article/article1379409974_Kudi%20et%20al.pdf

United Nations. (2015). *Transforming our world: the 2030 Agenda for Sustainable Development*. Retrieved August 29, 2016 from https://sustainabledevelopment.un.org/post2015/transformingourworld

UNSG. (2014). *Agriculture development, food security and nutrition*. Report of the UN Secretary-General, A/69/279, 7 August 2014.

USAID. (2014). *Land tenure in urban environments*. Retrieved August 27, 2016, from http://www.usaidlandtenure.net/sites/default/files/USAID_Land_Tenure_Urban_Brief_061214.pdf

Weitz, N., Nilsson, M., & Davis, M. (2014). A Nexus Approach to the Post-2015 Agenda: Formulating Integrated Water, Energy, and Food SDGs. *SAIS Review of International Affairs*, *34*(2), 37–50.

WFP. (2015). *World Food Programme, P4P Purchase for Progress - The Story: Connecting farmers to markets*. Retrieved August 27, 2016, from https://www.wfp.org/sites/default/files/the%20P4P%20Story%20-%20Connecting%20Farmers%20to%20Markets,%20Feb%202015.pdf

World Bank. (2014). *Employment in agriculture*. Retrieved August 28, 2016, from http://data.worldbank.org/indicator/SL.AGR.EMPL.ZS?page=2159Donor

World Bank. (2016). *Agriculture*. Retrieved August 28, 2016, from http://www.worldbank.org/en/topic/agriculture/overview

WRAP. (2015). *Waste and Resources Action Programme. Strategies to achieve economic and environmental gains by reducing food waste – Final Report*. Retrieved August 25, 2016, from http://static.newclimateeconomy.report/wp-content/uploads/2015/02/WRAP-NCE_Economic-environmental-gains-food-waste.pdf

Section 2

The Future of Agribusiness in the Digital Environment

Chapter 9
Determining the Role of Communication and Distribution Channels for Organic Foods

V. Aslihan Nasir
Bogazici University, Turkey

Suphan Nasir
Istanbul University, Turkey

ABSTRACT

While, the market for organic foods is growing; the proportion of consumers who buy organic foods is still considered low. The role of communication activities is very important for promoting the organic food consumption. In order to create awareness and generate demand for organic foods, companies need to use effective communication tools. Companies in the agribusiness sector try to take advantage of the information and communication technologies in the digital era with the purpose of communicating the value of their offer to consumers. Companies need to know which information sources (channels) are most influential in purchase decision while communicating with consumers. Thus, the aim of this study is to examine consumers' credibility perceptions of communication channels that are used to promote organic food. The great majority of the respondentsin this study mentioned that they had never seen organic food ads. However, a significant number of consumers who had seen organic food ads declared Internet as the medium they had exposed to organic food ads. Nevertheless, our study revealed that the respondents did not perceive Internet as a credible source of information about organic foods. Yet, online social networks were perceived as more reliable source of information about organic foods when compared to majority of traditional media such as radio and newspaper ads. The distribution channels that consumers prefer to purchase organic foods was also investigated; and it is found that a significant portion of the consumers choose supermarkets and neighborhood bazaars for their organic food shopping whereas Internet/online shops and pharmacy stores were shown as the least preferred shopping alternatives. Finally, a substantial majority of the consumers mentioned high prices and availability as the main barriers against buying organic foods.

DOI: 10.4018/978-1-5225-2107-5.ch009

INTRODUCTION

The "organic foods movement" began as a reaction to conventional agricultural methods, whichare heavily dependent on chemical pesticides, fertilizers, antibiotics, and growth hormones that are assumed to be harmful for human health. Organic agriculture is practiced in 172 countries, and 43.7 million hectares of agricultural land are managed organically by approximately 2.3 million farmers (FiBL-IFOAM survey, 2016).Organic food market continues to grow.American consumers' demand for organic was grown by double-digits nearly every year since the 1990s; and most impressively, organic sales was increased from \$3.6 billion in 1997 to \$43.3 billion in 2015 (OTA, 2016). According to the Organic Trade Association's (OTA) 2015 Organic Industry Survey,organic food sector in the United Stateswas grown at an average of 11% from 2014 to 2015 (OTA, 2015). Moreover, U.S. Families' Organic Attitudes & Beliefs 2015 Tracking Study indicated that 51% of families were buying more organic products than a year ago. There is a growing consumer interest for organic products; however consumers desire for transparency in the supply chain and to know where food comes from. Thus, the industry is coming together in collaborative ways to build a secure supply chain that can support demand. Although the demand for organic foods is driven by personal health and environmental reasons, organic food sales represented almost 5% of total U.S. food sales (OTA, 2015).In Denmark the share of organic food represented the 7.6% of total food market. The leader Denmark is followed by Switzerland (7.1%), Austria (6.5%), U.S (5%), and Germany (4.4%)(FIBL-IFOAM Survey, 2016).

Based on the 17th edition of The World of Organic Agriculture, published by the Research Institute of Organic Agriculture (FIBL) and IFOAM, the global market for organic food in 2014 have reached 80 billion US Dollars (more than 60 billion Euros). The United States was the leading market with 27.1 billion Euros, followed by Germany (7.9 billion Euros), France (4.8 billion Euros), and China (3.7 billion Euros).

Organic packaged food and beverage sales generated in the United States in 2014 through the mass merchandise retail channel amounted to about 15.9 billion U.S. dollars. On the other hand, in that year, organic packaged food and beverage sales generated through natural health farm retail channel and internet retailing channel was 14.8 and 1.2 billion U.S. dollars, respectively (Statista, 2016a).

In the light of the above statistical information, it is clear that organic food sector has great potential for further expansion. Therefore, companies in agribusiness need to persuade consumers to buy organic food through using various communication channels. While designing their communication strategies, companies also need to learn why consumers do not buy organic food. Companies in the agribusiness sector try to take advantage of the information and communication technologies in the digital era with the purpose of communicating the value of their offer to consumers. While communicating the value of the offer, companies spend a lot of money for marketing communication budget. However it is important to note that, sources of information and credibility of information source influence the purchase decision of consumers. Thus, companies need to know that which information sources (channels) are most influential in purchase decision while communicating with consumers. Another aim of this study is to examine consumers' credibility perceptions of communication channels that are used to promote organic food.This chapter tries to find out:

- Whether consumers have ever seen an organic food advertisement.
- In which communication channel(s) did consumers see the organic food advertisement.
- The credibility of the communication channels for receiving information about organic food.

- The role of online and conventional communication channels.
- From which distribution channel (online vs offline) do consumers prefer to buy organic food.
- Why consumers don't buy organic food.

The findings of this study enable organic agribusiness companies to determine information sources (channels) that are most influential in purchase decision and design effective communication strategies to increase demand for organic food. Furthermore, it seems extremely important to learn about the most preferred distribution channels by the actual organic food consumers. Likewise, this research also investigates why consumers are reluctant to purchase organic foods.

LITERATURE REVIEW ON ORGANIC FOOD

Societal recognition of organic food has led to a rise in organic food studies. Existing studies about organic food largely focused on *motivations in the purchase of organic food* (Lockie et al., 2002; Makatouni, 2002; McEachern&Mcclean, 2002; Zanoli&Naspetti, 2002; Baker et al., 2004; Lockie et al., 2004; Honkanen, Verplanken, & Olsen, 2006; de Magistris&Gracia, 2008; Seyfang, 2008; HamzaouiEssoussi&Zahaf, 2009; Chakrabarti, 2010; Aertsens et al., 2011; Ergönül&Ergönül, 2015;Teng& Wang, 2015). Moreover, aconsiderable amount of study in the literature hadexamined *perceptions and attitudes of consumers* toward organic food (Thompson, 1998; Magnusson et al., 2001; Burton et al., 2001; Harper &Makatouni, 2002; Lea & Worsley, 2005; Padel& Foster, 2005; Tarkiainen& Sundqvist, 2005; Chen, 2007; Michaelidou& Hassan, 2008; Shafie& Rennie, 2012; Paul & Rana; 2012;Sangkumchaliang& Huang, 2012; Zagata, 2012) Gracia& de Magistris, 2013; Lee & Yun, 2015; Xi et al., 2015).

Organic food literature had also investigated the *profile of organic foodbuyers* in order to find out who buy organic food (Davies, Titterington, & Cochrane, 1995; Fotopoulos, &Krystallis, 2002;Krystallis, Fotopoulos, &Zotos, 2006; Hughner, McDonagh, &Prothero, 2007; Onyango, Hallman, & Bellows, 2007; Zepeda & Li, 2007; Aertsens et al., 2009; Zakowska-Biemans, 2011; Dimitri, &Dettmann, 2012;Hamzaoui-Essoussi&Zahaf, 2012; Paul & Rana, 2012; Kesse-Guyot et al., 2013; Chen, Lobo, &Rajendran, 2014; Lobo, Mascitelli, & Chen, 2014). The role of *price and consumers' willingness to pay for organic food*were among topics that were alsoexplored in the literature (Govindasamy& Italia, 1999; Gil, Gracia, & Sanchez, 2000; Loureiro& Hine, 2002; Krystallis&Chryssohoidis, 2005; Krystallis, Fotopoulos, &Zotos, 2006; Batte et al., 2007; Didier &Lucie, 2008; Ureña, Bernabéu, &Olmeda, 2008; Adams &Salois, 2010; Nie& Zepeda, 2011; Van Doorn&Verhoef, 2011; Hamzaoui-Essoussi&Zahaf, 2012; Janssen & Hamm, 2012; Marian et al. 2014; Rödiger, & Hamm, 2015)

Although there was abundant literature on organic food, there is limited number of studies on promoting organic food as well as organic food distribution channels. The role of communication activities is very important in the marketing of products. In order to create awareness and generate demand for organic foods, companies need to use effective communication tools. It is vital to persuade customers that organic foods worth for the given money.

The study of Bodini, Richter, and Felder (2009) indicated that when purchasing organic fruit, consumers had preferred more straightforward sources of information for exploring product quality related information, such as obtaining information from the producers and sales people at the point of sale (at farm shops, markets or during promotional activities in supermarkets). By contrast, in the context of organic fruit purchase, consumers did not prefer to use the electronic media sources frequently as a source of information on quality issues. Thus, the authors proposed to use producers as multipliers or well-informed sales people as a source of authentic quality communications for organic food. Moreover, they proposed that in order to make organic food more attractive to organic consumers, the communication message strategy should focus more on quality- related issues such as informing consumers about the extra quality value inherent in organic food, and product's key quality attributes (Bodini, Richter, & Felder, 2009).

The market for organic foods is growing; however the proportion of consumers who buy organic foods is still considered low. Lack of information was considered as significant barrier for purchasing more organic foods (Kastberg, 2015). Therefore, some of the studies focused to analyse and develop ideal communication strategies for promoting organic foods. The study of Rousseau and Vranken (2013) contributed to a better understanding of the role information provision in expanding the market for organic food products. They investigated how policy makers can use information provision to develop the demand for organic food. The findings of the study had indicated that the information distributed to consumers through labelling can influence the willingness to pay of consumers for labelled organic food. Therefore, authors suggest that there is a role for policy makers and producers in providing more accurate and reliable information about socially responsible production processes. Since the effect of information provision was more pronounced for certain groups of consumers such as non-vegetarians, infrequent buyers of organic products and members of a nature protection organization, this study underlined the importance of taking the observed preference heterogeneity into account and tailor policies to specific consumer groups (Rousseau &Vranken, 2013). In another study, Loebnitz andAschemann-Witzel (2016) investigated the Chinese consumer reactions to organic food labels and explored whether Chinese consumers' fruit and vegetable quality inferences could be favourably influenced by communication efforts. The results indicated that communicating organic food quality was key to strengthen organic demand in China, and further efforts were needed to strengthen communication of organic food quality. The findings of the study also indicated that priming environmental values in communication increased product expectations for organic-labelled food items only for participants whose environmental concern was central to them. Thus, they suggested that organic market communication efforts should be directed differently to various consumer segments based on the importance they attach to environmental values, and that focus should be given to the target group with strong environmental values (Loebnitz&Aschemann-Witzel, 2016).

Although, decisions on food purchase are expected to be undertaken using limited information search; when confronted with ethical products such as organic foods, consumers often become more involved, and this results in a more extensive search for information. Zander and Hamm (2012) concentrated on the information search behaviour of European consumers with regard to organic food. The findings of the study revealed that majority of consumers use simplifying and selective search strategies when looking for information on organic food with additional ethical attributes. They found out that consumers tend to start information search in the top, left hand corner and move towards the bottom, right hand corner. Thus, they suggested that principal attributes should be placed in the upper left corner and be followed by the next most important and so on when presenting information on an array of different products (Zander & Hamm, 2012).

In the organic food industry, actors engage in gaining and maintaining traceability and communicating it to the consumers. Traceability in the organic food industry was seen as significant for food safety and quality. Lindh and Olsson (2010) examined the objectives of each actor for gaining and maintaining traceability throughout the supply chain. Food safety and quality, managing the supply chain and internal resources, and communication with consumers were identified as the objectives of for gaining and maintaining traceability. The findings illustrated that all actors want to engage in traceability; however they prioritized the objectives differently (Lindh & Olsson, 2010).

Jarossova and Mind'asova (2015) compared the application of marketing communication tools used by organic food producers in Slovakia and Austria. This study emphasized that organic food producers should make greater use of marketing communication tools in their business and develop distribution channels. They proposed to use of loyalty programs and public relations (PR) activities to keep up long-term relationships with customers, and employ targeted promotional activities to educational entities such as schools, and also in business entities and training courses on the use of organic food in the cooking process (Jarossova & Mind'asova, 2015).

In the literature, a few number of study examined the trust orientations in the organic food distribution channels. Companies are increasingly focus on supply chain management in order to ensure high quality, traceability, and supply continuity. Atănăsoaie (2011) explored distribution channels on organic foods markets, and identified advantages and disadvantages of each distribution channels. Findings of this study indicated that distribution has a key role in the development of the organic food. Market development depends on the ability of farmers to choose an optimal distribution channel for their products. In this study small farmers were recommended to keep a closer link with the final customer through using distribution channels directly, without intermediaries. Large farms, which produce crops that require special storage conditions, were recommended to use indirect distribution channels such as supermarkets, organic shops specialized, processors and various intermediaries (Atănăsoaie, 2011).Hamzaoui-Essoussi, Sirieix, and Zahaf (2013) identified and analysed factors related to the supply side that determine trust/mistrust in organic food products, and determined the distribution channel strategies to increase trust in organic food products. Within the context of this study, individual in-depth interviews were conducted in Canada and France with managers from superstores, specialty stores, farmers, markets, producers and certification bodies. The findings of this study highlighted that consumers who buy from shorter channels have specific needs and motivations to buy organic foods such as health and support of local farmers, conversely consumers buying from longer channels are looking for different shopping and consumption experiences. This study indicated that consumers buying from producers/farmers were clearly looking for proximity with the producer, fresh products and quality. They had a better understanding of the organic farming process, and preferred organic food for its impact on health and the environment. On the other hand, consumers using standard channels of distribution were looking for convenience, healthy products and taste. These consumers were seem to get confused between organic and natural products. Trust in organic food is significant in the organic food networks, and trust in organic food has been related to the organic certification labels, product labels, brands, traceability, advice, and store reputation (Hamzaoui-Essoussi, Sirieix, &Zahaf, 2013).

In another study, Liang (2014) examined the profiles of consumers who purchase organic food online and analyzed differences in their behaviours with regard to different food-related lifestyles. Consumers were segmented as traditional food, uninvolved food, and enthusiastic food shoppers based on their food-related lifestyles. Findings revealed that these three segments were statistically different from each other with respect to their online organic food purchasing profiles, and demographic variables (Liang, 2014).Olech and Kuboń (2016) focused to understand preferences of consumers in buying organic foods to facilitate development and functioning of existing distribution channels and formation of new ones. Research results indicated that consumers were interested in the purchase of products in small packaging and findings underlined that the purchase directly from a producer and through the Internet had become more popular form of shopping. Authors indicated that consumers, due to lack of time, appreciate more on-line shopping and the use of the so-called "Package from a Farmer", where products bought for a specific minimum amount were delivered by a producer to a consumer (Olech & Kuboń, 2016).

DATA COLLECTION AND ANALYSIS

A survey was developed and conducted in one of the largest metropolitan areas of a European city. Since this study took place in an emerging economy of Eastern Europe, grocery stores and shopping malls seem to be more accurate places to conduct this research in order to reach potential consumers of organic foods. The consumers, who were shopping in grocery stores or shopping malls, were asked to complete the questionnaire. For a whole week, three trained students were asked to stay at the exit doors of six different supermarkets. The supermarkets were chosen on the basis of visitor number and popularity as well as the representativeness of distinct regions of this cosmopolitan city. Actually, more than 1000 consumers were requested to participate the survey. However, some of the consumers refused to participate whereas some were left off without finishing the survey completely. Hence, at the end a total of 316 surveys were collected, and the data collected from this sample were analyzed by using SPSS. The 52% of the respondents were male, and the great majority of the participants were within the age range of 18-25 years (42%), followed by the age groups of 26-35 (29%), 36-45 (14 per cent) and 46 years and above (15%). It should be noted, however, that 50% of the general population in the country where the survey was conducted are below the age of 30. 40% of the respondents have $1,100 monthly income, 35% have a monthly income of $1,100-2,500 and 25% have a monthly income of $2,500. The majority of the respondents (50%) are considered to belong to the middle income group while a fairly high percentage (40%) are considered to come from a low-income group. Since the study was conducted in an emerging/developing country, these statistics are not unusual. Finally, 37% of the respondents had a university degree, whereas 18% has some high school degree.

In the other parts of the questionnaire, the respondents were asked whether they had ever exposed to advertisements about organic foods. Furthermore, those who mentioned that they'd seen organic food ads were more deeply explored by asking the media type that they had encountered with these ads. In addition, the credibility of the media type has been also examined. For this purpose, the respondents were asked to show their level of agreement through a five point Likert type scale (5=strongly agree,... 1=strongly disagree) for the credibility of each media type. Depending on the literature, a list of barriers were developed as the reasons of not purchasing organic foods. A five point Likert scale (5=strongly agree,... 1=strongly disagree) has been used to measure the main reasons of not purchasing organic foods.

FINDINGS

As it is illustrated in Table 2, approximately 41,8% of the respondents declared that they'd seen organic food advertisement. Later on, we asked a further question to learn in which media channel those consumers had seen organic food ads. Table 3 shows that more than 23,4% of the respondents mentioned Internet as the media that they'd exposed to organic food ads. Point of purchase displays was the second communication channel with almost 20,9%, and it was followed by magazines (18%) and TV (17,1%). Our findings indicate that from the perspective of consumers, billboards (3,2%), radio (1,9%), and direct

Table 1. Demographic profile of the respondents

Age	18-25	26-35	36-45	46-55	≥ 56
	130 (41%)	88 (28%)	46 (15%)	41 (13%)	11 (3%)
Gender	**Female**	**Male**			
	155 (49%)	161 (51%)			
Education	**Some High School Degree**	**University Student**	**University Graduate**	**Postgraduate Degree**	
	57 (18%)	94 (30%)	118 (37%)	47 (15%)	
Income	**≤ 1150 $**	**1151-2550 $**	**2551-4570 $**	**4571-5700 $**	**≥ 5701 $**
	126 (40%)	111 (35%)	46 (15%)	20 (6%)	13 (4%)

mail (1,6%) were the least attention-grabbing media for organic foods. Actually, this finding is in line with the decline or at least stability of certain traditional advertising channels and rise of Internet and digital advertising channels (PwC, 2016). Organic food has a unique nature as a product category, hence it is understandable that consumers who are aware of its existence may be those who search for it at most convenient channels such as Internet. However, it is really surprising that consumers still depend on what their friends and acquaintances recommend about organic foods (i.e. WOM). We've also asked respondents which media type/channel and to what extent they find credible while receiving information about organic foods. Actually, the mean values of a variety of media types demonstrate that the information gathered from friends and acquaintances (i.e. WOM) has the highest credibility among the consumers.

As presented in Table 4, WOM has been followed by magazine ads, online social networks, and TV ads in terms of credibility of the information about organic foods. Questions addressing WOM and e-WOM separately were not included in the questionnaire, hence it is not made clear whether the respondents understand e-WOM included in WOM or not. Furthermore, online social networks are perceived as more reliable source of information about organic foods when compared to majority of traditional media such as radio and newspaper ads, except magazine ads. Nevertheless, TV ads and online social networks are almost perceived as identical in terms of trustworthiness. On the other hand, even though the majority of the respondents (23,4%) stated that they'd seen organic food ads on Internet, they did not recognize Internet ads as a reliable information source for organic foods. Direct mail ads, Internet ads, and radio ads were realized as the least credible information source among the respondents.

We've also investigated which distribution channels consumers prefer to purchase organic foods. For this purpose, a variety of channels were listed in the survey where consumers can reach to purchase organic foods. It is revealed that a significant portion of the consumers choose supermarkets (65,3%) and neighborhood bazaars (36,3%) for their organic food shopping. Unexpectedly, Internet/online shops

Table 2. Have you ever seen an organic food advertisement?

	Yes%	No%
Have you ever seen organic food advertisement	41.8%	58.2%

Table 3. In which of the following media have you seen the organic food advertisement?

Media Channel Type	Organic Food Ad Noticed %
Internet	23.4%
Point of purchase displays	20.9%
Magazines	18%
TV	17.1%
Brochures/Inserts	14.2%
Newspapers	11.7%
Outdoor/Billboards	3.2%
Radio	1.9%
Direct Mail	1.6%

Source: Adapted from Kotler & Armstrong (2014).

Table 4. The credibility of the media type for receiving information about organic foods

	N	Min.	Max.	Mean	Std. Deviation
WOM	282	1.00	5.00	***3.74***	.97678
Magazine ads	281	1.00	5.00	***3.19***	.84595
On-line social networks	278	1.00	5.00	***3.06***	.93158
TV ads	285	1.00	5.00	***3.05***	.87560
Newspaper ads	279	1.00	5.00	3.04	.84244
Brochures/Inserts	279	1.00	4.00	3.01	.97647
Outdoor/Billboard ads	277	1.00	5.00	2.94	.82714
Radio ads	277	1.00	4.00	2.84	.76690
Internet ads	285	1.00	5.00	2.70	.83844
Direct-mail ads	277	1.00	5.00	2.69	.83688

Source: Adapted from Kotler & Armstrong (2014).

(3,2%) and pharmacy stores (1,9%) were least preferred shopping alternatives. Table 5 demonstrates which distribution channels were preferred by the respondents, but we've to note that the respondents were allowed to select more than one distribution channel as their organic food shopping destination.

Finally, we've explored the basic reasons of not purchasing organic foods as well. The survey questions developed by Nasir & Karakaya (2014, p. 304) to measure the reasons of not purchasing organic foods have been used. Table 6 shows the mean values of all the probable reasons of not buying organic foods. A substantial majority of the consumers mentioned high prices as a main barrier against buying organic foods. The second significant barrier against buying organic foods is the availability issue. Consumers participated to our study mentioned that "organic foods are hard to find and/or not available". Furthermore, the third issue stated by the respondents is about lack of variety of organic foods. Therefore, we can conclude that high prices, unavailability, and lack of variety are the basic difficulties for reaching to and purchasing of organic foods.

Table 5. Where do you acquire organic foods?

Distribution Channel Type	%
Supermarkets	65.3%
Neighborhood bazaars	36.3%
Selling points in shopping malls	27.4%
Herbal stores	21.3%
Specialty stores	16.9%
Directly from producer	15.3%
Other	6.7%
Internet/Online shops	3.2%
Pharmacy stores	1.9%

Table 6. The reason of not consuming organic food

	N	Mean	Std. Dev.
They are more expensive	313	**4.40**	.783
They are hard to find and/or not available	315	**3.66**	.971
It is difficult to find the variety of foods that I need/look for	310	**3.00**	1.116
I do not trust that they are really organic	312	2.97	1.039
The package sizes of organic foods do not meet the need of my family	310	2.93	1.082
I'm not the person who does food shopping	310	2.90	1.409
The institutions that give organic food certification are not reliable	311	2.72	.820
They don't have attractive physical appearance	312	2.52	1.156
They don't have delicious taste	311	1.97	.993
I do not know what organic products are	305	1.86	1.003

Source: Adapted from Nasir and Karakaya (2014, p. 304).

CONCLUSION

The findings of this study revealed that 58% of the respondents did not encounter with organic food advertisements. The majority of the consumers, who declared that they had seen organic food ads, mentioned Internet (23,4%) and point of purchase displays (20,9%) as the media where they had exposed to organic food ads. Actually, our finding is supported by the study of Bodini, Richter, and Felder (2009), in which the authors declared that obtaining information from the producers and sales people at the point of sale (at farm shops, markets or during promotional activities in supermarkets) is mostly preferred by the consumers. Yet, magazines (18%) and TV (17,1%) ads were other media channels that were mostly referred by the respondents. In addition, organic food ads that appear at billboards, radio, and direct mail ads were least likely noticed by the respondents. Furthermore, the credibility of the media type for organic foods has been also explored. It is found that WOM, magazine ads, and online social networks were perceived as most reliable sources of information for organic foods. Similarly, Chakrabarti (2010) mentioned that the importance of WOM was high in consumers' organic food purchase process. This study also investigated which distribution channels consumers prefer to purchase organic foods. It is shown that a greatratio of the consumers preferred supermarkets (65,3%) and neighborhood bazaars (36,3%) for their organic food shopping. However, Internet/online shops (3,2%) and pharmacy stores (1,9%) were least preferred shopping alternatives for organic foods.At first glance, our findings seem to be contradicted with those of Olech and Kuboń (2016), who asserted that the purchase directly at a producer and in the Internet has become more popular form of organic food shopping. However, in the same study Olech and Kuboń (2016) stated that consumers provide special shop racks in big stores and organic food stores – which function as separate places in big super-markets as basic places of shopping. Therefore, we can conclude that despite the incremental increase in alternative shopping places (such as Internet/online shops), consumers still prefer supermarkets as the most convenient shopping place for organic foods. Similar to the study of Olech and Kuboń (2016), this research also supports the importance of neighborhood bazaars where consumers can directly purchase from producers. Likewise, Hamzaoui-Essoussi, Sirieix, and Zahaf (2013)indicated that consumers buying from producers/farmers were clearly looking for proximity with the producer, fresh products and quality; whereas, consumers using standard channels of distribution were looking for convenience, healthy products and taste.

This study also examined the main barriers against purchasing organic foods. It is found that price, availability, and lack of variety were the three obstacles that were mentioned by the respondents as the reason of not purchasing organic foods. Past research supported this finding that price and availability were main barriers (Lea and Worsley, 2005; Padel and Foster 2005; Lockie et al., 2002; McEachern and McClean, 2002). For instance, Padel and Foster's study (2005) assumedthat the issue of price, access and availability, visual product quality and presentation, mistrust of organic food in supermarkets, eating habits and lack of cooking skills as the important barriers of not buying organic foods.In addition, it is seen that respondents were not sure whether the food that they had bought were really "organic", with a mean value of 2.97 (out of 5). Similarly, the respondents declared that they do not find the organic food certification institutions as reliable with a mean value of 2.72. Likewise, Lea and Worsley (2005) found that consumers mistrusted organic labels, that is to say, consumers thought organic food labelling was unreliable; whereas Padel and Foster (2005) mentioned mistrust of organic food in supermarkets as another barrier.

MANAGERIAL IMPLICATIONS

Actually, the findings of this study have intertwined relations with each other. This article revealed that the cost and logistics of organic foods were directly related with the consumption of organic foods. Hence, Internet/online shops should be more effectively and efficiently used in order to decrease distribution costs of organic foods. Furthermore, from the supply side it would be more convenient and economic to carry out a wide variety of organic foods if Internet shops were used as a distribution channel. Therefore, it can be asserted that an integrative production-distribution system should be developed and encouraged by using technology, particularly Internet as a ground.

The 58% of the respondents in this study mentioned that they have not seen any organic food advertisement. So as a part of communication strategy, marketers first of all should increase the awareness level of consumers by using informative advertisements. The ads that attract the attention of consumers about organic foods must be designed. What is more interesting in this research is that the ones who declare that they saw organic food ads stated that they had seen that ad on the Internet. That is to say 23,4% of the respondents expressed that they had seen organic food ads on the Internet. This may be attributed to the fact that consumers generally search for organic products on the Internet and therefore they notice and recognize the online ads about organic foods that theywere exposed to.However, when it is asked which media channel they find more credible to get information about organic foods, the ratio of participants mentioning Internet as a credible source is very low. Consecutively, WOM, magazine ads, and online social networks are found as reliable source of information for organic foods. Hence, from the perspective of marketers first of all attention-grabbing ads should be designed but to increase the credibility of the ads it can be recommended to use highly believable or likeable sources endorsing the organic product consumption. Similarly, organic food producers, in their ads, may show how they carefully grow crops and give detailed information about the differences between organically grown and traditionally grown agricultural products. Another method is related with labelling and certification institutions. The labels should inform the consumers and the trustworthiness of the certification institutions should be improved through effective communication strategies. Organic food producers can use online social networks more effectively so that consumer-generated content may help companies to generate greater consumer engagement and increase their trustworthiness. Furthermore, as stated by Chakrabarti (2010), health advisors like doctors and nutritionists may act as the professional opinion leaders for the organic foods and they may exert formal influences upon patients in WOM conversation.Likewise, as asserted by Olech and Kuboń (2016)organic food producers must pay special attention to a suitable form of promotion of their products and information on them with the use of specialized Internet websites or social networks.

It is clear that organic food sector has great potential for further expansion. Therefore, companies in agribusiness need to use information sources (channels) that are most influential in purchase decision and design effective communication strategies to persuade consumers to buy organic food.

REFERENCES

Adams, D. C., & Salois, M. J. (2010). Local versus organic: A turn in consumer preferences and willingness-to-pay. *Renewable Agriculture and Food Systems*, *25*(04), 331–341. doi:10.1017/S1742170510000219

Aertsens, J., Mondelaers, K., Verbeke, W., Buysse, J., & Van Huylenbroeck, G. (2011). The influence of subjective and objective knowledge on attitude, motivations and consumption of organic food. *British Food Journal*, *113*(11), 1353–1378. doi:10.1108/00070701111179988

Aertsens, J., Verbeke, W., Mondelaers, K., & Van Huylenbroeck, G. (2009). Personal determinants of organic food consumption: A review. *British Food Journal*, *111*(10), 1140–1167. doi:10.1108/00070700910992961

Atănăsoaie, G. (2011). Distribution channels on the organic foods market. *Journal of Horticulture. Forestry and Biotechnology*, *15*(3), 19–25.

Baker, S., Thompson, K. E., Engelken, J., & Huntley, K. (2004). Mapping the values driving organic food choice: Germany vs the UK. *European Journal of Marketing*, *38*(8), 995–1012. doi:10.1108/03090560410539131

Batte, M. T., Hooker, N. H., Haab, T. C., & Beaverson, J. (2007). Putting their money where their mouths are: Consumer willingness to pay for multi-ingredient, processed organic food products. *Food Policy*, *32*(2), 145–159. doi:10.1016/j.foodpol.2006.05.003

Bodini, A., Richter, T., & Felder, R. (2009). Quality related communication approaches for organic food. *Journal of Food Products Marketing, 15*(3), 364–377. doi:10.1080/10454440902942117

Burton, M., Rigby, D., Young, T., & James, S. (2001). Consumer attitudes to genetically modified organisms in food in the UK. *European Review of Agriculture Economics, 28*(4), 479–498. doi:10.1093/erae/28.4.479

Chakrabarti, S. (2010). Factors influencing organic food purchase in India-expert survey insights. *British Food Journal, 112*(8), 902–915. doi:10.1108/00070701011067497

Chen, J., Lobo, A., & Rajendran, N. (2014). Drivers of organic food purchase intentions in mainland China–evaluating potential customers attitudes, demographics and segmentation. *International Journal of Consumer Studies, 38*(4), 346–356. doi:10.1111/ijcs.12095

Chen, M. F. (2007). Consumer attitudes and purchase intentions in relation to organic foods in Taiwan: Moderating effects of food-related personality traits. *Food Quality and Preference, 18*(7), 1008–1021. doi:10.1016/j.foodqual.2007.04.004

Davies, A., Titterington, A. J., & Cochrane, C. (1995). Who buys organic food? A profile of the purchasers of organic food in Northern Ireland. *British Food Journal, 97*(10), 17–23. doi:10.1108/00070709510104303

deMagistris, T., & Gracia, A. (2008). The decision to buy organic food products in Southern Italy. *British Food Journal, 110*(9), 929–947. doi:10.1108/00070700810900620

Didier, T., & Lucie, S. (2008). Measuring consumers willingness to pay for organic and Fair Trade products. *International Journal of Consumer Studies, 32*(5), 479–490. doi:10.1111/j.1470-6431.2008.00714.x

Dimitri, C., & Dettmann, R. L. (2012). Organic food consumers: What do we really know about them? *British Food Journal, 114*(8), 1157–1183. doi:10.1108/00070701211252101

Ergönül, B., & Ergönül, P. G. (2015). Consumer motivation for organic food consumption. *Emirates Journal of Food and Agriculture, 27*(5), 416.

FiBL-IFOAM Survey. (2016). *The World of Organic Agriculture: Statistics & Emerging Trends 2016.* Retrieved November 05, 2016 from https://shop.fibl.org/fileadmin/documents/shop/1698-organic-world-2016.pdf

Fotopoulos, C., & Krystallis, A. (2002). Purchasing motives and profile of the Greek organic consumer: A countrywide survey. *British Food Journal, 104*(9), 730–765. doi:10.1108/00070700210443110

Gil, J. M., Gracia, A., & Sanchez, M. (2000). Market segmentation and willingness to pay for organic products in Spain. *The International Food and Agribusiness Management Review, 3*(2), 207–226. doi:10.1016/S1096-7508(01)00040-4

Govindasamy, R., & Italia, J. (1999). Predicting willingness-to-pay a premium for organically grown fresh produce. *Journal of Food Distribution Research, 30*, 44–53.

Gracia, A., & de Magistris, T. (2013). Organic food product purchase behaviour: A pilot study for urban consumers in the South of Italy. *Spanish Journal of Agricultural Research, 5*(4), 439–451. doi:10.5424/sjar/2007054-5356

Hamzaoui-Essoussi, L., Sirieix, L., & Zahaf, M. (2013). Trust orientations in the organic food distribution channels: A comparative study of the Canadian and French markets. *Journal of Retailing and Consumer Services*, *20*(3), 292–301. doi:10.1016/j.jretconser.2013.02.002

Hamzaoui Essoussi, L., & Zahaf, M. (2009). Exploring the decision-making process of Canadian organic food consumers: Motivations and trust issues. *Qualitative Market Research: An International Journal*, *12*(4), 443–459. doi:10.1108/13522750910993347

Hamzaoui-Essoussi, L., & Zahaf, M. (2012). Canadian organic food consumers profile and their willingness to pay premium prices. *Journal of International Food & Agribusiness Marketing*, *24*(1), 1–21. do i:10.1080/08974438.2011.621834

Harper, G. C., &Makatouni, A. (2002). Consumer perception of organic food production and farm animal welfare. *British Food Journal, 104*(3/4/5), 287-299.

Honkanen, P., Verplanken, B., & Olsen, S. O. (2006). Ethical values and motives driving organic food choice. *Journal of Consumer Behaviour*, *5*(5), 420–430. doi:10.1002/cb.190

Hughner, R. S., McDonagh, P., Prothero, A., Shultz, C. J., & Stanton, J. (2007). Who are organic food consumers? A compilation and review of why people purchase organic food. *Journal of Consumer Behaviour*, *6*(2-3), 94–110. doi:10.1002/cb.210

Janssen, M., & Hamm, U. (2012). Product labelling in the market for organic food: Consumer preferences and willingness-to-pay for different organic certification logos. *Food Quality and Preference*, *25*(1), 9–22. doi:10.1016/j.foodqual.2011.12.004

Jarossova, M. A., &Mind'asova, K. (2015). Comparison of the application of marketing communication tools used by organic food producers in Slovakia and Austria. *NaukiInżynierskieiTechnologie*, *4*(19).

Kastberg, P. (2015). Promoting communication, participation, and learning with regard to organic food products: A communication theoretical approach. *Ecology and Society*, *20*(1), 3. doi:10.5751/ES-07139-200103

Kesse-Guyot, E., Peneau, S., Mejean, C., de Edelenyi, F. S., Galan, P., Hercberg, S., & Lairon, D. (2013). Profiles of organic food consumers in a large sample of French adults: Results from the Nutrinet-Sante cohort study. *PLoS ONE*, *8*(10), e76998. doi:10.1371/journal.pone.0076998 PMID:24204721

Kotler, P., & Armstrong, G. (2014). *Principles of Marketing* (15th ed.). Pearson.

Krystallis, A., & Chryssohoidis, G. (2005). Consumers willingness to pay for organic food: Factors that affect it and variation per organic product type. *British Food Journal*, *107*(5), 320–343. doi:10.1108/00070700510596901

Krystallis, A., Fotopoulos, C., & Zotos, Y. (2006). Organic consumers profile and their willingness to pay (WTP) for selected organic food products in Greece. *Journal of International Consumer Marketing*, *19*(1), 81–106. doi:10.1300/J046v19n01_05

Lea, E., & Worsley, T. (2005). Australians organic food beliefs, demographics and values. *British Food Journal*, *107*(11), 855–869. doi:10.1108/00070700510629797

Lee, H. J., & Yun, Z. S. (2015). Consumers perceptions of organic food attributes and cognitive and affective attitudes as determinants of their purchase intentions toward organic food. *Food Quality and Preference*, *39*, 259–267. doi:10.1016/j.foodqual.2014.06.002

Liang, R. D. (2014). Enthusiastically consuming organic food:: an analysis of the online organic food purchasing behaviors of consumers with different food-related lifestyles. *Internet Research: Electronic Networking Applications and Policy, 24*(5), 3-4.

Lindh, H., & Olsson, A. (2010). Communicating imperceptible product attributes through traceability: A case study in an organic food supply chain. *Renewable Agriculture and Food Systems*, *25*(04), 263–271. doi:10.1017/S1742170510000281

Lobo, A., Mascitelli, B., & Chen, J. (2014). Opportunities for small and medium enterprises in the innovation and marketing of organic food: Investigating consumers purchase behaviour of organic food products in Victoria, Australia. *AI & Society*, *29*(3), 311–322. doi:10.1007/s00146-013-0457-z

Lockie, S., Lyons, K., Lawrence, G., & Grice, J. (2004). Choosing organics: A path analysis of factors underlying the selection of organic food among Australian consumers. *Appetite*, *43*(2), 135–146. doi:10.1016/j.appet.2004.02.004 PMID:15458800

Lockie, S., Lyons, K., Lawrence, G., & Mummery, K. (2002). Eating green: Motivations behind organic food consumption in Australia. *Sociologia Ruralis*, *42*(1), 23–40. doi:10.1111/1467-9523.00200

Loebnitz, N., & Aschemann-Witzel, J. (2016). Communicating organic food quality in China: Consumer perceptions of organic products and the effect of environmental value priming. *Food Quality and Preference*, *50*, 102–108. doi:10.1016/j.foodqual.2016.02.003

Loureiro, M. L., & Hine, S. (2002). Discovering niche markets: A comparison of consumer willingness to pay for local (Colorado grown), organic, and GMO-free products. *Journal of Agricultural and Applied Economics*, *34*(03), 477–487. doi:10.1017/S1074070800009251

Magnusson, M. K., Arvola, A., Koivisto Hursti, U.-K., Åberg, L., & Sjödén, P.-O. (2001). Attitudes towards organic foods among Swedish consumers. *British Food Journal*, *103*(3), 209–227. doi:10.1108/00070700110386755

Makatouni, A. (2002). What motivates consumers to buy organic food in the UK? Results from a qualitative study. *British Food Journal, 104*(3/4/5), 345-352.

Marian, L., Chrysochou, P., Krystallis, A., & Thøgersen, J. (2014). The role of price as a product attribute in the organic food context: An exploration based on actual purchase data. *Food Quality and Preference*, *37*, 52–60. doi:10.1016/j.foodqual.2014.05.001

McEachern, M. G., & Mcclean, P. (2002). Organic purchasing motivations and attitudes: Are they ethical? *International Journal of Consumer Studies*, *26*(2), 85–92. doi:10.1046/j.1470-6431.2002.00199.x

Michaelidou, N., & Hassan, L. M. (2008). The role of health consciousness, food safety concern and ethical identity on attitudes and intentions towards organic food. *International Journal of Consumer Studies*, *32*(2), 163–170. doi:10.1111/j.1470-6431.2007.00619.x

Nasir, V. A., & Karakaya, F. (2014). Underlying motivations of organic food purchase intentions. *Agribusiness*, *30*(3), 290–308. doi:10.1002/agr.21363

Nie, C., & Zepeda, L. (2011). Lifestyle segmentation of US food shoppers to examine organic and local food consumption. *Appetite*, *57*(1), 28–37. doi:10.1016/j.appet.2011.03.012 PMID:21477631

Olech, E., & Kuboń, M. (2016). Clients preferences and development of organic food distribution channels. *Agricultural Engineering*, *20*(1), 119–125. doi:10.1515/agriceng-2016-0012

Onyango, B. M., Hallman, W. K., & Bellows, A. C. (2007). Purchasing organic food in US food systems: A study of attitudes and practice. *British Food Journal*, *109*(5), 399–411. doi:10.1108/00070700710746803

Organic Trade Association. (2015). *2015 Organic Industry Survey*. Retrieved November 05, 2016 from https://ota.com/sites/default/files/indexed_files/PolicyConference2015_Infographic_8.5x11_1a_0.pdf

Organic Trade Association. (2016). *State of the Organic Industry 2016*. Retrieved November 05, 2016 from http://ota.com/sites/default/files/indexed_files/OTA_StateofIndustry_2016.pdf

Padel, S., & Foster, C. (2005). Exploring the gap between attitudes and behaviour: Understanding why consumers buy or do not buy organic food. *British Food Journal*, *107*(8), 606–625. doi:10.1108/00070700510611002

Paul, J., & Rana, J. (2012). Consumer behavior and purchase intention for organic food. *Journal of Consumer Marketing*, *29*(6), 412–422. doi:10.1108/07363761211259223

PwC. (2016). *Global entertainment and media outlook 2016-2020*. PwC.

Rödiger, M., & Hamm, U. (2015). How are organic food prices affecting consumer behaviour? A review. *Food Quality and Preference*, *43*, 10–20. doi:10.1016/j.foodqual.2015.02.002

Rousseau, S., & Vranken, L. (2013). Green market expansion by reducing information asymmetries: Evidence for labeled organic food products. *Food Policy*, *40*, 31–43. doi:10.1016/j.foodpol.2013.01.006

Sangkumchaliang, P., & Huang, W. C. (2012). Consumers' perceptions and attitudes of organic food products in Northern Thailand. *International Food and Agribusiness Management Review*, *15*(1), 87–102.

Seyfang, G. (2008). Avoiding Asda? Exploring consumer motivations in local organic food networks. *Local Environment*, *13*(3), 187–201. doi:10.1080/13549830701669112

Shafie, F. A., & Rennie, D. (2012). Consumer perceptions towards organic food. *Procedia: Social and Behavioral Sciences*, *49*, 360–367. doi:10.1016/j.sbspro.2012.07.034

Statista. (2016a). *Organic packaged food and beverage sales in the United States in 2014, by distribution channel*. Retrieved November 05, 2016 from https://www.statista.com/statistics/249231/share-of-organic-packaged-food-and-beverage-sales-in-the-us-by-distribution-channel/

Tarkiainen, A., & Sundqvist, S. (2005). Subjective norms, attitudes and intentions of Finnish consumers in buying organic food. *British Food Journal*, *107*(11), 808–822. doi:10.1108/00070700510629760

Teng, C. C., & Wang, Y. M. (2015). Decisional factors driving organic food consumption: Generation of consumer purchase intentions. *British Food Journal*, *117*(3), 1066–1081. doi:10.1108/BFJ-12-2013-0361

Thompson, G. D. (1998). Consumer demand for organic foods: What we know and what we need to know. *American Journal of Agricultural Economics*, *80*(5), 1113–1118. doi:10.2307/1244214

Ureña, F., Bernabéu, R., & Olmeda, M. (2008). Women, men and organic food: Differences in their attitudes and willingness to pay. A Spanish case study.*international Journal of consumer. Studies*, *32*(1), 18–26.

Van Doorn, J., & Verhoef, P. C. (2011). Willingness to pay for organic products: Differences between virtue and vice foods. *International Journal of Research in Marketing*, *28*(3), 167–180. doi:10.1016/j.ijresmar.2011.02.005

Xie, B., Wang, L., Yang, H., Wang, Y., & Zhang, M. (2015). Consumer perceptions and attitudes of organic food products in Eastern China. *British Food Journal*, *117*(3), 1105–1121. doi:10.1108/BFJ-09-2013-0255

Zagata, L. (2012). Consumers beliefs and behavioural intentions towards organic food. Evidence from the Czech Republic. *Appetite*, *59*(1), 81–89. doi:10.1016/j.appet.2012.03.023 PMID:22504401

Zakowska-Biemans, S. (2011). Polish consumer food choices and beliefs about organic food. *British Food Journal*, *113*(1), 122–137. doi:10.1108/00070701111097385

Zander, K., & Hamm, U. (2012). Information search behaviour and its determinants: The case of ethical attributes of organic food. *International Journal of Consumer Studies*, *36*(3), 307–316. doi:10.1111/j.1470-6431.2011.00998.x

Zanoli, R., & Naspetti, S. (2002). Consumer motivations in the purchase of organic food: A means-end approach. *British Food Journal*, *104*(8), 643–653. doi:10.1108/00070700210425930

Zepeda, L., & Li, J. (2007). Characteristics of organic food shoppers. *Journal of Agricultural and Applied Economics*, *39*(01), 17–28. doi:10.1017/S1074070800022720

Chapter 10
Segmenting Paddy Farmer's Attitude and Behavior:
A Study Towards the Green Fertilizer Technology Adoption Among Malaysian Paddy Farmers – Adoption of GFT

Nadia Adnan
Universiti Teknologi Petronas, Malaysia

Shahrina Md Nordin
Universiti Teknologi Petronas, Malaysia

Amir Noor Noor
London Metropolitan University, UK

ABSTRACT

Agriculture is the major driving force of Malaysian economic. The aim of this research study is to segment the behavior of paddy farmers in Malaysia and understand how they influence adoption, a green fertilizer technology (GFT). The first objective of this chapter is to establish the thinking which enables a society to bridge the gap between embracing GFT among paddy farmer in Malaysia. Furthermore, the study builds the conceptual framework and examine the relationship among the relevant construct of this conceptual framework which was found by critically examining the different agricultural innovation literature. To make this conceptual framework robust it is found in the literature that theory of planned behavior and theory of reasoned action play a major role in segment farmer's behavior towards the adoption of GFT. Policy implications and/or suggestions for future research are deliberated for each issue and factor that affecting the adoption of GFT.

DOI: 10.4018/978-1-5225-2107-5.ch010

INTRODUCTION

The market of the 21st century is a giant network, which covers the whole globe. The market is divided into hundreds of domestic markets and international markets. There are thousands of organizations and millions of consumers, a wide range of products and services to be offered. A great skill is needed to be able to manage these markets. The marketing science studies show markets from bottom to top, develops techniques in managing consumers, competitors, partners, dealing with products and advertising strategies. Every single company has a marketing strategy. It might be well-structured or not thought through at all, but yet there is one in every company. One of the first steps on the way to a marketing strategy is completing market segmentation. Nowadays it is very important for a company to have a crystal clear understanding of its marketing strategy. Any attempts in doing marketing and product positioning won't work as they should unless a company has an understanding of its clients and develops a marketing strategy.

Market segmentation is one of the most basic and essential parts of developing a marketing strategy. During the past hundred years, the market has grown enormously: there are lots of new products and consumers' needs getting more and more twisted and complicated over time. Technology and innovations encourage farmers' wants and companies' imagination, which is why markets become more complicated. With continuous growth and development of different market segments, it is important for a company to have a deep knowledge of consumers and markets. This study concentrates on market research, specifically on market segmentation: what kind of information and advantages can be gained out of market segmentation. The purpose of this study is to provide to exemplifying that how market segmentation can determine the right target (paddy farmers) farmers for GFT which is name as One BAJA.

In the beginning of the study, the researcher will explain the segmentation process of marketing and after that, it will follow by the analysis as per the questionnaire led and using respectively the Minerva model and the Mosaic model. In the first part of the study, the market segmentation process will be defined later the different types of market segmentation and the variables will explain. These types are significant when identifying the right target customer's farmers to a product (One BAJA). There are four major types of segmentation first is the demographic segmentation, which is considered the most shared one that deals with basic demographic factors such as age, income, gender etc. and divides the target farmers into segments based on these variables. Whereas, the geographic segmentation divides the target farmers into segments based on geographical areas such as nations, regions, cities, etc. The psychographic segmentation divides the farmers into segments according to their values and lifestyle. Finally, the behavioral segmentation divides the target customer farmers into segments based on their attitude toward product.

On the basis of the description of the types, it will be concluded which of the types are best suited when identifying the target farmers of respectively the landline telephone and the mobile telephone, in this case, being the demographic and the behavioral segmentation. The second step for (One BAJA) team is the segmentation process which is a matter of right target marketing. After identifying the target customers (farmers), the fertilizer company essentially chooses which segment to target. Subsequently the fertilizer company then must decide which market strategy to choose i.e. undifferentiated marketing, where the fertilizer company does not consider differences between the segments and targets the market with one offer, differentiated marketing, where the company targets many market segments with offers specially designed for each segment or concentrated marketing, where the fertilizer company chooses one or few markets. Furthermore, the third and final step of the segmentation process is the matter of right positioning.

The fertilizer company (One BAJA) will have identified and evaluated the target customer it is then essential to choose what position the (One BAJA) wants to occupy in the chosen segments. In order for the One BAJA to achieve a successful positioning i.e. when the (farmers) find that the product satisfies their expectations and desires, there are steps the (One BAJA) must follow. These include amongst others; the fertilizer company (One Baja) must understand what the farmers expect and believe to be most important when deciding on a purchase, the fertilizer company (One Baja) must develop a product which caters specifically for the farmers' needs and expectations different analysis.

LITERATURE REVIEW

Segmentation

Segmentation involves finding out what kinds of consumers with different needs exist. In the auto market, for example, some consumers demand speed and performance, while others are much more concerned about roominess and safety (Kotler, 2012). In general, it holds true that "You can't be all things to all people," and experience has demonstrated that firms that specialize in meeting the needs of one group of consumers over another tend to be more profitable (Adnan, Nordin, & Redza, 2017).Whereas, Figure 1 states the different level of segmentation. However, in the line of this research our target consumer is paddy farmers in Malaysia.

Why Segmentation?

When it comes to marketing strategies, most people spontaneously think about the 4P (Product, Price, Place, Promotion) –maybe extended by three more Ps for marketing services (People, Processes, Physical Evidence) (Kotler, 2012; Kotler, Armstrong, Saunders, & Wong, 2003). Market segmentation and the identification of the target market are an important element of each marketing strategy? They are the basis for determining, any particular marketing mix (Kotler, 2012; Kotler et al., 2003). Literature of (Kotler, 2012; McDonald & Wilson, 2011)suggested that. Whereas, Figure 2 depicts the marketing segmentation process.

MAIN FOCUS OF THE CHAPTER

Segmentation and Marketing Planning

One of the essential issues in the marketing strategy of a company is marketing research. Marketing research gathers information about markets which concern the company as well as its potential farmers / current farmers (Wittman, Beckie, & Hergesheimer, 2012). The overall purpose is to get a clear picture of what is a current market, which market segments are related to a company, what are their needs and wants (Kotler, 2012). An outcome of a marketing research is a development of a strategy or improvements to it for reaching out to those market segments and satisfying their needs, or a strategy for a marketing company's specific product (McDonald & Wilson, 2011). Kristoferson and Bokalders (2013) marketing research consists of information about actual and potential farmers, needs, products, technologies, and competitors.

Figure 1. Segmentation

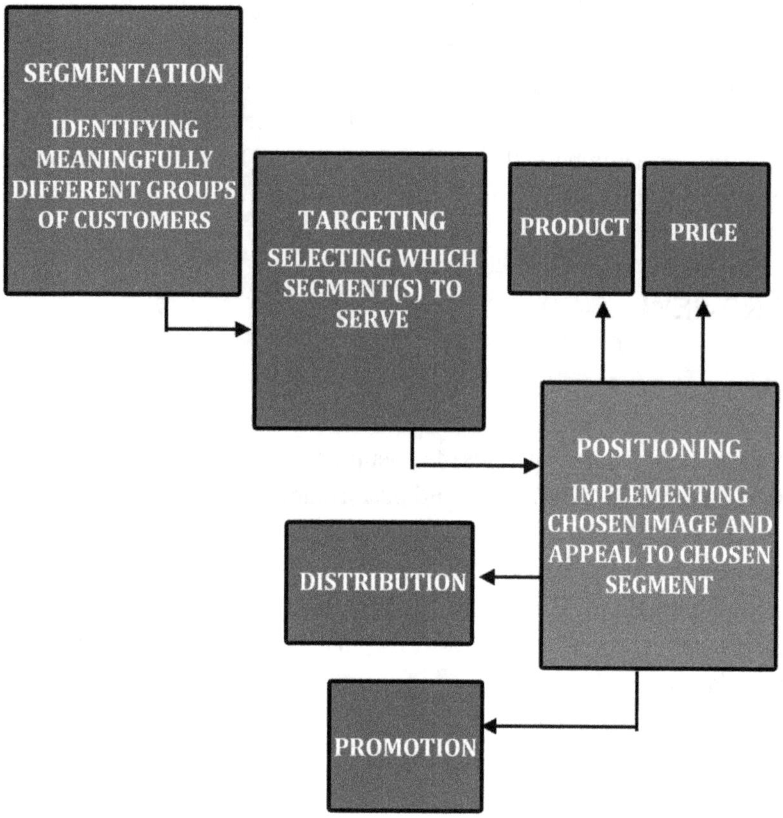

Figure 2. Market segmentation process

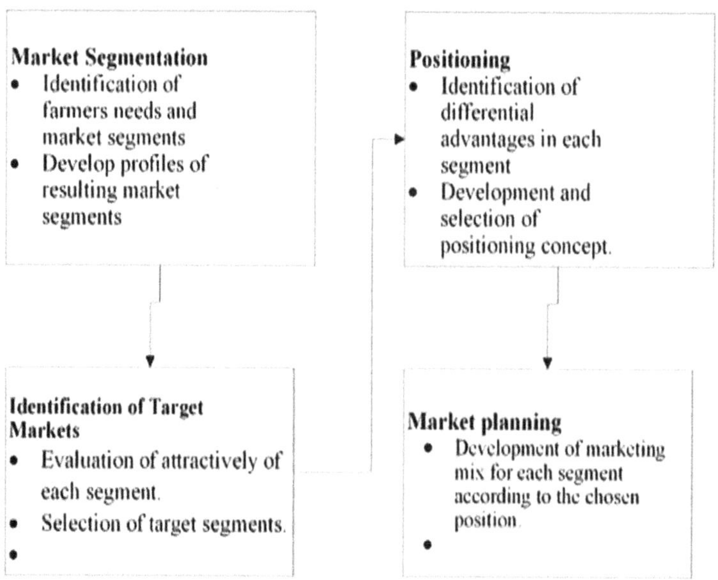

To collect information about markets, either primary or secondary sources should be used in different type of marketing strategies. Furthermore, (Armstrong, Adam, Denize, & Kotler, 2014; McDonald & Wilson, 2011) highlighted the point that marketing research consists of information about actual and potential farmers, needs, products, technologies, and competitors. To collect information about markets, either primary or secondary sources should be used. Usually, it is way easier and faster to collect information from secondary sources, which are: published articles in journals, newspapers, commercially published market researchers, government statistics, yearbooks, Internet, and other published materials. Primary sources are the ones which are collected from the original source: with the help of questionnaires, interviews, experiments or product tests with consumers.

A combined way of the research process is quite often used by companies – when there is not enough information from secondary sources, information is collected directly from primary sources (Sacks et al., 1996).There are several types of marketing research that are carried out by marketers:

- Customer research
- Promotion research
- Distribution research
- Sales research
- Marketing environment research

THEORY OF MARKET SEGMENTATION

It all started from early economic theory, where it was discovered that demand is connected to the level of competition and pricing (McDonald, 2012). Based on the article of Croft (1994) "The international consumer market segmentation managerial decision process", lots of information about the theory of market segmentation were discovered. The market segmentation concept itself has its roots starting from the 1930s when Chamberlin and Robinson proposed their theories of imperfect competition. Only in 1950s marketing writer Wendell Smith stated his point that the recognition of diversity or heterogeneity in demand and supply proposed the existence of two quite different marketing strategies, which are in use even in our modern world– product differentiation and market segmentation.

Market Segmentation

The market segmentation is mentioned as being one of the key elements of modern marketing and is, as mentioned, the process of dividing the market into several groups and/or segment(s) based on factors such as demographic, geographic, psychological and behavioral factors. By doing so the marketers will have a better understanding of their target audience and thereby make their marketing more effective (Furnham & Gunter, 2015).This is due to the fact that by using the analytical process that puts farmers first, the marketer will get more satisfied farmers and thereby gain a great advantage over competitors (Kotler, 2012).There are four levels defined by (Kotler, 2012; Kotler & Caslione, 2009) are segments, niches, local areas, and individuals. Therefore, the following marketing strategies arise segment marketing, niche marketing, local marketing and individual marketing, as well as mass marketing (Croft, 1994). Furthermore, in the line of this study, we segment the fertilizer industry in Malaysia.

- Mass marketing's advantage is that it has the biggest potential market, however, it becomes more difficult to practice mass marketing over time due to a huge range of products offered to consumers.
- Segment marketing isolates the broad market and adapts so that products match one or a few sectors' needs. Segment marketing's benefits are more efficient marketing and communications in terms of reaching out to consumers. It allows the company to create a more segment-adjusted product and price according to its audience. Also the ways of communicating to consumers and distributing products become easier.
- Niche marketing focuses on subgroups within segments of the market; the company knows its clients so well that it is able to offer very specific products to a consumer when other competitors cannot offer similar products. The client is ready to pay a premium price for that offer. Niche farmers usually have a very distinct set of wants and needs usually not satisfied. The niches of the market do not attract that many competitors due to their small size and specific expectations.
- Micromarketing is the practice of serving individual clients and their individual needs and wants, which goes beyond segment or niche marketing. Philip Kotler broadened this concept into a deeper one by dividing Micromarketing into local marketing and individual marketing.

MARKET SEGMENTATION BASES

According to M. Hutt and Speh (2012), two major categories for business to business market segmentation bases are a macro segmentation and micro-segmentation. Macro segmentation concentrates on the characteristics of an organization and its situation, such as size, geographic location, operating market, industry, etc. On the other hand, micro-segmentation requires a lot more detailed information and observations about companies, such as buying decision criteria, the importance of buying and attitudes within each macro segment defined. Philip Kotler and K. Keller ("Marketing Management", 2009 p. 355) on the other hand have their own vision on segmentation variables of business markets.

Even though it is pretty close to what has mentioned above, it just introduces a slightly different way of thinking. Kotler's and Keller's major segmentation variables are: demographic (industry, company size, location); operating variables (technology, user or nonuser status, customer capability); purchasing approaches (power structure, nature of existing relationships, general purchase policies, purchasing criteria); situational factors (urgency, specific application, size of order); personal characteristics (buyer-seller similarity, attitude towards risk, loyalty).Macro segmentation is a common choice of companies when doing market segmentation.

GEOGRAPHIC LOCATION

Geographic segmentation includes such variables as location, macro-economic factors, customer concentration, etc. It can be done by countries, states, towns or even continents such as,

- **Organization's Size:** The size of an organization can be measured by a few factors: the amount of employees or revenues. In different economies and different markets, there are different definitions for small to big companies, depending on the wealth of the economy, the size of the country and the size of the market.

- **Industry Classification:** Industry segmentation is a first one to do when dividing markets into groups. Moreover, it is one of the most complicated sub-variables due to a wide range of industries and sub-industries, and new ones appear with time and high technologies. One of complications related to industry segmentation is companies operating in many industries at the same time. Nowadays it is quite normal to operate in several industries and be successful in them.
- **Characteristics Relevant to the Organization:** In case a company finds some other factors related to the descriptive characteristics of its market, it can use those factors for the segmentation process. It can totally depend on the type of business, services and partnership history and other factors.
- **User/Nonuser Status:** It is important for a farming organization to know who its active users are, who are interested in its services and who are not. That will give useful insights on who was reached and who was not. Most of the companies keep track of their past activities and past results, that is where the information on segmentation upon user/nonuser can be received from.

IMPORTANCE AND BENEFITS OF MARKETING SEGMENTATION IN MALAYSIAN AGRI-INDUSTRY

Nowadays, the market segmentation concept is considered as a big and important part of marketing theory. Marketing is all about identifying and satisfying the needs of farmers, and the needs of possible farmers need to be identified as well as farmers need to be analysed. Market segmentation gives an agricultural industry a possibility to satisfy their farmers' needs in a smart way: to offer products

A STUDY ON AGRICULTURE INDUSTRY OF MALAYSIA

In history, Malaysia was an agricultural country, proceeding to its independence in 1957. The agriculture industry has been a backbone of the Malaysian economy (Hosseini & Wahid, 2013).Whereas, Dr. Mahathir Muhammad, the former prime minister of Malaysia with his Vision 2020, imagined Malaysia as an advanced industrialized country in the coming year of 2020 (Brooker,2012; Nagarajan, 2008). Furthermore, he mentioned in this vision that agriculture is the "sunset industry "for the nation (Bakar, 2009; Melchior, 2011). Moreover, agriculture remains an important part of the national budget in the new era for the gradually growing population with the challenge to deliver both food security and safety, and sustainable growth and wealth formation. As, (Ramli et al., 2013; M. Shaffril et al., 2010) describe the Malaysia's land area measures for agriculture is about 329,733 square km, most of the soil is acidic hence, infertile.

Additionally, the agriculture sector engaged as the backbone of the country's economy and had ominously contributed to the 11.6% GDP, till it remained outshined by contributions from the manufacturing sector as the country moves rapidly towards industrialization from the mid-1970's (Haris, Hamzah, Krauss, & Ismail, 2013). Therefore, the agriculture sector has persisted glumness previously, the government agencies focus on the manufacturing sector, but during the economic crisis in 1997, Malaysian industrial sector faces the severe downfall. Whereas, in the agricultural industry has been proven as an effective medium to overcome self-sufficiency (Shaffril, Asmuni, & Ismail, 2010).

The country's agricultural sector has experienced various phases of transformation (Azadi, Ho, &Hasfiati, 2011). Now, after all, these phases of transformation, the agriculture sector contributes more to Gross Domestic product (GDP) increased from RM17.01 billion in 1995 to 18.22 billion in 2000 (8th Malaysian Plan). It also gained government attention in the 9th Malaysian Plan (2005-2010), agriculture is declared as the 3rd largest income generator for Malaysian. In the Malaysian agriculture sector, there are several commodities like palm oil, cocoa, livestock's fisheries and paddy. But, the government gives more emphasis more on the paddy industry because it is the basic good for the nation (FAO, 2012). Approximately, it is projected that around 300,000 farmers are involved in paddy cultivation (Moeskops et al., 2010).

Paddy is the staple food for more than three billion people all around the world. At least 114 countries grow rice and more than 50 have an annual production of 100,000 tonnes or more (Dastagiri, Gajula, & Patil; Edwards, 2015). Rice is the main food for most countries in Asia and, about 90% of the global rice area, production and consumption are concentrated on(Noppers, Keizer, Bolderdijk, & Steg, 2014). At this time, when the world's population is already reeling from higher food prices, many countries have already banned or restricted their rice exports, which pushes up the price of rice even higher (Edwards, 2015). Paddy yields have been increasing since the 1960s, but since the 1990s, the increase in rice production has been slower than population growth. Indeed, it is anticipated that rice production will need to increase by 30% by 2025 in order to cater for the world's growing population(Othman, 2012).

The current study is to look at Malaysia's paddy fertilizer sectors. Malaysia's paddy fertilizer sector is considered unique in several ways or elements. The government of Malaysia pays special attention to paddy farming. Therefore, the food crop has strategic value where rice is grown in both Peninsular and East Malaysia which is shown in Figure 3 (Haris, Hamzah, Krauss, & Ismail, 2013). About 300,500ha in peninsular Malaysia and 190,000 ha in East Malaysia are developed for rice production (Ahmad & Tahar, 2014).

There are eight granaries area in Malaysia named as MADA, Kedah, KADA, Kelantan, and Northwest Selangor project, SeberangPerai IADA, Penang, Perak and Ketara. IADA is the main paddy producer that meets seventy two percent (72%) of the demand of Malaysia stated by the Ministry of Agriculture and Agro-based Industry, (MARDI, 2010). Furthermore, it is predicted by the lawful agency that paddy will produce RM 988 million in 2010 (DOAStat, 2013). Nevertheless, if we relate paddy with other commodities of agriculture, there is still lot need to be done to strengthen paddy as the main agriculture sector in Malaysia. As per the report of (FAOstat, 2009a) the yearly production of rice stands at 2.51 million metric tons which is not sufficient. Stability of paddy plantation is essential in Malaysia as it is recognized as staple food for the majority of Malaysian.

Paddy is considered as important and essential food for both its citizen and the government, which is grown on 672,000 hectares (1.66 million acres) and the production is estimated at 3.66 metric tons per hectare which is not up to the consumption mark (Othman, 2012). Resolve this issue of low production with sustainable mean government has started providing all paddy farmers subsidized fertilizer since 1970 (Noh, 2012). On average, a Malaysian consumes around 300kg of rice per capita (FAOstat, 2012). There are many strategies to increase production in sustainable contexts. Recently, the government of Malaysia had allocated RM 100 million in the 2013 budget for the development and management of paddy industry including the subsidized fertilizer (Noh, 2013a). Furthermore, the government has taken various efforts to strengthen the paddy and rice industry like establishing many organizations. For instance, paddy marketing accountability was taken over by National paddy and rice board (NPRB) which was established in 1971. In the year of 1996, NPRB got privatized and changed to (Plan, 1996)

Figure 3. Map of Peninsular Malaysia
Source:(Tey, 2013)

PadiBerasNasionalBerhad, by having a mission to strengthen the paddy industry (Fahmi, Samah, & Abdullah, 2013). The Government's rigorous effort continues to be one of the big agendas. During the review (NinthMalaysianplans, 2006) 70% of 2.5 billion was allocated to national food security. But currently, the country produces only 60-65% of its requirement. It meets the remaining through imports, mostly from Thailand (Haris et al., 2013). Inasmuch, substantial progress has been made in the doubling of paddy production in the last 30 years (Grassini, Eskridge, & Cassman, 2013).However, there is still a need to increase the production to meet the need of the population without unsuitable practice (Adnan, Nordin, & Noor, 2017) It is mostly a farmer's effort that contributes in large measures to paddy production in the country. A multitude of factors could be relevant this disparity of production and sustainability (Nhamo, Rodenburg, Zenna, Makombe, &Luzi-Kihupi, 2014).

The paddy plantation needs to be cultivated by new ideas and methods. Farmers need to adopt One BAJA fertilizer to increase the production without damaging the environment. This contributes toward the sustainable environment (Unescap, 2011), because it increase the crop farm production system that avoids the use of synthetic fertilizers and pesticides, hormones, antibiotics and takes measures to protect the environment and increase the production (Othman, 2012).The annual growth rate of rice production from year 2000 to 2013 was about 1.47% and the production of rice increased from 2.14 million metric tons (MT) in 2000 to 2.63 million MT in 2013. However, the Self-Sufficiency Level (SSL) of rice in Malaysia was 70% in 2000, 72% in 2007 and it increased to 74% in 2013 (Lee, Kim, &Vohs, 2011). Consistent with the National Agriculture Policy (NAP), the Malaysian government wishes to escalate the SSL of paddy. The SSL of paddy production can be achieved by effective fertilizer application. Malaysia is a dynamic player as a consumer and producer of fertilizer. But most of the fertilizer used in Malaysia is produced out of the country. The total value of fertilizer imported into Malaysia has been on the rise from RM 1.47 billion in 2003 to RM 3.4 billion in 2007 and this has caused huge foreign exchange losses to the country (Nordin, Noor, & bin MdSaad, 2014). This indicates that there is a large market potential for locally produced fertilizers as import substitute as well as for local producers to move up the value chain.

To enhance the agriculture production and to get optimum plant growth, nutrients must be accessible in sufficient and balanced amount (Chen, 2006). Nevertheless, the traditional method of agricultural farming in which chemical-based fertilizers are heavily used has contributed to increased global warming and greenhouse effect. Moreover, increasing soil destruction and immense use of biochemical fertilizer, particularly the rice cultivation made the Malaysian government to introduce the 'National Green Technology Policy' (NGTP) in 2009. NGTP put emphasis on boosting and implementing the green technology in agriculture (Vlek& Byrnes, 1986).

GREEN FERTILIZER TECHNOLOGY DEVELOPMENT

The worldwide concerns about worsening environmental conditions necessitate the need for eco-friendly agriculture products (Hashim&Ho, 2011; Roopan et al., 2013). Green technology policy refers to the development and application of products, equipment, and systems used to converse the natural environment and resources which minimize and reduces the negative impact of human activities (Hashim&Ho, 2011; Roopan et al., 2013).The continual application of traditional fertilizers in Malaysian paddy production has been creating adverse environmental impact with negative social consequences. The development of an eco-friendly fertilizer is hence timely and appropriate. Green fertilizer technology in the form of control release fertilizer (CRF) has numerous advantages. These advantages are:

- It increases fertilizer efficiency and crop yield,
- It reduces losses of nutrients through leaching, runoff, volatilization, and denitrification,
- It saves time, cost, and labor in reducing the frequency required for fertilizer application compared to the conventional method,
- It synchronizes the release of all macro- and micro-nutrient in the soil necessary for crop plantation.

CONTROL RELEASE FERTILIZER "ONE BAJA" RESEARCH PROGRAM

A group of researchers from several public and private universities in Malaysia has embarked on a research program named as "One Baja". The primary objective of the One Baja research program is to develop a green fertilizer technology by featuring a green production of ammonia as well as a controlled-release mechanism of urea for paddy farming fertilizer. This One Baja research program received a grant worth (Ringgit Malaysia) RM 12 million from the Ministry of Higher Education (MOHE) of Malaysia. The aims of this research program are:

- To model and synthesize ammonia and urea in a magnetic induction reaction zone with the presence of a catalyst and catalyst support;
- To design, multi-functional layers of biodegradable granulated urea and to develop a new coating formulation material to increase nutrient uptake;
- To systematically model the process design of Enhanced Efficiency Fertilizer (EEF) plant as well as to calculate the energy saving of the new system;
- To run efficacy test on the newly developed EEF products and to conduct some economic impact analysis via stakeholder communication.

NECESSITY OF THE STUDY FOR ONE BAJA

This remarkable achievement is a result of policy done by the Party and Government, which has paid attention to agricultural development. In addition, the agriculture has also applied the scientific and technical methods and effective usage of fertilizers in farming. In order to promote agriculture development, both in quantity and quality basis, the industry needs the support and contribution of all economic components, of which the important role of fertilizer production enterprises is undeniable. Fertilizers facilitate plants to have high productivity and quality products. Fertilizer companies must manufacture diversified and good products to meet the market needs. This would be a responsibility as well as a challenge for FELDA, RISDA, FELCRA and FOA when the firm provides urea and NPK fertilizers in the market. Before 2006, the production of fertilizer companies did not meet the need of the market; thus, the marketing activities were not paying enough attention and promoted by the fertilizer companies.

Background of Study

In average, Malaysia produces 2.67 ton of paddy each year (DOS, 2012). However, more than 60% of the production comes from the northern region where two of three largest granary areas in Malaysia are located. Furthermore, Muda Agriculture Development Authority (MADA) alone contains around 40,000 families who work as farmers (MADA, 2007). It makes the northern region of Malaysia as one of the main rice producers in Malaysia and contains the largest number of farmers compared to other regions. There are three major granary areas under MADA, IADA KADA. These regulatory bodies are responsible for managing farmers in terms of irrigation, planting, and technical advice. Under this management authority, there are several small groups of management team known as "Pertubuhan Peladang Kawasan" or PPK (Area Farmers Organization) which is the local union of farmers in the area. Their role is managing and planning for any errands for a single planting season. Furthermore, PPKs are also responsible for managing the purchase of non-subsidized fertilizer and storing the subsidized ones before the disbursement to the farmers. To this end, it is necessary for the company to carefully analyze the macro and industry environment and then map out the appropriate marketing solutions for NPK fertilizers of One BAJA. All the above-mentioned contents are covered in the specific study.

DATA SOURCES AND RESEARCH METHODOLOGY

The research methodology is used in this study is a case study and this methodology meets the research's objective. Data used in the study were taken from both reliable secondary data about One BAJA operation collected in the process of working with One BAJA and primary data collected from interviewing agents and customers in target market about its products, prices, places and promotions and from author's observation. There are a number of techniques used to conduct a case study research will be:

- Interviewing staff and experts
- Economic statistic method
- Methods of description and comparison

This Project management all in all is a methodology and a discipline that intends to bring benefits to its One Baja fertilizer: managing complex change, using prudent resources on the right projects, using creativity and knowledge of specifically chosen team members, setting quality objectives. Its success is driven by the project 7 on schedule, expenses within budget, meeting predefined quality goals, meeting the expectations, objectives, goals and mission of the customers, effective teamwork, new business improvements and productive outcome. However, market segmentation is an important asset of the project planning phase, too. It does help managers to develop a proper set of projects (products/services), target it to the right people with the right strategy technique.

MARKETING MIX (4P)

Product is the most basic element in the marketing mix. The product can have tangible features including quality, design, function, package and brand. It can also have intangible features such as delivery service, warranty, training, etc. Price is a no less important element of market mix, which includes: whole sale and retail prices, discount, credit. Price must be in consistence with the value customers received and be competitive. Place is one of the major elements of marketing mix. This refers to how an organization will distribute the product or service they are offering to the end user. The company must understand the importance of distribution, and select the right intermediary for effective distribution. Promotion is all activities to communicate the product's advantages and convince target customer to buy that product. Promotion may include many actives to reach the target market. Company must set up activities such as, advertising, promotion, public relation, direct marketing. Thus, the research on the basic 4 p's of marketing mix is to check the market strategies of One BAJA such as positioning, product promotion and price

Positioning Strategies

Positioning of NPK fertilizer basing on the analysis of One BAJA comparative advantage and the result of interviews with customers in target markets, One BAJA can position its NPK products in the target market as follows: There are a lot of different kinds of Malaysia NPK Fertilizer of different brands namely: This data show the major player of manufacturers for compound fertilizer. Malaysian NPK Fertilizer (MNF) is the largest single factory manufacturer. (FPM got two factories at PasirGudang). There is compaction, granulated (method). If bio granulation it means during compaction process, they put biofertilizer element during the compaction. We still import compound fertilizer, but the amount is not simply 3 million (requirement – manufactured) because some of the requirements involve straight fertilizer. MNF Sdn Bhd is major producers of paddy fertilizer. Table 1 highlights the Malaysian fertilizer producer. These brands have a remarkable reputation in the market, yet some companies own the backward technology, which makes poor product quality and unstable price. One BAJA is a new player with the advantage of modern technology in order to make products of superior quality. Also, the company considers the high quality product as the comparative advantage to penetrate and attract customers. The aim of market penetration is to offer diversified products and upgrade the product consumption. Whereas, Table 1 illustrated the Malaysian fertilizers producers.

Product Strategies

Product strategies study, design the suitable package for consumers and diversify the products with different types of packaging and weight, namely 10 kg, 25 kg, 50 kg, 100 kg, which will satisfy all demands of consumers. Frequently update and summarize the One BAJA market information, analyze

Table 1. List of Malaysian fertilizer producer

	Compound fertilizer manufacturer	Capacity (MT)	Type of process	Locations	
1	FPM Sdn Bhd	300,000	Liquified Urea	Pasir Gudang	Peninsula
2	Malaysian NPK Fertilizer Sdn Bhd	270,000	Liquified Urea	Gurun, Kedah	Peninsula
3	CCM Fertilizers Sdn Bhd	270,000	Liquified Ammonical	Shah Alam	Peninsula
4	CCM Fertilizers Sdn Bhd	230,000	Liquified Urea	Bintulu	Sarawak
5	Agromate Sdn Bhd	200,000	Liquified Urea	Lahad Datu	Sabah
6	Kemira-Kuok Fertilizers Sdn Bhd	120,000	Granulated	Port Klang	Peninsula
7	Narsco-Kuok Fertilizers Sdn Bhd (Agrifert / Kuok Brothers)	120,000	Granulated	Pasir Gudang	Peninsula
8	Agri-Bintulu Fertilizers Sdn Bhd (Agrifert / Kuok Brothers)	120,000	Granulated	Bintulu	Sarawak
9	United Compound Fertilizers Sdn Bhd	100,000	Compaction	Kuantan	Peninsula
10	All Cosmos Industries Sdn Bhd	80,000	Bio Compaction	Pasir Gudang	Peninsula
11	TM-Kay Fertilizers Sdn Bhd (Agrifert / Kuok Brothers)	80,000	Granulated	Port Klang	Peninsula
12	Agri-Sabah Fertilizers Sdn Bhd (Agrifert / Kuok Brothers)	50,000	Granulated	Sandakan	Sabah
13	PK Fertilizers Sdn Bhd	25,000	Granulated	Pasir Gudang	Peninsula
14	Twin Arrow Fertilizer Sdn Bhd	25,000	Granulated	Port Klang	Peninsula
15	Consolidated Fertiliser Corporation (CFC) Sdn Bhd	20,000	Granulated	Port Klang	Peninsula
16	Sasco Sdn Bhd	20,000	Granulated	Lahad Datu	Sabah
	Total MT	2,030,000			

Source:(Safitri&Andriyani, 2011)

and evaluate the information, identify the market trend, competitors' abilities, recommend solutions and strategies to bring into full play the products' competitive advantages. Frequently visit the primary consumers, farmers, who directly use the products, to assess their knowledge as well as the satisfaction from the products. For this reason, One BAJA should focus on high quality products, which have high demand, to convince the customers and distributors. One BAJA should invest in technology, which is more modern than that of other competitors, to create competitive advantages and superior products. One BAJA market is a fierce market; therefore, in order to survive and go far in the market, One BAJA need to invest in R&D activities in the long-term to differentiate their products with those of competitors.

Price Strategies

In terms of pricing a product, One BAJA is a newcomer; thus, the company needs a flexible pricing strategy, price determination must be taken into consideration of production costs, domestic and foreign prices and especially competitors' pricing. Being a fertilizer producer, the company deeply aware that its main customers are farmers, who are mostly living in poverty, underfunding and having a limited scientific understanding. To create favorable conditions for farmers to buy fertilizers on the installment plan, the company needs a late payment mechanism from 3 to 6 months so the farmers can purchase the fertilizer.

Promotion Strategies

Promotion strategy regarding the promotion and communication strategy, One BAJA should pay attention to, the following activities: The Company needs to hold seminars in the localities within the target market to introduce the product, which can help farmers and retailers aware of the products and the superior quality of NPK fertilizer against other competitors. In order to popularize as well as provide information about NPK products, the company needs to select media channels which are close to farmers such as agriculture programs broadcasted on radio and TV. The company should also take the agents in the urea nitrate market as a channel to introduce about NPK products. It is an important channel and popular with farmers. Besides that, One BAJA fertilizer awareness strategies through ICT should be proposed in the future as a promotional tool for the department of Agriculture, Malaysia organic farming.

The use of ICT can be used as a tool by the Department of Environment and the Ministry of Energy, Green Technology and Water for their sustainability promotion among Malaysian farmers. One Baja successful promotion with the help of ICT innovation will enhance the sustainability awareness through its recognition at the Malaysia Technology Expo (MTE) 2009. According to the National Green Technology Policy Malaysia, effective promotion and public awareness are two of the main factors that would affect the success of sustainable development among farmers through the adoption of One BAJA fertilizer. This is particularly significant as such adoption requires a change of mindset of the farmers through various approaches, including effective education and information dissemination to increase public awareness of sustainable agriculture and on ways to conserve the environment. This awareness of One BAJA fertilizer among paddy farmers in Malaysia with the help of ICT application with the element of captology, integrated into a persuasive learning environment can play an innovative role in promoting sustainable practices in agriculture. This research also aims to relate the issues of adoption of One BAJA fertilizer by using ICT in order to disseminate information about sustainable paddy farming practices in Malaysia. The government also plans to increase the usage of ICT in all sectors to increase productivity. Mastery of the use of information technologies will be emphasized to enhance the acquisition and dissemination of new knowledge and technologies, and to motivate greater youth participation in technology development and transfer. Advances in "expert systems" or computer simulated scenario analysis will be exploited to enhance on-farm advice and information exchange to agriculture producers through extension. The most innovative approach is the development of a new product or process to solve a new problem or usage. An example is the Digital Video Disk (DVD), which illustrates an old process with a new usage. DVDs employ the same basic technology as CDs; however, the means for compression and reading hardware are more advanced.

Another example of product innovation is an electronic mail security that involves virus protection software (Othman et al., 2012). However, almost as quickly as new software and processes are developed for protecting a firm's information, new problems emerge. It is a constant war of innovation nowadays, in Malaysia there are many innovations in ICT in learning and promoting sustainability issues, including learning through VR, sustainable agriculture in multimedia and VR application, persuasive technology, educational games, serious games, criteria in choosing the learning games and positive impact of computer games on education and training. According to the Planning and Information Technology Department of MADA, the usage of ICT was existing in their daily operations. This is especially true when they convey information to the public through their website (http://www.mada.gov.my/). It is similar in Kajang, Bandar Baru Tunjong, Tanjung Karang and Sabak Bernam Agriculture Departments. They have all developed their own web-based information systems for the purpose of disseminating information about innovation in the fertilizer industry.

DIFFERENT TYPE OF FERTILIZER IN MALAYSIA

Mineral Fertilizers

Mineral fertilizers account for more than 90 percent of fertilizers used by all types of farming systems in Malaysia. The main fertilizers are urea, ammonium sulphate, calcium ammonium nitrate, phosphate rock, super phosphates, ammonium phosphate, potassium chloride, potassium sulphate and NPK, NP and PK compound fertilizers. Due to the rapid expansion in crop production, especially of plantation crops (rubber, oil-palm, cocoa, and paddy) there has been a corresponding increase in fertilizer use. Potassium fertilizers have shown the largest increase and very effective for the crops.

Organic Fertilizers

The government is promoting the use of organic fertilizers in Malaysia for two main reasons. Firstly, organic agriculture is seen as important for the sustainable use and management of natural resources. Secondly, in the NAP3, organic agriculture is identified as a niche market opportunity for fruits and vegetables. In an effort to reduce the dependence on mineral fertilizers and to move towards more natural and healthier methods of food production, the government is promoting programmers that encourage the recycling and use of agricultural waste.

Table 2. Number and area of organic producers per State in 2001. Source: (Wai, 2001)

State	Number	Area(ha)
Selangor	4	10.8
Negri Sembilan	10	90
Melaka	2	1.1
Johor	2	3.5
Pahang	6	11.6
Sabah	2	12
Sarawak	1	2
Total	27	131

PRODUCTION AND MARKETING OF THAT FERTILIZER

Most of the fertilizer used in Malaysia is produced abroad. Urea, ammonium based and organic fertilizers are produced in large amounts, but the urea used in Malaysia for agriculture is not locally produced material. This is because the prilled urea produced in Malaysia fetches a high price in the international market and is therefore exported. The parent fertilizer producing company has two subsidiaries involved in the production of urea. These are one subsidiary was incorporated in 1980 and located in Bintulu, Sarawak. The plant has a capacity to produce 600 000 tonnes of granular urea and 420 000 tons of prilled urea per annum.The other came into operation in 1999 and is located in Gurun, Kedah. It has an annual capacity of 650 000 tonnes granular urea. This second plant exports about 65 percent of its products to Australia, Thailand and other South East Asian agricultural countries. The remainder is sold to the National Farmers' Association, Malaysia's largest urea consumer, for distribution in Malaysia. Most of the companies involved in fertilizer production engage in the mixing of straight fertilizers to produce compound fertilizers. The fertilizer industry in Malaysia is efficient and highly competitive. There are over 50 companies involved in the branded fertilizer trade with over 350 brands of various forms of fertilizers (Malaysian Agriculture Directory and Index, 2004). The normal channels of distribution are shown in Figure 4. The Government, in its drive to help farmers, particularly smallholders, to procure fertilizers has: stimulated fertilizer consumption through subsidy and credit schemes; facilitated the supply and distribution of fertilizers through FELDA, RISDA, FELCRA and FOA; stimulated the establishment of a large number of distribution points through farmers' cooperatives improved the cost-value ratio between fertilizers and agricultural produce by providing price support; provided extension services, research on fertilizer use and quality control. The government has also encouraged suppliers to market fertilizers in packages smaller than 50-kg bags, to allow farmers who cannot afford 50-kg bags to purchase lesser quantities.

FERTILIZER USAGE IN PADDY INDUSTRY MALAYSIA

In the context of Malaysian paddy industry, reliance on fertilizer usage to increase yield requires a good relationship fostered between the farmers as consumers, generally represented by farmer organizations, and manufacturers or retailers as suppliers (Ismaila, 2008; Zainal, Shamsudin, Mohamed, Adam, &

Figure 4. Marketing and distribution channel of fertilizers in Malaysia

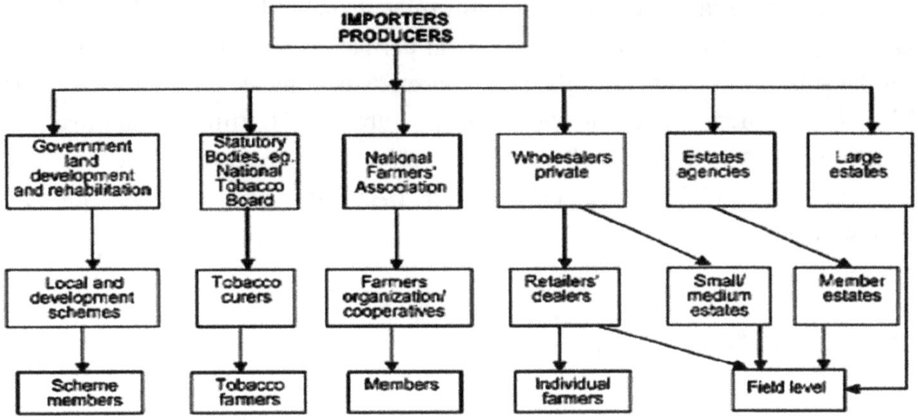

Kaffashi, 2014). The relationship does not merely involve purchasing and selling products as it also requires professional advice to the farmers. For instance, in the case of introductions to new technology or products would require feedback from the consumers. The government's effort to strengthen the paddy and rice industry continues to be one of the main agenda. The Ninth Malaysia Plan saw 70% of RM 2.5 billion, and later RM 3 million during the mid-review, allocated to national food security. The paddy sector in Malaysia is a heavily subsidized industry, which the summary of the input and fertilizer subsidies is as follows:

Despite the rigorous efforts and initiatives by the Government, Fahmiet al. (2013) mentioned that Malaysia's rice production is considered inefficient in terms of cost and production. It has also been noted that there are also issues and challenges specifically on the supply of fertilizer that has an impact on the yield of paddy. Some of the challenges in fertilizer supply revolve around the following issues (reported in BeritaTransformasiPertanian: Lahir Peladang Progres if, 2012): Ensuring that fertilizer

Figure 5. Input and fertilizer subsidy in paddy agriculture
Source:(A. Rizal et al., 2014)

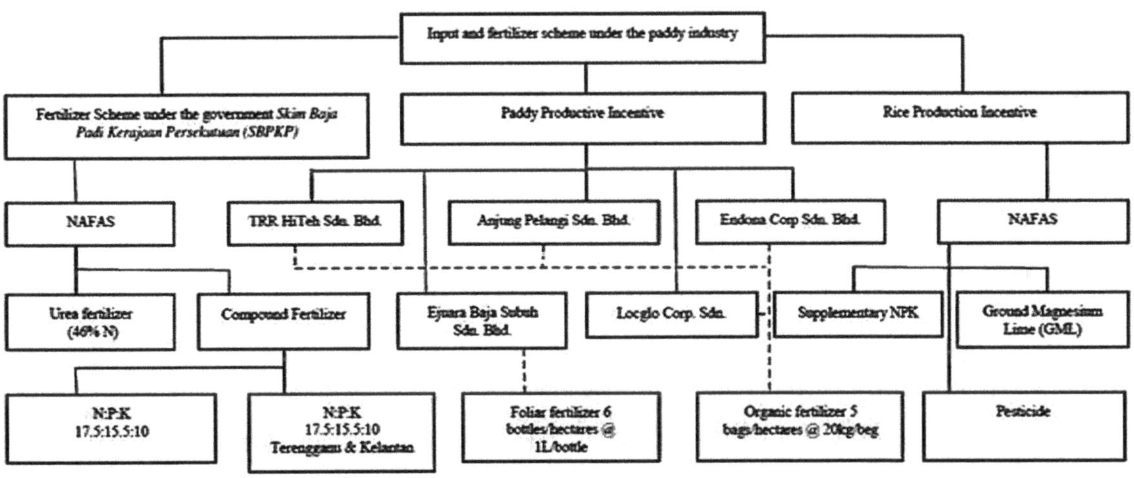

is supplied only to eligible farmers, more frequent visits to meet the farmers for monitoring, timely distribution of fertilizer to farmers. Malaysian government spent about 30 million USD annually on chemical fertilizer. As illustrated in Figure 5 National Farmer Association (NAFAS) has been the sole distributor of the subsidized fertilizer to the paddy farmer's nationwide (A. R. A. Rizal, Nordin, Saad, & Ismail, 2012).Hence, there is a need to ensure that the subsidized fertilizer is supplied to and received only by eligible farmers which require efficient communication management, specifically between the suppliers and consumers. In of this research study One BAJA need proper segmentation in order to be successful in the market.

FUTURE RESEARCH DIRECTIONS

This paragraph summarizes the review results and suggests future directions of One Baja fertilizer and procedures. The specific research field is relatively new and possible future perspectives have to be emphasized so that new innovation can be effective in order to increase the production.

Investment positive:

- Rising population increased the demand of chemical fertilizer usage and changing diets, putting a heavy demand on higher crop yields. With limited available acreage for planting, fertilizers will be increasingly needed to produce strong yields.
- Crop prices have most likely hit with upward potential in the near future Increases in crop prices will directly result in higher sales of fertilizers.
- Using One BAJA fertilizer is a strong an opportunity exist for paddy farmers in Malaysia to increase the production without damaging the environment.

Investment negative:

- Most fertilizer companies specialize in a specific fertilizer type since fertilizers are dependent on various conditions, companies can experience cyclicality. Investing in the wrong fertilizer farmers in a downward cycle could result in heavy losses.

Key to monitor:

- Crop prices will have the most immediate effect on fertilizer consumption. Farmers will be more likely to purchase and utilize fertilizer when they experience high margin. One Baja fertilizer is also known as control released fertilizer

CONCLUSION

The study has met the goal of mapping out some marketing solutions that are suitable for the NPK target market of One BAJA. To this end, the author has logically revised the marketing literature, including concepts relating to marketing, models to analyze macro and industry environment and marketing mix. Based on the literature review, the author has analyzed the situation of One BAJA NPK fertilizer. The study covers the macro and industry environment analysis of NPK fertilizer, major rivals' marketing mix. Then, the analysis of NPK market has laid the foundation of the marketing strategies to develop One BAJA NPK target market. With such a wide range of activities and a huge market to serve, it is important to have a structured marketing strategy and a marketing plan. In order to accomplish that, one needs to know the customers of the market. That is why market segmentation plays such a big role in marketing planning and marketing strategy. However, the results of One Baja research program validated that it is a value-added research program. Since the value of the obtained benefits is more than the funding devoted (RM12 million) on this particular research.

Overall, the marketing strategies of One Baja results indicated that the RM 12 million research program provided by the government can potentially generate benefits value of RM 1,824 million per year for the Malaysian fertilizer industry. In summary, the study concludes that the participants in research collaboration appear to get significant benefits and it also increase the paddy production. The most important benefit realized by the government is that the new green fertilizer technology in the form of CRF developed in this research program will have a high impact on the domestic paddy farming, the ecosystem as well as on the value proposition of the fertilizer industry. In a nutshell, this technology is economically efficient in the context of the fertilizer application frequency as compared to the conventional methods.

ACKNOWLEDGMENT

The authors would like to thank Universiti Teknologi PETRONAS (UTP) and the Ministry of Higher Education (MOHE), Malaysia, for the financial support under Long Research Grant Scheme (LRGS) and Management and Humanities to conduct this research.

REFERENCES

Adnan, N., Nordin, S. M., & Noor, A. N. (2017). *A Comparative Study on GFT Adoption Behaviour Among Malaysian Paddy Farmers. In Green Marketing and Environmental Responsibility in Modern Corporations* (pp. 239–263). IGI Global. doi:10.4018/978-1-5225-2331-4.ch012

Adnan, N., Nordin, S. M., & Redza, A. (2017). *Benefit of One Baja Fertilizer for Attaining Agricultural Sustainability among Malaysian Paddy Farmers: Agricultural Sustainability among Malaysian Paddy Farmers. In Business Infrastructure for Sustainability in Developing Economies* (pp. 143–160). IGI Global. doi:10.4018/978-1-5225-2041-2.ch008

Advisory, F. (1993). Development and Information Network for Asia and the Pacific (FADINAP). *Proceedings of the regional FADINAP seminar on fertilisation and the environment.*

Ahmad, S., & Tahar, R. M. (2014). Selection of renewable energy sources for sustainable development of electricity generation system using analytic hierarchy process: A case of Malaysia. *Renewable Energy*, *63*, 458–466. doi:10.1016/j.renene.2013.10.001

Ahmad, S., & Tahar, R. M. (2014). Selection of renewable energy sources for sustainable development of electricity generation system using analytic hierarchy process: A case of Malaysia. *Renewable Energy*, *63*, 458–466. doi:10.1016/j.renene.2013.10.001

Ahmed, O. H., Hanif, M., Husni, A., Abd Rahim, A., & Musa, M. H. (2001). Pineapple residue management practices and fertilizer regimes: Effects on P and K uptake, yield and some economic implications. *Pertanika. Journal of Tropical Agricultural Science*, *24*(2), 123–129.

Armstrong, G., Adam, S., Denize, S., & Kotler, P. (2014). *Principles of marketing*. Pearson Australia.

Azadi, H., Ho, P., & Hasfiati, L. (2011). Agricultural land conversion drivers: A comparison between less developed, developing and developed countries. *Land Degradation & Development*, *22*(6), 596–604. doi:10.1002/ldr.1037

Bakar, B. B. (2009). T*he Malaysian agricultural industry in the new millennium: Issues and challenges*. Academic Press.

Balbus, J. M., Boxall, A., Fenske, R. A., McKone, T. E., & Zeise, L. (2013). Implications of global climate change for the assessment and management of human health risks of chemicals in the natural environment. *Environmental Toxicology and Chemistry*, *32*(1), 62–78. doi:10.1002/etc.2046 PMID:23147420

Balogun, R. B., Ogbu, J. U., Umeokechukwu, E. C., & Kalejaiye-Matti, R. B. (2016). *Effective Microorganisms (EM) as Sustainable Components in Organic Farming: Principles, Applications and Validity. In Organic Farming for Sustainable Agriculture* (pp. 259–291). Springer.

Bean, J., & Hussey, L. (2012). *Marketing Public Sector Services: essential skills for the public sector*. HB Publications.

Brashear Alejandro, T., Javalgi, R. G., Gross, A. C., Benoy Joseph, W., & Granot, E. (2011). Assessing competitive advantage of emerging markets in knowledge intensive business services. *Journal of Business and Industrial Marketing*, *26*(3), 171–180. doi:10.1108/08858621111115895

Brooker, D. (2012). Build it and they will come? A critical examination of utopian planning practices and their socio-spatial impacts in Malaysias intelligent city. *Asian Geographer*, *29*(1), 39–56. doi:10.1080/10225706.2012.659192

Croft, M. J. (1994). *Market segmentation: A step-by-step guide to profitable new business*. Cengage Learning Emea.

Djellal, F., Francoz, D., Gallouj, C., Gallouj, F., & Jacquin, Y. (2003). Revising the definition of research and development in the light of the specificities of services. *Science & Public Policy*, *30*(6), 415–429. doi:10.3152/147154303781780227

FADINAP. (1984). Report. Marketing, distribution and use of fertilizer in Malaysia. FADINAP.

Fahmi, Z., Samah, B. A., & Abdullah, H. (2013). Paddy Industry and Paddy Farmers Well-being: A Success Recipe for Agriculture Industry in Malaysia. *Asian Social Science*, *9*(3), 177. doi:10.5539/ass.v9n3p177

FAO. (2012). Retrieved from http://www.fao.org/docrep/x0262e/x0262e06.htm

Fred van Raaij, W., & Verhallen, T. M. (1994). Domain-specific market segmentation. *European Journal of Marketing*, *28*(10), 49–66. doi:10.1108/03090569410075786

Furnham, A., & Gunter, B. (2015). *Corporate Assessment (Routledge Revivals): Auditing a Company's Personality*. Routledge.

Grassini, P., Eskridge, K. M., & Cassman, K. G. (2013). Distinguishing between yield advances and yield plateaus in historical crop production trends. *Nature Communications*, 4. PMID:24346131

Haris, N. B. M., Hamzah, A., Krauss, S. E., & Ismail, I. A. (2013). Relationship between Decision-Making Inputs and Productivity among Paddy Farmers" in Integrated Agriculture Development Areas (IADAs), in Malaysia. *International Journal on Advanced Science, Engineering and Information Technology*, *3*(1), 64–70.

Haris, N. B. M., Hamzah, A., Krauss, S. E., & Ismail, I. A. (2013). Relationship between Decision-Making Inputs and Productivity among Paddy Farmers" in Integrated Agriculture Development Areas (IADAs), in Malaysia. *International Journal on Advanced Science, Engineering and Information Technology*, *3*(1), 64–70.

Hashim, H., & Ho, W. S. (2011). Renewable energy policies and initiatives for a sustainable energy future in Malaysia. *Renewable & Sustainable Energy Reviews*, *15*(9), 4780–4787. doi:10.1016/j.rser.2011.07.073

Homburg, C., Kuester, S., & Krohmer, H. (2013). *Marketing management: A contemporary perspective*. McGraw-Hill Higher Education.

Hosseini, S. E., & Wahid, M. A. (2013). Feasibility study of biogas production and utilization as a source of renewable energy in Malaysia. *Renewable & Sustainable Energy Reviews*, *19*, 454–462. doi:10.1016/j.rser.2012.11.008

Hunter, W., Hutschemaekers, G. J., Unger, J.-P., de Paepe, P., Ghilbert, P., Soors, W., & Rissanen, P. etal. (2003). AP Nauta, J. von Grumbkow. *Policy*, *3*, 4.

Hutt, M., & Speh, T. (2012). *Business marketing management: B2B*. Cengage Learning.

Hutt, M. D., & Speh, T. W. (1984). The marketing strategy center: Diagnosing the industrial marketers interdisciplinary role. *Journal of Marketing*, *48*(4), 53–61. doi:10.2307/1251510

Ismaila, S. S. (2008). *Off farm employment among paddy farmers in the Muda and Kemasin Semarak granary areas*. Universiti Putra Malaysia.

Jenkins, M., & McDonald, M. (1997). Market segmentation: Organizational archetypes and research agendas. *European Journal of Marketing*, *31*(1), 17–32. doi:10.1108/03090569710157016

Kearney, D. (2013). Food Deprivations as Crimes against Humanity. *NYUJ Int'l L. & Pol.*, *46*, 253.

Kotler, P. (2012). *Kotler on marketing*. Simon and Schuster.

Kotler, P. (2012). *Kotler on marketing*. Simon and Schuster.

Kotler, P., Armstrong, G., Saunders, J., & Wong, V. (2003). *Principi di Marketing (Italian Translation of Principles of Marketing)*. ISEDE and Prentice Hall International.

Kotler, P., & Caslione, J. A. (2009). *Chaotics: the business of managing and marketing in the age of turbulence*. AMACOM Div American Mgmt Assn.

Kotler, P., & Keller, K. L. (2009). *Dirección de marketing*. Pearson educación.

Kristoferson, L. A., & Bokalders, V. (2013). *Renewable energy technologies: their applications in developing countries*. Elsevier.

Lee, K., Kim, H., & Vohs, K. D. (2011). Stereotype threat in the marketplace: Consumer anxiety and purchase intentions. *The Journal of Consumer Research*, *38*(2), 343–357. doi:10.1086/659315

Levina, O. (2010). *Business market segmentation: Segmenting market for AmCham Finland*. Academic Press.

McDonald, M. (2012). *Market Segmentation*. Wiley Online Library. doi:10.1002/9781119207863

McDonald, M., & Wilson, H. (2011). *Marketing plans: How to prepare them, how to use them*. John Wiley & Sons.

McDonald, M., & Wilson, H. (2011). *Marketing plans: How to prepare them, how to use them*. John Wiley & Sons.

Melchior, M. R. (2011). From Design Nations to Fashion Nations? Unpacking Contemporary Scandinavian Fashion Dreams. *Fashion Theory*, *15*(2), 177–200. doi:10.2752/175174111X12954359478681

Miorandi, D., Sicari, S., De Pellegrini, F., & Chlamtac, I. (2012). Internet of things: Vision, applications and research challenges. *Ad Hoc Networks*, *10*(7), 1497–1516. doi:10.1016/j.adhoc.2012.02.016

Moeskops, B., Buchan, D., Sleutel, S., Herawaty, L., Husen, E., Saraswati, R., & De Neve, S. etal. (2010). Soil microbial communities and activities under intensive organic and conventional vegetable farming in West Java, Indonesia. *Applied Soil Ecology*, *45*(2), 112–120. doi:10.1016/j.apsoil.2010.03.005

Nagarajan, S. (2008). Indians in Malaysia towards vision 2020. *Rising India and Indian communities in East Asia*, 375-398.

Nhamo, N., Rodenburg, J., Zenna, N., Makombe, G., & Luzi-Kihupi, A. (2014). Narrowing the rice yield gap in East and Southern Africa: Using and adapting existing technologies. *Agricultural Systems*, *131*, 45–55. doi:10.1016/j.agsy.2014.08.003

Noppers, E. H., Keizer, K., Bolderdijk, J. W., & Steg, L. (2014). The adoption of sustainable innovations: Driven by symbolic and environmental motives. *Global Environmental Change*, *25*, 52–62. doi:10.1016/j.gloenvcha.2014.01.012

Noppers, E. H., Keizer, K., Bolderdijk, J. W., & Steg, L. (2014). The adoption of sustainable innovations: Driven by symbolic and environmental motives. *Global Environmental Change*, *25*, 52–62. doi:10.1016/j.gloenvcha.2014.01.012

Nordin, S. M., Noor, S. M., & Saad, M. S. M. (2014). Innovation Diffusion of New Technologies in the Malaysian Paddy Fertilizer Industry. *Procedia: Social and Behavioral Sciences*, *109*, 768–778. doi:10.1016/j.sbspro.2013.12.542

Othman, Z. (2012). *Information and communication technology innovation as a tool for promoting sustainable agriculture: a case study of paddy farming in west Malaysia*. University of Malaya.

Patel, R. (2013). The long green revolution. *The Journal of Peasant Studies*, *40*(1), 1–63. doi:10.1080/03066150.2012.719224

Prabovo, N. E., Foster, H. L., Nelson, S., Sitepu, B., & Nelson, P. (2012). *Practical use of oil palm nutrient physiological efficiency with regard to nutrient recovery and agronomic efficiencies at different Sumatran sites*. Academic Press.

Ramli, N. S., Hassan, M. S., Samah, B. A., Ali, M. S. S., Shaffril, H. A. M., & Azaharian, Z. S. (2013). Problems in using agricultural information from television among farmers in Malaysia. *Asian Social Science*, *9*(9), 70. doi:10.5539/ass.v9n9p70

Rizal, A., Redza, A., Md Nordin, S., Saad, M., Shamsuri, M., & Wahab, H. (2014). Inter-Organization Communication Management between Organizations in a Subsidized Fertilizer Market in Malaysia. *Global Business and Management Research: An International Journal*, *6*(3), 229–236.

Rizal, A. R. A., Nordin, S. M., Saad, M. S., & Ismail, K. (2012). Engagement Strategies for Stakeholder Management in New Technology Development in the Fertilizer Industry-A Conceptual Framework. *Proceedings of World Academy of Science, Engineering and Technology*.

Rizal, A. R. A., Nordin, S. M., Saad, S. M., & Wahab, H. (2013). Communication Management between Organizations in a Subsidized Market: The Case of Paddy Fertilizer in Malaysia. *Asian Journal of Research in Social Sciences and Humanities*, *3*(12), 1.

Roopan, S. M., Madhumitha, G., Rahuman, A. A., Kamaraj, C., Bharathi, A., & Surendra, T. (2013). Low-cost and eco-friendly phyto-synthesis of silver nanoparticles using Cocos nucifera coir extract and its larvicidal activity. *Industrial Crops and Products*, *43*, 631–635. doi:10.1016/j.indcrop.2012.08.013

Sacks, F. M., Pfeffer, M. A., Moye, L. A., Rouleau, J. L., Rutherford, J. D., Cole, T. G., & Wun, C.-C. etal. (1996). The effect of pravastatin on coronary events after myocardial infarction in patients with average cholesterol levels. *The New England Journal of Medicine*, *335*(14), 1001–1009. doi:10.1056/NEJM199610033351401 PMID:8801446

Safitri, R., & Andriyani, A. (2011). Keefektifan Pemberian Posisi Semi Fowler Terhadap Penurunan Sesak Nafas Pada Pasien Asma di Ruang Rawat Inap Kelas III RSUD Dr. Moewardi Surakarta. *Gaster Jurnal Ilmu Kesehatan,* *8*(2), 783-792.

Shaffril, H. A. M., Asmuni, A., & Ismail, A. (2010). Menu. *Journal of International Social Research*, *3*, 450–458.

Shaffril, M., Azril, H., Uli, J., Asiah, M., Ali, N. A., Abu Samah, B., & Thomas, K. etal. (2010). Can quality of work life affect work performance among government agriculture extension officers? A case from Malaysia. *Journal of Social Sciences*, *6*(1), 64–73. doi:10.3844/jssp.2010.64.73

Taylor, S., & Todd, P. (1995). Decomposition and crossover effects in the theory of planned behavior: A study of consumer adoption intentions. *International Journal of Research in Marketing*, *12*(2), 137–155. doi:10.1016/0167-8116(94)00019-K

Tey, Y. S. (2013). *The Adoption of Sustainable Agricultural Practices: An Integrative Approach for Malaysian Vegetable Farmers*. The University of Adelaide.

Windeshausen, H. N. (1973). Market Segmentation: Concepts and Applications. *Journal of Marketing (pre-1986)*, *37*(4), 124.

Wittman, H., Beckie, M., & Hergesheimer, C. (2012). Linking local food systems and the social economy? Future roles for farmers markets in Alberta and British Columbia*. *Rural Sociology*, *77*(1), 36–61. doi:10.1111/j.1549-0831.2011.00068.x

Zainal, Z., Shamsudin, M. N., Mohamed, Z. A., Adam, S. U., & Kaffashi, S. (2014). Assessing the Impacts of Climate Change on Paddy Production in Malaysia. *Research Journal of Environmental Sciences*, *8*(6), 33.

ADDITIONAL READING

Chen, J. H. (2006). The combined use of chemical and organic fertilizers and/or bio fertilizer for crop growth and soil fertility. In *International Workshop on Sustained Management of the soil-rhizosphere system for efficient crop production and fertilizer*.

Fahmi, Z., Samah, B. A., & Abdullah, H. (2013). Paddy Industry and PaddyFarmers Well-being: A Success Recipe for Agriculture Industry inMalaysia. *Asian Social Science*, *9*(3), 177. doi:10.5539/ass.v9n3p177

Hashim, H., & Ho, W. S. (2011). Renewable energy policies and initiatives for a sustainable energy future in Malaysia. *Renewable & Sustainable Energy Reviews*, *15*(9), 4780–4787. doi:10.1016/j.rser.2011.07.073

Liew, Y. A., Omar, S. S., Husni, M. H. A., Abidin, M. Z., & Abdullah, N. A. P.(2010). Effects of Micronutrient Fertilizers on the Production of. Malaysianjournal of soil science, 14, 71-82.

Vlek, P. L., & Byrnes, B. H. (1986). 7. The efficacy and loss of fertilizer N inlowland rice. *Fertilizer Research*, *9*(1-2), 131–147. doi:10.1007/BF01048699

KEY TERMS AND DEFINITIONS

Business-to-Business: (B2B): Refers to a situation where one business makes a commercial transaction with another. This typically occurs when: A business is sourcing materials for their production process (e.g. a food manufacturer purchasing salt).

Marketing Research: "The process or set of processes that links the producers, customers, and end users to the marketer through information used to identify and define marketing opportunities and problems; generate, refine, and evaluate marketing actions; monitor marketing performance; and improve understanding of marketing as a process.

Market Segmentation: A marketing strategy which involves dividing a broad target market into subsets of consumers, businesses, or countries that have, or are perceived to have, common needs, interests, and priorities, and then designing and implementing strategies to target them.

One Baja Fertilizer: In Malaysia green fertilizer technology in the form of Control Release Fertilizer (CRF).

Segmentation Bases: The dimensions that can be used to segment a market. "A segmentation basis is defined as a set of variables or characteristics used to assign potential customers to homogeneous groups.

NOMENCLATURE

CRF Control released Fertilizer
PPK Pertubuhan PeladangKawasan
NAFAS National Farmer Association
BERNAS Padiberas NasionalBerhad (National Rice Company)
BLS Barat Laut Selangor (North West)
DOA Department of Agriculture
GAP Good Agricultural Practice
IRRI International Rice Research Institute
IADA Integrated Agriculture Development Area
KADA Kemubu Agricultural Development Authority (Kelantan)
KETARA Northern Terengganu
KOREF Kahang Organic Rice Eco Farm
KSM Kerian Sungai Manik (Perak)
MADA Muda Agricultural Development Authority
MARDI Malaysian Agricultural Research and Development Institute

Chapter 11
Multiple Exploration of Entrepreneurs' Suggestions for Agricultural Development of Local Regional Units in Greece

Odysseas Moschidis
University of Macedonia, Greece

Vasileios Ismyrlis
Greek Statistical Authority, Greece

ABSTRACT

The purpose of the present article is to evaluate the factors which are considered to be important for the agribusiness development of a local economy, with data derived from the entrepreneurs' perspective. For this purpose, an appropriate methodology was designed, in order to include the most of the aforementioned factors. Emphasis was given to questions which can illustrate the level of technological innovation with actions and initiatives like digital marketing, innovative ability and others. Therefore, a questionnaire was created and was then applied to many regions in northern Greece. In respect of data analysis, the contribution of Correspondence Analysis (CA), a method from the multidimensional statistics field, was crucial because it easily revealed the characteristics that intensively differentiated themselves. The above methodologies and their special characteristics facilitated also the implementation of SWOT analysis. In the case of the Regional Units examined in the current research, the positive and negative factors-points were easily revealed and presented.

DOI: 10.4018/978-1-5225-2107-5.ch011

INTRODUCTION

The agribusiness sector is facing many challenges worldwide, as the globalization affects its structure and access to markets. Having already many additional problems due to the complexities and uncertainties linked to the sector, it is nowadays even more essential to try to cope with and implement new methods and technologies. One major example of these complexities is the global production networks (Hampton et al., 2007).

The present chapter is the beginning of a new research with a principle goal to help the administrations of the local authorities or/and central government evaluate their performance in aiming to pursue rural economic development. In their effort to contribute to this field, one of the most important actions is to implement initiatives that can assist the agricultural enterprises to develop and stabilize strong entrepreneurship values. Another important activity in the same direction, is to evaluate and record the current situation in their local regional unit. This can be succeeded with the utilization of suitable instruments that collect information from the entrepreneurs themselves.

The above suggestion to collect information, can be realized with an effort to measure the factors that enhance agricultural development from the entrepreneurs' perspective. In order to collect the relevant information, the research presented here, is based on data obtained by a questionnaire survey. The questionnaire that created by the authors, contained a section about demographics and another one concerning the entrepreneurs' perception of the existed local agricultural development's actions-factors. The main goal is to provide an important tool that administrators of the local authorities could use to make decisions, with an aim to improve the economic climate, and furthermore to pursue development.

For the analysis of the data, Correspondence Analysis, a multidimensional statistical methodology, is mainly used, as the most suitable for discovering correspondences (Benzecri, 1992) between the variables. It is an exploratory methodology of data analysis that does not assume any distribution of the data and puts forward possible trends that exist in the data graphically (Greenacre, 2007), as well. The results are presented on graphs that represent the configuration of points in projection planes formed by the first principal axes (Lebart et al., 1984, p. 44). This approach enables the researcher not only to analyze the phenomenon in a more holistic way, but also to highlight potential issues and questions that have not been previously identified. Two proposed tables, which enable the evaluation of ordinal data in a different aspect, will also be utilized (Moschidis, 2006; 2009).

The methodology utilized for the determination of the relationships of the different characteristics, is S.W.O.T analysis. The specific technique is used to evaluate the Strengths, Weaknesses/Limitations, Opportunities, and Threats involved in any business project (Piercy & Giles, 1989). It involves specifying the objective of the business venture or project and identifying the internal and external factors that are favorable and unfavorable to achieve that objective (Helms & Nixon, 2010). In the present case it can be used as a measure for evaluating and presenting the factors that can contribute to the economic development of a local regional unit.

The research had started from regional units from Northern Greece, with a view to be generalized in national level. The study concluded in some interesting results; firstly, in some distinguished factors for the agricultural growth that should be taken into consideration by the authorities, and secondly in the realization that there appeared to be no differences in the opinions of entrepreneurs with different demographic backgrounds. The methodology used, seemed to be perfectly suitable to extract the elements that distinguished as from the present chapter.

BACKGROUND

Agricultural Economic Development and Innovation

Agriculture is a business sector of the economy, but it seems to be a strategic one as it handles food safety and quality, it produces interdependencies with other sectors of the economy and deals with most of the natural resources of a country's economy (Ariza et al., 2013). The global demand for food will increase by 70% until 2050, but the productivity of the agricultural sector has decelerated the last years (GSRT, 2013;Moschidis & Arabatzis, 2013). For all the above factors, governments are obliged to pay a close attention to this part of the economy. For example, in the unified Europe, there are some common targets and policies for the agricultural policy.

As referred to the ''European Commission rural development program for 2014-2020'' (European Commission, 2014), some of the priorities to be followed are:

- Fostering knowledge transfer
- Enhancing competitiveness
- Promoting food chain organization and risk management
- Restoring, preserving and enhancing ecosystems
- Promoting resource efficiency and supporting the shift toward a low-carbon and climate-resilient economy
- Promoting social inclusion, poverty reduction and economic development in rural areas.

From all the above priorities it can be concluded that the agricultural sector inside the European Union should expand to fields like innovation or/and technology, that they may be seem peculiar (especially to Greek farmers) and perhaps a new addition for this sector, but they are implemented to all other business sectors and activities for many years and they seem to be essential to enhance competitiveness and to survive the demanding global environment. In addition, many researches have highlighted that topics as innovation, entrepreneurship and the learning organization have a linkage with market orientation (Johnson et al., 2009). Market orientation is also a key factor for the appropriate connection with the markets and the customer and this aspect has already been examined in the agribusiness field (van Duren et al., 2003).

Development of agricultural entrepreneurship has been an important policy in order to increase the value of agricultural production and open up the sector for businesses which is a clear departure from what obtained in the past, when oil prices were at their peak (Olawa & Olawa, 2015). Agribusinesses should compete not only on domestic markets, but on global as well (Esterhuizen, 2006) as competition and customer orientation have further increased (Dlamini & Kirsten, 2014). Developing entrepreneurial skills of the farmers seem to be essential to expand the competitive ness of their enterprises, many times needed just to survive or to find chances to expand more. A contribution to this effort can always be an effective regional policy (Polyzos & Arabatzis, 2005), which involves the increase of economic productivity. The productivity of the agricultural sector can be further associated with factors like good rural infrastructure (Llanto, 2012) and access to appropriate technology (Pinstrup-Andersen & Shimokawa, 2007). Investment in infrastructure contributes to the reduction in transport and marketing costs and therefore producers are better linked to markets (Ashok & Balasubramanian, 2006). Surely, governments through public investment can further contribute to the establishment of the above factors (Nadeem et al., 2011).

Innovation seems to be another key factor for economic growth and development, both for enterprises and countries. Innovation can be implemented in the fields of product, processes, organization and market (Śledzik, 2013) and can include and can be facilitated by technological aspects and actions (Papaioannou et al, 2015). Many of these actions are examined in the present questionnaire.

Greece's Status in Agricultural Entrepreneurship, Innovation, and Technology

Agriculture is the fourth most important economic activity (see economic sectors in ''Key terms and Definitions'' section) within the structure of Greek GDP(Gross Domestic Product) and represents 3.7% of the total (ELSTAT, 2016), which makes it essential for the Greek economy, although there was a decline in this percentage from 6.08% in 2000 to 3.7% in 2013. Certainly, the economic turmoil existed since 2008 has a serious negative impact to the Greek food sector (Chatzipetrou & Moshidis, 2016). In terms of employment, agriculture accounts for 13.2% of the total in the country (European Commission, 2016). The Northern Greece regions, which are examined in the present study, include those of Central, Eastern and Western Macedonia, Hepirus and Thrace. All together they assemble the 29.4% of the total number of Greek agricultural enterprises and the 26.7% of the added values of agricultural production (ELSTAT, 2009).

It seems that globalization has affected seriously the Greek agricultural sector, as it comes up afainst faces a situation with intense competitiveness with many other economies/countries that are in a position to offer much better prices and very often products of better quality. On the other hand, this globalization is always a big opportunity to enter into new markets and countries. Nevertheless, the Greek agricultural sector is confronts many negative issues, as the great number of small farms (European Commission, 2014) and the high rate of economically active people employed by the sector (Polyzos & Arabatzis, 2005).

Concerning the innovation implemented in this sector, there is a lag in innovation measurement for agricultural firms (Ariza et al., 2013) all over the world and therefore data from two big surveys, the Global Competitiveness and the Global Innovation were also utilized. These surveys present specific indices to express many aspects of economic activities, but surely the most important ones, like competitiveness and innovation (Cornell University, INSEAD & WIPO, 2015; World Economic Forum, 2013). Greece seems to have achieved notable results in the last years (Cornell University, INSEAD & WIPO, 2015; World Economic Forum, 2015), concerning the Competitiveness and Innovation indices and certain pillars (Graduates in science, Quality research institutes, Ease of protecting investions) have distinguished. However, these results refer to the overall Greek economy and entrepreneurship and there is no specific data for the agricultural sector.

METHODOLOGY

Questionnaire

The fifty-five (55) questions of the current questionnaire (Table 11, Appendix), were taken from the relevant bibliography and included topics concerning competitiveness, innovation and other aspects of entrepreneurship.

More specifically the fifty-five questions (E1-E55), could be grouped in the following categories (in the parentheses some of them are referred):

- Economic factors (ease of getting credit, economic climate, strength of investors protection, effects of taxation, funding from ESPA program, supports new workplaces, agritourism's expansion).
- Marketing factors (effective advertising, consumer satisfaction (examination of trends), possibility of exports, local demand, intensity of local competition)
- Quality of products and processes (quality of local suppliers, possession of quality standards as ISO 9001, 22000 or Agrocert)
- Human capital-Training-Skills (education, extent of staff training, skills possession)
- Infrastructure Logistics-Transports: (quality of roads, connection with big urban centers)
- Internet access and usage: (ICT Access, ICT Use)
- Information and knowledge diffusion-technology: (knowledge diffusion)
- Environment and Energy (ecological sustainability, photovoltaic parks, natural resource protection, protection of biodiversity)
- Innovation in production and marketing: capacity for innovation, use of digital marketing, number of new products, patents.

There were also four questions about demographics: age, level of education, profession and sex. This study covers the results from a research started in the beginning of 2015 and completed in the April of the same year in all the Regional Units of Northern Greece. The questionnaire was distributed to a broad sample of entrepreneurs in the local economy over eighteen years old, covered all professions and education levels. The question for the entrepreneurs was: "To what degree do you think the following characteristics exist in your business sector or local economy?", and the possible answers were, 1:"not at all", 2:"little", 3:"somewhat", 4:"much", 5:"a great deal". Consequently, higher scores on this scale indicate the strong existence of the characteristic. Moreover, there was a definite distinction to which factors referred to the local economy and which to the local business sector.

The main subject to be answered was to seek the most and least featured factors that existed in the local economy or business sector. Afterwards, merely the questions from the agricultural entrepreneurs where distinguished and are presented from now on.

SWOT

The S.W.O.T. analysis is a methodology from the field of management science, which is used to evaluate the Strengths, Weaknesses/Limitations, Opportunities, and Threats existed in any business project. It involves specifying the objective of the business venture or project and identifying the internal and external factors that are favorable and unfavorable to achieve that objective. The final step of the method is a presentation of a table, which includes all the factors for the Strengths, Weaknesses/Limitations, Opportunities, and Threats that exist in the current project.

In the case of the regional public administration, SWOT analysis is a means of reviewing and evaluating on the performance and potentials of the local economy. With the execution of this process, to identify the local, regional unit economic performance, SWOT analysis can be a valuable evaluation tool for the management of the regional administration.

Correspondence Analysis and Special Tables of Coincidences

C.A. is an exploratory technique of the data analysis field, which does not assume any distribution of the data and is putting forward intensively differentiated trends that exist in the data, graphically as well. (Moschidis et al., 2009).

In this project C.A. is applied to two proposed tables of coincidences:

1. The table of evaluation: this table displays the distribution of the n_1 individuals of group A in 3 grades of the three-grade scale (we assume for convenience that the p questions E1, E2,..., Ep are formulated in a three-grade scale (not at all(1)-moderate(2)- very much(3)), therefore the table of evaluation has the following form (Table 1):

For the meaning of the numbers $K_{ij,}$ we note that number, e.g. K_{23} equals the number of the individuals of group A that chose for the question E2 the grade 3.

2. The table of comparative evaluation (Moschidis, 2006, 2009, 2015), which is defined in the comparative evaluation of the questions E1, E2,..., Ep from the two (or more generally) groups A and B. This table derives from the horizontal union of the tables of evaluations of groups A and B, therefore the table of comparative evaluation has the following form (Table 2):

Subsequently, implementing the correspondence analysis to the table of comparative evaluation and the points-columns e.g. A2, B3 are close to the point-line E2 in the first factorial space, it transpires that group A evaluated the criterion E2 with the grade "moderate"(2), while the group B with the grade "very much" (3). With this aspect, the comparison of the views of the different groups A and B, for the criteria E1, E2,..., Ep is realized (Moschidis, 2006).

Table 1. The table of evaluation of the group A

	A			Sum
	A1	A2	A3	
E1	K_{11}	K_{12}	K_{13}	n_1
E2	K_{21}	K_{22}	K_{23}	n_1
Ep	K_{p1}	K_{p2}	K_{p3}	n_1
Sum	K_1	K_2	K_3	pn_1

Table 2. The table of comparative evaluation of groups A, B

	A			B			Sum
	A1	A2	A3	B1	B2	B3	
E1	K_{11}	K_{12}	K_{13}	K'_{11}	K'_{12}	K'_{13}	$n_1+n_2=n$
E2	K_{21}	K_{22}	K_{23}	K'_{21}	K'_{22}	K'_{23}	n
Ep	K_{p1}	K_{p2}	K_{p3}	K'_{p1}	K'_{p2}	K'_{p3}	n
Sum	$K_{.1}$	$K_{.2}$	$K_{.3}$	$K'_{.1}$	$K'_{.2}$	$K'_{.3}$	pn

ANALYSIS

Descriptive Statistics

A brief presentation of some descriptive statistics of the sample analysed, are presented below:

The average grade of the responses in the fifty-five questions, of all the agricultural entrepreneurs in the 5-point scale was 2.65, which means that the economic characteristics examined in this research, are not very well implemented in the regions. For example the percentage of the entrepreneurs that replied "somewhat" was 32.72% (Table 4).

Application of the Proposed Methodology

Multivariate Analysis

Firstly, for the implementation of the Correspondence analysis, the table of evaluation for the fifty-five questions was created (see Table 5). This table, is the realized application of the first one of the proposed tables, which were referred above in the methodology section.

For the better understanding of the numbers in the above cells, an example is given. Number 50 in question E7, corresponds to the grade 2 of the scale, which means that fifty respondents have answered "little" in this question.

The results after the implementation of the correspondence analysis, to the table of evaluation are following.

Firstly, the table of eigenvalues is presented, where the total inertia is 0.25167 (Table 6).

- The first factorial axis (first main trend) interprets with a percentage of 72.66 the researched issue.
- The second factorial axis (second main trend) has a 19.79 interpretation percentage.

The first factorial space interprets data with a percentage of 92.45, which is a very good and adequate percentage for further analysis.

From the results of correspondence analysis, which are presented in Table 7, we will use the most important interpretation indicator of point (characteristic) towards axis, which is Contribution (CTR), as it expresses its contribution's percentage, in axis construction. The points with large CTR towards the axis, construct and many times highlight its physical importance.

Table 3. Demographics of the respondents (agricultural entrepreneurs)

Gender	%	Age	%	Education Level	%
Male	65.2	18-30	36.1	Not attended	1.0
Female	34.8	30-44	28.1	Primary	33.5
		>45	35.8	Secondary	44.6
				Univ.graduate	19.2
				Post graduate	1.7

Table 4. Percentages of the responses in the 5-grade scale

Response-Grade	Percentage-%
1:not at all	18.00
2: little	27.00
3:somewhat	32.72
4:much	16.93
5:a great deal	5.35

Table 5. Table of evaluation of the fifty-five questions

Questions	Grades					Questions	Grades					Questions	Grades				
	1	**2**	**3**	**4**	**5**		**1**	**2**	**3**	**4**	**5**		**1**	**2**	**3**	**4**	**5**
E1	20	51	78	29	2	**E19**	34	60	60	23	3	**E38**	76	50	40	9	5
E2	28	66	55	28	3	**E20**	20	45	71	37	7	**E39**	37	58	50	24	11
E3	20	50	76	28	6	**E21**	19	40	50	57	14	**E40**	24	62	67	22	5
E4	18	49	75	33	5	**E22**	23	50	79	27	1	**E41**	12	36	48	65	19
E5	19	58	71	29	3	**E23**	61	55	51	13	0	**E42**	12	27	46	63	32
E6	57	53	51	19	0	**E24**	28	52	75	22	3	**E43**	37	47	63	27	6
E7	44	**50**	69	16	1	**E25**	24	45	69	37	5	**E44**	50	64	53	11	2
⋮	⋮	⋮	⋮	⋮	⋮	⋮	⋮	⋮	⋮	⋮	⋮	⋮	⋮	⋮	⋮	⋮	⋮
E15	72	62	37	7	2	**E33**	16	12	34	51	67	**E52**	25	56	71	25	3
E16	35	22	57	44	22	**E34**	23	45	85	20	7	**E53**	29	53	71	25	2
E17	30	28	49	43	30	**E35**	11	48	66	45	10	**E54**	16	46	54	48	16
E18	71	52	42	14	1	**E36**	45	62	50	21	2	**E55**	22	56	76	26	0

Table 6. Eigenvalues-Inertia for the table of evaluation

TOTAL INERTIA		0.25167		
AXIS	**INERTIA**	**%EXPLAN**	**SUM**	**SCREE PLOT**
1	0.1828533	72.66	72.66	**
2	0.0498098	19.79	92.45	***********
3	0.0120358	4.78	97.23	***
4	0.0069669	2.77	100.00	**

Using F indicator (coordinate) we define the side of the axis in which the point (characteristic) is represented. Therefore, the points with positive coordinate are situated on the right side and on the other side are those with negative. We note that the average CTR is 1000:55= 18.18, where 55 is the number of the points-questions. Therefore, we can consider points of high contribution in axis construction, those with CTR values above average (CTR values over 20 are in bold for the first axis and over 25 for the second). Having in mind the above estimations and the help of the visualization of the axes (see Fig. 1, 2, 3), we conclude in the following results (Greenacre, 2007).

The first axis (Figure 1), opposes characteristics, that have proved to exist in a great extent in the local economy or agricultural sector, the following: "Productivity of the sector" (E29), ''Connection with big urban centers (E32), "Unemployment" (E33), "Sectorial unemployment" (E37), "Tourist sector's development perspectives" (E41) "Secondary production perspectives" (E42), "Green development" (E17), '' Connection with other sectors'' (E26), to characteristics that their existence have been realized to be negligible, like: "Investments from outsiders" (E15) and "Innovative companies in the sector" (E47), ''Digital marketing in the sector'' (E14) and ''Photovoltaic parks (E18).

Table 7. Interpretation indicators: Co-ordinates (#F1), Projections-correlations (COR), Contributions (CTR) of the first two axes

	#F11	COR	CTR	#F22	COR	CTR			#F1	COR	CTRR	#F22	COR	CTR
E1	-92	87	1	-281	825	**28**		**E28**	-86	192	1	171	763	10
E2	-166	415	3	-103	159	4		**E29**	1342	953	**179**	281	41	**28**
E3	-24	8	1	-219	758	17		**E30**	212	322	5	-272	529	26
E4	3	0	1	-264	942	25		**E31**	77	72	1	-246	719	22
E5	-90	100	1	-251	780	22		**E32**	647	843	**41**	-105	21	4
E6	-382	768	14	175	162	11		**E33**	1350	816	**181**	604	163	**133**
E7	-309	835	10	9	0	1		**E34**	-57	30	1	-185	323	12
E8	-409	849	16	157	125	10		**E35**	207	353	5	-273	607	27
E9	-167	453	3	-44	30	1		**E36**	-301	857	9	86	70	3
E10	190	263	4	-270	528	**26**		**E37**	907	854	**81**	253	66	23
E11	-319	912	10	-57	29	2		**E38**	-442	426	19	493	532	**88**
E12	-381	775	14	106	60	5		**E39**	-69	155	1	116	454	5
E13	-305	709	10	35	9	1		**E40**	-136	286	2	-139	300	8
E14	-481	765	**22**	240	191	21		**E41**	527	754	**27**	-149	59	9
E15	-525	585	**27**	435	405	**69**		**E42**	774	965	**59**	32	1	1
E16	375	690	13	142	99	8		**E43**	-106	722	2	-8	3	1
E17	518	824	**26**	232	165	19		**E44**	-396	838	15	156	130	9
E18	-463	567	**21**	370	364	**50**		**E45**	-427	807	18	172	131	10
E19	-209	856	5	-49	45	1		**E46**	-96	318	1	-76	201	3
E20	67	84	1	-218	888	17		**E47**	-566	629	**31**	422	350	**65**
E21	339	621	11	128	87	6		**E48**	-118	421	2	-86	223	3
E22	-135	182	2	-263	697	25		**E49**	-369	709	13	165	144	10
E23	-446	763	19	233	209	19		**E50**	117	417	2	-134	539	7
E24	-162	384	3	-158	368	10		**E51**	-198	394	4	-220	491	17
E25	11	2	1	-199	909	14		**E52**	-140	321	2	-181	542	11
E26	594	913	**35**	-119	36	6		**E53**	-168	494	3	-157	428	9
E27	311	812	10	131	143	7		**E54**	315	772	9	-114	100	5
								E55	-172	268	2	-270	663	**26**

Figure 1. First factorial axis

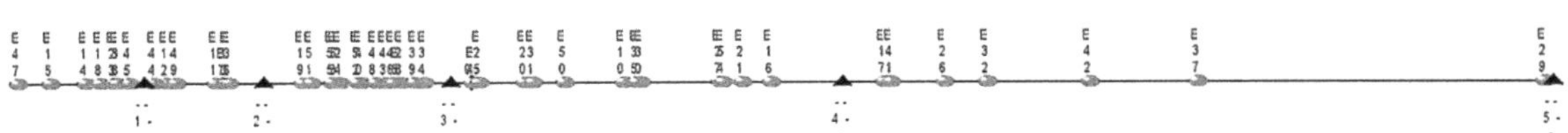

Figure 2. Second factorial axis

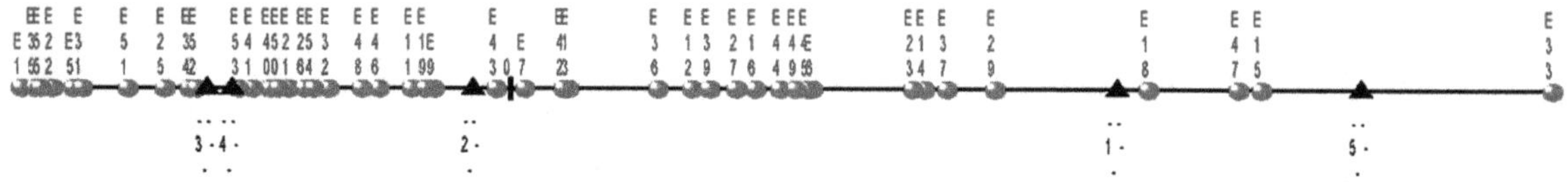

The second axis (Figure 2) opposes factors that were ranked with 1 or 5, as: the "Investments from outsiders" (E15), "Photovoltaic parks" (E18), "Productivity in the sector" (E29), "Unemployment" (E33), "Knowledge diffusion in the sector" (E38), "Innovative companies in the sector" (E47) with factors that got the mediate values (3,4) like, "Care for biodiversity" (E1), "Access to internet" (E10), "Possession of quality certificates" (E55).

From the preceding analysis and Figure 3, we can conclude (Greenacre, 2007), that there are many factors that seem to strongly differentiate themselves. Firstly, the positive ones (that contribute positively to the economic environment and entrepreneurship) are: "Productivity of the sector" (E29), ''Connection with big urban centers (E32), "Tourist sector development perspectives" (E41) "Secondary production perspectives" (E42), "Green development" (E17), "Connection with other sectors" (E26). From the opposing point of view, the negative ones are presented: "Investments from outsiders" (E15), ''Innovative companies" (E47), ''Digital marketing" (E14), ''Knowledge diffusion" (E38), ''Connection with other sectors" (E26), ''Unemployment in the sector" (E37), "Unemployment" (E33), ''General state of the economy" (E5). From the above results, important conclusions can be drawn from the SWOT analysis, which can be seen in Tables 8,9 for the local economy and the agricultural sector, respectively.

Figure 3. First factorial space

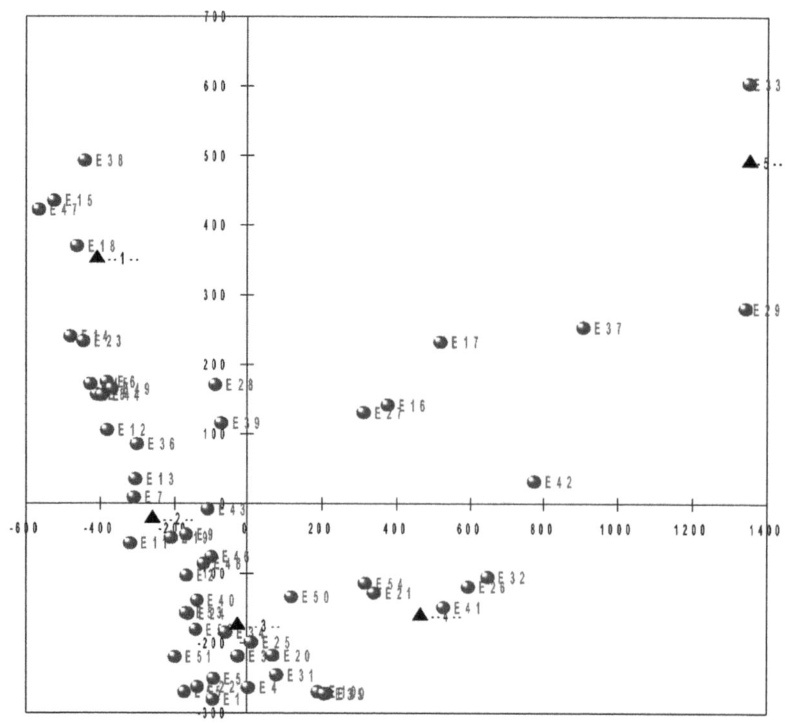

Table 8. SWOT analysis of the local economy

Strengths	Weaknesses	Opportunities	Threats
Green development	Investments from outsiders	Secondary production perspectives	Unemployment
	Photovoltaic parks	Tourism development perspectives	General state of economy
		Connection with big urban centers	

Table 9. SWOT analysis for the agricultural sector of the local economy

Strengths	Weaknesses	Opportunities	Threats
Productivity	Digital marketing	Connection with other sectors	Unemployment in the sector
	Innovative companies		
	Knowledge diffusion		

Consequently, it is obvious that unemployment (inside the sector or not) from the perspective of the entrepreneurs, is a serious threat in the regions examined. On the other hand, there is an opportunity for the secondary production as well the tourist sector to further expand. Both previous factors are considered very important for the economic prosperity of any region. The agricultural sector of the regions examined, seems to experience some other problems too, like the lack of: innovation, knowledge diffusion and digital marketing, all of which are entrepreneurship friendly. On the other hand, the local agricultural sector seems to have an adequate connection with other sectors, and its productivity is well established. The general state of the economy is not well and there existed few investions from outside the local area. However, some other factors like green development, connecting with big urban centers, seem to be sufficiently featured in the regions examined.

Comparative Evaluation

Except from the above analysis, the effect of all the demographic variables on the factors was also examined. Here, the table of comparative evaluation, which gives the possibility to compare the performance of different groups is applied. The findings showed that the opinions of the citizens do not change when they have different demographic characteristics, a result that means the evaluation from an entrepreneur's perspective, is not affected by other factors.

As an example, the comparative table of the fifty-five questions and the level of education, was also analyzed, after the implementation of Correspondence Analysis in the table of comparative evaluation (Table 10). The symbolization "Sec" belongs to entrepreneurs having completed education below bachelor (secondary or primary or no) and "Univ" to entrepreneurs with bachelor or even higher education. Certainly, Sec1 is the first grade of the category "Sec", Sec2 is the second grade of the category and so on.

The results derived, concluded that there is no difference in the opinions between entrepreneurs with a different level of education. This is also obvious in the visualization of the results, which are displayed in the first factorial space (Figure 4). The entrepreneurs with different level of education 1:Sec, and 2:Univ, have given similar responses, which can be resulted by the closeness of the points. For example, it can be seen from the points 15 and 25 (in the circle), where 15 is the fifth grade of education level 1 and 25 is the fifth grade of education level 2.

Table 10. Table of comparative evaluation of questions and level of education

Ind	Sec1	Sec2	Sec3	Sec4	Sec5	Univ1	Univ2	Univ3	Univ4	Univ5
E1	10	25	44	10	2	10	26	34	19	0
E2	15	30	33	12	1	13	36	22	16	2
E3	9	26	40	13	3	11	24	36	15	3
E4	8	30	36	14	3	10	19	39	19	2
E5	15	23	38	14	1	4	35	33	15	2
E6	32	28	24	7	0	25	25	27	12	0
E7	23	21	40	6	1	21	29	29	10	0
E8	30	30	25	5	1	21	35	25	8	0
E9	21	24	37	8	1	15	22	37	11	4
E10	9	18	37	25	2	9	19	30	26	5
⋮	⋮	⋮	⋮	⋮	⋮	⋮	⋮	⋮	⋮	⋮
E50	13	28	32	13	5	9	19	30	26	5
E51	14	27	43	7	0	13	25	34	17	0
E52	18	27	36	10	0	7	29	35	15	3
E53	17	31	32	11	0	12	22	39	14	2
E54	11	30	25	18	7	5	16	29	30	9
E55	15	31	34	11	0	7	25	42	15	0

Figure 4. First factorial space of the table of comparative evaluation

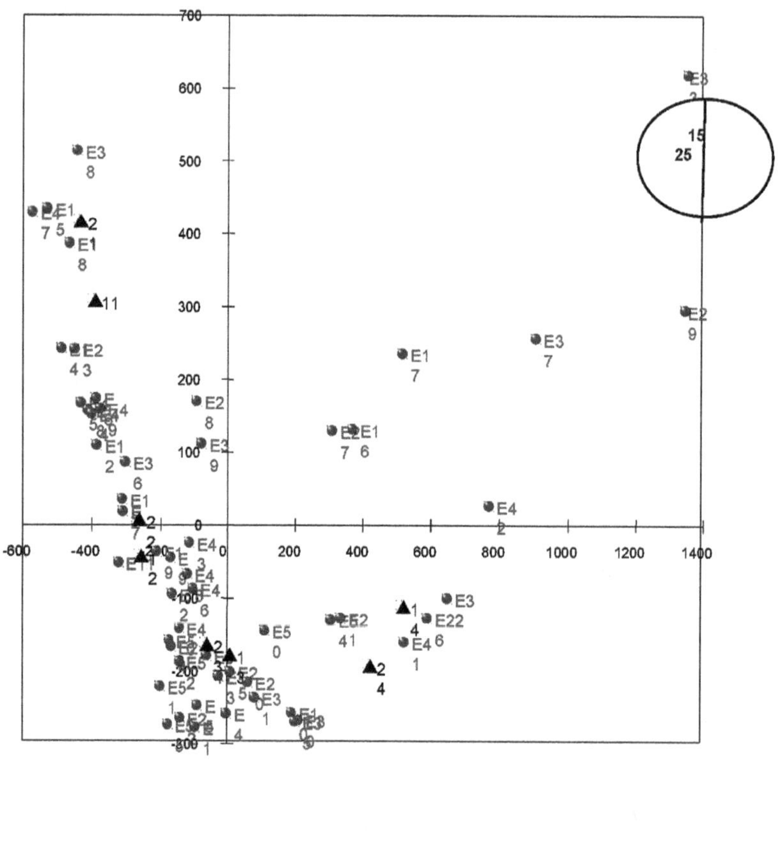

FUTURE RESEARCH DIRECTION

The methodology presented in this research could be generalized in all Greek regions and therefore this offers a possibility to compare the economic situation in the different regions.

More questions could be included in the questionnaire, in order to explore the possible reasons for the delay of the economic development and moreover the implementation of technology innovations in the agribusiness sector in Greece.

Local and central governments can adopt measures to improve the situation regarding the economic climate and the same research could be repeated after some years in the same regions, in order to make longitudinal comparisons.

CONCLUSION

The questionnaire designed for the purpose of the present study and which was pilot tested in regions of Northern Greece, has achieved with great ease to extract the most important factors of the economic environment and economic performance, assisted in this effort by a method of the multidimensional statistics field. Regarding the final results, about the local entrepreneurship's opinions of the factors appropriate for economic development in all the regions, differences were detected in some factors. This result makes it possible for the researcher to discover the strengths and weaknesses of municipality's economic performance and the possible potentials, for further improvement. In addition, it can be concluded that the general perception of the current situation in the regions examined is not influenced by the social characteristics of the entrepreneurs like level of education, age and sex. Concerning the innovation and technological aspects that companies could use, it can be deducted that the agricultural sector in the regions examined in this study, did not manage to implement most of them.

Moreover, the present questionnaire can be an important tool to evaluate factors that contribute to the local economic development and can be used to extract the crucial factors that SWOT analysis requires. The correspondence analysis method seems to be perfectly suited for the extraction of the elements needed to perform the SWOT analysis. Particularly, with the contribution of the two proposed tables of evaluation, the final choice of the crucial factors was precise and immediate. This proposed method, can also be used as an instrument to measure and compare different regions' performance, with the help of the indicators (questions) presented in the current study. This project can provide a measurement tool for Central Administration-Government, in order to compare and evaluate the economic performance and environment to different regions. It is a way to implement benchmarking, with the creation of a network of public administrations' organizations which have common practices, and to develop a peer evaluation among them (Cappelli et al., 2011). This performance can be a criterion for another evaluation; the appropriate exploitation of the resources, that municipalities have at hand, in order to achieve their goals.

REFERENCES

Ariza, C., Rugeles, L., Saavedra, D., & Guaitero, B. (2013). Measuring Innovation in Agricultural Firms: A Methodological Approach. *Electronic Journal of Knowledge Management*, *11*(3), 185–198.

Ashok, K. R., & Balasubramanian, R. (2006). *Role of Infrastructure in Productivity and Diversification of Agriculture, A Research Report*. Islamabad, Pakistan: SANEI.

Benzecri, J. P. (1992). *Correspondence Analysis Handbook*. New York: Taylor & Francis.

Cappelli, L., Guglielmetti, R., Mattia, G., Merli, R., & Renzi, M. F. (2011). Peer evaluation to develop benchmarking in the public sector. *Benchmarking an International Journal*, *4*(4), 490–509. doi:10.1108/14635771111147605

Chatzipetrou, E., & Moschidis, O. (2016). Quality costing: A survey in Greek supermarkets using multiple correspondence analysis. *International Journal of Quality & Reliability Management*, *33*(5), 615–632. doi:10.1108/IJQRM-01-2014-0004

Cornell University, INSEAD, & WIPO. (2015). *The Global Innovation Index 2015: Effective Innovation Policies for Development*. Authors.

Dlamini, B., Kirsten, J., & Masuku, M. (2014). Factors Affecting the Competitiveness of the Agribusiness Sector in Swaziland. *Journal of Agricultural Studies*, *2*(1).

Esterhuizen, D. (2006). *An Evaluation of the Competitiveness of the South African Agribusiness Sector*. University of Pretoria. Available from UPeTD: Retrieved June 12, 2016, from http://upetd.up.ac.za/thesis/ available/ etd-12082006144349/ unrestricted/ 00front.pdf,9/3/2010

European Commission. (2014). *Rural development programme*. Retrieved June 10, 2016, from http://ec.europa.eu/agriculture/rural-development-2014-2020/

European Commission. (2016). *Factsheet on 2014-2020 Rural development programme for Greece*. Retrieved July 12, 2016, from http://ec.europa.eu/agriculture/rural-development-2014-2020/country-files/el/factsheet-greece_en.pdf

General Secretariat of Research and Technology-GSRT. (2013). *Research and Innovation in the Agro-bioalimentary cluster for the 2014-2020 period* (in Greek). Retrieved July 12, 2016, from http://www.gsrt.gr/Financing/Files/ProPeFiles48/ΤΟ%20ΑΓΡΟΒΙΟΔΙΑΤΡΟΦΙΚΟ%20ΣΥΜΠΛΕΓΜΑ%202014-2020_ΕΣΠΕΚ_10-7-2013_A_FINAL.pdf

Greenacre, M. (2007). *Correspondence Analysis in Practice*. Boca Raton, FL: Chapman and Hall/CRC. doi:10.1201/9781420011234

Hampton, R., Fromm, I., & Nyhodo, B. (2007). *International Trade, Consumer Behaviour and Trust: Factors Affecting Agribusinesses in Developing Countries*. Marketing Department Faculty Publications. Retrieved June 20, 2016), from http://digitalcommons.unl.edu/marketingfacpub/6

Hellenic Statistical Authority (ELSTAT). (2009). *Results of Agricultural-Livestock Census 2009*. Retrieved from http://www.statistics.gr/el/statistics/-/publication/SPG31/-

Johnson, A., Dibrell, C., & Hansen, E. (2009, Spring/Fall). Market Orientation, Innovativeness, and Performance of Food Companies. *Journal of Agribusiness*, *27*(1/2), 85–106.

Hellenic Statistical Authority (ELSTAT). (2016). *Gross value added by industry (A10) - NACE REV.2, the sector of Agriculture, Forestry*. Retrieved June 10, 2016, from http://www.statistics.gr/el/statistics/-/publication/SEL45/-

Helms, M. M., & Nixon, J. (2010). Exploring SWOT analysis - where are we now? A review of academic researchfrom the last decade. *Journal of Strategy and Management*, *3*(3), 215–251.

Llanto, G. (2012). *The impact of infrastructure on agricultural productivity*. Philippine Institute for Development Studies, Series: Discussion paper series (Philippine Institute for Development Studies), no. 2012-12.

Moschidis, O. (2006). A proposal of comparative evaluation with the correspondence analysis [in Greek]. *Journal SPOUDAI*, *56*(3), 95–113.

Moschidis, O. (2015). Unified coding of qualitative and quantitative variables and their analysis with ascendant hierarchical classification. *International Journal of Data Analysis Techniques and Strategies*, *7*(2), 114–128. doi:10.1504/IJDATS.2015.068745

Moschidis, O., & Arabatzis, G. (2013). SHP stations and integrated rural development: A multivariate statistical approach. *International Journal of Green Economics*, *7*(4), 333–347. doi:10.1504/IJGE.2013.058561

Moschidis, O., Spathis, C., & Floropoulos, I. (2009). Methodological approach to Multidimensional exploratory evaluation of Taxis (Taxation information system) usefulness:Greek economy in Perspective. *Journal of Financial Management and Analysis:International Review of Finance*, *22*(2), 1–12.

Moschidis, O. E. (2009). A different approach to multiple correspondence analysis (MCA) than that of specific MCA. *Mathématiques et sciences humaines. Mathematical Social Sciences*, (186), 77–88.

Nadeem, N., Mushtaq, K., & Javed, M. I. (2011). Impact of social and physical infrastructure on agricultural productivity in Punjab, Pakistan - A production function approach. *Pakistan Journal of Life and Social Sciences*, *9*(2).

Olowa O.W., & Olowa, O.A. (2015). Factors Affecting Entrepreneurship Development in Agribusiness Enterprises in Lagos State. *Global Journal of Management and Business, 15*(7).

Papaioannou, E., Georgiadis, C. K., Moschidis, O., & Manitsaris, A. (2015). Factors Affecting Customers Perceptions and Firms Decisions Concerning Online Fast Food Ordering. *International Journal of Agricultural and Environmental Information Systems*, *6*(1), 48–78. doi:10.4018/ijaeis.2015010104

Piercy, N., & Giles, W. (1989). Making SWOT Analysis Work. *Marketing Intelligence & Planning*, *7*(5/6), 5–7. doi:10.1108/EUM0000000001042

Pinstrup-Andersen, P., & Shimokawa, S. (2007). Rural Infrastructure and Agricultural Development. In F. Bourguignon & B. Pleskovic (Eds.), *Annual World Bank Conference on Development Economics – Global 2007: Rethinking Infrastructure for Development* (pp. 175-203). Washington, DC: The World Bank.

Polyzos, S., & Arabatzis, G. (2005). *Labour Productivity of the Agricultural Sector in Greece: Determinant Factors and Interregional Differences Analysis*. Department of Planning and Regional Development, School of Engineering, University of Thessaly, Discussion Paper Series.

Śledzik, K. (2013). Schumpeter's view on innovation and entrepreneurship. In Management Trends in Theory and Practice. University of Zilina. doi:10.2139/ssrn.2257783

van Duren, E., Sparling, D., Turvey, C., & Lake, L. (2003). An assessment of the strategies and strengths of medium-sized food processors. *Agribusiness: International Journal (Toronto, Ont.)*, *19*(1), 115–132.

World Economic Forum. (2013). *The Global Competitiveness Report 2013–2014*. Geneva: World Economic Forum.

World Economic Forum. (2015). Retrieved from http://www.weforum.org/reports/global-competitiveness-report-2014-2015

KEY TERMS AND DEFINITIONS

Central Administration-Government: The government of a nation-state and this is a characteristic of a unitary state.

Digital Marketing: The advancement of items or brands through one or more types of electronic media. For instance, publicizing mediums that may be utilized as a major aspect of the advanced advertising system of a business could incorporate limited time endeavors made by means of the Internet, online networking, cellular telephones and electronic boards, and in addition through computerized and TV and radio stations.

Green Development: It is an area use arranging idea that incorporates thought of group wide or local natural ramifications of improvement, and site-particular green building ideas. This incorporates city arranging, natural arranging, engineering, scene design and group building.

Infrastructure: Relatively lasting and foundational capital venture of a nation, firm, or venture that underlies and makes conceivable all its monetary movement. It incorporates authoritative, information transfers, transportation, utilities, and waste expulsion and preparing offices. A few definitions additionally incorporate instruction, human services and innovative work.

Innovation: The way toward deciphering a thought or development into a decent or administration that makes esteem or for which clients will pay. In business, advancement frequently comes about when thoughts are connected by the organization with a specific end goal to facilitate fulfill the necessities and desires of the clients.

Knowledge Diffusion: The way toward imparting exploration, developments as well as information to people, gatherings or associations.

Learning Organization: An association that gains learning and enhances sufficiently quick to survive and flourish in a quickly evolving environment. Learning associations (1) make a society that empowers and backings nonstop representative learning, basic considering, and hazard bringing with new thoughts, (2) permit mix-ups, and esteem worker commitments, (3) gain for a fact and test, and (4) disperse the new information all through the association for fuse into everyday exercises.

Market Orientation: A business methodology or logic that spotlights on distinguishing and meeting the expressed or shrouded needs or needs of clients.

Productivity: A measure of the productivity of a man, machine, processing plant, framework, and so on., in changing over inputs into valuable yields. Productivity is processed by partitioning normal yield per period by the aggregate costs caused or assets (capital, vitality, material, faculty) devoured in that period.

Sector (Economy/Business): The economic-business sector or corporate sector is a part of the economy made up by companies with the same characteristics. The three-sector theory in the economics, subdivides them into: The Primary Sector (Raw Materials): Agriculture, Fishery, Forestry. The Secondary Sector (Manufacturing): Manufacturing, Construction. The Tertiary Sector (Sales and Services): Services, Trade.

Technology: The intentional use of data in the configuration, generation, and usage of products and administrations, and in the association of human exercises.

APPENDIX

The Table 11 below, presents the number of entrepreneurs per question and grade, the averages of the responses per question and the analytical description of the questions.

Table 11. Grades and averages of the 55 questions

Number	Gr1	Gr2	Gr3	Gr4	Gr5	Aver.	Question	Gen/Sec*
E1	20	51	78	29	2	2.7	Care for the protection of biodiversity	General
E2	28	66	55	28	3	2.5	Care for the protection of natural resources	General
E3	20	50	76	28	6	2.7	Diffusion of knowledge and information	Sectorial
E4	**18**	**49**	**75**	**33**	**5**	**2.8**	Satisfaction from income	General
E5	**19**	**58**	**71**	**29**	**3**	**2.7**	Effective advertising	Sectorial
E6	57	53	51	19	0	2.2	Expectations of the agricultural policy implemented	Sectorial
E7	44	50	69	16	1	2.3	Intense business activity in your sector	Sectorial
E8	51	65	50	13	1	2.2	Encouragement of private initiative from local or central government	General
E9	36	46	74	19	5	2.5	Cooperation of local authority and entrepreneurs	General
E10	18	37	67	51	7	3.0	Access to internet and WWW	Sectorial
E11	37	61	65	17	0	2.3	Exploitation of European funding programs (ESPA)	General
E12	45	72	46	16	1	2.2	New technology and informatics usage	Sectorial
E13	40	70	46	23	1	2.3	Innovative agricultural companies	Sectorial
E14	56	70	45	8	1	2.0	Usage of digital marketing	Sectorial
E15	72	62	37	7	2	**1.9**	**Investments from outsiders**	General
E16	35	22	57	44	22	3.0	Environment awareness	Sectorial
E17	**30**	**28**	**49**	**43**	**30**	**3.1**	**Green development**	Sectorial
E18	**71**	**52**	**42**	**14**	**1**	**2.0**	**Creation of Photovoltaic parks**	General
E19	34	60	60	23	3	2.5	Support from government	General
E20	20	45	71	37	7	2.8	Contribution of tourism to the local economy	General
E21	19	40	50	57	14	**3.0**	Contribution of agricultural sector to the local economy	Sectorial
E22	23	50	79	27	1	2.6	Market knowledge	Sectorial
E23	**61**	**55**	**51**	**13**	**0**	**2.1**	**General state of economy**	General
E24	28	52	75	22	3	2.6	*Cost of using infrastructure*	General
E25	24	45	69	37	5	2.7	Access to internet and WWW	Sectorial
E26	12	27	58	60	23	**3.3**	**Connection with other sectors**	Sectorial
E27	27	43	52	35	23	2.9	Quality products	Sectorial
E28	45	49	50	26	10	2.5	Education and training of your sector	Sectorial
E29	3	10	45	64	58	**3.9**	**Productivity of your sector**	Sectorial
E30	16	39	66	51	8	3.0	Impact of Taxation	General
E31	22	39	71	43	5	2.8	Connection with the rest Greece	General
E32	14	23	52	68	23	**3.4**	**Connection with big near urban centres**	General

continued on following page

Table 11. Continued

Number	Gr1	Gr2	Gr3	Gr4	Gr5	Aver.	Question	Gen/Sec*
E33	16	12	34	51	67	**3.8**	**Unemployment**	General
E34	23	45	85	20	7	2.7	Number of Associations, Partnerships, Consortiums	Sectorial
E35	11	48	66	45	10	3.0	Number of Tourists	General
E36	45	62	50	21	2	2.3	Number of new patents, products	General
E37	11	25	57	40	47	**3.5**	Seasonal unemployment	General
E38	76	50	40	9	5	2.0	Knowledge diffusion in your sector	Sectorial
E39	37	58	50	24	11	2.5	The interest of the young people to become businessmen in the region	General
E40	24	62	67	22	5	2.6	Connection of primary sector with secondary	General
E41	**12**	**36**	**48**	**65**	**19**	**3.2**	**Developmental perspectives of the tourist sector**	General
E42	12	27	46	63	32	3.4	**Secondary sector perspectives**	General
E43	37	47	63	27	6	2.5	Existence of companies with export activity	General
E44	50	64	53	11	2	2.2	Agrotourism expansion	General
E45	51	69	47	12	1	2.1	Investors' protection	General
E46	28	56	66	23	7	2.6	Reduction of local products	Sectorial
E47	75	57	42	6	0	1.9	**Innovative companies**	**Sectorial**
E48	30	59	56	31	4	2.6	Special skills of the staff	Sectorial
E49	58	47	56	19	0	2.2	Support new workplaces	General
E50	22	47	62	39	10	2.8	Well-trained staff	Sectorial
E51	27	52	77	24	0	2.5	Quality of roads	General
E52	25	56	71	25	3	2.6	**Satisfied with the state of economy**	General
E53	29	53	71	25	2	2.5	Companies are close to the consumer	Sectorial
E54	16	46	54	48	16	3.0	Intensity of local competition	Sectorial
E55	22	56	76	26	0	2.6	Possession of ISO or other quality certificates	Sectorial

* Gen=General(for the local economy), Sec=Sectorial(for the specific sector)

Chapter 12
Does Nonfarm Income Affect Agricultural Income and Investment in Pakistan?

Zia Ullah Khan
University of Swabi, Pakistan

Zahoor ul Haq
Abdul Wali Khan University Mardan, Pakistan

Khalid Khan
Higher Education Department Peshawar, Pakistan

Muhammad Ishaq
Pakistan Agricultural Research Council, Pakistan

Fazli Wahid
University of Waterloo, Canada

ABSTRACT

The study investigates the impact of nonfarm income (NFI) on agricultural income and investment using the Pakistan Social and Living Measurement survey data for the year 2005-06. Results show that NFI negatively affects agricultural income and investment whenever it is statistically significant; and these effects are not same across the four provinces of Pakistan. The one to one comparison between the four provinces of the country shows that the effects of NFI on agricultural income and investment differ across provinces. The policy implication is that as compared to other sectors of the economy, agriculture generates low returns and consequently NFI is invested in other more productive sectors of the economy.

INTRODUCTION

Not all people receive their earnings from a single source, hold their wealth as one asset and employ their labor in one activity. Multiple motives encourage families and individuals to diversify their assets and income generating activities (Barett and Reardon, 2001). Participating in the nonfarm income (NFI) generating activities are one of the ways for rural households to diversify their earning sources and increase their gross income. In return, these earnings affect farm productivity by enhancing investment in farming. Studies show that NFI has positive effects on farm investment (Heartz, 2009) and increase

DOI: 10.4018/978-1-5225-2107-5.ch012

expenditure on inputs (Kilic et al.,2009; Oseni & Winters, 2009; Pfeiffer et al., 2009). Consequently, farm productivity increases (Huang et al., 2009) and poverty reduces (Kijima et al., 2006; Ruben & Van Den Berg, 2001). NFI can also finance longer term on-farm capital investment such as construction of irrigation channels, purchase of machinery which can positively affect farm productivity (Barett and Reardon, 2001). Little et al. (2006) found that farm households diversify their earning sources to improve insurance against the risks of agro-climatic natural shocks, help overcome credit constraints and stabilize aggregate income flows. However, contrary to studies cited above, Pfeiffer et al. (2009) show that NFI negatively affects crop production, but positively effect the purchase of inputs. Hence, NFI has divergent effects on production and inputs use.It is the focus of this research in Pakistan, where agriculture is the second largest sector of country's economy contributing 21 percent to the GDP and provides livelihoods to 40 percent of the population.

It is generally believed that only poor households may diversify their earning sources to increase their aggregate income. However, this may not be true. Rich rural households diversify their earning sources to further maximize their profit while poor diversify to minimize risk, stabilize income and secure food access (Kilic et al., 2009). Haggblade et al. (2010) and Davis et al. (2007) identified the growth linkages between the agriculture sector and rural nonfarm employment. These linkages are: 1) the increase in income, increasing effective consumption of nonfarm products, affecting nonfarm employment; 2) the effect of demand-induced changes on downstream production linkages from processing and distribution; and 3) the changes in input-demand and its effect on production. This study focuses on the third linkage that is the nonfarm-income-induced demand for agricultural inputs and its effect on agricultural productivity. The selection is motivated by our lack of knowledge of the effect of NFI on agriculture sector in Pakistan. In an emerging economy like that of Pakistan, it is important to understand that why some farmer perform better than others? Does NFI create positive spillover effects on agriculture and livestock investment and consequently income? The four provinces of the country are agriculturally very different. Punjab and Sindh produce cash crops like cotton, rice and sugarcane while Baluchistan and Khyber Pakhtunkhwa have small natural resource base supporting livestock keeping. These differences raise the question that whether NFI have divergent effects on agriculture and livestock income and ivestment across these regions. This study estimates the effect of NFI on expenditure made on farm inputs and consequently farm income in Pakistan. Alternatively, the study investigates whether farmers who diversify their income have higher farm incomes as compared to other producers.

The study makes a three important contributions to the existing literature. First, it develops empirical models illustrating the effect of NFI on agricultural income and expenditures on crops and livestock raising in the country. Second, these models are used to test specific hypotheses about the effect of NFI on agricultural income and expenditures on crops and livestock raising across the four provinces of Pakistan. It is also important to mention that understanding behavior of agricultural households with respect to income and investment is important to analyze the effects of government interventions (e.g., pricing policies, investment projects) and external changes in market conditions on the rural economy and livelihoods. Such knowledge becomes more important for a country like Pakistan where agriculture is the second largest sector of the economy. Third, the study provides empirical evidence on the effect of NFI on agricultural income and expenditures on crops and livestock raising which can help in developing the relevant policies for creating and promoting opportunities of earning NFI.

The next section presents the empirical model used to estimate the effect of NFI on agricultural income and investment, followed by discussion about data used in the analysis in section three. The estimated results are presented and discussed in section four, followed by conclusions given in section five.

The Agricultural Sector in Pakistan

Pakistan has two seasons, namely the '*kharif*' and '*rabi*'. *Kharif* begins in April-June and ends during October-December when *rabi* begins which ends in April-May. *Kharif* crops include rice, sugarcane, cotton, maize, mong and mash while *rabi* crops include wheat, gram, lentil, tobacco, rapeseed, barley and mustard. The agriculture sector is divided into crops, livestock, forestry and fishing sub-sectors. Crops sector is further divided into important crops, other crops and cotton ginning. Important crop sector includes cotton, sugarcane, wheat, rice and maize while the other crop sector consists of lentils, peas, potatoes, onions and chilies. The Agriculture sector accounts for 20.9 percent of the country's Gross Domestic Product (GDP) in 2015. The crop sector accounts for 43.7 percent of the 20.9 percent agricultural sector contributions to a country's GDP in 2015. In the crop sector, important crop sub-sector contributes 64.6 percent to the value added in the crops sector, followed by other crops (28.1%) and cotton ginning (7.3%). The livestock sector contributes 11.8 percent to the country's GDP and 56.3 percent to the agricultural GDP (GoP, 2011). The collective contribution of forestry and fishing to a country's GDP is less than one percent.

The Empirical Model

Consider an agrarian household endowed with land, capital, and inputs. The household faces decision variables of consumption, investment, and purchase of inputs for production. The household is assumed to maximize profit as producer and utility as a consumer. In case of capital constraint, the household can get any amount of credit from a perfect financial market. However, given the imperfect financial market in a developing country like Pakistan, household faces credit constraint. In such a case, household production and consumption decisions are inseparable (Taylor and Martin, 2001; de Janvry and Sadoulet, 2003). Hence, the household sacrifices its leisure time and participate in the NFI generating activities to earn more income and overcome the credit constraint. The NFI is then available for investment in the crops and livestock sectors. Consequently, agricultural (collectively crops and livestock) investment and income are expected to increase. Our specification of the econometric model is based on Kilic et al. (2009), and Osenia and Winters (2009). The empirical model postulates that the variable of interest (Y_i)in thousand rupees[1] is the function of socioeconomic variables, NFI of household i .($NFIii$.) in thousand rupees and regional dummies (RD_i .). Regional dummies (RDi .) represent the four provinces of Pakistan such that RD_{i1} . represents Punjab, RD_{i2} . represents Sind, RD_{i3} . represents Khyber Pakhtunkhwa and RD_{i4} . represents Baluchistan. The variable of interest includes agricultural income and expenditure made on crops production and livestock keeping. In the empirical model, farm characteristics such as land, poverty status (Evans and Nagau, 1991), education (Ellis, 2000), family size and marital status of the respondents are included. Marital status and family size are included in the model because these factors directly affect the supply of labor to NFI generating activities. Larger families can release more labor from farm production towards NFI generating activities.

$$Y_i = \begin{array}{l} \alpha_0 + \alpha_1 lNFI_i + \alpha_2 RD_{i1} + \alpha_3 RD_{i2} + \alpha_4 RD_{i3} + \alpha_5 RD_{i3} + \\ \alpha_6 Poor_i + \alpha_7 lAge_i + \alpha_8 lFSize_i + \alpha_9 lFArea_i + \alpha_{10} Urban_i + \\ \alpha_{11} lExpc_i + \alpha_{12} lExpl_i + \alpha_{13} Married_i + \alpha_{14} Literate_i + \varepsilon_i \end{array} \quad (1)$$

Where i. indexes households, Age_i represents age of the household head in years, $FSize_i$ is family size measured as number of family members, $FArea_i$ is farm area in hectares, $Urban_i$ is dummy, one representing urban areas zero otherwise, $Expc_i$ represents expenditure on crops in thousand rupees, $Expl_i$ is expenditure on livestock in thousand rupees, $Married_i$ is dummy, one for married household head zero otherwise, *Literatei* is a dummy, one for literate households zero otherwise, ε_i. represents the random error assumed to be distributed normally with mean zero and variance σ^2 ., l stands for logarithm and α_i. are the parameters to be estimated. The parameter α_1. indicates the effect of NFI (in thousand rupees) on the variable of interest in Pakistan. It is interpreted as the elasticity of NFI with respected to either agricultural income or farm investment.

Following Haq and Meilke (2010), and in order to compare the effects of NFI_i. on Y_i. across the provinces of Pakistan, regional (provincial) slope-shifters of NFI_i. were derived using the following relationship. These slopes allow testing regional specific hypotheses about the effect of NFI_i. on variable of interest.

$$NFI_{iP} = Y_i * RD_{i1}.NFI_{iS} = Y_i * RD_{i2}. \tag{2}$$

$$NFI_{iK} = Y_i * RD_{i3}.NFI_{iB} = Y_i * RD_{i4}.$$

where NFI_{iP}. represents nonfarm income of household i. in Punjab, NFI_{iS}. represents nonfarm income of household i .in Sind, NFI_{iK} . represents nonfarm income of household i .in KP and NFI_{iB} . represents nonfarm income of household i .in Baluchistan. Augmenting equation (1) with the regional NFI_i. shifters yields equation (3).

$$Y_i = \begin{matrix} \delta_0 + \delta_1 lNFI_{ip} + \delta_2 lNFI_{iS} + \delta_3 lNFI_{iK} + \delta_4 lNFI_{iB} + \\ \delta_5 Poor_i + \delta_6 lAge_i + \delta_7 lFSize_i + \delta_8 lFArea_i + \delta_9 Urban_i + . \\ \delta_{10} lExpc_i + \delta_{11} lExpl_i + \delta_{12} Married_i + \delta_{13} Literate_i + \varepsilon_i \end{matrix} \tag{2}$$

Consider Y_i Y_i to be agricultural income than the following hypotheses are tested using equation (3).

H_1: Nonfarm income does not determine agricultural income in the provinces of Pakistan.
H_2: The effect of NFI on agricultural income is similar across the four provinces of the country.

Similarly other hypotheses are tested using the same equation, results of which are compiled in tables 5 and 6.

DATA

This study uses the cross-sectional data collected under the Pakistan Social and Living Measurement Survey (PSLM) for the year 2005-06. The survey was carried in all the four provinces of the country. The survey is conducted by Federal Bauru of Statistics (FBS) bi-annually. FBS has developed its own

Table 1. Distribution of sample across provinces

Province	Count	Percent
Punjab	1631.0	44.0
Sindh	844.0	22.8
NWFP	796.0	21.5
Balochistan	433.0	11.7
Pakistan	3704.0	100.0

sampling frame for urban and rural areas. Each frame is split into enumeration blocks. Each enumeration block consists of 200 to 250 households. Each block is divided into lower, middle and high income groups. FBS obtains list of villages from the Population Census Organization of the country. Urban domain consists of big cities like Islamabad, Lahore, Gujranwala, Faisalabad, Rawalpindi, Multan etc. Stratum from urban areas also classified according to income levels. After excluding population of the large sized cities, the remaining urban population in each defunct Division in all the provinces has been grouped together to form a stratum. Each district in the provinces are grouped to constitute a stratum, whereas defunct administrative division has been treated as stratum in Baluchistan province.

FBS determined a sample size of 15512 households in the country. This sample size is obtained from 1113 sample enumeration blocks. Our sample includes only those households who cultivate land. In this way the sample reduces to 3704 households. Detail of the sample frame across the four provinces is given in Table 1.

RESULTS AND DISCUSSION

This first part presents the effect of NFI on agricultural income and investment (Table 2), followed by similar effects estimated for crops (Table 3) and livestock (Table 4) sectors. Each table consists of four models, two each for expenditure and income both distinguished by the inclusion/exclusion of regional dummies.

Table 2. OLS estimates of the effect of nonfarm income on agricultural income and farm investment

Variables	Model-1	Model-2	Model-3	Model-4
	Agricultural Income		Agricultural Investmen	
Poverty (Poor is 1, otherwise 0)	0.048	0.167***	0.004	0.410***
	(0.066)	(0.030)	(0.098)	(0.044)
Age (Years)	0.107	0.066	-0.005	0.037
	(0.104)	(0.048)	(0.145)	(0.073)
Family Size (No. of persons)	0.089	0.159***	0.402***	0.337***
	(0.057)	(0.026)	(0.087)	(0.050)

continued on following page

Table 2. Continued

Variables	Model-1	Model-2	Model-3	Model-4
	Agricultural Income		Agricultural Investmen	
Farm Area (Hectares)	0.055*	0.017	0.301***	0.447***
	(0.028)	(0.015)	(0.040)	(0.026)
Urban (Urban is 1, otherwise 0)	-0.253*	0.025	-0.319*	-0.097
	(0.141)	(0.055)	(0.172)	(0.102)
Expenditure on Crops (Rs/Year)	0.414***	0.485***	----	----
	(0.029)	(0.014)	----	----
Expenditure on Livestock (Rs/Year)	0.278***	0.214***	----	----
	(0.042)	(0.016)	----	----
Nonfarm Income (Rs (000)/Year)	-0.032	----	0.178***	----
	(0.032)	----	(0.048)	----
Punjab (Punjab is 1, otherwise 0)	0.068	----	0.730***	----
	(0.111)	----	(0.147)	----
Sind (Sind is 1, otherwise 0)	0.345**	----	0.898***	----
	(0.119)	----	(0.169)	----
KP (KP is 1, otherwise 0)	0.202*	----	-0.478**	----
	(0.121)	----	(0.163)	----
Married (Married is 1, otherwise 0)	-0.09	-0.078	0.293	0.015
	(0.125)	(0.069)	(0.196)	(0.111)
Literate (Literate is 1, otherwise 0)	0.159	0.322	0.635	0.221
	(0.337)	(0.233)	(0.626)	(0.314)
Nonfarm Income-Punjab (Rs (000)/Year)	----	-0.042***	----	-0.005
	----	(0.009)	----	(0.017)
Nonfarm Income-Sind (Rs (000)/Year)	----	0.023	----	0.009
	----	(0.017)	----	(0.031)
Nonfarm Income-KP (Rs (000)/Year)	----	-0.019*	----	-0.249***
	----	(0.012)	----	(0.016)
Nonfarm Income-Baluchistan (Rs (000)/Year)	----	-0.058	----	-0.275***
	----	(0.039)	----	(0.046)
R-squared	0.668	0.728	0.399	0.309
Number of observations	3704	3704	3704	3704
F-Statistics	75.01***	373.2***	41.56***	88.6***

Note: *, ** and *** denote variables significant at 0.1, 0.05 and 0.01 level of probability respectively. All the standard errors are corrected for heteroscadasticity.

Table 3. OLS estimates of the effect of nonfarm income on crops' income and expenditure

Variable	Dependent Variable			
	Log of Crops Expenditure		Log of Crops Income	
Poverty (Poor is 1, otherwise 0)	0.115	0.530***	0.057	0.085***
	(0.112)	(0.050)	(0.056)	(0.025)
Age (Years)	-0.153	-0.045	-0.033	0.01
	(0.154)	(0.080)	(0.077)	(0.039)
Family Size (No. of persons)	0.336***	0.294***	0.129**	0.177***
	(0.096)	(0.050)	(0.052)	(0.023)
Farm Area (Hectares)	0.337***	0.577***	0.060**	0.065***
	(0.050)	(0.027)	(0.022)	(0.011)
Urban (Urban is 1, otherwise 0)	-0.252	-0.044	-0.158	0.039
	(0.160)	(0.103)	(0.106)	(0.042)
Expenditure on Crops (Rs/Year)		----	0.715***	0.711***
		----	(0.027)	(0.010)
Nonfarm Income (Rs (000)/Year)	0.182**	----	-0.052*	----
	(0.057)	----	(0.029)	----
Punjab (Punjab is 1, otherwise 0)	0.442**	----	-0.160*	----
	(0.169)	----	(0.087)	----
Sind (Sind is 1, otherwise 0)	0.789***	----	0.071	----
	(0.188)	----	(0.100)	----
KP (KP is 1, otherwise 0)	-1.023***	----	-0.008	----
	(0.185)	----	(0.098)	----
Nonfarm Income-Punjab (Rs (000)/Year)	----	-0.019	----	-0.052***
	----	(0.020)	----	(0.007)
Nonfarm Income-Sind(Rs (000)/Year)	----	0.024	----	0.012
	----	(0.034)	----	(0.018)
Nonfarm Income-KP (Rs (000)/Year)	----	-0.310***	----	-0.028**
	----	(0.017)	----	(0.010)
Nonfarm Income-Balochistan (Rs (000)/Year)	----	-0.223***	----	0.012
	----	(0.057)	----	(0.028)
Married (Married is 1, otherwise 0)	0.487**	0.189	0.097	-0.037
	(0.206)	(0.120)	(0.102)	(0.054)
Literate (Literate is 1, otherwise 0)	0.815	0.074	0.498**	0.347**
	(0.863)	(0.457)	(0.177)	(0.138)
Summary Statistics				
R-squared	0.449	0.385	0.774	0.818
Number of observations	3704	3704	3704	3704
F-Statistics	51.94***	137.50***	142.77***	778.36***

Note: *, ** and *** denote variables significant at 0.1, 0.05 and 0.01 level of probability respectively. All the standard errors are corrected for heteroscadasticity.

Table 4. OLS estimates of the effect of nonfarm income on livestock income and expenditure

Variable	Dependent Variable			
	Log of Livestock Expenditure		**Log of Livestock Income**	
Poverty (Poor is 1, otherwise 0)	-0.04	0.367***	-0.051	0.145**
	(0.110)	(0.046)	(0.108)	(0.049)
Age (Years)	0.151	0.123*	0.112	0.177**
	(0.143)	(0.073)	(0.143)	(0.078)
Family Size (No. of persons)	0.210**	0.233***	-0.014	0.090**
	(0.088)	(0.047)	(0.086)	(0.042)
Farm Area (Hectares)	0.200***	0.287***	0.019	-0.050**
	(0.034)	(0.019)	(0.033)	(0.019)
Urban (Urban is 1, otherwise 0)	0.326**	0.087	-0.417*	-0.079
	(0.125)	(0.088)	(0.226)	(0.090)
Expenditure on Livestock (Rs/Year)			0.601***	0.638***
			(0.081)	(0.030)
Nonfarm Income (Rs (000)/Year)	0.232***		0.031	
	(0.044)		(0.046)	
Punjab (Punjab is 1, otherwise 0)	0.723***		0.288*	
	(0.173)		(0.170)	
Sind (Sind is 1, otherwise 0)	0.574**		0.554**	
	(0.243)		(0.177)	
KP (KP is 1, otherwise 0)	-0.108		0.446**	
	(0.186)		(0.166)	
Nonfarm Income-Punjab (Rs (000)/Year)	0.079	-0.084	-0.208	-0.228**
	(0.176)	(0.102)	(0.169)	(0.091)
Nonfarm Income-Sind(Rs (000)/Year)	-0.126	0.347	0.177	0.409
	(0.449)	(0.300)	(0.432)	(0.374)
Nonfarm Income-KP (Rs (000)/Year)		0		0
		(0.016)		(0.015)
Nonfarm Income-Balochistan (Rs (000)/Year)		-0.058		0.079***
		(0.061)		(0.024)
Married (Married is 1, otherwise 0)		-0.170***		0.019
		(0.016)		(0.015)
Literate (Literate is 1, otherwise 0)		-0.255**		-0.06
		(0.079)		(0.059)
Summary Statistics				
R-squared	0.262	0.209	0.339	0.341
Number of observations	3704	3704	3704	3704
F-Statistics	21.99***	52.13***	19.95***	64.97***

Note: *, ** and *** denote variables significant at 0.1, 0.05 and 0.01 level of probability respectively. All the standard errors are corrected for heteroscadasticity.

In case of agricultural income and investment, results are compiled in table 2. Model-1 shows the effect of NFI on agricultural income while Model-2 shows the same effect for the four provinces of Pakistan. Model-3 and 4 shows the effect of NFI on farm investment in the country and across the four provinces respectively. Results show that all the estimated models fit the data as well as the coefficient of determination ranges from 30.9% for model-4 to 72.8 for model-1. F-statistics show that all the estimated models are statistically significant, indicating that the hypotheses that the coefficients of regression models except the intercept are zero, is rejected at the 0.01 level of significance.

Table 2 further shows that as compared to poor households, non-poor households invest more in agricultural production and generate more agricultural income. Non-poor invest 50.7[2] percent of agricultural production and generate 18.2 percent more agricultural incomes, keeping other variables constant. However, with the inclusion of provincial slope shifters in the model (Model-1 and 3), this effect fades away. Age of the household head does not statistically significantly affect agricultural income and farm investment. Family size is statistically significant determinant when NFI provincial slope shifters are included in the model (Model-2). The effect of farm area on agricultural income and farm investment is statistically significant, but inelastic implying that increase in the farm area increases investment cost more, but yields proportionately less income. Its coefficient in model-4 shows that a one percent increase in the farm area increases farm investment by 0.45 percent, keeping other variables constant. However, its effect on farm investment is high as compared to its effect on agricultural income. Households located in the urban areas invest 37.6 percent less (model-3) and generate 28.7 percent more income. The effect of marriage and literacy on agricultural income as well as investment is statistically insignificant in all the models. Further, the effect of NFI on agricultural income is statistically not significant (Model-1), however, it is an important determinant of the farm investment (Model-3). The regional dummies show that as compared to Baluchistan, agricultural income is high by 41.2 percent in Sind and 22.4 percent in Khyber Pakhtunkhwa. Similarly, as compared to Baluchistan, production costs are higher by 107.5 percent in Punjab, 145.5 percent in Sind and lower by 38 percent in Khyber Pakhtunkhwa. Hence, production costs and agricultural income vary across the provinces.

The table also shows that the effect of Niño agricultural income in Punjab and Khyber Pakhtunkhwa (KP) is statistically significant while its effect on agricultural investment is also statistically significant and negative in KP and Baluchistan. The estimated results show that a ten percent increase in NFI decreases agricultural income by 0.4 percent in Punjab and by 0.2 percent in Khyber Pakhtunkhwa. Similarly, a ten farm increase in nonfarm in Khyber Pakhtunkhwa and Baluchistan decreases farm investment by 2.5 and 2.8 percent, respectively. The collective implications of these results are that returns in agriculture sector as compared to other sectors of the economy are low. Further, NFI may not be readily available to the farm sector for investment due to many reasons including the consumption, social and financial requirements of rural households. Also, the effect of the NFI on farm income and investment may be different not only across the crops and livestock sectors but also across time and regions. Hence, farm households do not invest the additional income in agriculture production and consequently its effect on agricultural income is negative.

Agricultural income and investment are then split into crops and livestock income and investments and the effect of the NFI is separately estimated on these sectors. Estimated results for the crops and livestock sectors are given in tables 3 and 4 respectively. However, separately estimating the models for crops and livestock sectors did not change the effect of most of the variables on both income and investment. For both the sectors (Tables 3 and 4), the effect of NFI on the variable of choice is negative when statistically significant with the exception of livestock sector in Balochistan (Table 4). In Balochistan, a ten percent increase in the NFI marginally (0.8%) increases investment in livestock sector. Hence, on an overall, the direction of the results presented in table 2 hold for both crops and livestock sectors.

The estimated parameters of NFI are then used to estimate a number of hypotheses about agricultural income and investment and results are compiled in Tables 5 and 6, respectively.

Results show that NFI is an important determinant of both agricultural investments (Table 5) and income (Table 6). The impact of NFI on agricultural investment (Table 5) and income (table 6) differs across the four provinces of the country. Its effect on agricultural income between Punjab and Sindh, Punjab and KP, Sind and KP and Sind and Baluchistan are statistically different. Similarly, the effect of NFI on agricultural investment is also statistically significant (Table 6). Its effect on agricultural investment between Punjab and KP, Punjab and Baluchistan and Sind and KP and Sind and Baluchistan are statistically different.

Table 5. Test of the hypotheses about the role of nonfarm income in farm investment

Hypothesis	F-Statistics
Nonfarm income does not affect agricultural production cost.	71.19***
The effect of nonfarm income on agricultural production cost is same across the provinces of Pakistan	56.93***
The effect of nonfarm income on agricultural production cost in Punjab is similar to its effect in Sind	0.17
The effect of nonfarm income on agricultural production cost in Punjab is similar to its effect in Khyber Pakhtunkhwa.	135.14***
The effect of nonfarm income on agricultural production cost in Punjab is similar to its effect in Baluchistan	31.39***
The effect of nonfarm income on agricultural production cost in Sind is similar to its effect in Khyber Pakhtunkhwa.	58.33***
The effect of nonfarm income on agricultural production cost in Sind is similar to its effect in Baluchistan.	26.88***
The effect of nonfarm income on agricultural production cost in Khyber Pakhtunkhwa is similar to its effect in Baluchistan.	0.28

Note: *, ** and *** denote variables significant at 0.1, 0.05 and 0.01 levels, respectively.

Table 6. Test of the hypotheses about the role of nonfarm income in agricultural income

Hypothesis	F-Statistics
Nonfarm income does not affect agricultural income.	7.20***
The effect of nonfarm income on agricultural income is same across the provinces of Pakistan	4.44**
The effect of nonfarm income on agricultural income in Punjab is similar to its effect in Sind	11.66**
The effect of nonfarm income on agricultural income in Punjab is similar to its effect in Khyber Pakhtunkhwa.	2.83*
The effect of nonfarm income on agricultural income in Punjab is similar to its effect in Baluchistan	0.17
The effect of nonfarm income on agricultural income in Sind is similar to its effect in Khyber Pakhtunkhwa.	4.22*
The effect of nonfarm income on agricultural income in Sind is similar to its effect in Baluchistan.	3.74*
The effect of nonfarm income on agricultural income in Khyber Pakhtunkhwa is similar to its effect in Baluchistan.	0.93

Note: *, ** and *** denote variables significant at 0.1, 0.05 and 0.01 levels, respectively.

CONCLUSION

Nonfarm income can be used to timely purchase crops and livestock inputs and undertake scheduled farm production and harvesting activities. However, there is no evidence suggesting that NFI has positive effects on agricultural income and investment in Pakistan. This study underscores the existing literature by providing the evidence on the effect of NFI on agricultural productivity in Pakistan using the Pakistan Social and Living Measurement survey.

The econometric analysis investigating the effect of NFI on agricultural income and expenditure shows that nonfarm is an important determinant of agricultural income. The effect of NFI on agricultural income is not same across the four provinces of Pakistan. Its effect on agricultural income between Punjab and Sindh, Punjab and KP, Sind and KP and Sind and Baluchistan are statistically different. Similarly, the effect of NFI on agricultural investment is also statistically significant. Its effect on agricultural investment between Punjab and KP, Punjab and Baluchistan and Sind and KP and Sind and Baluchistan are statistically different.

The authors highlighted that if there are negative linkages between NFI and agricultural production, then understanding the nature of these linkages could prove useful in designing programs to facilitate agricultural households' adjustment to rural economic change.

LIMITATIONS AND FUTURE DIRECTION

The study uses PSLM data in the analysis. PSLM data provides enterprise specific output data but does not provide similar data for inputs. Hence, using PSLM data it is not possible to determine the effect of NFI on major crops raised in different agro-ecological zones of the country. Hence, an investigation of the effect of NFI on major crops in different agro-geographical regions of the country is required. Further, since agriculture is one of the biggest sectors of the economy and is a source of livelihood for about one-half of the population, it is important to understand the effect of NFI on poverty in the country. This study also ignored the effect of NFI on variability and distribution of rural household's income.

REFERENCES

Barett, C. B., & Reardon, T. (2001). *Asset, activity, and income diversification among African agriculturists: Some practical issues*. UASID Basis CRSP.

Davis, B., Winters, P., Carletto, C., Covarrubias, K., Quinones, E., & Zezza, A. (2007). Rural income generating activities: Across country comparison. *World Development*, 119–123.

deJanvry, A., & Sadoulet, E. (2003). *Progress in the modeling of rural households' behavior under market failures. Chapters in Honor of Erik Thorbecke*. New York: Kluwer Publishing.

Ellis, F. (2000). *Rural Livelihoods and Diversity in Developing Countries*. Oxford, UK: Oxford University Press.

Evans, H. E., & Ngau, P. (1991). Rural-urban relations, household income diversification and agricultural productivity. *Development and Change*, *22*(3), 519–545. doi:10.1111/j.1467-7660.1991.tb00424.x

GoP. (2011). *Economic survey of Pakistan. Economic Advisory Wing. Finance division*. Islamabad: MINFAL Pakistan.

Haggblade, S., Hazell, P., & Reardon, T. (2010). The Rural Non-farm Economy: Prospects for Growth and Poverty Reduction. *World Development*, *38*(10), 1429–1441. doi:10.1016/j.worlddev.2009.06.008

Haq, Z., & Meilke, K. (2010). Do the BRICs and Emerging Markets Differ in their Agrifood Trade? *Journal of Agricultural Economics*, *61*(1), 1–14. doi:10.1111/j.1477-9552.2009.00229.x

Hertz, T. (2009). The effect of nonfarm income on investment in Bulgarian. *Agricultural Economics*, *2*(40), 161–176. doi:10.1111/j.1574-0862.2009.00367.x

Huanga, J., Wu, Y., & Rozelle, S. (2009). Moving off the farm intensifying agricultural production in Shandong: A case study of rural labor market linkages in China. *Agricultural Economics*, *2*(40), 203–218. doi:10.1111/j.1574-0862.2009.00370.x

Kijima, Y., Matsumoto, T., & Yamano, T. (2006). Nonfarm employment, agricultural shocks, poverty dynamics: Evidence from rural Uganda. *Agricultural Economics*, 459-467.

Kilic, T., Carletto, C., Miluka, J., & Savastano, S. (2009). Rural nonfarm income its effect on agriculture: Evidence from Albania. *Agricultural Economics*, *2*(40), 139–160. doi:10.1111/j.1574-0862.2009.00366.x

Little, P., Stone, P., Mogues, T., Castro, P., & Negatu, W. (2006). Moving in place drought and poverty dynamics in South Wollo, Ethiopia. *Development Studies,* (42), 200-225.

Oseni, G., & Winters, P. (2009). Rural nonfarm activities and agricultural crop. *Agricultural Economics*, *2*(40), 189–201. doi:10.1111/j.1574-0862.2009.00369.x

Pfeiffer, L., Feldman, L. A., & Taylor, J. (2009). Is off-farm income reforming the farm? *Evidence from Maxico.*, *2*(40), 125–138.

Ruben, R., & Van Den Berg, P. (2001). Nonfarm employment and poverty alleviation of rural farm households in Honduras. *World Development*, *3*(29), 549–560. doi:10.1016/S0305-750X(00)00107-8

Taylor, J. E., & Martin, P. (2001). Human capital: Migration and rural population change. In B. Gardener & G. Rausser (Eds.), *Handbook of Agricultural Economics* (Vol. 1, pp. 457–511). Amsterdam: Elsevier.

ENDNOTES

1. One average, one US\$ = 85.5 Rupees during the study period.
2. Proportionate effect of a dummy variable is calculated as (Exponent (α)-1)*100. Hence, (Exponent (0.41)-1)*100=50.68 $\approx$ 50.7.

Chapter 13
The Risk Parity Approach Applied to Agricultural Commodities:
A Different Approach to the Risk

Denis Veliu
Canadian Institute of Technology, Albania

ABSTRACT

The recent years were hard for commodities, with most suffering of high losses. The uncertainty of the financial markets after the 2008 crisis has pushed in the interest of finding new way of diversification. With the Risk Parity or Equally Weighted Risk Contribution strategy, Maillard, Roncalli, and Teiletche (2008) suggested a method that maximize the diversification. These authors have applied this strategy to the volatility (standard deviation). In this chapter, the author describes how to apply Risk Parity to the Conditional Value at Risk using historical data estimation. Passing to CVaR, a coherent measure, the model can benefit from its properties with the needed assumptions. As a special case, the author has applied this method to an agricultural portfolio, compared the Risk Parity strategies with each other and with the Mean Variance and Conditional Value at Risk. An important part is the analysis of the riskiness, the diversification and the turnover. A portfolio with a certain numbers of agricultural commodities may have particular specified that an investor requires.

INTRODUCTION

In recent years, the financial markets have been afflicted by high volatility, both equity and bond markets. After Markowitz (1952) with his first milestone work in modern portfolio theory, a number of other portfolio optimization models have been proposed in the literature. (Sharpe,1964) tried to linearize the portfolio optimization model. (Konno and Yamazaki 1991) introduced the Mean-Absolute Deviation (MAD), a different risk mcasurc using linear programming model instead of a quadratic programming model.

DOI: 10.4018/978-1-5225-2107-5.ch013

The MiniMax approach, introduced by Young (1998), minimizes the worst-case scenario, which is used as risk measure. Risk Metrics introduced methods to quantify market risks, such as $VaR_\alpha(x)$ which is defined as the maximum potential change in value of a portfolio with a given probability over a certain horizon. Risk Management has used this instrument for many years, in order to evaluate the performance and regulatory requirements, and to develop methodologies to provide accurate estimates. $VaR_\alpha(x)$ does not allow diversification. There are many works on the alternative risk measure $CVaR_\alpha(x)$ from the authors such as Andersson, Mausser and Uryasev (2000) that show why this is more preferred to $VaR_\alpha(x)$.

The most important properties are that $CVaR_\alpha(x)$ is a coherent and a convex measure shown in the model presented by (Artzner,1999), a model that allows diversification.

All these models have one problem in common: they need as an input the estimation of expected return for the assets. Models that need to estimate expected returns, produce extreme weights and have significant fluctuation over time. The Mean-Variance model is too sensitive to the input parameters, specially to the expected returns (Merton,1980). Thus, a significant variation of the input parameters can significantly change the composition of the portfolio, like in the Mean Variance portfolio. Models that rely on expected returns tend to be very concentrated on few assets and perform poorly out of sample (Merton,1980). The Black&Litterman model can be obtained using a Bayesian approach to change the estimated returns (Black, Litterman, 1990). With the passing of time, more sophisticated and advanced models were developed for the market forecasting, introducing different techniques and simulations. Thus, investors continue to use such simple allocation rules for allocating their capital across assets.

This chapter focuses on the models of portfolio selection under the Risk Parity criteria. More attention was focused on these models after the financial crisis of 2008 in the way they distribute the risk among the assets that compose the financial portfolio. The idea was introduced by Qian (2005) and it led to the creation of Risk Parity Portfolios, where we allocate an equal amount of risk to stocks and bonds in order to capture long-term risk premium embedded within various assets. Risk Parity portfolios are more efficient than the traditional 60/40 portfolios and they are truly balanced in terms of risk allocation.

The first authors formulate and discuss this argument were Maillard, Roncalli and Teiletche (2009). They showed that the volatility of Risk Parity is located between that of the minimum variance and of the equally weighted portfolio. Also, they prove the uniqueness and the existence of the Risk Parity portfolio.

Risk Parity approaches are frequently used to allocate the risk of a portfolio by decomposing the total portfolio risk into the risk contribution of each component in the same quantity. One of the biggest advantages of the Risk Parity approach is that it does not require the estimation of the expected returns. A crucial point of the thesis is the risk decomposition. Using the properties of the coherent and convex measures defined by Artzner(1999), it is possible to use the Euler decomposition for first order homogeneous functions.

In the Risk Parity models used in the literature, the measure of risk is the standard deviation of the financial portfolio. In this chapter is described that is possible to apply the Risk Parity approach to a different risk measure, the Conditional Value at Risk ($CVaR_\alpha(x)$), which is a coherent and convex risk measure, that allows to apply the Euler decomposition for first order homogeneous functions. The decomposition requires the calculation of the derivatives of risk measure. In the literature this model is used under the hypothesis that the returns are distributed like a multivariate normal for the calculation of the optimal weights with historical simulation. This hypothesis is less credible due to the lack of reality. Another author, Stefanovits (2009) in his master thesis, applies the equally risk contribution to the expected shortfall in case of standardized multivariate distribution, using a Gaussian kernel estimation. He implemented Risk Parity approach to Expected Shortfall assuming normally or t-student data in a parametric approach. In this chapter the Risk Parity model with $CVaR_\alpha(x)$ as a risk measure can be applied any (real) return distribution. This is possible with approximation methods in the calculation of the partial derivatives of the Conditional Value at Risk (Tasche, 2000). The author compares the Risk Parity strategies with different risk measures (standard deviation and Conditional Value at Risk). The results are very similar but the time of computation of Risk Parity with Conditional Value at Risk is significantly shorter. Starting from the studies of Colucci (2013), the author creates a Risk Parity with Conditional Value at Risk which has no true diversification, in order to compare it with Risk Parity with $CVaR_\alpha(x)$.

The models have been applied to daily and weekly frequencies in order to have a good approximation of Risk Parity with $CVaR_\alpha(x)$. Since the last year the commodities market had been afflicted from high volatility and a negative trend, the focus will be on how the portfolio created will perform in these cases.

The algorithms for the optimization are developed in Matlab 2012b ©, which is very effective in the calculation of portfolios with a large number of assets. For the Risk Parity with $CVaR_\alpha(x)$, we use an interior point algorithm with a defined number of iterations.

RISK BUDGETING

The Risk Budgeting Approach

In this section are derived the theoretical properties of the risk budgeting portfolios. Starting from the work of Maillard and Roncalli (2008), we formulate the general case of the Risk Budgeting approach a risk measure.

We create a portfolio with *n* assets, each weight x_i and $\mathcal{R}(x)$ as a risk measure for the portfolio $x = (x_1, x_2, \ldots, x_n)$. Using the Euler decomposition, for positive homogenous risk measures, we know that:

$$\mathcal{R}(x) = \sum_{i=1}^{n} x_i \frac{\partial \mathcal{R}(x)}{\partial x_i}$$

The Risk Budgeting approach uses the following marginal and total risk contribution of each asset:

$$MRC_i(x) = \frac{\partial \mathcal{R}(x)}{\partial x_i}$$

$$TRC_i(x) = x_i \frac{\partial \mathcal{R}(x)}{\partial x_i}$$

We consider the vector of risk budgets of all asset, $b = (b_1, b_2, ..., b_n)$., where b_i is the amount of risk in percentage of the total risk. We set $b_1 \geq 0$ and $\sum_{i=1}^{n} b_i = 0$.

If $b_1=0$ it means that the asset has no risk. We do not include risk free assets in our portfolio construction, so each asset will contribute to the total risk.

For a given risk budget b, the mathematical problem for the case with no short selling and no leverage can be summarized as follows:

$$x^* \in \left\{ x \in [0,1] : \sum_{i=1}^{n} x_i = 1, TRC_i(x) = b_i \mathcal{R}(x), \forall i \right\}$$

The difference between a risk budgeting portfolio and an optimized portfolio is that the first one does not try to maximize the utility function and the expected performance of the portfolio, but it just considers the risk dimension (Maillard, Roncalli, 2012).

The Risk Parity method is a particular case of risk budgeting when each total risk contribution is equal: in other words when $b_i=b_j=1/n$

$$TRC_i(x) = TRC_j(x), \forall i, j$$

$$x_i \frac{\partial \mathcal{R}(x)}{\partial x_i} = x_j \frac{\partial \mathcal{R}(x)}{\partial x_j} \forall i, j$$

then:

$$\mathcal{R}(x) = \sum_{i=1}^{n} x_i \frac{\partial \mathcal{R}(x)}{\partial x_i} = \sum_{i=1}^{n} x_i TRC_i(x) = n TRC_i(x)$$

In other words:

$$TRC_i(x) = \frac{\mathcal{R}(x)}{n}$$

In this way the risk is divided in the same proportion for each asset that composes the portfolio. A problem for this model consists in calculating the partial derivative of the risk $\mathcal{R}(x)$ respect to the weights x_i:

The mathematical problem for the Risk parity case can be summarized as follows:

$$x^* \in \left\{ x \in [0,1]^n : \sum_{i=1}^{n} x_i = 1, TRC_i(x) = TRC_j(x), \forall i, j \right\}$$

In this chapter, the author will apply Risk Parity to the standard deviation and to Conditional Value at Risk. In both cases, the models are in equal conditions: same starting points in the algorithms and with no short selling or no possibility to leverage.

Risk Parity Applied to Standard Deviation

In the literature, the most common use of Risk Parity is the case with the standard deviation as risk measure.

For a portfolio with n assets and weights $x = (x_1, x_2, \ldots, x_n)$, the standard deviation is:

$$\mathcal{R}(x) = \sigma_P(x) = \sqrt{\sum_{i=1}^{n}\sum_{j=1}^{n} x_i x_j \sigma_{ij}} = \sqrt{x'\Omega x}$$

where Ω is the covariance matrix.

The marginal risk contribution of the i asset:

$$MRC_i(x) = \frac{\partial \sigma_P(x)}{\partial x_i} = \frac{\partial \sigma_i^2 + \sum_{j=1}^{n} x_i \sigma_{ij}}{\sigma_P(x)} = \frac{(\Omega x)_i}{\sqrt{x'\Omega x}}$$

And the total risk contribution:

$$TRC_i(x) = x_i \frac{\partial \sigma_P(x)}{\partial x_i} = x_i \frac{\partial \sigma_i^2 + \sum_{j=1}^{n} x_i \sigma_{ij}}{\sigma_P(x)} = x_i \frac{(\Omega x)_i}{\sqrt{x'\Omega x}}$$

It is easy to show that:

$$\sum_{i=1}^{n} TRC_i(x) = \sum_{i=1}^{n} x_i \frac{(\Omega x)_i}{\sqrt{x'\Omega x}} = \sqrt{x'\Omega x} = \sigma_P(x)$$

Recall that the solutions Mean-Variance model enjoys the following problem:

$$\frac{\partial \sigma_P(x)}{\partial x_i} = \frac{\partial \sigma_P(x)}{\partial x_j}$$

In other words it to equalizes the marginal risk contributions, instead of the total risk contributions as in case of the Risk Parity:

$$TRC_i(x) = TRC_j(x) \forall i, j$$

The Risk Parity model can be formulated as the following optimization problem:

$$x^* = \arg\min \sum_{i=1}^{n}\sum_{j=1}^{n}\left(TRC_i(x) - TRC_j(x)\right)^2$$
$$\sum_{i=1}^{n} x_i = 1$$
$$x \geq 0$$

Since $TRC_i(x) = x_i \dfrac{\partial \sigma_P(x)}{\partial x_i}$ and $TRC_i(x) = \dfrac{\sigma_P(x)}{n}$, the problem can be written as follow:

$$x^* = \arg\min \sum_{i=1}^{n}\sum_{j=1}^{n}\left(TRC_i(x) = TRC_j(x)\right)^2$$
$$\sum_{i=1}^{n} x_i = 1$$
$$x \geq 0$$

An important point is proving the existence and, after the uniqueness of the Risk Parity portfolio, that Maillard S., Roncalli T., Teiletche J (2012) had proved in their working paper. This is possible, by using the properties of the covariance matrix Ω which is positive-definite. Thus, this becomes a problem which requies the minimization of a convex quadratic function with convex constraints, where can be applied the first order Khun-Tucker conditions for the Lagrangian. An important conclusion of their work is the mathematical proof where they showed that the standard of the Risk Parity portfolio is in between the Mean- Variance Model and the naive portfolio (the portfolio that take equal weights 1/n).

$$\sigma_{MV} \leq \sigma_{RP} \leq \sigma_{\frac{1}{n}}$$

Chaves D., Hsu J., Li F., Shakernia O.(2012) introduced two simple iterative algorithms to calculate the portfolio weights for a risk parity strategy. These algorithms presented require only simple computations and quickly converge to the optimal solution. They put the constraints of no short selling and no possibility to leverage. The first algorithm starts from an equal weight, calculate the betas of each asset and finishes when the condition of Risk Parity is satisfied. This method does not have a mathematical proof of convergence to a solution, but in many numerical applications one finds that the weights are the right one to guarantee the Risk Parity. The algorithms based on covariance are less efficient in terms of computation time, do not guarantee convergence to a solution, but are easier to implement using nonlinear optimization.

The second algorithm is an application of Newton's method for solving a system of nonlinear equations. This method converges faster than the first one and we just have to deal with operations such as inverse matrix. It tends to be more robust, reaching the optimal solution even when the first algorithm fails in particular situations. Both algorithms compute the same "optimal" risk parity solution as the original Maillard, Roncalli and Teiletche (2008) Risk Parity solution using nonlinear optimization.

RISK PARITY APPLIED TO CVAR

Derivatives of the Conditional Value at Risk

To guarantee the existence of the partial derivatives of $CVaR_\alpha(x)$ we need to impose some assumptions on the distribution of the random vector $R = (r_1, r_2, \ldots\ldots, r_n)$. Starting from the work of Tasche (2000), the first problem to deal with is differentiating the quaintile function $q_\alpha(X)$ and form that, the expression of $CVaR_\alpha(x)$ partial derivatives.

We present sufficient conditions for quaintile of the portfolio return $X = R'x = \sum_{i=1}^{n} x_i r_i$ to be differentiable respect to the weights x_i. These conditions rely on the existence of a conditional probability density function (pdf) of the i-th asset return r_i given the others which is measured as follow:

$$r_{i,t+1} = \frac{P_{i,t+1} - P_{i,t}}{P_{i,t}}$$

Definition 1: For the random vector $R = (r_1, r_2, \ldots\ldots, r_n)$, r_1 has a conditional density given $(r_2, \ldots\ldots, r_n)$ if it exits a measurable function $\theta : \mathbb{R}^n \rightarrow [0, \infty)$ such for that all $A \in \mathcal{B}(\mathbb{R})$ we have

$$P[r_1, \in A | r_2, \ldots\ldots, r_n] = \int_A \theta(u, r_2, \ldots\ldots, r_n)\, du$$

The existence of a joint pdf of R implies the existence of the conditional pdf but not necessarily the vice versa is true.

Lemma 1: Assume that r_1 has a conditional density θ given given $(r_2, \ldots\ldots, r_n)$, where $(r_1, r_2, \ldots\ldots, r_n)$ is a an $\mathbb{R}^n \mathbb{R}^n$-valued random vector. For any weight vector, $x = (x_1, x_2, \ldots\ldots, x_n) \in \mathbb{R} \setminus \{0\} \times \mathbb{R}^{n-1}$ we have:

The random variable $X = \sum_{i=1}^{n} x_i r_i$ has a pdf given by the following absolutely continuous functions

$$f_X(u) = \frac{1}{x_1} E\left[\theta\left(\frac{u - \sum_{j=2}^{n} x_j r_j}{x_1}, r_2, \ldots\ldots, r_n\right)\right] \tag{1.1}$$

if $f_X(u)>0$ we have almost surely for $i=2,\ldots\ldots,n$, and for $u \in \mathbb{R}$

$$E\left[r_i \mid \sum_{j=1}^{n} x_j r_j = \mathrm{u}\right] = \frac{E\left[r_1\theta\left(\frac{1}{x_1}\left(u - \sum_{j=2}^{n} x_j r_j\right), r_2, \ldots\ldots, r_n\right)\right]}{E\left[\theta\left(\frac{1}{x_1}\left(u - \sum_{j=2}^{n} x_j r_j\right), r_2, \ldots\ldots, r_n\right)\right]} \tag{1.1 a}$$

if $f_X(u)>0$ we have almost surely for $i=2,\ldots\ldots,n$, and for $u \in \mathbb{R}$

$$E\left[r_1 \mid \sum_{j=1}^{n} x_j r_j = \mathrm{u}\right] = \frac{E\left[\frac{u - \sum_{i=2}^{n} x_i r_i}{x_1}\theta\left(\frac{1}{x_1}\left(u - \sum_{j=2}^{n} x_j r_j\right), r_2, \ldots\ldots, r_n\right)\right]}{E\left[\theta\left(\frac{1}{x_1}\left(u - \sum_{j=2}^{n} x_j r_j\right), r_2, \ldots\ldots, r_n\right)\right]} \tag{1.1 b}$$

The point 1 of the Lemma says that if there is a conditional density of r_1 given the other component, then subject of the condition $x\neq 0$ the distribution $X = R'x = \sum_{i=1}^{n} x_i r_i$ is absolutely continuous with a density of point 1.

Proof:

1. Consider $x_1>0$, then we can write:

$$P\left[X \le u\right] = E\left[1_{\{X\le u\}}\right] = E\left[E\left[1_{\{X\le u\}}\right] \mid r_2, \ldots\ldots, r_n\right] = E\left[\int_{-\infty}^{\frac{u-\sum_{i=2}^{n} x_i r_i}{x_1}} \theta\left(v, r_2, \ldots\ldots, r_n\right) dv\right]$$

$$= E\left[\int_{-\infty}^{u} \frac{1}{x_1}\theta\left(\frac{v - \sum_{i=2}^{n} x_i r_i}{x_1}, r_2, \ldots\ldots, r_n\right) dv\right] = \int_{-\infty}^{u} E\left[\frac{1}{x_1}\theta\left(\frac{v - \sum_{i=2}^{n} x_i r_i}{x_1}, r_2, \ldots\ldots, r_n\right)\right] dv$$

$$= \frac{1}{x_1} E\left[\theta\left(\frac{v - \sum_{i=2}^{n} x_i r_i}{x_1}, r_2, \ldots\ldots, r_n\right)\right]$$

In the last step, it is required to apply the Fubini Theorem in order to change the order of integration. For $x_1 < 0$ we proceed in the same way.

2. $$E\left[r_i \mid \sum_{j=1}^{n} x_j r_j = \mathrm{u}\right] = \frac{E\left[r_i 1_{\{X\le u\}}\right]}{P\left[X = u\right]} = \lim_{\delta\to 0}\lim_{\delta\to 0} \frac{\delta^{-1}E\left[r_i 1_{\{u<X<u+\delta\}}\right]}{\delta^{-1}P\left(u < X < u+\delta\right)} = \frac{\frac{\vartheta}{\vartheta u}E\left[r_i 1_{\{X\le u\}}\right]}{f_X\left(u\right)}$$

where $f_X(u)>0$ (1.2)

Furthermore, we have:

$$\frac{\partial}{\partial u}E\left[r_i 1_{\{X\le u\}}\right]=\frac{\partial}{\partial u}E\left[Er_i\left[1_{\{X\le u\}}\right]\mid,r_2,\ldots..,r_n\right]=\frac{1}{x_1}E\left[r_i\theta\left(\frac{u-\sum_{i=2}^{n}x_ir_i}{x_1},r_2,\ldots..,r_n\right)\right] \quad (1.3)$$

Substituting (1.1) and (1.3) in (1.2) we obtain (1.1a)

3. We can write the expression (1.1a) and obtain (1.1b)

$$E\left[r_i\mid\sum_{j=1}^{n}x_jr_j=u\right]=E\left[\frac{u-\sum_{i=2}^{n}x_ir_i}{x_1}\mid\sum_{j=1}^{n}x_jr_j=u\right]$$

These are possible only for these assumptions of the conditional density θ: For more see the work of Tasche (2000).

Assumptions:

1. For fixed $r_2 \ldots r_n$, the mapping $t \rightarrow (t, r_2 \ldots r_n)$ is continuous in t.
2. The map $(t,x)\rightarrow E\left[\theta\left(\frac{u-\sum_{i=2}^{n}x_ir_i}{x_1},r_2,\ldots..,r_n\right)\right]$ is finite value and continuous.
3. For i=2,.....,n the mapping $E\left[r_i\theta\left(\frac{u-\sum_{i=2}^{n}x_ir_i}{x_1},r_2,\ldots..,r_n\right)\right]$ is finite value and continuous.

Applying these assumptions, Tasche (2000) gives the conditions for the partial differentiation with respect to the weights.

Theorem 1: Assume that the distribution of the returns is such that there exists a conditional density of r_i given $r_2 \ldots r_n$, satisfying the above Assumptions in some open set $H\subset\mathbb{R}\setminus\{0\}\times\mathbb{R}^{n-1}$ and that $f_X\left(q_\alpha(X)\right)>0$. Then $q_\alpha(X)$ is partially differentiable with respect to each weight x_i as follows:

$$\frac{\partial q_\alpha(X)}{\partial x_i}=E\left[r_i\mid R'x=q_\alpha(X)\right]$$

Proof. Applying Lemma 1 the random variable $X=R'x=\sum_{i=1}^{n}x_ir_i$ has a continuous pdf conditional density of r_1 given $r_2 \ldots r_n$ as follow

$$f_X(u) = \frac{1}{x_1} E\left[\theta\left(\frac{u - \sum_{j=2}^{n} x_j r_j}{x_1}, r_2, \ldots\ldots, r_n\right)\right] \forall x \text{ with } x_i \geq 0$$

$$\alpha = P\left[X \leq q_\alpha(X)\right] = E\left[\int_{-\infty}^{\frac{q_\alpha(X) - \sum_{i=2}^{n} x_i r_i}{x_1}} \theta\left(v, r_2, \ldots\ldots, r_n\right) dv\right] \tag{1.4}$$

Differentiating expression (1.4) with respect x_i for $i=2,\ldots\ldots,n$, we have:

$$0 = \frac{1}{x_1} E\left[\theta\left(\frac{u - \sum_{j=2}^{n} x_j r_j}{x_1}, r_2, \ldots\ldots, r_n\right)\right] = f_X(u) \tag{1.5}$$

Solving (1.5) for $\frac{\partial q_\alpha(X)}{\partial x_i}$ and applying the Lemma 1, we find the result of Theorem 1:

$$\frac{\partial q_\alpha(X)}{\partial x_i} = E\left[r_i \mid R'x = q_\alpha(X)\right]$$

Note that $VaR_\alpha(x) = -q_\alpha(X)$ then we can write:

$$\frac{\partial VaR_\alpha(x)}{\partial x_i} = -E\left[r_i \mid R'x = -VaR_\alpha(x)\right]$$

Applying to $VaR_\alpha(x)$ the Euler decomposition we have:

$$VaR_\alpha(x) = \sum_{i=1}^{n} \frac{\partial VaR_\alpha(x)}{\partial x_i} = -\sum_{i=1}^{n} E\left[r_i \mid R'x = -VaR_\alpha(x)\right]$$

The calculation of the partial derivatives for the Value at Risk is crucial for the of the partial derivatives of the Conditional Value at Risk. Indeed, be the definition of $CVaR_\alpha(x)$ (Uryasev, 2000) we have

$$CVaR_\alpha(x) = \frac{1}{\alpha} \int_0^\alpha VaR_v(x)\,\mathrm{d}v \tag{1.6}$$

Thus, using the Assumption 1 and differentiating (1.6) we obtain that:

$$\frac{\partial CVaR_\alpha(x)}{\partial x_i} = \frac{1}{\alpha}\int_0^\alpha \frac{\partial CVaR_\alpha(x)}{\partial x_i} \mathrm{d}v = -\frac{1}{\alpha}\int_0^\alpha E\left[r_i \mid -R'x = VaR_\alpha(x)\right]\mathrm{d}v$$
$$= -\frac{1}{\alpha}\int_0^\alpha E\left[r_i \mid X = q_\alpha(X)\right]\mathrm{d}v = -E\left[r_i \mid X \le -VaR_\alpha(x)\right] \tag{1.7}$$

The same result starting from the Expected shortfall $ES_\alpha(x)$, which is equivalent to the $CVaR_\alpha(x)$, as Tasche (2000) and Stefanovits (2010) showed in their work, under the condition that $E[X^-]<\infty$.

The Total Risk contribution for each asset i of a portfolio is given by the following expression:

$$TRC_i^{CVaR_\alpha(x)}(x) = x_i \frac{\partial CVaR_\alpha(x)}{\partial x_i}$$

The expression in case of continuous returns distribution is the following:

$$TRC_i^{CVaR_\alpha(x)}(x) = -x_i E\left[r_i \mid X \le -VaR_\alpha(x)\right]$$

$$CVaR_\alpha(x) = \sum_{i=1}^{n} TRC_i^{CVaR_\alpha(x)}(x) = -\sum_{i=1}^{n} x_i E\left[r_i \mid X \le -VaR_\alpha(x)\right]$$

NUMERICAL APPROXIMATION FOR ESTIMATING $VaR_\alpha(x)$ AND $CVaR_\alpha(x)$ RISK PARITY USING HISTORICAL DATA

In this section is described how to compute the $VaR_\alpha(x)$ and $CVaR_\alpha(x)$ using historical scenarios of assets returns.

Suppose that the *i*-th asset return r_i consist of *T* number outcomes r_{ji} with *i*=1,.....,*n* and *j*=1,...,T. For each portfolio $x \in \mathbb{R}^n$ where *n* is the number of assets in the market, the vector of the observed portfolio returns is $R_p = (r_{p1}, \ldots, r_{pT})$ where:

$$r_{pj} = x' r^j \text{ with } \mathrm{j} = 1, \ldots, \mathrm{T}$$

where $r^j = (r_{j1}, \ldots, r_{jT})$

If the number of observation *T* is large enough, we can apply the Law of Large Numbers for the numerical approximation of the empirical distribution of the historical portfolio return:

$$P(R_p \le y) \approx \frac{\#\left(j = 1, \ldots, T \mid r_{p1} \le y\right)}{T}$$

Therefore, we compute the $VaR_\alpha(x)$ and $CVaR_\alpha(x)$ of portfolio returns as follows:

$$VaR_{\alpha}\left(x\right) \approx -r^{\text{sorted}}_{\lfloor p\alpha T \rfloor}$$

$$CVaR_{\alpha}\left(x\right) \approx -\frac{1}{\alpha T}\sum_{j=1}^{\lfloor \alpha T \rfloor} r^{\text{sorted}}_{p\,j}$$

where α is a specified significance level and $r^{\text{sorted}}_{p\,j}$ are the sorted portfolio returns that satisfy

$$r^{sorted}_{p1} \leq r^{sorted}_{p2} \leq \ldots r^{sorted}_{p\,j} \leq \ldots \leq r^{sorted}_{p\,j}$$

Using historical data, from (1.4) the approximation of the partial derivatives $CVaR_{\alpha}\left(x\right)$ for each asst *i* becomes:

$$\frac{\partial CVaR_{\alpha}\left(x\right)}{\partial x_{i}} \approx -\frac{1}{\lfloor \alpha T \rfloor}\sum_{k=1}^{\lfloor \alpha T \rfloor} r^{\text{sorted}}_{k\,i} \quad \forall\ i = 1,\ldots,n$$

and then the total risk contribution of asset *i* is

$$TRC_{i}^{CVaR_{\alpha}\left(x\right)}\left(x\right) = x_{i}\frac{\partial CVaR_{\alpha}\left(x\right)}{\partial x_{i}} \approx -\frac{1}{\lfloor \alpha T \rfloor}x_{i}\sum_{k=1}^{\lfloor \alpha T \rfloor} r^{\text{sorted}}_{k\,i}$$

where $r^{\text{sorted}}_{k\,i}$ are the corresponding returns of asset *i* to the sorted portfolio returns. This method was suggested by Stefanovits (2010) in his master thesis, where he applies the equally risk contribution in case of standardized multivariate distribution, using a Gaussian kernel estimation. He implemented Risk Parity approach to Expected Shortfall assuming normally or t-student data in a parametric approach.

THE RISK PARITY PORTFOLIO FOR THE CVAR WORST CASE SCENARIO

In this section is provided a naive method to compute the Risk Parity portfolio weights when $CVaR_{\alpha}\left(x\right)$ is the risk measure. Starting from the work of Colucci (2011), this method does not require any optimization approach and it uses the $CVaR_{\alpha}\left(x\right)$ convexity property.

Let us consider the vector of portfolio weights $x = (x_1, x_2, \ldots, x_n)$ and $R = (r_1, r_2, \ldots, r_n)$ the vector of asset returns. Combining the property of sub-additivity and positive homogeneity we obtain $CVaR_{\alpha}\left(x\right)$ is a convex function:

$$CVaR_{\alpha}\left(R'x\right) \leq x_{1}CVaR_{\alpha}\left(r_{1}\right) + x_{2}CVaR_{\alpha}\left(r_{2}\right) + \ldots + x_{n}CVaR_{\alpha}\left(r_{n}\right) \tag{2.1}$$

where the first member (right hand side) of (2.1) represents the $CVaR_\alpha(x)$ worst case scenario when $x_i \geq 0$ and $\sum_{i=1}^{n} x_i = 1$. Thus we denote the absolute contribution of asset i to the maximum total risk as follows:

$$AC_i = x_i CVaR_\alpha(r_i) \tag{2.2}$$

Then, the Risk Parity portfolio can be found by the following steps:

1. Start with a uniform portfolio 1/n;
2. Find the portfolio upper bound risk $CVaR^U = \sum_{i=1}^{n} x_i CVaR_\alpha(r_i)$ that corresponds to the worst case scenario;
3. We find the absolute contribution equal for every asset that belongs to the portfolio. For a fixed $CVaR^U$ compute the value of the absolute contribution of each asset in case of equality among the assets:

$$AC^U = \frac{CVaR^U}{n}$$

4. From (2.2) the Risk Parity portfolio weights are obtained by setting:

$$x_i^{**} = \frac{AC^U}{CVaR_\alpha(r_i)} \quad \forall i=1,\dots,n$$

and normalizing the weights to get:

$$x_i^{*} = \frac{x_i^{**}}{\sum_{k=1}^{n} x_k^{**}}$$

We call this method *Naive Risk Parity CVaR*, as it is not the true diversification.

It is possible to show that the weights of the Naive Risk Parity CVaR portfolio are proportional to the inverse of the $CVaR_\alpha(r_i)$:

$$x_i^{**} = \frac{AC^U}{CVaR_\alpha(r_i)} = \frac{CVaR^U}{nCVaR_\alpha(r_i)}$$

thus normalizing the portfolio weights we obtain:

$$x_i^* = \frac{x_i^{**}}{\sum_{k=1}^{n} x_k^{**}} = \frac{\frac{CVaR^U}{nCVaR_\alpha(r_i)}}{\sum_{k=1}^{n} \frac{CVaR^U}{nCVaR_\alpha(r_{ik})}} = \frac{CVaR_\alpha(r_i)^{-1}}{\sum_{k=1}^{n} CVaR_\alpha(r_k)^{-1}}$$

The total risk contribution of asset *i* for $\sum_{i=1}^{n} x_i = 1$ and is:

$$AC^U = TRC_i^{CVaR_\alpha(x)}(x) = x_i CVaR_\alpha(r_i) = \frac{1}{\sum_{k=1}^{n} CVaR_\alpha(r_k)^{-1}}$$

The worst case scenario of CVaR

$$CVaR^U = \frac{\text{n}}{\sum_{k=1}^{n} CVaR_\alpha(r_k)^{-1}}$$

The same result can be provided if the returns are arranged in a non decreasing order, instead to the corresponding sorted portfolio return.

Colucci (2011), in his working paper proposes to work with maximum portfolio risk, as the wiser choice if we take in consideration the co-dependence when the market crashes (2008 crisis). Thus, we have an upper bound of the risk.

DIVERSIFICATION MEASURES

Consider a portfolio $x = (x_1, x_2, \ldots, x_n)$ satisfying the budget constraint $\sum_{i=1}^{n} x_i = 1$ with short sales not allowed $x_i \geq 0$. The first naive diversification measure is the Herfindal index:

$$D_{Her} = 1 - xx'$$

which takes the value 0 if the portfolio is concentrated in one asset and the maximum value $1 - \frac{1}{n}$ for the equally weighted (or naive) portfolio.

We introduce the measure proposed by Bera and Park (Bera, Park, 2004), for strategies with no short selling.

This diversification measure can be interpreted as the probability of each weight measure in terms of entropy:

$$D_{BP} = -\sum_{i=1}^{n} x_i \log(x_i) = \sum_{i=1}^{n} x_i \log(\frac{1}{x_i})$$

The D_{BP} takes value between 0 (fully concentrated in one asset) and log(n) for the naive portfolio.

Another index of diversification based on the weights that compose the portfolio has been proposed by Hannah and Kay:

$$D_{HK}^{\alpha} = -\left(\sum_{i=1}^{n} x_i^{\alpha}\right)^{\frac{1}{\alpha-1}}$$

For all α>0α>0. It is easy to verify that $D_{Hk}{}^2 = D_{Her} - 1$.

These three quantities represent diversification only in terms of capital invested and do not take into account that assets contribute differently to the total portfolio volatility.

Another useful index for estimating transaction costs, is the turnover of the portfolio:

$$TO = \sum_{i=1}^{n} | x_i^{t+1} - x_i^{t} |$$

where x_i^t denotes the weight of asset i at time t.

STRUCTURES OF THE ANALYSIS AND DEFINITION OF THE INDICES FOR THE BENCHMARK PORTFOLIOS

Suppose that the i-th asset return r_{ji} consist of T outcomes with i=1,....,n and j=1,...,T. For each portfolio $x \in \mathbb{R}^n$ where n is the number of assets, the vector of the observed portfolio returns portfolio $x \in \mathbb{R}^n$ where n is the number of assets in the market, the vector of the observed portfolio returns is $R_P = (r_{p1},.....,r_{pT})$ where:

$$r_{pj} = x' r^j \text{ with j=1,....,T}$$

where $r^j = (r_{j1},.....,r_{jT})$ as we described in the section of the numerical approximation.

In the analysis we choose an in-sample period ΔL and an out of sample period of ΔH which are shorter than the ΔL using, generally daily time series.

The holding (or out of sample) period represents the investment horizon of the selected portfolio. The daily average portfolio return is:

$$\mu(R_P) = \frac{1}{T}\sum_{j=1}^{T} r_{pj}$$

The annualized mean portfolio return for the daily observation:

$$\mu\left(R_P\right)_{ann} = \left(1+\mu\left(R_P\right)\right)^{252} - 1$$

In this way mean returns are going to be used in order to quantify relationships between portfolio risk and return. To quantify the total gain of the strategy the author computes for $k = 1,....,T$ the compounded return:

$$\mu_k^c\left(R_P\right) = \prod_{j=1}^{k}\left(1 + r_{pj}\right) - 1$$

so that $\mu_T^c\left(R_P\right)$ is the compounded return over the whole period (terminal compound return).

As measures of risk, the author computes the sample volatility, $VaR_\alpha\left(x\right)$ and $CVaR_\alpha\left(x\right)$ of the daily returns over the period.

$$\sigma\left(R_P\right) = \left(\frac{1}{T}\sum_{j=1}^{T}\left(r_{pj} - \mu\left(R_P\right)\right)^2\right)^{\frac{1}{2}}$$

$$VaR_\alpha\left(x\right) \approx - r_{p\lfloor\alpha T\rfloor}^{sorted}$$

$$CVaR_\alpha\left(x\right) \approx -\frac{1}{\lfloor\alpha T\rfloor}\sum_{j=1}^{\lfloor\alpha T\rfloor} r_{pj}^{sorted}$$

Note that in order to have a good approximation the tail of the observations αT, it is suggested choose a longer period of estimation, the so called in sample period L, or, in order to reflect better and more recently the fluctuation of the market, a larger α, for instance α=10%.

These represent daily risks and can be annualized by multiplying them with $\sqrt{252}$

For the annualized risks, the notations are the following σ_{ann}, VaR_{ann} and. $CVaR_{ann}$ and the performance ratios $S_\sigma = \frac{\mu\left(R_P\right)_{ann}}{\sigma_{ann}}$, $S_{VaR_{ann}} = \frac{\mu\left(R_P\right)_{ann}}{VaR_{ann}}$ and $S_{CVaR_{ann}} = \frac{\mu\left(R_P\right)_{ann}}{CVaR_{ann}}$ also with the Sortino ratio (which is a variation of Sharpe ratio that takes in consideration only positive parts of the returns) with risk free rate equal to zero and the Rachev ratio (to measure the fund's upside potential as measured by expected return in the right tail and expected loss in the left) at the confidence level to α=5%

$$SortR = \frac{\mu\left(R_P\right)}{\sigma_d}, \qquad RaR_{\alpha,\beta} = \frac{CVaR_\alpha\left(R_P - r_f\right)}{CVaR_\beta\left(r_f - R_P\right)}$$

PORTFOLIO OPTIMIZATION WITH AGRICULTURAL COMMODITIES

In this part of the chapter is described the optimization and the performance of the models using a group of commodities in different sectors, mostly agricultural (see Table 1).

We have excluded from the quoted in the exchange market from the grains group rough rice and palm oil, from the softs the cocoa and rubber, and all the livestock (Live cattle, feeder cattle and lean hogs) for the missing and discontinuities data. There is a difference between the two elements that compose the metals group: while gold can be considered as a safe asset, the silver is heavily used in industry.

Two of the most common commodity indexes are the Goldman Sachs Commodities Index (GSCI) and the Dow Jones-USB Commodity Index (DJ-USBCI).The price of them is quoted in United State Dollars, except gold and silver the rest refers to the futures market index. After the 2008 crisis the market of commodities had been fluctuating, till 2014, where the price of commodities futures started decreasing. The year 2015 was tough for commodities, with most suffering colossal losses. The GSCI index, lost 33% in 2015. There are many factors that contributed to the precipitate the decline in commodities over the year, including a strong American currency policy. Commodities prices usually rise when inflation is increasing and so they be considered, as protection from the effects of inflation.

Historically, commodities have a negative correlation to stocks and bonds, for that, the diversification of the portfolio will be better if the investor decides to allocate some part of the capital invested.

To fulfill the conditions of our models for the numerical approximation, the period of data should be large enough, in this from 02/01/2014-10/06/2016, with highly (daily) frequencies. Considering that the market is opened for trading 252 days a year, we have 614 observation for these 13 assets.

To have a clear idea of the assets that compose our portfolio, it is better to take a glance at the characteristics of the distribution of the returns (mean, median, range, skewness and kurtosis) for the daily case.

The purpose of this analysis is to see the kind of distribution of each asset that composes the portfolio and if we can apply other models of optimization that require particular conditions for the distribution.

So each of the return distributions has a leptokurtic distribution compared to a normal distribution, due the fact of the decreasing trend of the market starting from 2014, and continuing for all 2015.

RISK PARITY WITH STANDARD DEVIATION APPLIED TO COMMODITIES

To have a complete view of the Risk Parity strategies, the first implementation will be the optimization of the portfolio strategy using the standard deviation as risk measure. To see each contribution to the

Table 1. The composition of the portfolio

ENERGY	METALS	GRAINS	SOFTS
Crude oil	Gold	Corn	Coffee
Coal	Silver	Wheat	Sugar
Gasoline		Soybean	Cotton
			Ethanol
			Orange juice

Figure 1. The returns of 13 future prices of the commodities

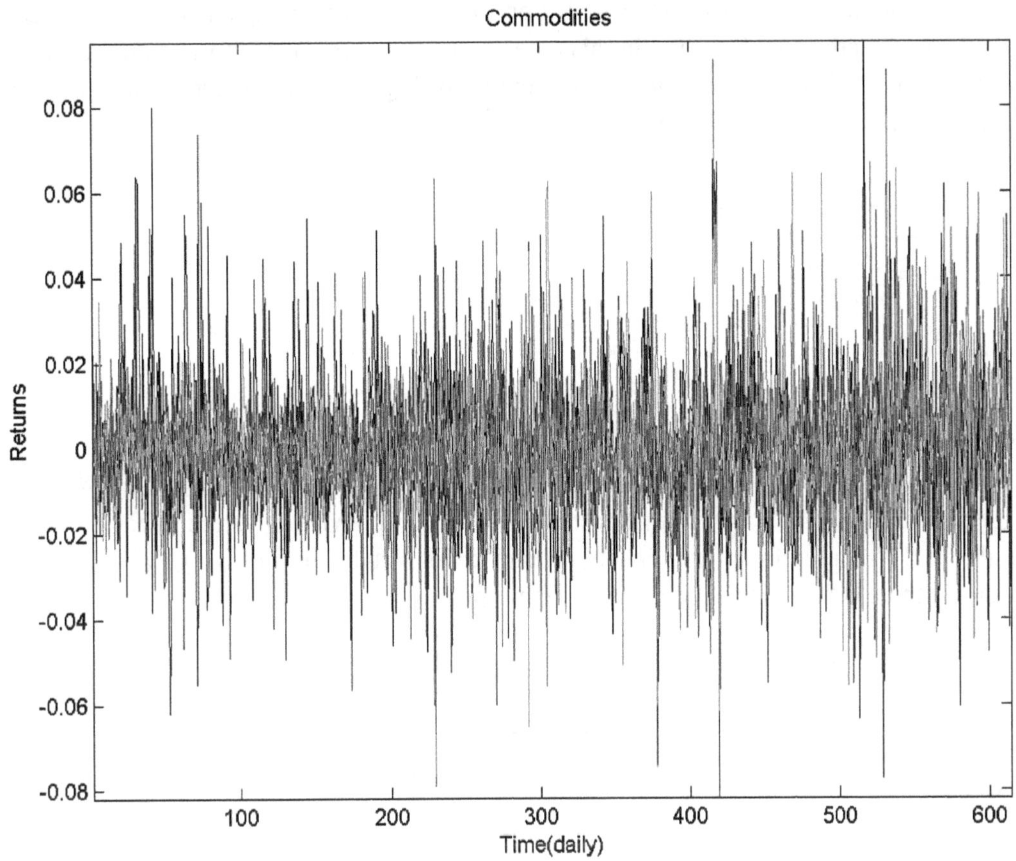

Table 2. The distribution analyses of data

Commodities	Mean	Median	Range	Skewness	Kurtosis
Crude oil	-0,000820	-0,0002352	0,1727379	0,27461797	6,4470912
Coal	-0,001012	-0,0009639	0,1750453	0,30994380	6,9936932
Gasoline	-0,000706	-0,0010145	0,1546692	0,13929234	5,8167876
Gold	0.000058	-0,0003987	0,0743328	0,20781726	4,5867796
Silver	-0,000241	-0,0002048	0,1299112	-0,07549043	5,6117836
Corn	-0,000175	0	0,1044446	0,01203753	5,5049178
Wheat	-0,000397	-0,0003836	0,1146884	0,04687712	3,7115257
Soybean	0,000103	0	0,0935916	0,38200273	5,7915782
Coffee	-0,000086	0,0005306	0,1420502	0,22778999	3,9163795
Sugar	0,000008	-0,0005998	0,1099828	0,53165481	6,2513187
Cotton	-0,000325	0	0,0727474	0,18188836	5,1493391
Ethanol	0.000060	0	0,0869296	-0,19067671	4,0056354
Orange juice	0,000221	-0,0003892	0,1318472	0,44318187	4,9861287

risk, in this case to standard deviation, we consider the data for the same period of time from 2/1/2014 to 10/06/2016 for daily frequencies (128 weeks or 29 months). The sample period is large enough to apply the Law of Large Numbers and in this case we get the maximum information for the range. In order to have an equal risk contribution we use the following optimization model with no short sales and no leveraged positions.

$$x^* = \arg\min \sum_{i=1}^{n}\sum_{j=1}^{n}\left(TRC_i(x) - TRC_j(x)\right)^2$$
$$\sum_{i=1}^{n} x_i = 1$$
$$x \geq 0$$

The algorithm was made by Ph.D. Farid Moussaoui using Matlab optimization (c) .

The sum of the total risk contribution $TRC_i(x)$ gives the standard deviation of the portfolio, 0.006534.

You can notice that the Risk Parity strategy with standard deviation as risk measure, in this case choose almost the same amount for the energy commodities (crude oil, coal and gasoline). For the other commodities, there is a significant difference in the choice.

Table 3. The composition of portfolio with Risk Parity STD and Mean-Variance models

	Risk Parity with STD			Mean-Variance
Commodities	x_i	$\frac{\partial \sigma_P(x)}{\partial x_i}$	$TRC_i(x)$	x_i
Crude oil	0.050994	0.0124460	0.0006346796	0
Coal	0.051228	0.0123321	0.0006317446	0
Gasoline	0.060883	0.0104735	0.0006376612	0.028101
Gold	0.085082	0.0035815	0.0003047227	0.242514
Silver	0.080371	0.0079013	0.0006350339	0
Corn	0.084303	0.0062398	0.0005260313	0.0189846
Wheat	0.084785	0.0067162	0.0005694359	0.0134039
Soybean	0.084447	0.0050205	0.0004239658	0.1586902
Cofee	0.075681	0.0097735	0.0007396617	0.0001589
Sugar	0.085259	0.0044051	0.0003755771	0.1190228
Cotton	0.085124	0.0028290	0.0002408164	0.2714717
Ethanol	0.083806	0.0057310	0.0004802905	0.0784745
Orange juice	0.088035	0.0037964	0.0003342115	0.0691776
Sum	1	std	0.006534	1

The Mean-Variance model without the return constraint (i.e, we find the minimum value of the variance portfolio) the results of the last column: the portfolio in concentrated in a significant weights in cotton 27,14%, gold 24,25%, soybean 15,87%, sugar 11,9% ethanol 7,84% and orange juice 6,92%.

It has excluded crude oil and coal, and chosen gasoline in a small amount. The variance obtained is 2.5656e-05 or standard deviation of 0.0050651. If we show the marginal risk contribution it should be equal for each asset selected and the total risk contribution is 0 where the assets are not selected. We also compute the Naive portfolio for comparative reasons with the weights $x_{\frac{1}{n}} = \frac{1}{n} = \frac{1}{13}$ and get the standard deviation $\sigma_{\frac{1}{n}} = 0.007107964$.

$$\sigma_{MV} \leq \sigma_{RP} \leq \sigma_{\frac{1}{n}}$$

Due to the negative trend of the period of reference, the return indexes and the ratios have a negative value if we compared the returns of the portfolios without changing the weights invested (re calibrating the portfolio). From the point of view of riskiness, the risk parity strategy is between Mean-Variance and the uniform portfolio even compared from other risk measures.

RISK PARITY WITH $CVaR_{\alpha}(x)$ APPLIED TO COMMODITIES

The Risk Parity strategy using the Conditional Value at Risk as risk measure, can benefit from the benefits of a coherent risk measure, as described by Artzner (1999). For this, an algorithm is needed, elaborated in Matlab, to compute the total risk contribution of each asset to Conditional Value at Risk, using the same time series of the case of the standard deviation as a risk measure (i.e. the period of time

Table 4. Performance analyses for the maximum range of data

	Risk Parity with STD	M-V	Uniform
$\mu(R_p)$ (%)	-0.0001937	-0.0000569	-0.0002493
$\mu(R_p)_{ann}$ (%)	-0.0030759	-0,0009040	-0.0039578
σ (%)	0.0065338	0.0050651	0.0071079
$VaR_{10\%}$ (%)	0.0164073	0.0162976	0.0227050
$CVaR_{10\%}$ (%)	0.0242521	0.0224701	0.0311667
σ_{ann} (%)	0.1037214	0.0804067	0.1128354
$VaR_{10\%ann}$ (%)	0.2604578	0.1949965	0.2709837
$CVaR_{10\%ann}$ (%)	0.3849896	0.2877766	0.4002862
S_{σ}	-0.0296557	-0.0112476	-0.0350763
$S_{VaR_{ann}}$	-0.0118097	-0.0034957	-0.0098875

from 2/1/2014 to 10/06/2016 for daily frequencies). To have a comparative measure for the true diversification, another method can be introduced, like the Naive Risk Parity $CVaR_{\alpha}(x)$, with no true diversification, to see the difference between these models in the contribution of the risk. We choose a confidence level of 10%, in order to have a fast convergence of the model.

For the daily frequencies we have the following tables with weights, marginal risk contribution and total risk contribution for Naive Risk Parity $VaR_{\alpha}(x)$ and Risk Parity $CVaR_{\alpha}(x)$ (see Table 5).

You can notice that the total risk contribution is bigger in case of Risk Parity Portfolio Naive than in case of Risk Parity CVaR. As a consequence the R.P Naive CVaR is riskier than R.P CVaR. Another interesting result is that the marginal risk contribution of Risk Parity Portfolio Naive is higher than the corresponding assets marginal risk contribution of Risk Parity in each case. Also the CVaR is highly concentrated in Gold, Soybean and Cotton while excluding completely crude oil coal and silver.

The minimum risk of the portfolio from the point of view of $CVaR_{10\%}(x) = 0.008815$. In this case the portfolio con concentrated mainly in Gold, soybean, sugar and cotton.

For the Naive portfolio the Conditional Value at Risk $CVaR_{\frac{1}{n}}(x) = 0.031167$. From the following order:

Table 5. The composition of portfolio with Risk Parity strategies and CVaR

Commodities	Risk Parity $CVaR_{10\%}(x)$ - Naive			Risk Parity $CVaR_{10\%}(x)$			$CVaR_{10\%}(x)$
	x_i	$\frac{\partial CVaR(x)}{\partial x_i}$	$TRC_i(x)$	x_i	$\frac{\partial CVaR(x)}{\partial x_i}$	$TRC_i(x)$	$TRC_i(x)$
Crude oil	0,049189	0,03575	0,001758	0,04020	0,020294	0,000816	0
Coal	0,048915	0,03595	0,001758	0,04000	0,019933	0,000797	0
Gasoline	0,056267	0,03125	0,001758	0,04766	0,016718	0,000797	0,00706
Gold	0,105505	0,01667	0,001758	0,11334	0,007186	0,000815	0,26236
Silver	0,062240	0,02825	0,001758	0,05766	0,014452	0,000833	0
Corn	0,086073	0,02043	0,001758	0,07268	0,010867	0,000788	0,02223
Wheat	0,072646	0,02421	0,001758	0,07489	0,010894	0,000816	0,02172
Soybean	0,123295	0,01426	0,001758	0,09282	0,008662	0,000804	0,18116
Cofee	0,048209	0,03648	0,001758	0,06880	0,011423	0,000786	0,01317
Sugar	0,088615	0,01984	0,001758	0,09250	0,008743	0,000809	0,11270
Cotton	0,111240	0,01581	0,001758	0,13690	0,005737	0,000785	0,23038
Ethanol	0,085970	0,02054	0,001758	0,07428	0,010745	0,000798	0,08599
Orange juice	0,061833	0,02840	0,001758	0,08830	0,009103	0,000804	0,06323
Sum	1	CVaR	0.02286112	1	CVaR	0.0104491	1

$$CVaR_{10\%}(x) < CVaR_{RP-CVaR}(x) < CVaR_{RP-CVaR\ naive}(x) < CVaR_{\frac{1}{n}}(x)$$

the Risk Parity strategy is a good tradeoff between the minimum $CVaR_{10\%}(x)$ and the naive portfolio.

In the next session are comparing the performance of all these models.

COMPARISON BETWEEN MODELS

In the comparison between models we have to distinguish between 2 strategies. The first one consists in a passive strategy when, after we decide the optimal weights with each strategy, we don't change them during the holing period. In this case, without turnover, for the period 2/1/2014 to 10/06/2016 (614 observations), the results in Table 6 are obtained.

Mean-Variance and CVaR perform better from the point of view of performance The Risk Parity strategies have almost the same indicators of performance. The naive strategy will perform worse than the others.

Table 6. The Performance analyses of all models (Passive method)

Indexes	R.P. -STD	M-V	R.P. CVaR N.	R.P. CVaR	$CVaR_{10\%}(x)$	Uniform (Naive)
$\mu(\%)$	-0,019376	-0,005697	-0,01842	-0,016643	-0,003163	-0,024932
$\mu_{ann}(\%)$	-1,002617	-0,295817	-0,95307	-0,861793	-0,164348	-1,288260
$\mu^c(\%)$	-12,371881	-4,193826	-11,75916	-10,720170	-2,702779	-15,514390
Median	-0,027887	-0,010965	-0,037593	-0,028979	-0,002397	-0,037492
$\sigma(\%)$	0,653383	0,506515	0,62638	0,604405	0,510284	0,710796
$VaR_{10\%}(\%)$	0,863358	0,627662	0,82356	0,791882	0,600460	0,888762
$CVaR_{10\%}(\%)$	1,140183	0,893956	1,09510	1,044900	0,883309	1,254896
$\sigma_{ann}(\%)$	4,711614	3,652530	4,51687	4,358429	3,679710	5,125630
$VaR_{10\%ann}(\%)$	13,705400	9,963830	13,07365	12,570700	9,532000	14,108700
$CVaR_{10\%ann}(\%)$	18,099800	14,191100	17,38417	16,587417	14,022000	19,920800
S_{σ}	-0,212797	-0,080989	-0,21100	-0,197730	-0,044660	-0,251337
$S_{VaR_{ann}}$	-0,347750	-0,143062	-0,34687	-0,326769	-0,083292	-0,431672
$S_{CVaR_{ann}}$	-0,263320	-0,100446	-0,26086	-0,247641	-0,056621	-0,305726
Sortino Ratio	-0,041711	-0,016006	-0,04137	-0,038817	-0,008903	-0,049203
Rachev Ratio	1,057818	1,030872	1,07860	1,053800	1,045660	1,064300

The second method, the active, consists in recalibrating the portfolio, for example in regular interval of time.

This is a better to compare the performance between the models. It is possible to create a rolling window starting from 2/1/2014 to 10/06/2016 with in sample period of L=1 year (252 days) and out of sample of H=5 days, so recalibration of each week (considering the time lapse when the stock market is open). In this way we have to recalculate the values for about 72 times. So, the out of sample period starts for the next year, 1/1/2015 till 10/06/2016.

It is easy to notice a better performance compared to the passive method for all models. Even in this decreasing trend of the market, the Mean-Variance model has a positive performance in term of returns. The interesting fact is that the Risk Parity with CVaR, perform better than CVaR itself.

In most of the cases, the investors have to compare the performance during the period of interest. For that, he can compare the cumulated return in each time.

From the graph of the compound returns, the study can be divided in two parts: in the first half we can see that the Risk Parity strategies perform better that the others. In the second half, there is a regain of the return of the Mean Variance portfolio; the other models perform better than the Naive portfolio.

Another study consists in the riskiness of the portfolios out of sample. If the volatility is calculated, the Mean-Variance is advantaged since his purpose is to minimize the variance, that without the return

Table 7. The performance of portfolios with Rolling window L=252 days H=5 days

Indexes	R.P. -STD	M-V	R.P. CVaR N.	R.P. CVaR	$CVaR_{10\%}(x)$	Uniform
$\mu(\%)$	-0,002631	0,007121	-0,002644	-0,002401	-0,009310	-0,006563
μ_{ann} (%)	-0,660819	1,810511	-0,664241	-0,603300	-2,319016	-1,640261
μ^c (%)	-1,9622	2,002069	-1,852127	-1,712605	-3,890597	-3,507745
Median	-0,016972	-0,004089	-0,011355	-0,021354	-0,020437	-0,026115
σ (%)	0,759411	0,569029	0,715113	0,693490	0,585983	0,820656
$VaR_{10\%}$ (%)	0,975616	0,712137	0,956385	0,907248	0,729627	1,077318
$CVaR_{10\%}$ (%)	1,2925	0,967594	1,200329	1,174454	1,028446	1,370611
σ_{ann} (%)	12,05528	9,033068	11,352058	11,008816	9,302194	13,027510
$VaR_{10\%ann}$ (%)	15,48740	11,30483	15,182136	14,402130	11,582472	17,102000
$CVaR_{10\%ann}$ (%)	20,51775	15,36008	19,054632	18,643871	16,326083	21,757780
S_σ	-0,054816	0,200431	-0,058513	-0,054802	-0,249298	-0,125900
$S_{VaR_{ann}}$	-0,042668	0,160154	-0,043752	-0,041890	-0,200218	-0,095911
$S_{CVaR_{ann}}$	-0,032210	0,117870	-0,034850	-0,032359	-0,142044	-0,075387
Sortino Ratio	-0,004965	0,018138	-0,005314	-0,004942	-0,022418	-0,011505
Rachev Ratio	1,1047111	1,061681	1,107324	1,083027	1,014862	1,118365

Figure 2. Compound returns of the portfolios

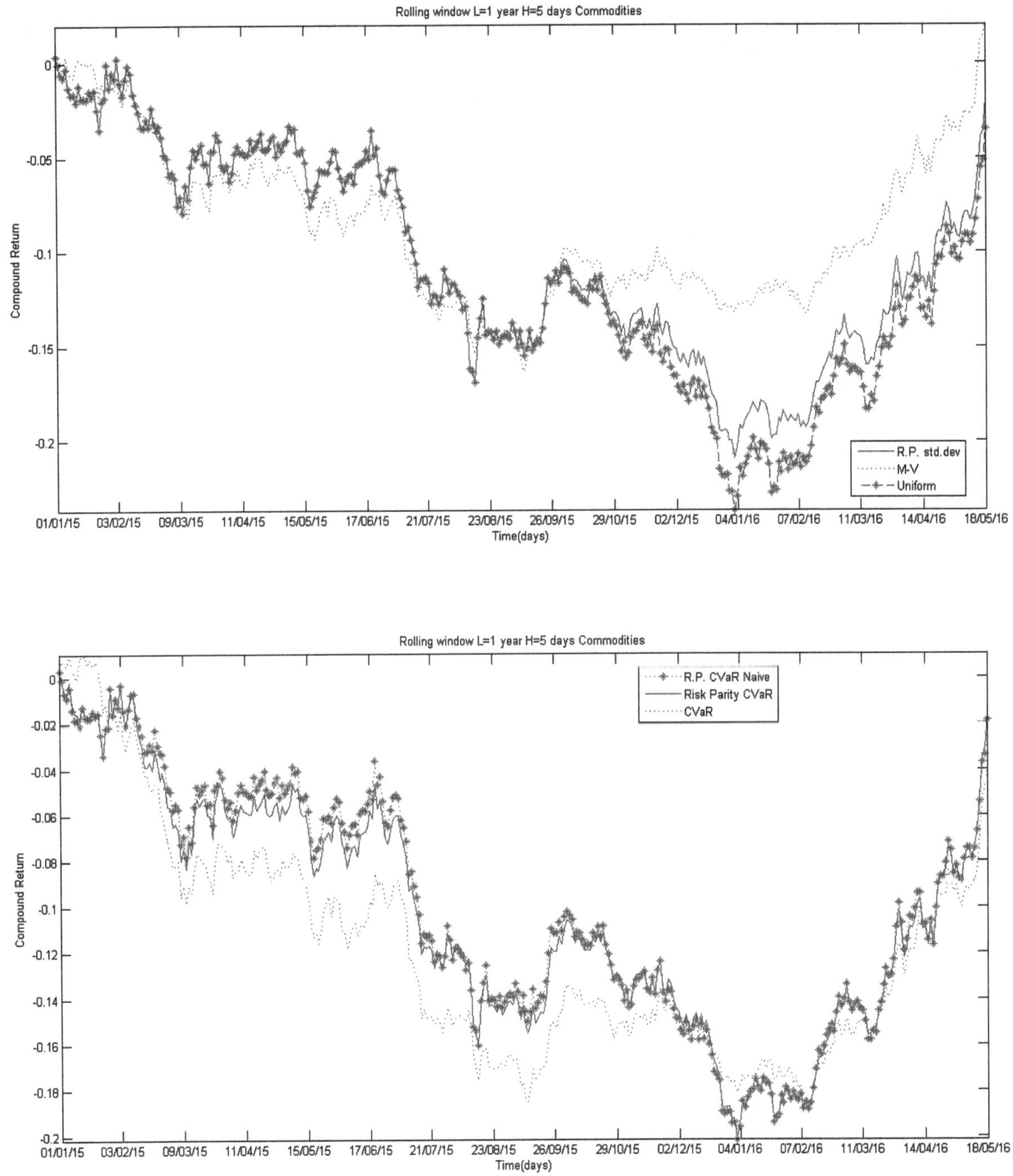

constrain gives the global minimum riskiness. The same reasoning will be with CVaR. Although, this measure is good to have a clear idea of the Risk Parity strategies.

The Risk Parity with CVaR is less riskier than the corresponding with a standard deviation as a risk measure in both cases.

Figure 3. The Volatility and the CVaR of the portfolios

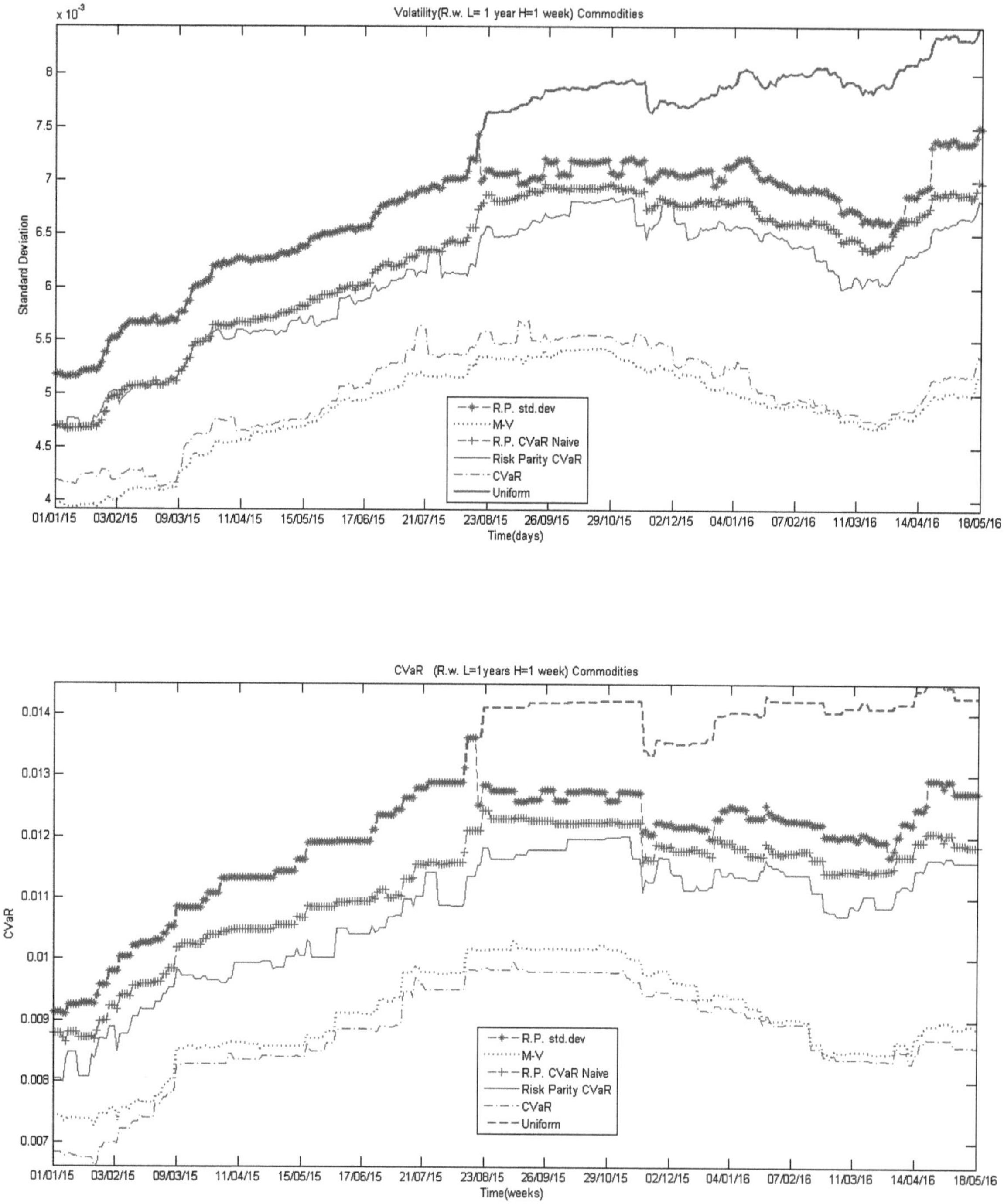

To get closer to the real markets when measuring performance, we have to deal with the transaction cost, fixed or variable in all cases. For that we must consider the portfolio turnover for each period where we recalculate the optimal weights. As we know the CVaR model and the Mean Variance model are concentrated in small groups of assets and for that they suffer from high turnover.

The other models have lower turnover but for that, the amount of capital invested and the fixed costs.

In Figure 4 we show the portfolio turnover for each model:

As the turn over index is measured in absolute values, some of the proportion of the amount invested should be decreased (sell asset) and some should be increased.

In Table 8 we show the average turnover for each period of rebalancing (each week).

The problem with the average is that in some cases the portfolios do not change the composition so we have to take a closer look. As we described in section 1.5.1, the Herfindal index takes the value 0 if the portfolio is concentrated in one asset and the maximum value 1-1/n for the naive portfolio. So for the naive (or Uniform) portfolio, we have the maximum value 0.9231 for the Herfindal Index. The more the portfolio is concentrated, like CVaR and Mean-Variance, the lower is the index.

Another way to study diversification is to apply the Bera Park Index, which is similar to the Herfindal index. The only problem to deal with using this measure is when the portfolio assumes the position 0 for a certain asset and there we have to adapt the quantities equal to 0 in a way to apply the index. As we see from the graph, the most concentrated are the CVaR and the Mean Variance.

As the last point we consider the number of assets that each model selected with a reasonable quantity (we do not consider the weights smaller than 10^{-6}. Since Risk Parity models and the naive portfolio,

Figure 4. Portfolio turnover of commodities

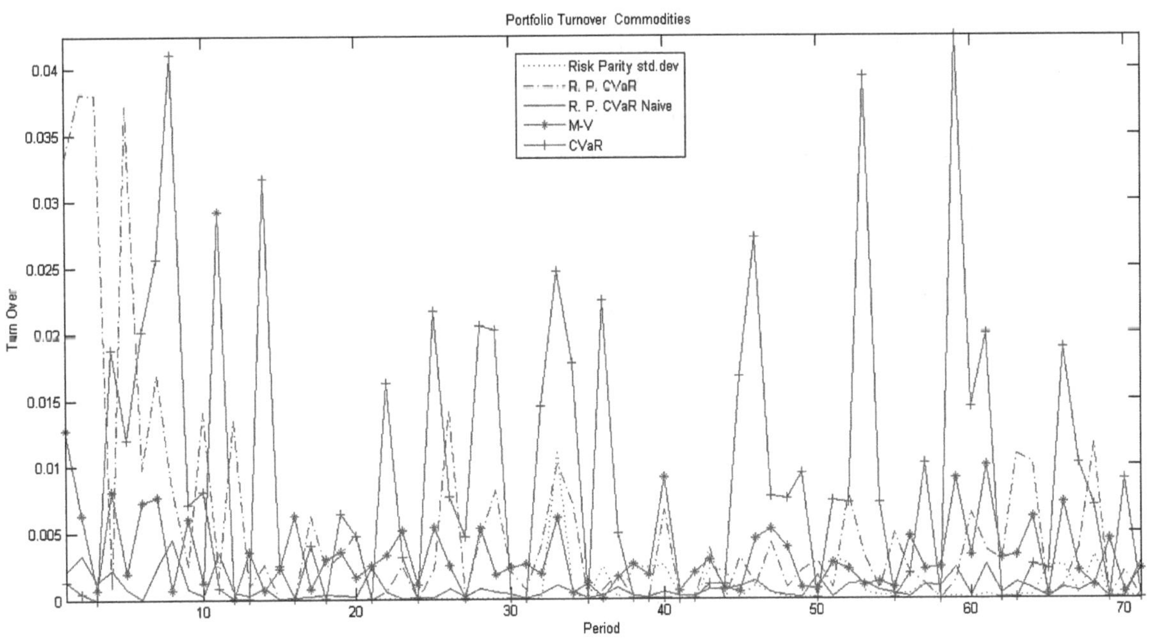

Table 8. The Turnover analyses

	R.P. -STD	M-V	R.P. CVaR N.	R.P. CVaR	CVaR
Average Turnover (%)	0,0632	0,35536	0,08221	0,55304	0,893552
Total Turnover(%)	4.484	39.265	25.2309	5.8368	63.4422

Figure 5. Diversification indexes

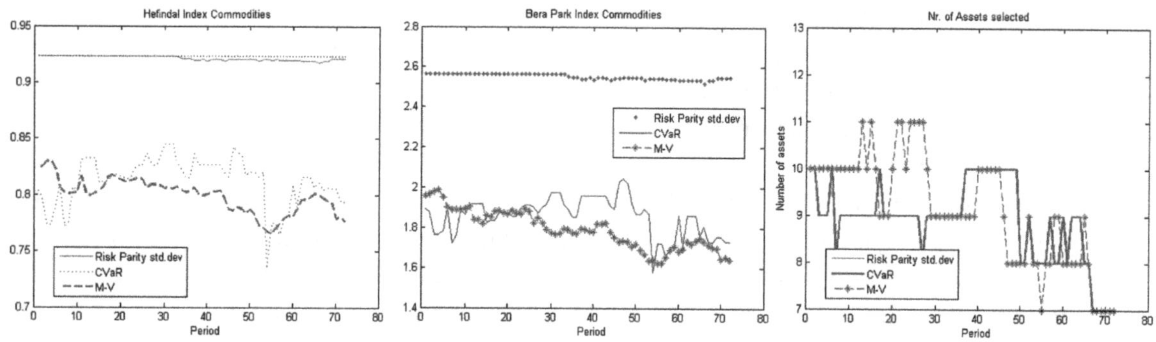

consider all the assets we will show just one of them. If we show the weights, in some cases the Mean Variance and the CVaR is concentrated in almost 30% in one commodity. If that commodity is Oil, that had a decreasing value and, in this case the investor should sell each week quantities in order to balance the portfolio.

RISK PARITY STRATEGIES IN OTHER ASSET ALLOCATIONS

In this part of the chapter the author describes other case studies of the Risk Parity strategies in order that the reader creates a good idea of what happens if the frequencies and the number of the elements that compose the portfolio changes. To guarantee a good converge of the models with the weekly frequencies of data, it is necessary to extend the period of study. The author chooses a period of observation from 1/1/2000 to 4/7/2014 consisting of 756 weeks or 174 months (14.5 years), without including all titles because of missing data or interrupted series.

Choosing from the indexes in Table 9, the groups are selected with different numbers of assets in order to study how Risk Parity strategies perform out of sample.

Using weekly data, in other to guarantee the conditions, the rolling time window is extended with in sample period of 4 past years (L= 4 years or 208 weeks) and out of sample period of one month (4 weeks).

Since the Risk Parity strategies take into consideration all the assets of the portfolio in a significant way, their performance tends to be between that of the Mean Variance and of the naive portfolio. If the investor tend to select a smaller subsets of assets, he can use different strategies but he cannot we cannot apply the cardinality constraints for the optimization model.

For each of these groups of assets, the author applies the Mean Variance and the Conditional Value at Risk without the expected return constrain, obtaining in this way the global minimum risk with the respective measure of risk. For rolling window, the average number of asset selected is shown in Table 10.

Table 9. Number of elements selected from each index

Indexes	DAX30	CAC40	Eurostoxx50	FTSE100	NIKKEI225
Nr. assets selected/Total	26/30	32/40	44/50	77/100	188/225

Table 10. The concentration of conditional value at risk and mean-variance models

Average number of assets	Concentration/Number of possible assets				
Portfolio Model	DAX30	CAC40	Eurostoxx50	FTSE100	NIKKEI225
CVAR	8/26	7/32	12/44	20/77	18/188
Mean-Variance	11 /26	12/32	8/44	12/77	20/188

Passing the data to monthly time series the concentration will be higher (in less assets). We only consider assets that have weights higher than 10^{-6}. The Risk Parity strategies take in consideration all the assets in a significant way. From the other side we have an higher turnover (see Table 11).

In the beginning, it may look a small number, but this is the average for the (756-208)/4=137 iterations.

From the point of view of riskiness, we have the same situation as in the case of commodities portfolio. So in each case the following inequalities are respected:

$$\sigma_{MV} \leq \sigma_{RP-std} \leq \sigma_{\frac{1}{n}}$$

$$CVaR_{10\%}(x) \leq CVaR_{RP-CVaR}(x) \leq CVaR_{RP-CVaR\ naive}(x) \leq CVaR_{\frac{1}{n}}(x)$$

The graphs of the CVaR and Variance (standard deviation) show the same dominance as in the case of commodities portfolio.

The performance of the models can be described in two parts: The first is before the subprime crisis of 2008 and second after the crisis. We notice that Mean-Variance and CVaR, that are heavily concentrated, in the first part have the same trajectory and after the crisis the Mean Variance dominates all the model in the performance. Let's take an example the Eurostock50 case.

Since CVaR and Mean-Variance have accumulated more wealth before the 2008 crisis, they still remain in the lead in terms of cumulated wealth. For a better understanding, Table 12 completes the case.

The Risk Parity with standard deviation and with CVaR are almost identical in the performance.

From the compounded return graph, It is easy to notice that the risk parity group is almost in the same area. This is due to the fact that they take in consideration all the 44 assets, and some of these had a poor performance, yet better than the naive portfolio.

Table 11. The average turnover (%) of the portfolios

Portfolio Model	DAX30	CAC40	Eurostoxx50
CVAR	0.2356	0.546	0.4568
Mean-Variance	0.1893	1.065	0.1546
Risk Parity with St.d.	0.0526	0.053	0.0392
Risk Parity with CVaR	0.0847	0.105	0.0530
Risk Parity with CVaR-Naive	0.0438	0.052	0.0267

Figure 6. The Cumulated wealth of each portfolio created from assets of Eurostock 50

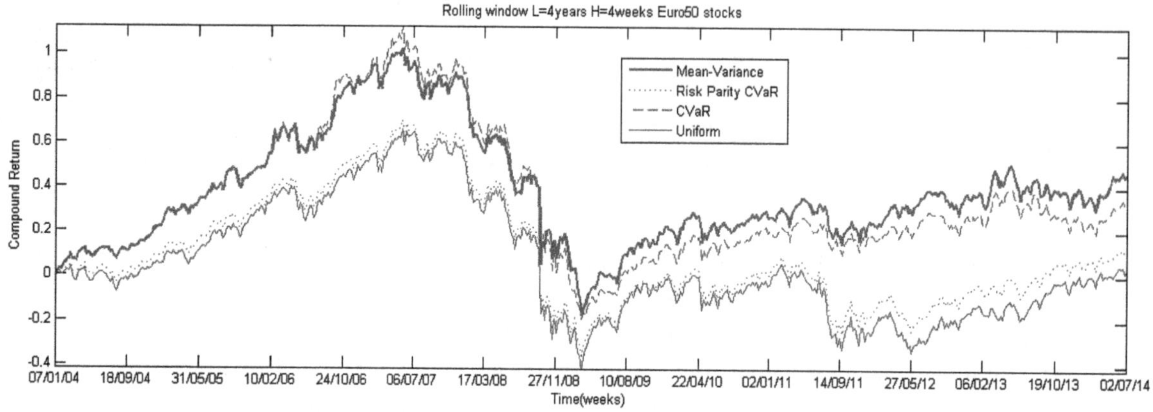

Table 12. The performance of the portfolios with assets of Eurostock50

	R.P. -STD	**M-V**	**R.P. CVaR N.**	**R.P. CVaR**	$CVaR_{10\%}(x)$	**Uniform (Naive)**
μ(%)	0.0645	0.0980	0.0595	0.0635	0.0818	0.0570
μ_{ann} (%)	3.4114	5.2281	3.1439	3.3565	4.341	3.0058
μ^c (%)	12.5300	46.0789	8.8183	12.4130	33.6318	3.6832
Median	0.4084	0.2874	0.3634	0.4102	0.2897	0.3956
σ (%)	2.8835	2.3551	2.9222	2.8555	2.363	3.1253
$VaR_{10\%}$ (%)	3.2348	2.5195	3.3222	3.1785	2.5093	3.4154
$CVaR_{10\%}$ (%)	5.4675	4.2722	5.541	5.4211	4.3819	5.9139
σ_{ann} (%)	20.7932	16.9825	21.0724	20.5913	17.0399	22.5367
$VaR_{10\%ann}$ (%)	23.3262	18.1683	23.9567	22.9207	18.0952	24.6288
$CVaR_{10\%ann}$ (%)	39.4267	30.8073	39.9567	39.0920	31.5983	42.6455
S_σ	0.1641	0.3079	0.1492	0.163	0.2548	0.1334
$S_{VaR_{ann}}$	0.1462	0.2878	0.1312	0.1464	0.2399	0.1220
$S_{CVaR_{ann}}$	0.0865	0.1697	0.0787	0.0859	0.1374	0.0705
Sortino Ratio	0.0291	0.0539	0.0265	0.0289	0.0452	0.0239
Rachev Ratio	0.7713	0.8119	0.7699	0.7709	0.7989	0.7877

The Risk Parity strategies take into consideration every asset of the market in order to contribute to the risk in the same quantity. We cannot choose a smaller subset of assets, applying the cardinality constraints.

This pushes us to develop other methods of selection of a subset of assets.

Using the same rolling time window, we choose the subset selected with Mean Variance and apply the Risk Parity with the standard deviation, and from the subset of CVaR apply Risk Parity with the CVaR and R.P. CVaR- Naive. This is just a matter of selection of a subset from all possible assets in order to have minimum risk with benefits of diversification.

From the Table 13, the Risk Parity with standard deviation, like in the Mean-Variance model, has a significant improvement in order of terminal compounded returns and performance ratios. From the graphical representation it is easy to notice that Mean-Variance still performs better than the others but, in this case, the Risk Parity with standard deviation is getting closer.

An interesting fact is that the Risk Parity with standard deviation and Risk parity, with CVaR now have a higher turnover than the corresponding measure of risk.

Another interesting case is a portfolio created with 4 commodities (Gold, silver, Heat Oil, Oil) and 4 foreign currency respect to the dollar (Euro, British Pound, Australian Dollar, New Zealand Dollar). The foreign currency can be allocated as a monetary reserve in case of necessity of the financial institution or the private investment, that may be adjusted each time for the currency exchange. Applying the portfolio selection as in Eurostock50 case we have the following inverted results for the weekly frequencies (see Figure 8).

In Figure 7, the performance is inverted if compared to the other cases. This urges the investors to combine a portfolio with stocks and bonds, and with some commodities in order to have the benefits of the Risk Parity strategies, with low turnover and also a good performance of the accumulated wealth.

As last environment, we consider a mixed portfolio with stocks, bonds, and commodities.

The target of this study is to show the behavior of the Risk Parity strategies for a set of assets with different classes of risk.

We consider the period from January 2000 to December 2013 for the following assets:

Figure 7. The cumulated wealth of the 2 times selection

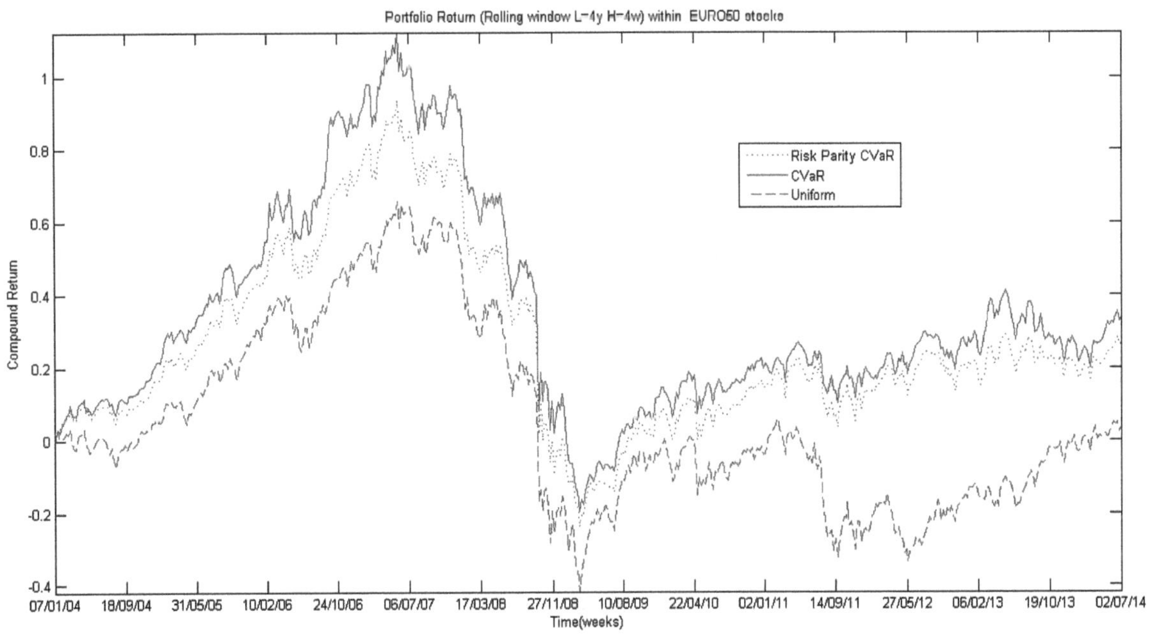

Table 13. The performance of the portfolios with the 2 times selection for the asset of Eurostock50

	R.P. -STD	M-V	R.P. CVaR N.	R.P. CVaR	$CVaR_{10\%}(x)$	Uniform (Naive)
$\mu(\%)$	0.0865	0.098	0.0766	0.0718	0.0818	0.057
μ_{ann} (%)	4.6002	5.2281	4.0626	3.8002	4.3410	3.0058
μ^c (%)	36.6637	46.0789	30.5511	26.9905	33.6318	3.6832
Median	0.3349	0.2874	0.2218	0.2445	0.2897	0.3956
σ (%)	2.388	2.3551	2.3246	2.3313	2.363	3.1253
$VaR_{10\%}$ (%)	2.5316	2.5195	2.4905	2.5020	2.5093	3.4154
$CVaR_{10\%}$ (%)	4.4633	4.2722	4.3908	4.4128	4.3819	5.9139
σ_{ann} (%)	17.2201	16.9825	16.7626	16.8111	17.0399	22.5367
$VaR_{10\%ann}$ (%)	18.2554	18.1683	17.9595	18.0425	18.0952	24.6288
$CVaR_{10\%ann}$ (%)	32.1850	30.8073	31.6629	31.8215	31.5983	42.6455
S_σ	0.2671	0.3079	0.2424	0.2261	0.2548	0.1334
$S_{VaR_{ann}}$	0.252	0.2878	0.2262	0.2106	0.2399	0.122
$S_{CVaR_{ann}}$	0.1429	0.1697	0.1283	0.1194	0.1374	0.0705
Sortino Ratio	0.0469	0.0539	0.0425	0.0396	0.0452	0.0239
Rachev Ratio	0.7785	0.8119	0.7684	0.7659	0.7989	0.7877

Figure 8. The cumulated wealth of 8 commodities

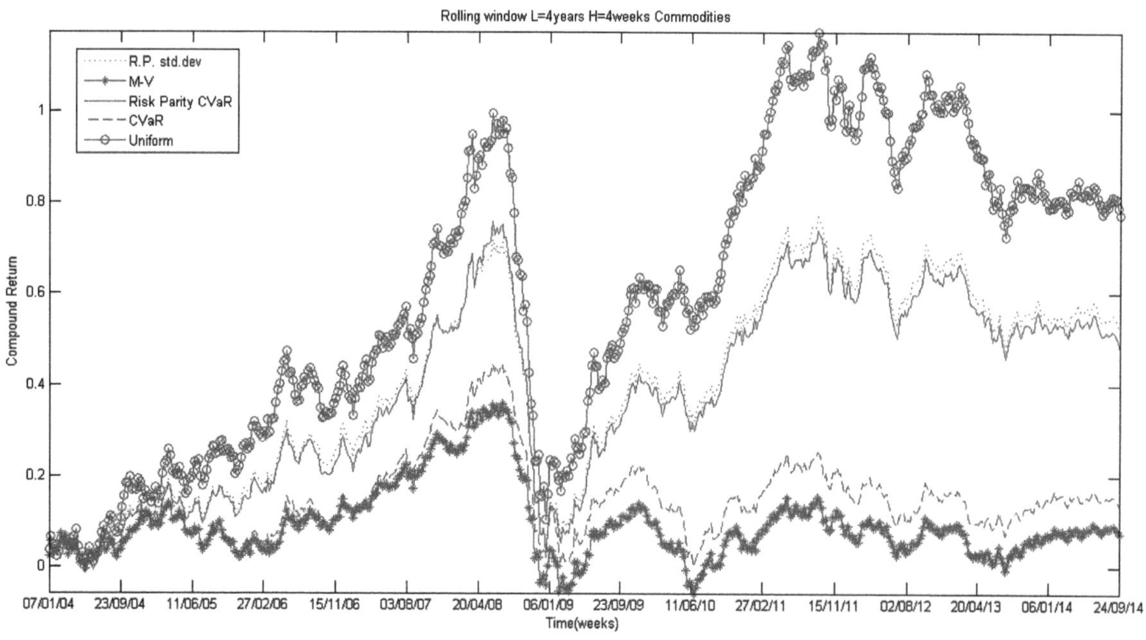

- 26 stocks of DAX30
- 9 Euro Government Bond
- Gold
- Silver

The portfolio statistics, returns, volatilities and total turnover diagrams of the R.P. strategy and the usual benchmarks are explained in Table 14.

The terminal compounded return is better for the Risk Parity with CVaR but in terms of performance ratio Mean-Variance is better than the others. Since the Mean Variance will be more concentrated to risk free assets, will have a lower turnover. The RP-CVaR almost the same turnover as the CVaR (see Box 1).

CONCLUSION

The previous years had been tremendous for commodities as an asset class. This had urged to consider different methods to allocate the amount of the investment in commodities in financial portfolio. Some particular investor may collect significant amounts of agricultural commodities in their financial portfolio, for different purposes. The traditional models, such as the Markowitz model, focus only on the riskless

Figure 9. The cumulated wealth for Risk Parity with CVaR and mean Variance model

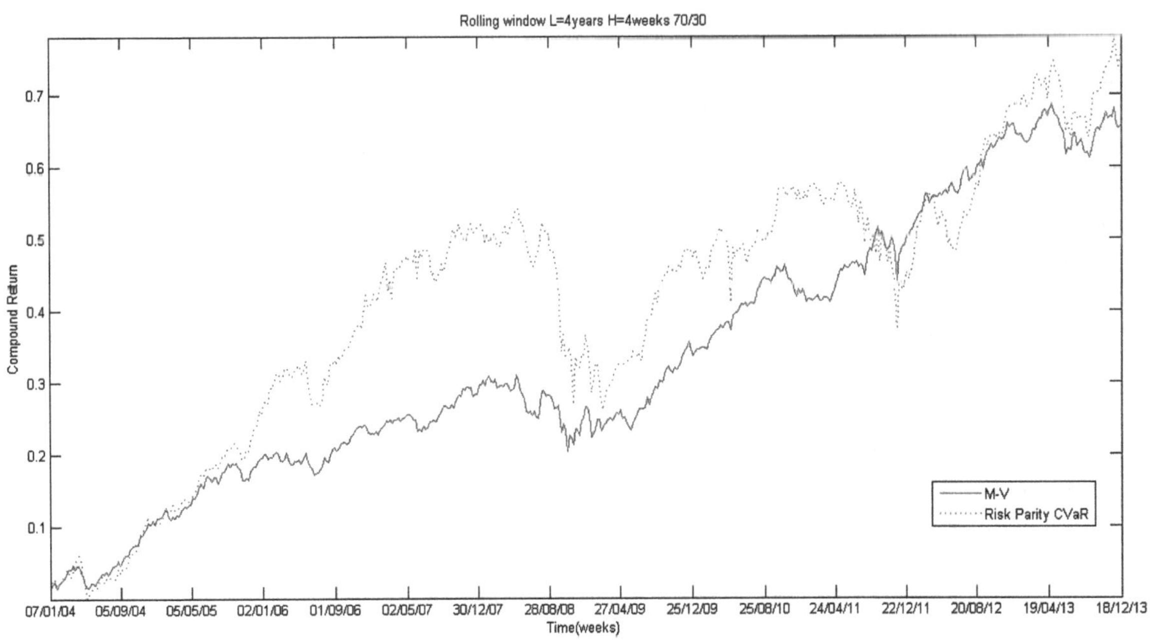

Box 1.

	RP-Std	M-V	RP-CVaR Naive	RP-CVaR	CVaR
Average Turnover(%)	0.1985	0.3166	0.0387	0.8762	0.8449

Table 14. The performance analyses of the mixed portfolio

	R.P. -STD	M-V	R.P. CVaR N.	R.P. CVaR	$CVaR_{10\%}(x)$	Uniform (Naive)
μ(%)	0.1142	0.0990	0.1059	0.1149	0.0881	0.1193
μ_{ann} (%)	6.1136	5.2819	5.6559	6.1508	4.6878	6.3970
μ^c (%)	72.7785	65.8115	69.4239	76.1627	56.5337	64.885
Median	0.2258	0.1328	0.1680	0.1552	0.1224	0.3251
σ (%)	1.3300	0.5899	0.9376	1.0863	0.6215	2.1310
$VaR_{10\%}$ (%)	1.2320	0.6609	0.8287	1.0212	0.6750	2.1818
$CVaR_{10\%}$ (%)	2.4233	1.0247	1.6459	1.938	1.0849	4.0388
σ_{ann} (%)	9.5905	4.2540	6.7611	7.8331	4.4818	15.367
$VaR_{10\%ann}$ (%)	8.8839	4.7655	5.9761	7.3643	4.8677	15.733
$CVaR_{10\%ann}$ (%)	17.4749	7.3894	11.8687	13.9753	7.8235	29.124
S_{σ}	0.6375	1.2416	0.8365	0.7852	1.0460	0.4163
$S_{VaR_{ann}}$	0.6882	1.1084	0.9464	0.8352	0.9630	0.4066
$S_{CVaR_{ann}}$	0.3499	0.7148	0.4765	0.4401	0.5992	0.2196
Sortino Ratio	0.1151	0.2512	0.1583	0.1528	0.2122	0.0745
Rachev Ratio	0.8364	1.0102	0.8694	0.9480	0.9918	0.7986

assets in we find the global minimum of riskiness. Thus, this high concentration will have also high transaction costs if the investor will try recalibrate the portfolio. Also, relying the in the expected returns during a negative trend of the economy, will bring unrealistic and pessimistic allocation of the assets.

For all this, after we have described the properties and conditions of the Risk Parity strategies, we have implemented a Risk Parity to Conditional Value at Risk.

In the portfolio created with agricultural commodities, we have created a rolling window for daily frequencies with in sample of one year (252 days) and out of the sample of 1 week (5 days). From the point of view of the performance, there is no significant difference between the Risk Parity strategies with different risk measures, but passing to CVaR, the investor can benefit from the properties of a coherent risk measure (Artzner, 1999). Also the riskiness of the Risk parity strategies are almost the same, allocated between the CVaR (or Mean Variance) and the naïve portfolio.

The portfolio created with agricultural commodities using the Risk Parity criteria presented a better diversification (tested with Herfindal and Bera Park indexes), less concentration in high weights compared to Conditional Value at Risk and Mean Variance. For this, there is less cost to recalibrate the portfolio, if of course the cost are variable. If the investor has the purpose to maintain different and significant quantities of each agricultural commodity, applying the Risk Parity strategies, he may cover the exposed quantity by keeping risk free assets in the same amount (for instance the same amount of bonds) in the case of CVaR. In the last part of the chapter, the author has applied the risk parity strategies to different groups to study their behavior from the point of view of performance.

In all these cases, the Risk Parity strategies are a good trade off between the traditional CVaR and Mean-Variance and the Naïve, from the point of view of performance and riskiness. In impossibility to apply the cardinality constraints in order to have a smaller group of assets, the author have applied a two stage optimization: In the first stage we have applied the Mean-Variance without the returns constrain (or the CVaR, with the same condition) and in the second stage, to these selected assets, apply the Risk Parity strategies. With these procedures, the compounded return and the performance increase in a significant proportion, and also we have the benefits of diversification.

Considering foreign currency as a commodity (such as currency reserves), brought an inverse tendency of the strategies from the performance but not of riskiness. Also, combining stocks with bonds and commodities, using the Risk Parity strategies, the portfolio obtained can have a better performance that the traditional ones, even more with the benefits of diversification.

From the analysis of different cases, there is still a question to make: What is the right number of assets that compose the portfolio to apply the Risk Parity strategies? These strategies allocate in a significant proportion, there is still to define the right number of assets.

Since the global demand for agricultural commodities will grow if the price is low, and with the intervention of the main producers of regulators (i.e. the recent interventions of Organization of the Petroleum Exporting Countries) the investors may increase the stockpile for future use. For all the properties described in this chapter, The Risk Parity approach will create stable portfolios, with less drawdown and good diversification, due to fact that the number of agricultural commodities is small.

For future research, it remains to study these models with the transaction cost, fixed and or variable.

REFERENCES

Acerbi, C., & Tasche, D. (2002). On the coherence of expected shortfall. *Journal of Banking & Finance, 26*(1487).

Aftalion, F., & Poncet, P. (1991). Les mesures de performance des OPCVM: Problemes et solutions. *Revue Banque*, *517*, 582–588.

Artzner, P., Delbaen, F., Eber, J.-M., & Heath, D. (1999). Coherent measures of risk. *Journal of Mathematical Finance*, *9*(3), 203–228. doi:10.1111/1467-9965.00068

Andersson, F., Mausser, H., & Uryasev, S. (2000). Credit Risk Optimization With Conditional Value at Risk Criterion. *Journal of Mathematical Programming*.

Bacon, C. R. (2008). *Practical Portfolio Performance Measurement and Attribution* (Vol. 546). Wiley.

Bertsimas, D., Lauprete, G., & Samarov, A. (2004). Shortfall as a risk measure: Properties and optimization. *Journal of Economic Dynamics & Control*, *28*(7), 1353–1381. doi:10.1016/S0165-1889(03)00109-X

Bera, A. K., & Park, S. Y. (2004). *Optimal portfolio diversification using maximum entropy*. Working Paper.

Black, F., & Litterman, R. (1990). *Asset Allocation: Combining Investors Views with Market Equilibrium*. Fixed Income Research. Goldman, Sachs & Company.

Biglova, A., Ortobelli, S., Rachev, S. T., & Stoyanov, S. (2004). Different Approaches to Risk Estimation in Portfolio Theory. *Journal of Portfolio Management*, *31*, 103–112.

Bollerslev, T. (1986). Generalized Autoregressive Conditional Heteroskedasticity. *Journal of Econometrics*, *31*(3), 307–327. doi:10.1016/0304-4076(86)90063-1

Caporin, M., & Lisi, F. (2011). Comparing and Selecting Performance Measures using Rank Correlations *Economics. The Open-Acces, Open-Assessment E-journal.*, *5*, 31.

Caporin, M., Lisi, F., & Janin, M. (2012). *A survey on Four Families of Performance Measures*. Working Papers Series.

Chaves, D., Hsu, J., Li, F., & Shakernia, O. (2012). *Efficient Algorithms for Computing Risk Parity Portfolio Weights*. Working Paper.

Colucci, S. (2013). *A quick introduction to quantitative models that discard estimation of expected returns for portfolio construction*. Working Paper.

Danielsson, J., & de Vries, C. (1997). Tail Index and Quantile Estimation with Very High Frequence Data. *Journal of Empirical Finance*, *4*(2-3), 241–257. doi:10.1016/S0927-5398(97)00008-X

Danielsson, J., & de Vries, C. (n.d.). *Beyond the Sample: Extreme Quantile and Probability Estimation.* London School of Economics, Discussion Paper 298.

Dowd, K. (2000). Adjusting for Risk: An Improved Sharpe Ratio. *International Review of Economics & Finance*, *9*(3), 209–222. doi:10.1016/S1059-0560(00)00063-0

Embrechts, P., Kluppelberg, C., & Mikosch, T. (1997). *Modelling External Events for Insurance and Finance*. Springer.

Fama, E. (1965). The behavior of Stock Market Prices. *The Journal of Business*, *38*(1), 34–105. doi:10.1086/294743

Konno, H., & Yamazaki, H. (1991). Mean-Absolute Deviation Portfolio Optimization Model and Its Applications to Tokyo Stock Market. *Management Science*, *37*(5), 519–531. doi:10.1287/mnsc.37.5.519

McNeil, A. J., & Frey, R. (2000). Estimation of Tail-Related Risk Measures for Heteroscedastic Financial Time Series: An Extreme Value Approach. *Journal of Empirical Finance*, *7*(3-4), 271–300. doi:10.1016/S0927-5398(00)00012-8

Manganelli, S., & Engle, R. F. (2001). *Value at Risk Models in Finance*. European Central Bank (ECB) Research Paper Series Working Papers.

Maillard, S., Roncalli, T., & Teiletche, J. (2009). On the properties of equally weighted risk contributions portfolios. *Journal of Portfolio Management*.

Maillard, S., Roncalli, T., & Teiletche, J. (2012). *Managing Risk Exposures using Risk Budgeting Approach*. Working Paper.

Markowitz, H.M. (1952). Portfolio selection. *Journal of Finance, 7*, 77-91.

Markowitz, H. (1987). *Mean-Variance analysis in portfolio choice and capital markets*. Basil Blackwell.

Merton, R. C. (1972). An Analytic derivation of the e¢ cient portfolio frontier. *Journal of Financial and Quantitative Analysis*, *7*(4), 1851–1872. doi:10.2307/2329621

Merton, R. C. (1980). On estimating the expected return on the market. *Journal of Financial Economics*, *8*(4), 323–361.

Morey, M. R., & Vinod, H. D. (2001). A Double Sharpe Ratio. *Advances in Investments Analysis and Portfolio Management*, *8*, 57–61.

Ortobelli, S., Rachev, S. T., Biglova, A., & Stoyanov, S. (2010). Portfolio Selection based on a Simulated Copula. *Journal of Applied Functional Analysis*, *5*, 177–194.

Qian, E. (2005). *Risk Parity Portfolios*. PanAgora Asset Management.

Risk Metrics. (1996). *Technical Document*. Morgan Guaranty Trust Company of NewYork.

Rockfellar, R.T., & Uryasev, S. (2000). Optimization of Conditional Value at Risk. *The Journal of Risk*.

Sharpe, W. (1964). Capital Asset Prices: A Theory of Market Equilibrium under Conditions of Risk. *The Journal of Finance*, *19*, 425–442.

Sharpe, W. (1966). Mutual Fund Performance. *The Journal of Business*, *39*(S1), 119–138. doi:10.1086/294846

Stefanovits, D. (2010). *Equal Contributions to Risk and Portfolio Construction* (Master Thesis). ETH Zurich.

Simaan, Y. (1997). Estimation Risk in Portfolio Selection: The Mean Variance Model Versus the Mean Absolute Deviation Model. *Management Science*, *43*(10), 1437–1446. doi:10.1287/mnsc.43.10.1437

Tasche, D. (2000). *Conditional expectation as quantile derivative. Risk Contributions and Performance Measurement*. Department of Mathematics, TU-Munchen.

Young, M. (1998). A Minimax Portfolio Selection Rule with Linear Programming Solution. *Management Science*, *44*(5).

Yitzhaki, S. (1982). Stochastic Dominance, Mean Variance and Gini's Mean Difference. *The American Economic Review*, *72*, 178–185.

Chapter 14
Risk Management in Agriculture:
Production and Technical Risk Management

Youssef Mohamed Hamada
Agriculture Research Center, Egypt

ABSTRACT

Risk has always been part of the business of agriculture. It's an industry built on the unpredictable forces of nature. What looks like a promising crop or herd can suddenly fall victim to the weather, insects or disease. Farmers are continually developing new ways to manage risk, from the use of hardier and higher yielding crop varieties and animal breeds to the application of new technologies on the farm to innovative marketing strategies. Smart agricultural policy has also evolved toward risk management programming that helps farmers deal with short-term income fluctuations as a result of risks largely outside their control. But the risks in agriculture today are greater and more complicated than ever before. International competition is fierce. Technological improvements are increasing world production and driving down real commodity prices. Public demand for higher food safety standards and better environmental practices requires new investments in the food system. Advances in science and technology are raising moral and ethical questions about the way food can and should be produced. At the same time, Smart agriculture itself has never been more diverse, ranging from specialty crops planted in small plots to grain farms covering thousands of hectares. In between being livestock operations of all sizes, greenhouses, organic farms and a growing number of agricultural businesses catering to unique consumer demands? It's an environment that is demanding new approaches to how business is conducted on the farm and consequently, how governments conduct agricultural policy.

DOI: 10.4018/978-1-5225-2107-5.ch014

INTRODUCTION

Farming is risky. Farmers live with risk and make decisions every day that affect their farming operations. Many of the factors that affect the decisions that farmers make cannot be predicted with 100 percent accuracy: weather conditions change; prices at the time of harvest could drop; hired labor may not be available at peak times; machinery and equipment could break down when most needed; draught animals might die; and government policy can change overnight. All of these changes are examples of the risks that farmers face in managing their farm as a business. All of these risks affect their farm profitability.

While farmers have always faced risk, farming has over the years, as a result of market liberalization and globalization, become increasingly risky. Small holder farmers have become especially vulnerable. A casual approach to farming, even if it is for household food consumption, is no longer viable. Farmers need to acquire more professional skills, not only in basic production, but also in farm business management. Among these are risk management skills.

Skillful farmers and other business people generally do not become involved in risky situations unless there is a chance of making money. Higher profits are usually linked with higher risks. These risky but potentially profitable situations need to be managed as carefully as possible. Good risk management involves anticipating potential problems and planning to reduce their detrimental effects. Simply reacting to unfavorable events after they occur is not good risk management.

Understanding risk will help extension workers to advise farmers on how to assess risk and to choose risk management strategies.

BACKGROUND

Agricultural systems, especially those of least developed countries face new challenges: to produce more to feed a growing population, especially in sub-Saharan Africa, to adapt to climate change and its variability but also to mitigate its contribution to emissions of greenhouse gases (the three pillars of Climate Smart Agriculture). These issues lead to a necessary adaptation for all agricultural systems whose diversity is proven. Among the pathways that can lead to such a transition, a better mobilization of ecological processes is the focus of research and development activities. This is primarily to increase production and improve its stability despite other changes, allowing farmers to improve their nutrition and generate income. It is also necessary to increase the other performances of agricultural systems, in particular by reducing their negative environmental externalities and increase resource use efficiency. This transition should contribute to reduce or to substitute chemical inputs and transformation of environments by techniques / practices that instead valorize the biodiversity and positive ecological processes within cultivated systems. In this approach, the specificity of local contexts in their biophysical, climatic and socioeconomic or cultural dimensions is a major element to be considered (AGRICORA, 2016).

This axis aims to better characterize, valorize and use this diversity (usable resources, biodiversity, production contexts, and local practices). System performance should thus be evaluated through their ability to better use natural resources such as solar energy, major nutrients (nitrogen -N-, -P- phosphorus) and water resources. Thus, they must promote facilitation processes between plant species for access to nutrients (e.g. inorganic P), the use of species that do not compete but instead have access to different nutrients tanks (nitrogen fixing, not fixers for example, surface water, deep water) and / or associating crops and livestock allowing nutrient transfer (crop residues, manure). Similarly, this approach requires consideration of articulated spatial scales of the plot, the cultivated system or the entire agro-ecosystem to better explore a wide range of available resources at these different scales. The continuum of land / farm / landscape is therefore of major interest (AGRICORA, 2016).

This adaptive transition goal requires a multidisciplinary approach between biophysical and social sciences of agriculture. Prospective work may be conducted on the feasibility of such alternatives and such a transition, for example by reasoning on climate change scenarios, the economic environment of agriculture, agricultural policies under the effect of demography, and global demand. The prospective nature of this work led to methods based on models in the broadest sense, whether mathematical models as role play games or any other favorable approach to the representation of multiple view points on complex problems. We will ensure the involvement of stakeholders in the validation of models and the discussion of their assumptions and results (AGRICORA, 2016).

Risk occurs whenever the consequences of a decision are not entirely known at the time a decision is made. There are a number of ways to manage risk. Farmers may try to prevent an unfavorable event from happening or they may take actions to reduce the adverse consequences should the unfavorable event happen. If possible, they should do both. Risk management strategies can be categorized as production, marketing, financial, human and institutional, while considered individually in this chapter they are often used in combination with another. Farmers choose and combine strategies based on their goals, attitudes towards risk and their personal and financial situations. Each individual farm family needs to find their own ways of coping with risk. Farmers' responses to risk are as diverse as the risks that affect their farms. One must bear in mind that whatever strategy is chosen, it will come at a price. The price could be a direct cost such as making insurance payments, or an indirect cost such as giving up a potential gain, or more time spent managing the farm (Kahan, 2008).

RISK SOURCES

Risk affects production such as changes in the weather and the incidence of pests and diseases. Equipment breakdown can be a risk as can market price fluctuations. Borrowing money can also be risky with sudden changes in interest rates. Risk also occurs as a result of changes in government policies. Such risks often have a major impact on farm income. Finally, there are risks related to the health and well being of the farmer and his family and the supply of labor for the farm. At the start of a season, farmers decide to grow different crops. They decide what to plant, how much to plant and when to plant. These decisions may appear simple, but for each decision there are many possible consequences. There will be only one outcome; only one result. But at the time the decision is made, the outcome is uncertain. When the chance or probability of an outcome is known in advance, this is called risk. When the chance of an outcome is not known in advance, this is called uncertainty (Kahan, 2008).

The most common sources of risk in farming can be divided into five areas: production, marketing, financial, institutional, human and interrelation of risks.

PRODUCTION AND TECHNICAL RISK

Crop and livestock performance depend on biological processes that are affected by the weather, and by pests and diseases. Low rainfall or drought may lead to low yields. Hail or heavy rains could damage or even wipe out crops. Outbreaks of pests or diseases could also cause major yield losses in crops and livestock. When farmers plant seeds and fertilize their land they do not know for certain how much rain will fall, or whether there will be a hail storm. They do not know if there will be a problem with pests or diseases. But still they must decide whether they are going to plant their crops or raise their livestock. The resources they spend to plough, plant and fertilize their crops or to care for their livestock may not be recovered. This is why there is a risk. Farmers produce without complete certainty about what will happen to their production. Another source of production risk is equipment. A farmer's tractor may break down during the production season, resulting in an inability to harvest in time, thus affecting yields. Similarly, if the farmer uses shared or hired traction or other equipment, will it be available when needed? If the farmer is using a new technology, will it perform as expected? Will it actually reduce costs and/ or increase yields? If seeds do not germinate and day old chicks die what will be the impact on production and farm family income? The farmer can never be completely certain (Kahan, 2008).

Agricultural systems, especially those of least developed countries face new challenges: to produce more to feed a growing population, especially in sub-Saharan Africa, to adapt to climate change and its variability but also to mitigate its contribution to emissions of greenhouse gases (the three pillars of Climate Smart Agriculture). These issues lead to a necessary adaptation for all agricultural systems whose diversity is proven. Among the pathways that can lead to such a transition, a better mobilization of ecological processes is the focus of research and development activities. This is primarily to increase production and improve its stability despite other changes, allowing farmers to improve their nutrition and generate income. It is also necessary to increase the othe performances of agricultural systems, in particular by reducing their negative environmental externalities and increase resource use efficiency (AGRICORA, 2016).

Production risk stems from the uncertainty regarding the factors that affect the quantity and quality of farm produce (e.g. weather, disease, pests). It also arises with the introduction of new technologies. Several strategies can be used to reduce production risk (Kahan, 2008).

MARKETING RISK: PRICES AND COSTS

Changes in prices are beyond the control of any individual farmer. The price of farm products is affected by the supply of a product, demand for the product, and the cost of production:

- Supply of a product is affected by a combination of production decisions made by farmers as a group and by the weather and other factors that influence yields;
- Demand for a product is affected by consumer preference, consumers' level of income, the strength of the general economy, and the supply and price of competing products;
- Cost of production of a unit of product depends on both input costs and yield. This makes it highly variable. Although input costs tend to be less variable than output prices, when combined with yield varying the cost of production becomes a serious source of risk.

Sometimes price movements follow seasonal or cyclical trends that can be predicted. Many times, however, supply or demand will change unexpectedly and, in turn, affect the market price. When farmers plant crops or commit resources to raising livestock, they do not know for certain what prices they will obtain for their products. In situations of low rainfall, production of crops is often reduced and, as a result, prices rise (Kahan, 2008).

Marketing risk exists because of the variability of product prices and the uncertainty of future market prices that the farmer faces when making the decision to produce a commodity. Several methods can be used to reduce price variability or to set a satisfactory price before the crops or livestock are ready for sale. These are discussed below.

FINANCIAL RISK

Financial risk occurs when money is borrowed to finance the operation of the farm business. This risk is caused by uncertainty about future interest rates and repayment schedules, changes in the loan collateral, and the ability of the farm to generate the cash flow necessary for credit repayments. In some countries small farmers have become bankrupt as a result of indebtedness. Farmers may purchase expensive inputs on credit, but with the failure of rainfall and consequent low yields may be unable to repay their loans. The failure to assess the financial risks associated with lending has a direct impact on their livelihoods. In some cases farmers have even committed suicide. This emphasizes the risk of farming and the need for extension workers and farmers to be aware of the need for appropriate risk management (Kahan, 2008).

The three aspects that need to be considered in managing financial risk are listed here:

- The availability and cost of credit and the repayment schedule;
- The farmer's liquidity or ability to generate cash flow;
- The farmer's ability to maintain and increase capital.

INSTITUTIONAL RISK

Institutional risk refers to unpredictable changes in the provision of services, such as the supply of credit and purchased inputs, and information from both formal and non-formal institutions. It also refers to uncertainties concerning government policies that affect farming. There are a number of strategies to manage institutional risk.

HUMAN AND PERSONAL RISK

Human risk refers to the risks to a farm business caused by illness and the personal situation of the farm family. It also covers issues that relate to hired workers. Human resource management An aspect of managing risk for larger farmers is good human resource management. This includes (Kahan, 2008):

- Selecting casual workers with suitable skills and experience;
- Ensuring workers are employed according to the relevant law (including fair pay);
- Regular communication;
- Ensuring the safety of workers;
- Providing adequate supervision and discipline.

INTERRELATION OF RISKS

Production, marketing, financial, institutional and human risks exist on most farms. They are frequently interrelated. The ability to repay debts depends on levels of production and the prices received for produce sold. Financing of production depends on the ability to borrow capital and the ability of the lender to supply capital in time. The different types of risk often need to be considered together (Kahan, 2008).

All farming systems seek to fully satisfy the water needs of all of their activities (cropping, livestock). Faced with low water resources, erratic rainfall and climate change, they develop (or will develop) a set of techniques to adapt and to make better use of this resource. One can speak of an engineering based on empirical knowledge which can occur at various levels of the agro-ecosystem to adapt to changing environments and conditions or resources or even extreme. The scales at which adaptation to hydro climatic risk applies are multiple: community practices and land development (including watersheds), the local landscape management, organization of land use, organization and management of the exploitation, permanent improvements of land, crop management (timing cycles, soil crop residues and crop residue management) or plant material choices. These choices at each level can affect the distribution of water and consequently the available resource or modify the efficiency of the use of this resource (AGRICORA, 2016).

Climate risk management is an emerging discipline based on the use of climate information in planning and decision making to not only protect producers in case of adverse weather hazards but also to create opportunities in case of favorable weather conditions. In addition, it will be necessary to develop innovative tools including satellite imagery and / or weather prediction and forecasting yields in early warning systems. Get real benefits for farmers also needs to take into account their needs, their practices and how they manage agricultural risks (including climate) and make decisions to address them. This understanding of current practices is a prerequisite for the design of appropriate and effective tools and methods and targeted communication with rural stakeholders so that climate information can be used for decision support (AGRICORA, 2016).

RISK TYPES

Risk can be categorized as production, marketing, financial, human and institutional. Production risk stems from uncertainty of factors that affect the quantity and quality of farm produce. Marketing risk exists because of the variability of product prices and the uncertainty of future market prices. Financial risk occurs when money is borrowed to finance the farm business. Institutional risk occurs because of unpredictable changes in the provision of services. And human risk refers to the risks to the farm business caused by human illness and the personal situation of the farm family (Kahan, 2008).

Farmers may be divided into three types: risk-neutral; risk-takers and risk-adverse. The risk-adverse farmers try to avoid taking risks. They tend to be more cautious individuals with preferences for less risky sources of income. In general, they will sacrifice some amount of income to reduce the chance of low income and losses. A risk averted does not refuse to accept any risk at all. However, the risk adverse farmer would seek to be compensated for the risk taken by receiving a higher return than would normally be obtained if there were no risk. Risk-takers are people who are open to more risky business options. Unlike the risk-adverse, risk-takers choose the alternative that gives some chance of a higher outcome, even though they may have to accept a lower outcome. When faced with the choice, risk-taking farmers tend to prefer to take the chance to make gains rather than protecting themselves from potential losses. Even so, risk-taking farmers are still influenced by the return they could receive. Risk-neutral lies between the risk-adverse and risk-taking positions. It is useful for the farmers and those who provide support services to know their attitudes towards risk. In this way, they are more conscious of the motivation behind the risk management decisions made. While most farmers tend to be risk averse, attitude concerning risk is not fixed. Many factors influence it. Thus, in one situation a farmer may be risk averse, and in another situation the same farmer may be a risk-taker (Kahan, 2008).

The following are some of the factors that may influence a farmer's attitude towards risk:

- Farmers who operate under subsistence conditions tend to be the most risk-averse. The provision of food for their dependents is an overriding priority for many of them. Activities with a monetary reward are frequently sacrificed in favor of meeting the objective of producing their own food (Kahan, 2008);
- Market-oriented farmers who are not willing or able to withstand the possible financial losses associated with a risk also tend to be more risk-averse. This is often true for smallholder farmers. In effect the relationship between the input costs and the value of output from the farm influences the farmer's attitude toward risk (Kahan, 2008);
- Family commitments and responsibilities can also play a role in attitudes toward risk. A person without family commitments may be more willing to take risks. Similarly, older people are likely to take fewer risks (Kahan, 2008);
- Past experience may also influence a farmer's decisions. The effects of particularly good or bad years in the past influence decisions to be made today. Again, this may be related to age; a younger person may not yet have had many experiences on which to base decisions (Kahan, 2008).

RISK MANAGEMENT STRATEGIES

The relative importance of the different sources of risk depends on the nature and circumstances of the individual farmer and the farm household. This includes the resource base of the farm, its physical location, the enterprise combinations chosen, the specific production processes practiced by the farm family and the attitude of the farmer towards risk (Kahan, 2008).

Decision-making is the principal activity of management. Early in the cropping season, farmers must make decisions about what crops to plant, and what seeding rates and fertilizer levels to use. The yield and prices obtained will not be known with certainty for several months, or even several years in the case of perennial crops and livestock. In only a few cases are farmers certain of the outcome of their decisions. This usually occurs when the decision is easy and there is only a single outcome.

For example, if farmers decide to take short-term loans, they know what will occur; banks will charge them interest at a specific rate. In this case, farmers know exactly the consequences of their decisions.

In most situations, however, the outcome of a decision cannot be predicted, as there is more than a single possible outcome. Farmers often find that their decisions turn out to be less than perfect because of changes that take place between the time the decision is made and the time the outcome of that decision is finalized. It may be that the outcomes themselves depend on the decisions of others and on future events that lie beyond the control of the farmer. For effective decisions to be taken, farmers must have all the necessary information regarding input prices, output prices and yields, as well as other technical data (Kahan, 2008).

Farmers who produce rain fed crops are likely to have good yields if the rainfall is adequate. But it is not certain whether it will rain, how much rain will fall or whether that rain will fall at the right time. These farmers are uncertain of the crop yield because of the risks of weather. If farmers plant their crop and an average amount of rainfall occurs, yields could be high and the crop could generate a satisfactory profit for the farmer. But if rainfall is not adequate, farmers may suffer low yields and low or non-existent profit. The pattern and amount of rainfall directly affect yields and the level of production of these crops. High rainfall results in good yields, but run the risk of increasing production among all farmers, resulting in price decreases. The combined effects of changes in production and the impact on price level of profit that can be earned. The risks associated with rain-fed farming are usually more complex than those encountered under irrigation (Kahan, 2008).

Farmers often have a basic understanding of how their crops will perform under dry, average and wet conditions. Some may have a formal record of the annual rainfall in the vicinity of their farm, while others may just remember the pattern over the years. Some farmers may have a feeling about the likelihood of a dry or wet year occurring before they decide on a cropping pattern for the season. Often, farmers think about the possible consequences of a decision to plant and grow their crops and then they decide what to do. Sometimes the risk may be so small that one does not give it any consideration. This is particularly the case when there is a long history of a consistent relationship between the decisions taken and the outcome.

For example, if a farmer has a long, successful working relationship with his or her hired labor force and the situation has not changed significantly, there will be little risk of labor not being available when needed.

Sometimes the risk may be very great and the farmer will need to give it careful consideration.

For example, being the first farmer to adopt a new seed or livestock variety may create a wide range of risks, each of which could potentially bring about losses or gains.

Before deciding to apply the new technology the farmer should take time to investigate and understand the nature of the risks and the degree of risk involved. Where there is little or no risk, decisions are generally easier to make. The greater and more complex the risk, the more difficult it becomes to make an informed decision. It is helpful to consider the fact that farmers do not only make active decisions to do something. Refusing to choose or to make a particular decision is, in itself, a decision that has outcomes and consequences.

So it is important that the farmer understands risk and how it affects his farming business. This puts the farmerin greater command of the factors that influence the household, farming and livelihood systems (Kahan, 2008).

Risk Management Strategies: The Need for Combinations

All of the strategies described in the previous sections are aimed at generating greater security for the farmer. However, it is for the farmer to decide if the benefits gained outweigh the direct or implied cost of the strategy (Kahan, 2008).

Making this decision involves a number of steps (Kahan, 2008):

- Identify the most appropriate strategy;
- Determine the degree to which risk may be reduced (i.e. the benefit);
- Identify the cost of implementing the strategy;
- Weigh up the costs and benefits and decide whether or not to implement the strategy.

It is important to note that some aspects of costs tend to be more important than others in choosing between risk management strategies. Also the various kinds of risk to which a farmer is exposed do not occur in isolation. Over the years a farmer will experience the consequences of many risks. It is therefore necessary to consider all of the risks involved and develop an integrated approach to manage them. Often a risk management strategy to cover one risk may increase another. Take, for example, a farmer who decides to diversify his crop to cover production risks. The choice of crops could increase marketing risks because, although the new crop is likely to be profitable,there may be a higher risk of price fluctuations. Farmers should develop a broad range of strategies that take into account the advantages and disadvantages (benefits and costs) of each risk management option individually and in combination (Kahan, 2008).

Key questions that can help a farmer make these decisions (Kahan, 2008):

- What risks is the farmer facing? What is the likelihood of these unfavorable events occurring? What are the consequences of these risks?
- What risk management strategies are available to the farmer?
- What effect does the risk management strategies have on the events or consequences faced?
- What are the economic benefits and costs of the options?
- How do these benefits and costs vary for each of the farm enterprises?
- How do the best risk-reducing options fit together?
- How do they affect one another in terms of costs and benefits and in terms of creating new risks?

When farmers explore these and similar questions, they will be in a better position to decide on the range of risk management strategies that are most effective for their farm. These strategies should take into account their household and farm goals, their attitude toward risk and their unique family, household and farm situations (Kahan, 2008).

MATHEMATICAL MODEL OF RISK MANAGEMENT

Risk has always been part of the business of agriculture. It's an industry built on the unpredictable forces of nature. What looks like a promising crop or herd can suddenly fall victim to the weather, insects or disease. Farmers are continually developing new ways to manage risk, from the use of hardier and higher yielding crop varieties and animal breeds to the application of new technologies on the farm to innovative marketing strategies. Agricultural policy has also evolved toward risk management programming that helps farmers deal with short-term income fluctuations as a result of risks largely outside their control. However, the risks in agriculture today are greater and more complicated than ever before. International competition is fierce. Technological improvements are increasing world production and driving down real commodity prices. Public demand for higher food safety standards and better environmental practices requires new investments in the food system. Advances in science and technology are raising moral and ethical questions about the way food can and should be produced. At the same time, agriculture itself has never been more diverse, ranging from specialist crops planted in small plots to grain farms covering thousands of hectares. In between being livestock operations of all sizes, greenhouses, organic farms and a growing number of agricultural businesses catering to unique consumer demands? It's an environment that is demanding new approaches to how business is conducted on the farm. Consequently, how governments conduct agricultural policy (APF, 2003). For some countries, the approach has been to increase subsidies to agriculture. New agriculture is taking a different approach. While working to reduce unfair, trade-distorting international subsidies, some countries is moving forward with a strategy that focuses on the sector's ability to increase profitability. The Government of some countries, along with provincial and territorial governments, is working with the agriculture and food industry and some countries on a national plan, the Agricultural Policy Framework (APF), to make some countries the world leader in food safety, innovation and environmental protection. A key element of this plan is to provide farmers the tools to manage and reduce risks that threaten the profitability of their operations, with programs cost-shared by governments and producers (APF, 2003).

In this chapter, 'production and technical risk management' refers to the probability of harmful consequences or expected losses resulting from the interaction of climate hazards with vulnerable conditions (RMS, 2016). 'Climate hazard' refers to a potentially damaging hydro meteorological event or phenomenon that can be characterized by its location, intensity, frequency, duration and probability of occurrence. This chapter considers both events with an identifiable onset and termination, such as a storm, flood or drought, and more permanent changes, such as a trend or transition from one climatic state to another, as hazards (Lim et al., 2005). Figure 1 shows structure model of Risk Management System (RMS) in Egypt.

REGIONAL DEVELOPMENT CONDITIONS, TRENDS AND CHALLENGES

Egypt is in the northeast part of Africa and is on two continents, Africa and the Sinai Peninsula in Asia. Its territory one million square kilometers (386.6 square miles) of which less than 4% is cultivated for crop production, bordered by a Gaza-trip to the northeast, Libya to the west, Sudan and Chad to the southeast, and the Mediterranean Sea to north. The 79 million people live on 40.000 square kilometers Figure 1. To

Figure 1. Structure model of Risk Management System (RMS) in Egypt

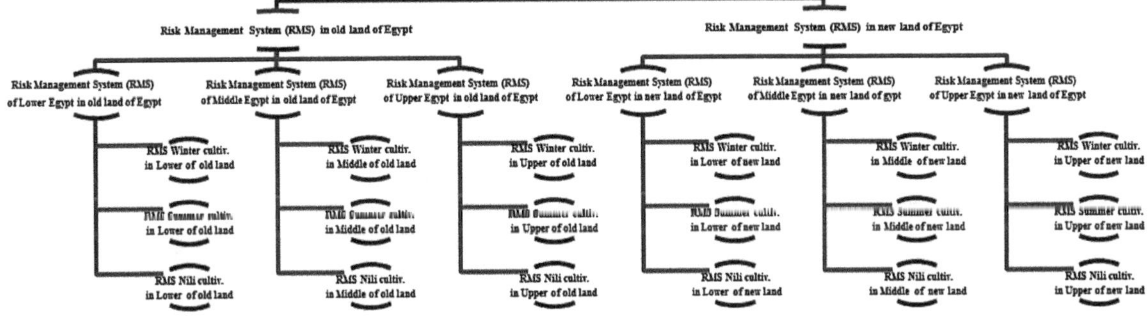

discuss Egypt is to discuss the Nile. It is formed by three tributaries, the White Nile, the Blue Nile, and the Atbara. Figure 2 shows map of the Arab Republic of Egypt-Locations of the governors in the zones

CASE STUDY OF RISK MANAGEMENT STRATEGIES

Production and technical risk management is the systematic approach and practice of incorporating production-related events, trends and projections into development decision making to maximize benefits and minimize potential harm or losses. Events are altering the trends of production and technical risk, increasing uncertainty and forcing us to re-evaluate conventional production and technical risk management practices. Historical experience with events, hazards may no longer be a sound basis for

Figure 2. Map of Egypt-locations of the governor rates in the zones

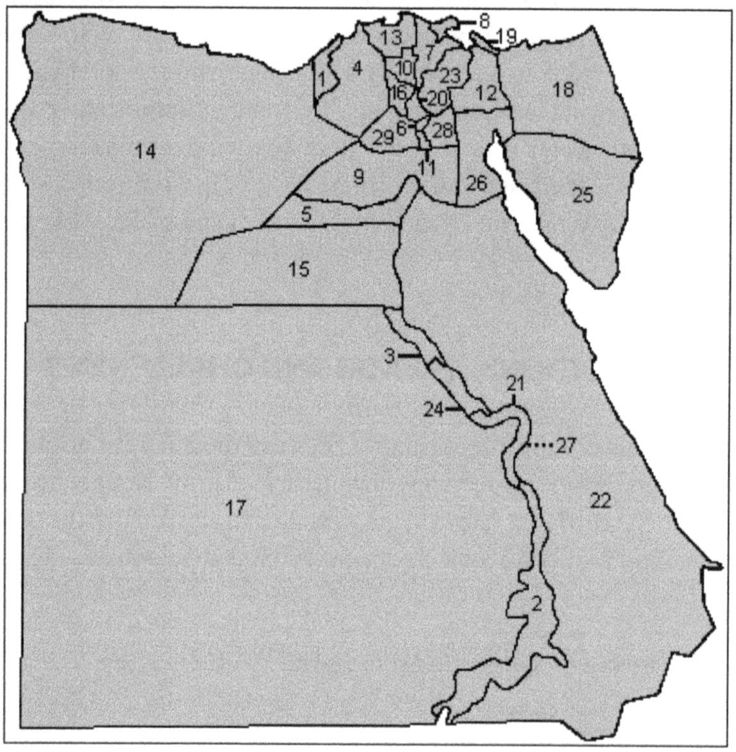

Outside the Valley	Upper Egypt	Middle Egypt	Lower Egypt	
17. New Valley	3. Assuit	9. Giza	19. Port Said	1. Alexandria
14. Matruh	24. Sohag	5. BeniSuef	20. Sharkia	10. Gharbia
25. South Sinai	21. Qena	29. Fayum	23. Damietta	16. Menoufia
18. North Sinai	27Luxor	15. Mania	26. Suez	12. Ismailia
27. Noubaria	2. Aswan		4. Behera	13. Kafr-El Sheikh
			11. Cairo	6. Qalyoubia
				7. Dakahlia

Source: (Hamada Youssef, 2014)

evaluating risk: observable trends and long-term, model-generated projections must also be taken into account if development is to be truly sustainable. Three key principles guide the implementation of the Risk Management System (RMS) in Egypt as a case study. First, the project builds on existing production and technical risk information and aims to fill critical knowledge gaps. Second, the main research phase focuses on one key sector, and potentially case study areas, in order to produce useful and concrete recommendations. Third, with a view to building capacity to identify, prioritize and manage production and technical risk, RMS works closely with in-country partners, which execute important parts of the chapter. These principles are put into practice in each country through a generic six-step implementation process.

RISK MANAGEMENT SYSTEM (RMS) IN EGYPT

Production and technical risk management in Egypt can be written as the following.

Minimize

$$\sum_{p=1}^{P}\sum_{r=1}^{R}\sum_{o=1}^{O}\sum_{d=1}^{D}\sum_{u=1}^{U}\sum_{c=1}^{C}\sum_{t=1}^{T}\sum_{i=1}^{I}\sum_{o=1}^{O}\sum_{n=1}^{N}\sum_{t=1}^{T}\sum_{e=1}^{E}\sum_{c=1}^{C}\sum_{h=1}^{H}\sum_{n=1}^{N}\sum_{i=1}^{I}$$

$$\sum_{c=1}^{C}\sum_{a=1}^{A}\sum_{l=1}^{L}\sum_{r=1}^{R}\sum_{i=1}^{I}\sum_{s=1}^{S}\sum_{k=1}^{K}\sum_{m=1}^{M}\sum_{a=1}^{A}\sum_{n=1}^{N}\sum_{a=1}^{A}\sum_{g=1}^{G}\sum_{e=1}^{E}\sum_{m=1}^{M}\sum_{e=1}^{E}\sum_{n=1}^{N}\sum_{t=1}^{T}$$

$$A_{\text{Production \& technical risk management}}\ B_{\text{Production \& technical risk management}}$$

$$\sum_{p=1}^{P}\sum_{r=1}^{R}\sum_{o=1}^{O}\sum_{d=1}^{D}\sum_{u=1}^{U}\sum_{c=1}^{C}\sum_{t=1}^{T}\sum_{i=1}^{I}\sum_{o=1}^{O}\sum_{n=1}^{N}\sum_{t=1}^{T}\sum_{e=1}^{E}\sum_{c=1}^{C}\sum_{h=1}^{H}\sum_{n=1}^{N}\sum_{i=1}^{I}$$

$$\sum_{c=1}^{C}\sum_{a=1}^{A}\sum_{l=1}^{L}\sum_{r=1}^{R}\sum_{i=1}^{I}\sum_{s=1}^{S}\sum_{k=1}^{K}\sum_{m=1}^{M}\sum_{a=1}^{A}\sum_{n=1}^{N}\sum_{a=1}^{A}\sum_{g=1}^{G}\sum_{e=1}^{E}\sum_{m=1}^{M}\sum_{e=1}^{E}\sum_{n=1}^{N}\sum_{t=1}^{T}$$

$$C_{\text{Production \& technical risk management}}\ D_{\text{Production \& technical risk management}}$$

$$\sum_{p=1}^{P}\sum_{r=1}^{R}\sum_{o=1}^{O}\sum_{d=1}^{D}\sum_{u=1}^{U}\sum_{c=1}^{C}\sum_{t=1}^{T}\sum_{i=1}^{I}\sum_{o=1}^{O}\sum_{n=1}^{N}\sum_{t=1}^{T}\sum_{e=1}^{E}\sum_{c=1}^{C}\sum_{h=1}^{H}\sum_{n=1}^{N}\sum_{i=1}^{I}$$

$$\sum_{c=1}^{C}\sum_{a=1}^{A}\sum_{l=1}^{L}\sum_{r=1}^{R}\sum_{i=1}^{I}\sum_{s=1}^{S}\sum_{k=1}^{K}\sum_{m=1}^{M}\sum_{a=1}^{A}\sum_{n=1}^{N}\sum_{a=1}^{A}\sum_{g=1}^{G}\sum_{e=1}^{E}\sum_{m=1}^{M}\sum_{e=1}^{E}\sum_{n=1}^{N}\sum_{t=1}^{T}$$

$$E_{\text{Production \& technical risk management}}\ F_{\text{Production \& technical risk management}}$$

$$\sum_{p=1}^{P}\sum_{r=1}^{R}\sum_{o=1}^{O}\sum_{d=1}^{D}\sum_{u=1}^{U}\sum_{c=1}^{C}\sum_{t=1}^{T}\sum_{i=1}^{I}\sum_{o=1}^{O}\sum_{n=1}^{N}\sum_{t=1}^{T}\sum_{e=1}^{E}\sum_{c=1}^{C}\sum_{h=1}^{H}\sum_{n=1}^{N}\sum_{i=1}^{I}$$

$$\sum_{c=1}^{C}\sum_{a=1}^{A}\sum_{l=1}^{L}\sum_{r=1}^{R}\sum_{i=1}^{I}\sum_{s=1}^{S}\sum_{k=1}^{K}\sum_{m=1}^{M}\sum_{a=1}^{A}\sum_{n=1}^{N}\sum_{a=1}^{A}\sum_{g=1}^{G}\sum_{e=1}^{E}\sum_{m=1}^{M}\sum_{e=1}^{E}\sum_{n=1}^{N}\sum_{t=1}^{T}$$

$$G_{\text{Production \& technical risk management}}\ H_{\text{Production \& technical risk management}}$$

$$\sum_{p=1}^{P}\sum_{r=1}^{R}\sum_{o=1}^{O}\sum_{d=1}^{D}\sum_{u=1}^{U}\sum_{c=1}^{C}\sum_{t=1}^{T}\sum_{i=1}^{I}\sum_{o=1}^{O}\sum_{n=1}^{N}\sum_{t=1}^{T}\sum_{e=1}^{E}\sum_{c=1}^{C}\sum_{h=1}^{H}\sum_{n=1}^{N}\sum_{i=1}^{I}$$

$$\sum_{c=1}^{C}\sum_{a=1}^{A}\sum_{l=1}^{L}\sum_{r=1}^{R}\sum_{i=1}^{I}\sum_{s=1}^{S}\sum_{k=1}^{K}\sum_{m=1}^{M}\sum_{a=1}^{A}\sum_{n=1}^{N}\sum_{a=1}^{A}\sum_{g=1}^{G}\sum_{e=1}^{E}\sum_{m=1}^{M}\sum_{e=1}^{E}\sum_{n=1}^{N}\sum_{t=1}^{T}$$

$$I_{\text{Production \& technical risk management}}\ J_{\text{Production \& technical risk management}}$$

$$\sum_{p=1}^{P}\sum_{r=1}^{R}\sum_{o=1}^{O}\sum_{d=1}^{D}\sum_{u=1}^{U}\sum_{c=1}^{C}\sum_{t=1}^{T}\sum_{i=1}^{I}\sum_{o=1}^{O}\sum_{n=1}^{N}\sum_{t=1}^{T}\sum_{e=1}^{E}\sum_{c=1}^{C}\sum_{h=1}^{H}\sum_{n=1}^{N}\sum_{i=1}^{I}$$

$$\sum_{c=1}^{C}\sum_{a=1}^{A}\sum_{l=1}^{L}\sum_{r=1}^{R}\sum_{i=1}^{I}\sum_{s=1}^{S}\sum_{k=1}^{K}\sum_{m=1}^{M}\sum_{a=1}^{A}\sum_{n=1}^{N}\sum_{a=1}^{A}\sum_{g=1}^{G}\sum_{e=1}^{E}\sum_{m=1}^{M}\sum_{e=1}^{E}\sum_{n=1}^{N}\sum_{t=1}^{T}$$

$$K_{\text{Production \& technical risk management}}\ L_{\text{Production \& technical risk management}} \tag{1}$$

MIN_A$_p$; C$_p$; E$_p$; G$_p$; I$_p$; K$_p$: Minimum absolute production and technical risk for planting, crop *p* in sub-zone *s* (2)

MAX_B$_p$; D$_p$; F$_p$; H$_p$; J$_p$; L$_p$: Maximum land area available for planting, crop *p* in sub-zone *s* (3)

MIN_e$_p$: Minimum total water consumption for planting, crop *p* in sub-zone *s* (4)

MIN_m$_p$: Minimum total kerosene fuel consumption for planting, crop *p* in sub-zone *s* (5)

MIN_e$_p$: Minimum total water consumption for planting, crop *p* in sub-zone *s* (6)

MAX_u$_p$: Maximum main crop yield in planting, crop *p* in sub-zone *s* (7)

MAX_c_p: Maximum main crop price in planting, crop *p* in sub-zone *s* (8)

MAX_t_p: Maximum Secondary crop yield in planting, crop *p* in sub-zone *s* (9)

MAX_o_p: Maximum labor wages cost in planting, crop *p* in sub-zone *s* (10)

MIN_n_p: Minimum draft animals cost in planting, crop *p* in sub-zone *s* (11)

MIN_t_p: Minimum machinery cost in planting, crop *p* in sub-zone *s* (12)

MIN_e_p: Minimum irrigation cost in planting, crop *p* in sub-zone *s* (13)

MIN_c_p: Minimum Seeds cost in planting, crop *p* in sub-zone *s* (14)

MIN_h_p: Minimum manure coast in planting, crop *p* in sub-zone *s* (15)

MIN_n_p: Minimum fertilizers coast in planting, crop *p* in sub-zone *s* (16)

MIN_i_p: Minimum insecticides coast in planting, crop *p* in sub-zone *s* (17)

MIN_c_p: Minimum laser land leveling coast in planting, crop *p* in sub-zone *s* (18)

MIN_a_p: Minimum other expenses coast in planting, crop *p* in sub-zone *s* (19)

MIN_l_p: Minimum rent cost in planting, crop *p* in sub-zone *s* (20)

MIN_n_p: Minimum crop emission in planting, *p* in sub-zone *s* (21)

Variables

$A_{\text{Production \& technical risk management}}$: Estimated production and technical risk for planting crop *p* in sub-zone (Lower Egypt in old land of Egypt) *r* in sub-season *o* in sub-soil type *d* by total main crop yield *u* with main crop price *c* and secondary crop yield *t* with secondary crop price *i* minus of it total crop production cost include labor wages cost *o* draft animals cost *n* machinery cost *t* include it irrigation cost *e* and seeds cost *c* manure coast *h* fertilizers coast *n* insecticides coast *i* laser land leveling *c* other expenses coast *a* and rent cost *l* by total energy consumptions include energy consumption for irrigation *r* energy consumption for labor *i* energy consumption for draft animal *s* energy consumption for land preparation *k* energy consumption for seed planting *m* energy consumption for manure *a* energy consumption for fertilization *n* energy consumption for insecticide *a* energy consumption for laser land leveling *g* and energy consumption for other expenses *e* by kerosene fuel *m* crop water consumption *e* crop emission *n* to give main and secondary crop yield *t*.

$B_{\text{Production \& technical risk management}}$: Estimated land area allocated for planting crop *p* in sub-zone (Lower Egypt in old land of Egypt) *r* in sub-season *o* in sub-soil type *d* by total main crop yield *u* with main crop price *c* and secondary crop yield *t* with secondary crop price *i* minus of it total crop production cost include labor wages cost *o* draft animals cost *n* machinery cost *t* include it irrigation cost *e* and seeds cost *c* manure coast *h* fertilizers coast *n* insecticides coast *i* laser land leveling *c* other expenses coast *a* and rent cost *l* by total energy consumptions include energy consumption for irrigation *r* energy consumption for labor *i* energy consumption for draft animal *s* energy consumption for land preparation *k* energy consumption for seed planting *m* energy consumption for manure *a* energy consumption for fertilization *n* energy consumption for insecticide *a* energy consumption for laser land leveling *g* and energy consumption for other expenses *e* by kerosene fuel *m* crop water consumption *e* crop emission *n* to give main and secondary crop yield *t*.

$C_{\text{Production \& technical risk management}}$: Estimated production and technical risk for planting crop *p* in sub-zone (Middle Egypt in old land of Egypt) *r* in sub-season *o* in sub-soil type *d* by total main crop yield *u* with main crop price *c* and secondary crop yield *t* with secondary crop price *i* minus of it total crop production cost include labor wages cost *o* draft animals cost *n* machinery cost *t* include it irrigation cost *e* and seeds cost *c* manure coast *h* fertilizers coast *n* insecticides coast *i* laser land leveling *c* other expenses coast *a* and rent cost *l* by total energy consumptions include energy consumption for irrigation *r* energy consumption for labor *i* energy consumption for draft animal *s* energy consumption for land preparation *k* energy consumption for seed planting *m* energy consumption for manure *a* energy consumption for fertilization *n* energy consumption for insecticide *a* energy consumption for laser land leveling *g* and energy consumption for other expenses *e* by kerosene fuel *m* crop water consumption *e* crop emission *n* to give main and secondary crop yield *t*.

$D_{\text{Production \& technical risk management}}$: Estimated land area allocated for planting crop *p* in sub-zone (Middle Egypt in old land of Egypt) *r* in sub-season *o* in sub-soil type *d* by total main crop yield *u* with main crop price *c* and secondary crop yield *t* with secondary crop price *i* minus of it total crop production cost include labor wages cost *o* draft animals cost *n* machinery cost *t* include it irrigation cost *e* and seeds cost *c* manure coast *h* fertilizers coast *n* insecticides coast *i* laser land leveling *c* other expenses coast *a* and rent cost *l* by total energy consumptions include energy consumption for irrigation *r* energy consumption for labor *i* energy consumption for draft animal *s* energy consumption for land preparation *k* energy consumption for seed planting *m* energy consumption for manure *a* energy consumption for fertilization *n* energy consumption for insecticide *a* energy consumption for laser land leveling *g* and energy consumption for other expenses *e* by kerosene fuel *m* crop water consumption *e* crop emission *n* to give main and secondary crop yield *t*.

$E_{\text{Production \& technical risk management}}$: Estimated production and technical risk for planting crop *p* in sub-zone (Upper Egypt in old land of Egypt) *r* in sub-season *o* in sub-soil type *d* by total main crop yield *u* with main crop price *c* and secondary crop yield *t* with secondary crop price *i* minus of it total crop production cost include labor wages cost *o* draft animals cost *n* machinery cost *t* include it irrigation cost *e* and seeds cost *c* manure coast *h* fertilizers coast *n* insecticides coast *i* laser land leveling *c* other expenses coast *a* and rent cost *l* by total energy consumptions include energy consumption for irrigation *r* energy consumption for labor *i* energy consumption for draft animal *s* energy consumption for land preparation *k* energy consumption for seed planting *m* energy consumption for manure *a* energy consumption for fertilization *n* energy consumption for insecticide *a* energy consumption for laser land leveling *g* and energy consumption for other expenses *e* by kerosene fuel *m* crop water consumption *e* crop emission *n* to give main and secondary crop yield *t*.

$F_{\text{Production \& technical risk management}}$: Estimated land area allocated for planting crop *p* in sub-zone (Upper Egypt in old land of Egypt) *r* in sub-season *o* in sub-soil type *d* by total main crop yield *u* with main crop price *c* and secondary crop yield *t* with secondary crop price *i* minus of it total crop production cost include labor wages cost *o* draft animals cost *n* machinery cost *t* include it irrigation cost *e* and seeds cost *c* manure coast *h* fertilizers coast *n* insecticides coast *i* laser land leveling *c* other expenses coast *a* and rent cost *l* by total energy consumptions include energy consumption for irrigation *r* energy consumption for labor *i* energy consumption for draft animal *s* energy consumption for land preparation *k* energy consumption for seed planting *m* energy consumption for manure *a* energy consumption for fertilization *n* energy consumption for insecticide *a* energy consumption for laser land leveling *g* and energy consumption for other expenses *e* by kerosene fuel *m* crop water consumption *e* crop emission *n* to give main and secondary crop yield *t*.

$G_{\text{Production \& technical risk management}}$: Estimated production and technical risk for planting crop *p* in sub-zone (Lower Egypt in new land of Egypt) *r* in sub-season *o* in sub-soil type *d* by total main crop yield *u* with main crop price *c* and secondary crop yield *t* with secondary crop price *i* minus of it total crop production cost include labor wages cost *o* draft animals cost *n* machinery cost *t* include it irrigation cost *e* and seeds cost *c* manure coast *h* fertilizers coast *n* insecticides coast *i* laser land leveling *c* other expenses coast *a* and rent cost *l* by total energy consumptions include energy consumption for irrigation *r* energy consumption for labor *i* energy consumption for draft animal *s* energy consumption for land preparation *k* energy consumption for seed planting *m* energy consumption for manure *a* energy consumption for fertilization *n* energy consumption for insecticide *a* energy consumption for laser land leveling *g* and energy consumption for other expenses *e* by kerosene fuel *m* crop water consumption *e* crop emission *n* to give main and secondary crop yield *t*.

$H_{\text{Production \& technical risk management}}$: Estimated land area allocated for planting crop *p* in sub-zone (Lower Egypt in new land of Egypt) *r* in sub-season *o* in sub-soil type *d* by total main crop yield *u* with main crop price *c* and secondary crop yield *t* with secondary crop price *i* minus of it total crop production cost include labor wages cost *o* draft animals cost *n* machinery cost *t* include it irrigation cost *e* and seeds cost *c* manure coast *h* fertilizers coast *n* insecticides coast *i* laser land leveling *c* other expenses coast *a* and rent cost *l* by total energy consumptions include energy consumption for irrigation *r* energy consumption for labor *i* energy consumption for draft animal *s* energy consumption for land preparation *k* energy consumption for seed planting *m* energy consumption for manure *a* energy consumption for fertilization *n* energy consumption for insecticide *a* energy consumption for laser land leveling *g* and energy consumption for other expenses *e* by kerosene fuel *m* crop water consumption *e* crop emission *n* to give main and secondary crop yield *t*.

$I_{\text{Production \& technical risk management}}$: Estimated production and technical risk for planting crop *p* in sub-zone (Middle Egypt in new land of Egypt) *r* in sub-season *o* in sub-soil type *d* by total main crop yield *u* with main crop price *c* and secondary crop yield *t* with secondary crop price *i* minus of it total crop production cost include labor wages cost *o* draft animals cost *n* machinery cost *t* include it irrigation cost *e* and seeds cost *c* manure coast *h* fertilizers coast *n* insecticides coast *i* laser land leveling *c* other expenses coast *a* and rent cost *l* by total energy consumptions include energy consumption for irrigation *r* energy consumption for labor *i* energy consumption for draft animal *s* energy consumption for land preparation *k* energy consumption for seed planting *m* energy consumption for manure *a* energy consumption for fertilization *n* energy consumption for insecticide *a* energy consumption for laser land leveling *g* and energy consumption for other expenses *e* by kerosene fuel *m* crop water consumption *e* crop emission *n* to give main and secondary crop yield *t*.

$J_{\text{Production \& technical risk management}}$: Estimated land area allocated for planting crop *p* in sub-zone (Middle Egypt in new land of Egypt) *r* in sub-season *o* in sub-soil type *d* by total main crop yield *u* with main crop price *c* and secondary crop yield *t* with secondary crop price *i* minus of it total crop production cost include labor wages cost *o* draft animals cost *n* machinery cost *t* include it irrigation cost *e* and seeds cost *c* manure coast *h* fertilizers coast *n* insecticides coast *i* laser land leveling *c* other expenses coast *a* and rent cost *l* by total energy consumptions include energy consumption for irrigation *r* energy consumption for labor *i* energy consumption for draft animal *s* energy consumption for land preparation *k* energy consumption for seed planting *m* energy consumption for manure *a* energy consumption for fertilization *n* energy consumption for insecticide *a* energy consumption for laser land leveling *g* and energy consumption for other expenses *e* by kerosene fuel *m* crop water consumption *e* crop emission *n* to give main and secondary crop yield *t*.

$K_{\text{Production \& technical risk management}}$: Estimated production and technical risk for planting crop *p* in sub-zone (Upper Egypt in new land of Egypt) *r* in sub-season *o* in sub-soil type *d* by total main crop yield *u* with main crop price *c* and secondary crop yield *t* with secondary crop price *i* minus of it total crop production cost include labor wages cost *o* draft animals cost *n* machinery cost *t* include it irrigation cost *e* and seeds cost *c* manure coast *h* fertilizers coast *n* insecticides coast *i* laser land leveling *c* other expenses coast *a* and rent cost *l* by total energy consumptions include energy consumption for irrigation *r* energy consumption for labor *i* energy consumption for draft animal *s* energy consumption for land preparation *k* energy consumption for seed planting *m* energy consumption for manure *a* energy consumption for fertilization *n* energy consumption for insecticide *a* energy consumption for laser land leveling *g* and energy consumption for other expenses *e* by kerosene fuel *m* crop water consumption *e* crop emission *n* to give main and secondary crop yield *t*.

$L_{\text{Production \& technical risk management}}$: Estimated land area allocated for planting crop *p* in sub-zone (Upper Egypt in new land of Egypt) *r* in sub-season *o* in sub-soil type *d* by total main crop yield *u* with main crop price *c* and secondary crop yield *t* with secondary crop price *i* minus of it total crop production cost include labor wages cost *o* draft animals cost *n* machinery cost *t* include it irrigation cost *e* and seeds cost *c* manure coast *h* fertilizers coast *n* insecticides coast *i* laser land leveling *c* other expenses coast *a* and rent cost *l* by total energy consumptions include energy consumption for irrigation *r* energy consumption for labor *i* energy consumption for draft animal *s* energy consumption for land preparation *k* energy consumption for seed planting *m* energy consumption for manure *a* energy consumption for fertilization *n* energy consumption for insecticide *a* energy consumption for laser land leveling *g* and energy consumption for other expenses *e* by kerosene fuel *m* crop water consumption *e* crop emission *n* to give main and secondary crop yield *t*.

SOLUTIONS AND RECOMMENDATIONS

Results of the Model and Discuss It

Risk Management System (RMS) is a model formulated as an analytical tool to apply production and technical risk management in the three zones of old and new land of Egypt in agriculture after laser land leveling did in some fields in the zone and in the season in the sector under the limitations of water sources in Egypt (Figure 1). In addition, the model has the flexibility of introducing production and technical risk management as the systematic approach and practice of incorporating production-related events, trends and projections into development decision making to maximize benefits and minimize potential harm or losses, achieving efficiency and equity in agriculture sector under the global financial and climate change, and as an analytical tool to reduce cost to become competitive in the world market, reduce water consumption and reduce the social cost of pollutants on farming crops. The economic, financial, risk, and the annual internal rate of return analysis of crop production are also investigated. Several steps were followed to implement the RMS model: first step was the optimum cropping pattern for each season for every zone in old and new land of Egypt, second step simulated optimum cropping pattern for the three zones (North, Middle and South Egypt), third step simulated optimum cropping pattern for the three zones with existing cropping pattern (2011/2012-2013/2014) to reallocate crop acreage according to production and technical risk management. To populate the model, field data reported by farmers was used. The data required was collected through a comprehensive survey of production and technical risk management and other inputs to crop fields on a seasonal basis, and included a comprehensive data set relating to the farm enterprise and associated socioeconomic conditions. Cropped area, yield, and cost data were obtained from the MALR (2016). Data on water consumption were collected from the MWRI (2016). Necessary data pertaining to the cropping pattern input of the respective production system were collected from primary sources and converted into corresponding cropping pattern values. Greenhouse gas emissions were calculated and represented per unit of the energy input. Data presented in this study are representative of typical and/or average data recorded over the three consecutive years of 2011/2012-2013/2014. The existing cultivation and it's an economic evaluation in Egypt in the three regions and seasons in old and new land is presented in Tables 1 and 2. Remaining base year data are available from the authors and Tables 3 and 4 places of crops in a larger view, showing place crops and its area as well as cultivating from its source (ECAPMS, 2016).

Optimal Solutions

Risk Management System (RMS) is a model should be used. In order to apply production and technical risk management could be reallocated to increase farm income, the model adjusted whatever change in land was needed to accompany the changes in soil type and water after making laser land leveling in old and new land of the Nile valley. The model structure to optimal cultivation based on suitable soil type and water in Egypt is given in Figure 1. Moreover, data in Tables 1 and 2 indicate that crop yields of

Table 1. Changes area in old land zones flow values from mean (2011/2012-2013/2014) to RMS (boold is values that have increased, gray is values that have decreased)

Winter Cultivation	Winter Cultivation in Lower Egypt				Winter Cultivation in Middle Egypt				Winter Cultivation in Upper Egypt				Winter Cultivation in Egypt			
	Mean	AWSCS	Change	%	Mean	AWSCS	Change	%	Mean	AWSCS	Change	%	Mean	AWSCS	Change	%
Wheat	666456.0	**671994.5**	**5538.5**	**0.8**	218744.8	211935.4	-6809.5	-3.1	182755.9	179789.0	-2966.9	-1.6	#######	#######	-4237.80	-0.40
Broad Beans	22649.8	20153.7	-2496.1	-11.0	1865.2	1594.3	-270.9	-14.5	3321.4	**4477.6**	**1156.3**	**34.8**	27836.3	26225.6	-1610.70	-5.79
Barley	4258.4	**4258.4**	**0.0**	**0.0**	1894.6	**2068.5**	**173.9**	**9.2**	197.0	**233.5**	**36.5**	**18.6**	6350.0	**6560.4**	**210.42**	**3.31**
Lentil	179.8	**179.8**	**0.0**	**0.0**	0.0	**688.0**	**688.0**	**0.0**	153.7	**688.0**	**534.2**	**347.5**	333.5	**1555.7**	**1222.20**	**366.50**
Fenugreek	23.9	**23.9**	**0.0**	**0.0**	1042.9	**1971.1**	**928.2**	**89.0**	648.9	**1133.6**	**484.7**	**74.7**	1715.7	**3128.6**	**1412.88**	**82.35**
Chick Peas	0.0	**0.0**	**0.0**	**0.0**	245.3	**388.9**	**143.6**	**0.0**	1189.9	**2776.6**	**1586.8**	**0.0**	1435.1	**3165.5**	**1730.40**	**120.57**
Lupine	337.7	**337.7**	**0.0**	**0.0**	5.0	**64.3**	**59.2**	**1175.0**	124.3	**138.2**	**13.9**	**11.1**	467.0	**540.1**	**73.08**	**15.65**
Flax	4054.7	**4054.7**	**0.0**	**0.0**	0.0	**0.0**	**0.0**	**0.0**	0.0	**0.0**	**0.0**	**0.0**	4054.7	**4054.7**	**0.00**	**0.00**
Onion	24357.5	24141.6	-215.9	-0.9	9533.2	8095.1	-1438.1	-15.1	5966.1	4234.9	-1731.2	-29.0	39856.7	36471.5	-3385.20	-8.49
clover	353264.9	**353264.9**	**0.0**	**0.0**	115989.3	**129571.7**	**13582.4**	**11.7**	61279.7	**72838.1**	**11558.4**	**18.9**	530533.9	**555674.7**	**25140.78**	**4.74**
Clover Tahreesh	98081.8	**98081.8**	**0.0**	**0.0**	20372.1	19936.1	-436.0	-2.1	5836.3	5741.8	-94.5	-1.6	124290.2	123759.7	-530.46	-0.43
Garlic	1351.1	**1351.1**	**0.0**	**0.0**	8181.2	7990.5	-190.7	-2.3	807.2	**851.8**	**44.5**	**5.5**	10339.6	10193.4	-146.16	-1.41
Sugar Beet	90047.2	**90047.2**	**0.0**	**0.0**	29574.3	**30867.9**	**1293.6**	**4.4**	2504.0	2048.8	-455.3	-18.2	122125.5	**122963.8**	**838.32**	**0.69**
Tomato	17569.0	**17569.0**	**0.0**	**0.0**	14487.1	**15586.6**	**1099.6**	**7.6**	9916.6	9453.8	-462.8	-4.7	41972.7	**42609.4**	**636.72**	**1.52**
Vegetables	116413.1	75637.4	-40775.7	-35.0	32676.0	27087.5	-5588.5	-17.1	4009.7	**4105.1**	**95.3**	**2.4**	153098.8	106829.9	-46268.9	-30.22
Others	1717.0	**797815.6**	**796098.7**	**46366.8**	13635.7	**15688.3**	**2052.5**	**15.1**	1247.8	**1531.7**	**283.9**	**22.8**	16600.5	**815035.6**	**798435.1**	**4809.71**
Summer Cultivation	**Summer Cultivation in Lower Egypt**				**Summer Cultivation in Middle Egypt**				**Summer Cultivation in Upper Egypt**				**Summer Cultivation in Egypt**			
	Mean	**AWSCS**	**Change**	**%**	**Mean**	**AWSCS**	**Change**	**%**	**Mean**	**AWSCS**	**Change**	**%**	**Mean**	**AWSCS**	**Change**	**%**
Cotton	99240.1	**192531.4**	**93291.2**	**94.0**	15632.4	11771.8	-3860.6	-24.7	4776.7	1445.2	-3331.4	-69.7	119649.2	**205748.3**	**86099.16**	**71.96**
Rice	558272.8	**549007.6**	**9245.5**	**1.7**	119.7	**637.1**	**481.3**	**402.1**	0.0	**0.0**	**0.0**	**0.0**	558392.5	549644.8	-8747.76	-1.57
Sorghum	0.0	**0.0**	**0.0**	**0.0**	45231.1	**48860.3**	**5050.9**	**11.2**	101077.2	**79094.0**	**8016.5**	**9.6**	146308.3	127954.3	#######	-12.54
Maize	310878.1	**296384.8**	**9834.7**	**3.3**	184075.5	**220580.9**	**57306.1**	**0.0**	94768.0	119527.8	-15493.0	-13.7	589721.6	**636493.5**	**46771.90**	**7.93**
Corn	71796.1	42100.8	-4075.3	-9.7	15164.1	20107.8	-1553.6	-10.2	32386.2	**44095.4**	**1381.8**	**3.6**	119346.4	106304.0	#######	-10.93
Soya Bean	113.4	316.7	-152.5	0.0	8556.2	7883.8	-2093.3	-24.5	607.3	**520.8**	**45.4**	**7.9**	9277.0	8721.3	-555.66	-5.99

continued on following page

Table 1. Continued

Sugar Cane	1046.2	1476.3	-188.6	-12.8	17719.0	11525.6	-118.4	-0.7	100868.5	101831.9	-194.5	-0.2	119633.6	114833.9	-4799.76	-4.01
Peanut	10928.0	12878.5	-4229.8	-32.8	5124.4	4376.0	-968.5	0.0	1727.5	**982.0**	**252.8**	**0.0**	17779.9	**18236.4**	**456.54**	**2.57**
Potatoes	38255.3	47717.5	-10971.2	-23.0	5625.9	**6470.7**	**1718.2**	**30.5**	749.7	**173.9**	**198.7**	**115.1**	44630.9	**54362.0**	**9731.12**	**21.80**
Onion	735.4	**1593.5**	**366.7**	**23.0**	2686.7	1653.4	-1648.5	-67.6	0.0	**0.0**	**0.0**	**0.0**	3422.2	3246.9	-175.28	-5.12
Sesame	3976.1	8012.8	-2999.2	-37.4	6423.5	5332.5	-1862.3	-29.0	2873.6	**1359.5**	**1755.6**	**84.5**	13273.3	**14704.8**	**1431.50**	**10.78**
Sunflower	1701.8	**2011.8**	**236.9**	**11.8**	3027.4	2276.1	-794.2	-26.2	1033.6	**367.5**	**1098.7**	**84.7**	5762.8	4655.4	-1107.40	-19.22
Tomato	26646.5	**23124.8**	**5036.6**	**21.8**	12975.5	**12368.3**	**978.2**	**7.5**	3510.4	2990.8	-950.0	-25.2	43132.3	38483.9	-4648.42	-10.78
Vegetables	151271.0	117352.6	-178.9	-0.2	32109.0	**27837.3**	**8826.3**	**27.5**	7726.7	**8126.2**	**3508.7**	**59.2**	191106.7	153316.1	#######	-19.77
Others	128733.4	**162360.7**	**111952.3**	**68.9**	45746.4	**42075.3**	**17284.7**	**61.7**	12865.4	**12081.7**	**2166.4**	**16.3**	187345.2	**216517.7**	**29172.50**	**15.57**
Nili Cultivation	**Nili Cultivation in Lower Egypt**				**Nili Cultivation in Middle Egypt**				**Nili Cultivation in Upper Egypt**				**Nili Cultivation in Egypt**			
	Mean	**AWSCS**	**Change**	**%**	**Mean**	**AWSCS**	**Change**	**%**	**Mean**	**AWSCS**	**Change**	**%**	**Mean**	**AWSCS**	**Change**	**%**
Rice	0.00	**0.00**	**0.00**	**0.00**	70.6	29.40	-164.22	-100.00	0.00	**0.00**	**0.00**	**0.00**	70.6	29.4	-41.16	-58.33
Sorghum	0.00	**0.00**	**0.00**	**0.00**	802.6	1425.50	-1036.98	-100.00	18.90	8.82	-8.82	-100.00	821.5	**1434.7**	**613.20**	**74.64**
Maize	30690.66	**42814.38**	**6878.34**	**19.14**	45719.1	58225.02	-2782.08	-6.10	8687.70	**17455.62**	**25358.76**	**145.28**	85097.5	**118495.0**	**33397.56**	**39.25**
Corn	22646.40	**19212.90**	**1015.98**	**5.58**	0.0	**0.00**	**19074.72**	**13804.26**	3494.40	**1647.66**	**17565.24**	**1066.07**	26140.8	20860.6	-5280.24	-20.20
Onion	0.00	**0.00**	**0.00**	**0.00**	3120.2	4019.52	-2405.34	-100.00	733.32	808.92	-808.92	-100.00	3853.5	**4828.7**	**975.24**	**25.31**
Potatoes	4730.46	**4709.04**	**186.06**	**4.11**	15716.4	16604.70	#######	-72.93	776.16	**702.66**	**4006.38**	**570.17**	21223.0	**22016.4**	**793.38**	**3.74**
Tomato	1785.00	**2134.86**	**271.32**	**14.56**	10699.5	9896.04	-8449.98	-79.83	700.56	**1420.02**	**714.84**	**50.34**	13185.1	**13450.9**	**265.86**	**2.02**
Vegetables	17823.54	**19054.56**	**6347.46**	**49.95**	11636.9	**11008.62**	**6196.68**	**48.19**	754.74	**1538.46**	**17516.10**	**1138.55**	30265.2	**31601.6**	**1336.44**	**4.42**
Others	6433.56	**16462.32**	**1807.68**	**12.34**	6046.7	**12608.82**	**6791.82**	**70.23**	1343.16	**1751.40**	**14710.92**	**839.95**	13823.5	**30822.5**	**16999.08**	**122.97**

Data source: (1) MALR (2016); (2) RMS model (2016)

Table 2. Changes area in new land zones flow values from mean (2011/2012-2013/2014) to RMS (boold is values that have increased, gray is values that have decreased)

Winter Cultivation	Winter Cultivation in Lower Egypt				Winter Cultivation in Middle Egypt				Winter Cultivation in Upper Egypt				Winter Cultivation in Egypt			
	Mean	AWSCS	Change	%	Mean	AWSCS	Change	%	Mean	AWSCS	Change	%	Mean	AWSCS	Change	%
Wheat	68931.7	**84000.0**	**15068.3**	**21.9**	20480.5	18557.3	-1923.2	-9.4	42232.7	**54406.8**	**12174.1**	**28.8**	131644.8	**156964.1**	**25319.28**	**19.23**
Broad Beans	4197.1	**6062.7**	**1865.6**	**44.5**	25.6	**281.4**	**255.8**	**998.4**	1097.0	821.5	-275.5	-25.1	5319.7	**7165.6**	**1845.90**	**34.70**
Barley	7095.1	**9802.4**	**2707.3**	**38.2**	371.7	**963.5**	**591.8**	**159.2**	4621.7	881.2	-3740.5	-80.9	12088.4	11647.0	-441.42	-3.65
Lentil	0.0	**0.0**	**0.0**	**0.0**	0.0	**6.3**	**6.3**	**0.0**	6.3	0.0	-6.3	-100.0	6.3	**6.3**	**0.00**	**0.00**
Fenugreek	24.4	**27.3**	**2.9**	**12.1**	113.0	**364.1**	**251.2**	**222.3**	59.6	**142.8**	**83.2**	**139.4**	197.0	**534.2**	**337.26**	**171.22**
Chick Peas	0.0	**0.0**	**0.0**	**0.0**	0.0	**0.0**	**0.0**	**0.0**	0.4	0.0	-0.4	0.0	0.4	0.0	-0.42	-100.00
Lupine	276.4	262.9	-13.4	-4.9	0.4	0.0	-0.4	-100.0	26.0	10.5	-15.5	-59.7	302.8	273.4	-29.40	-9.71
Flax	31.1	**105.8**	**74.8**	**240.5**	0.0	**0.0**	**0.0**	**0.0**	0.0	**0.0**	**0.0**	**0.0**	31.1	**105.8**	**74.76**	**240.54**
Onion	1977.8	1885.8	-92.0	-4.7	3363.4	**4286.9**	**923.6**	**27.5**	5161.8	**5936.3**	**774.5**	**15.0**	10502.9	**12109.0**	**1606.08**	**15.29**
clover	25668.7	**35782.7**	**10114.0**	**39.4**	2853.1	**6085.4**	**3232.3**	**113.3**	5397.8	**6792.7**	**1394.8**	**25.8**	33919.6	**48660.8**	**14741.16**	**43.46**
Clover Tahreesh	10106.9	**10594.9**	**488.0**	**4.8**	105.0	0.0	-105.0	-100.0	0.0	**0.0**	**0.0**	**0.0**	10211.9	**10594.9**	**383.04**	**3.75**
Garlic	301.6	220.1	-81.5	-27.0	42.0	**113.8**	**71.8**	**171.0**	148.3	**226.4**	**78.1**	**52.7**	491.8	**560.3**	**68.46**	**13.92**
Sugar Beet	32658.4	**45939.6**	**13281.2**	**40.7**	2049.6	1217.6	-832.0	-40.6	96.6	**363.3**	**266.7**	**276.1**	34804.6	**47520.5**	**12715.92**	**36.54**
Tomato	10579.4	10077.9	-501.5	-4.7	7540.3	**10166.1**	**2625.8**	**34.8**	10235.8	**14159.0**	**3923.2**	**38.3**	28355.5	**34403.0**	**6047.58**	**21.33**
Vegetables	20891.2	**30476.5**	**9585.2**	**45.9**	3690.5	3137.8	-552.7	-15.0	3993.4	3155.9	-837.5	-21.0	28575.1	**36770.2**	**8195.0**	**28.68**
Others	3058.0	2448.6	-609.4	-19.9	1934.9	**2632.6**	**697.6**	**36.1**	483.8	**707.3**	**223.4**	**46.2**	5476.8	**5788.4**	**311.6**	**5.69**
Summer Cultivation	**Summer Cultivation in Lower Egypt**				**Summer Cultivation in Middle Egypt**				**Summer Cultivation in Upper Egypt**				**Summer Cultivation in Egypt**			
	Mean	**AWSCS**	**Change**	**%**	**Mean**	**AWSCS**	**Change**	**%**	**Mean**	**AWSCS**	**Change**	**%**	**Mean**	**AWSCS**	**Change**	**%**
Cotton	9256.4	1641.8	-7614.6	-82.3	0.0	**0.0**	**0.0**	**0.0**	0.0	**0.0**	**0.0**	**0.0**	9256.4	1641.8	-7614.60	-82.26
Rice	35411.5	**41065.9**	**9245.5**	**1.7**	0.0	**0.0**	**481.3**	**402.1**	0.0	**0.0**	**0.0**	**0.0**	35411.5	**41065.9**	**5654.46**	**15.97**
Sorghum	0.0	**0.0**	**0.0**	**0.0**	703.9	**777.0**	**5050.9**	**11.2**	7863.7	**9372.7**	**8016.5**	**9.6**	8567.6	**10149.7**	**1582.14**	**18.47**
Maize	22347.8	**21755.2**	**9834.7**	**3.3**	8649.1	**10954.9**	**57306.1**	**0.0**	1803.9	4834.6	-15493.0	-13.7	32800.7	**37544.6**	**4743.90**	**14.46**
Corn	6848.9	10243.0	-4075.3	-9.7	777.8	420.8	-1553.6	-10.2	3307.1	**6192.1**	**1381.8**	**3.6**	10933.9	**16855.9**	**5922.00**	**54.16**
Soya Bean	8.4	0.0	-152.5	0.0	47.5	7.6	-2093.3	-24.5	0.0	**13.4**	**45.4**	**7.9**	55.9	21.0	-34.86	-62.41

continued on following page

Table 2. Continued

Sugar Cane	10.9	0.0	-188.6	-12.8	45.8	0.0	-118.4	-0.7	16340.1	18136.4	-194.5	-0.2	16396.8	**18136.4**	**1739.64**	**10.61**
Peanut	7968.2	11678.1	-4229.8	-32.8	2872.0	3869.0	-968.5	0.0	1839.2	**1748.0**	**252.8**	**0.0**	12679.4	**17295.2**	**4615.80**	**36.40**
Potatoes	3041.6	4279.4	-10971.2	-23.0	189.0	**186.9**	**1718.2**	**30.5**	0.0	**3444.0**	**198.7**	**115.1**	3230.6	**7910.3**	**4679.64**	**144.85**
Onion	115.5	**10.5**	**366.7**	**23.0**	614.0	447.3	-1648.5	-67.6	0.0	**0.0**	**0.0**	**0.0**	729.5	457.8	-271.74	-37.25
Sesame	2897.2	5509.1	-2999.2	-37.4	1866.5	2644.7	-1862.3	-29.0	2062.2	**3290.3**	**1755.6**	**84.5**	6825.8	**11444.2**	**4618.32**	**67.66**
Sunflower	120.1	**204.5**	**236.9**	**11.8**	209.5	260.4	-794.2	-26.2	0.0	**1.3**	**1098.7**	**84.7**	329.7	**466.2**	**136.50**	**41.40**
Tomato	18412.4	**23539.3**	**5036.6**	**21.8**	3065.6	**2682.5**	**978.2**	**7.5**	551.5	1752.7	-950.0	-25.2	22029.4	**27974.5**	**5945.10**	**26.99**
Vegetables	35025.9	62701.4	-178.9	-0.2	4957.3	**6444.9**	**8826.3**	**27.5**	3785.5	**5060.2**	**3508.7**	**59.2**	43768.6	**74206.4**	**30437.82**	**69.54**
Others	17227.1	**25309.2**	**111952.3**	**68.9**	930.3	**2793.8**	**17284.7**	**61.7**	7382.3	**10526.9**	**2166.4**	**16.3**	25539.8	**38629.9**	**13090.14**	**51.25**
Nili Cultivation	**Nile Cultivation in Lower Egypt**				**Nili Cultivation in Middle Egypt**				**Nili Cultivation in Upper Egypt**				**Nili Cultivation in Egypt**			
	Mean	**AWSCS**	**Change**	**%**	**Mean**	**AWSCS**	**Change**	**%**	**Mean**	**AWSCS**	**Change**	**%**	**Mean**	**AWSCS**	**Change**	**%**
Rice	0.00	**0.00**	**0.00**	**0.00**	0.0	0.00	-164.22	-100.00	0.00	**0.00**	**0.00**	**0.00**	0.0	0.0	0.00	0.00
Sorghum	0.00	**0.00**	**0.00**	**0.00**	0.0	0.00	-1036.98	-100.00	12.60	0.00	-8.82	-100.00	12.6	0.0	-12.60	-100.00
Maize	835.30	**2332.68**	**6878.34**	**19.14**	590.5	609.00	-2782.08	-6.10	1153.74	**2062.62**	**25358.76**	**145.28**	2580.1	**5004.3**	**2424.24**	**93.96**
Corn	2232.30	**3558.24**	**1015.98**	**5.58**	0.0	**0.00**	**19074.72**	**13804.26**	281.40	**2241.96**	**17565.24**	**1066.07**	2513.7	**5800.2**	**3286.50**	**130.74**
Onion	21.00	**75.60**	**0.00**	**0.00**	383.0	216.30	-2405.34	-100.00	513.66	844.20	-808.92	-100.00	917.7	**1136.1**	**218.40**	**23.80**
Potatoes	397.32	**1300.74**	**186.06**	**4.11**	184.8	184.80	-12686.10	-72.93	0.00	**0.00**	**4006.38**	**570.17**	582.1	**1485.5**	**903.42**	**155.19**
Tomato	756.42	**2424.66**	**271.32**	**14.56**	972.7	1250.76	-8449.98	-79.83	656.88	**650.58**	**714.84**	**50.34**	2386.0	**4326.0**	**1939.98**	**81.31**
Vegetables	1887.06	**4362.12**	**6347.46**	**49.95**	739.2	**1543.50**	**6196.68**	**48.19**	847.14	**550.62**	**17516.10**	**1138.55**	3473.4	**6456.2**	**2982.84**	**85.88**
Others	20.16	**1877.40**	**1807.68**	**12.34**	0.0	**361.20**	**6791.82**	**70.23**	395.64	**230.16**	**14710.92**	**839.95**	415.8	**2468.8**	**2052.96**	**493.74**

Data source: (1) MALR (2016); (2) RMS model (2016)

Table 3. Changes in Old land area, unit values and aggregate zones flow values from mean (2011/2012-2013/2014) to RMS (boold is values that have increased, gray is values that have decreased)

Lower Egypt	Winter Cultivation				Summer Cultivation				Nili Cultivation				Cultivation in Lower Egypt			
	Mean	AWSCS	Change	%	Mean	AWSCS	Change	%	Mean	AWSCS	Change	%	Mean	AWSCS	Change	%
Irrigated area of crop in old land	2922.49	**2922.49**	**0.00**	**0.00**	2253.22	**2253.22**	**0.00**	**0.00**	243.54	**243.54**	**0.00**	**0.00**	5419.25	**5419.25**	**0.00**	**0.00**
Soil type	11.93	14.61	2.67	22.38	9.06	8.61	-0.45	-4.97	1.11	**1.17**	**0.06**	**5.24**	22.10	**24.38**	**2.28**	**10.31**
Main crop yield	103.12	**136.90**	**33.78**	**32.76**	78.22	**96.06**	**17.85**	**22.82**	6.94	**10.49**	**3.55**	**51.14**	188.28	**243.45**	**55.18**	**29.31**
Secondry crop yield	30.41	**34.30**	**3.89**	**12.79**	35.58	**43.73**	**8.15**	**22.90**	2.59	**3.47**	**0.88**	**33.84**	68.58	**81.50**	**12.92**	**18.83**
Main crop price	19253.37	**20964.03**	**1710.66**	**8.88**	6603.15	**8465.56**	**1862.41**	**28.20**	726.43	**766.33**	**39.90**	**5.49**	26582.95	**30195.92**	**3612.97**	**13.59**
Secondry crop price	391.47	**426.56**	**35.10**	**8.97**	119.11	**128.59**	**9.48**	**7.96**	7.13	**10.40**	**3.27**	**45.91**	517.70	**565.55**	**47.85**	**9.24**
Total crop production cost	23410.93	26607.35	3196.42	13.65	24969.43	26684.42	1714.99	6.87	2381.39	2484.74	103.35	4.34	50761.75	55776.52	5014.76	9.88
*Labor Wages	3476.34	**7151.47**	**3675.13**	**105.72**	5525.14	**7397.02**	**1871.88**	**33.88**	491.20	**715.56**	**224.36**	**45.68**	9492.68	**15264.05**	**5771.38**	**60.80**
Draft Animals	30.77	31.77	1.00	3.24	179.90	**79.08**	**-100.81**	**-56.04**	4.19	**2.71**	**-1.49**	**-35.45**	214.86	**113.55**	**-101.31**	**-47.15**
Machinery	2189.90	2775.31	585.41	26.73	3046.96	3307.32	260.36	8.54	238.21	249.01	10.79	4.53	5475.07	6331.63	856.56	15.64
Seeds Cost	1035.28	1442.39	407.11	39.32	1689.77	1802.09	112.32	6.65	200.69	213.32	12.63	6.29	2925.74	3457.79	532.05	18.19
Manure	350.22	478.36	128.15	36.59	762.16	**702.74**	**-59.42**	**-7.80**	100.16	110.63	10.47	10.45	1212.54	1291.73	79.19	6.53
Fertilizers	1869.94	2644.19	774.25	41.40	2678.07	2738.88	60.81	2.27	308.06	**302.20**	**-5.85**	**-1.90**	4856.08	5685.28	829.20	17.08
Insecticides	369.80	492.54	122.74	33.19	439.82	504.20	64.38	14.64	67.20	**59.40**	**-7.80**	**-11.60**	876.82	1056.14	179.32	20.45
Laser land leveling cost	153.77	**153.77**	**0.00**	**0.00**	112.62	**109.56**	**-3.06**	**-2.72**	8.53	10.49	1.96	23.00	274.92	**273.82**	**-1.10**	**-0.40**
Other Expenses	874.16	1225.88	351.73	40.24	1498.35	1608.22	109.86	7.33	139.12	150.99	11.86	8.53	2511.63	2985.09	473.45	18.85
Rent	7206.27	9629.35	2423.1	33.62	8071.75	8435.33	363.58	4.50	652.71	670.44	17.73	2.72	15930.73	18735.12	2804.4	17.60
Crop revenue	157470.67	**257548.65**	**#######**	**63.55**	57960.85	**76210.05**	**18249.20**	**31.49**	4664.35	**8856.01**	**4191.66**	**89.87**	#######	**#######**	**122518.8**	**55.67**
Crop profit	139919.85	**226132.71**	**86212.86**	**61.62**	36205.85	**49525.62**	**13319.77**	**36.79**	3142.74	**4370.17**	**1227.43**	**39.06**	#######	**#######**	**100760.1**	**56.21**
Crop water consumption	16487.21	**9828.19**	**-6659.01**	**-40.39**	27955.68	**17438.51**	**#######**	**-37.62**	1824.92	**1271.68**	**-553.24**	**-30.32**	46267.81	**28538.38**	**-17729.4**	**-38.32**
Kerosene fuel Liter	4421.94	**3683.80**	**-738.14**	**-16.69**	2003.59	**1739.89**	**-263.70**	**-13.16**	198.78	**173.87**	**-24.91**	**-12.53**	6624.31	**5597.56**	**-1026.8**	**-15.50**
Energy consumption in cultivation MJ	96355.39	**81642.18**	**#######**	**-15.27**	#######	**94563.56**	**#######**	**-13.84**	10847.58	**9424.51**	**-1423.08**	**-13.12**	#######	**#######**	**-31324.0**	**-14.44**
Middle Egypt	**Winter Cultivation**				**Summer Cultivation**				**Nili Cultivation**				**Cultivation in Middle Egypt**			
	Mean	**AWSCS**	**Change**	**%**	**Mean**	**AWSCS**	**Change**	**%**	**Mean**	**AWSCS**	**Change**	**%**	**Mean**	**AWSCS**	**Change**	**%**
Irrigated area of crop in old land	473.53	**473.53**	**0.00**	**0.00**	423.76	**423.76**	**0.00**	**0.00**	113.82	**113.82**	**0.00**	**0.00**	1011.11	**1011.11**	**0.00**	**0.00**
Soil type	2.12	**2.12**	**0.00**	**0.20**	2.71	2.46	-0.25	-9.11	0.51	0.65	0.14	27.26	5.34	**5.24**	**-0.10**	**-1.95**

continued on following page

Table 3. Continued

Main crop yield	21.61	25.89	4.28	19.80	16.25	21.77	5.53	34.00	3.24	4.69	1.45	44.82	41.10	52.36	11.26	27.39
Secondry crop yield	6.97	7.58	0.61	8.82	6.78	9.31	2.53	37.39	1.13	1.47	0.34	30.25	14.87	18.36	3.49	23.47
Main crop price	1646.95	2004.20	357.25	21.69	874.55	1178.96	304.41	34.81	327.01	348.28	21.27	6.50	2848.51	3531.44	682.93	23.98
Secondry crop price	67.35	83.96	16.60	24.65	20.94	25.64	4.70	22.44	3.11	4.41	1.30	41.82	91.40	114.00	22.60	24.73
Total crop production cost	3866.07	4731.81	865.74	22.39	4536.64	4584.14	47.51	1.05	1163.69	1234.15	70.46	6.05	9566.39	10550.09	983.70	10.28
*Labor Wages	775.31	1057.49	282.17	36.40	954.12	1320.89	366.76	38.44	251.09	354.78	103.69	41.30	1980.52	2733.15	752.63	38.00
Draft Animals	6.24	7.14	0.90	14.47	24.06	11.81	-12.24	-50.89	1.93	1.34	-0.58	-30.37	32.22	20.30	-11.93	-37.01
Machinery	473.98	528.18	54.20	11.44	514.20	521.66	7.46	1.45	112.25	118.22	5.97	5.32	1100.43	1168.06	67.63	6.15
Seeds Cost	236.30	262.27	25.97	10.99	315.89	308.61	-7.28	-2.31	115.23	125.31	10.08	8.75	667.42	696.19	28.77	4.31
Manure	78.28	94.91	16.64	21.25	182.52	168.04	-14.47	-7.93	44.07	48.83	4.77	10.81	304.86	311.79	6.93	2.27
Fertilizers	420.64	432.51	11.87	2.82	521.92	495.43	-26.50	-5.08	158.31	156.93	-1.38	-0.87	1100.88	1084.87	-16.01	-1.45
Insecticides	81.31	88.51	7.19	8.85	60.88	78.50	17.62	28.94	33.52	30.42	-3.10	-9.25	175.72	197.43	21.71	12.36
Laser land leveling cost	28.63	28.63	0.00	0.00	22.13	19.34	-2.79	-12.59	4.27	5.12	0.86	20.05	55.03	53.10	-1.93	-3.51
Other Expenses	193.59	218.85	25.25	13.04	271.69	275.54	3.86	1.42	68.36	76.45	8.09	11.83	533.64	570.84	37.20	6.97
Rent	1557.19	1715.70	158.5	10.18	1410.91	1384.31	-26.60	-1.89	305.92	316.74	10.82	3.54	3274.02	3416.75	142.7	4.36
Crop revenue	30891.89	38839.05	7947.16	25.73	10071.19	13621.41	3550.22	35.25	2545.26	4337.22	1791.96	70.40	43508.34	56797.68	13289.3	30.54
Crop profit	27025.05	33748.77	6723.72	24.88	6582.90	9037.27	2454.36	37.28	1903.64	2346.35	442.71	23.26	35511.60	45132.39	9620.8	27.09
Crop water consumption	2877.61	1757.29	-1120.32	-38.93	4390.49	2728.08	-1662.41	-37.86	837.91	620.48	-217.43	-25.95	8106.01	5105.85	-3000.2	-37.01
Kerosene fuel Liter	651.80	562.60	-89.20	-13.69	380.39	304.79	-75.60	-19.87	95.15	84.38	-10.77	-11.32	1127.34	951.76	-175.6	-15.57
Energy consumption in cultivation MJ	15848.04	13759.43	-2088.60	-13.18	20193.10	16816.86	-3376.24	-16.72	5193.57	4574.87	-618.70	-11.91	41234.70	35151.16	-6083.5	-14.75

Upper Egypt	Winter Cultivation				Summer Cultivation				Nili Cultivation				Cultivation in Upper Egypt			
	Mean	AWSCS	Change	%	Mean	AWSCS	Change	%	Mean	AWSCS	Change	%	Mean	AWSCS	Change	%
Irrigated area of crop in old land	290.04	290.04	0.00	0.00	372.60	372.60	0.00	0.00	25.33	25.33	0.00	0.00	687.97	687.97	0.00	0.00
Soil type	1.14	1.57	0.44	38.35	1.46	1.44	-0.02	-1.48	0.15	0.11	-0.04	-27.30	2.75	3.12	0.37	13.56
Main crop yield	14.46	17.82	3.36	23.22	23.48	28.94	5.46	23.24	0.61	1.22	0.61	100.88	38.55	47.98	9.43	24.46
Secondry crop yield	5.02	5.65	0.63	12.55	5.79	7.16	1.37	23.72	0.25	0.47	0.22	86.99	11.06	13.29	2.22	20.08
Main crop price	735.39	956.43	221.04	30.06	439.55	507.86	68.31	15.54	50.50	51.48	0.99	1.95	1225.44	1515.78	290.34	23.69
Secondry crop price	56.62	69.27	12.65	22.34	17.55	19.61	2.06	11.76	0.69	1.41	0.72	104.35	74.85	90.29	15.43	20.62
Total crop production cost	2477.07	2992.23	515.15	20.80	4388.47	4563.56	175.09	3.99	227.15	238.13	10.98	4.83	7092.70	7793.92	701.22	9.89

continued on following page

Table 3. Continued

*Labor Wages	477.96	**659.77**	**181.81**	**38.04**	913.00	**1248.70**	**335.71**	**36.77**	34.45	**68.78**	**34.33**	**99.65**	1425.40	**1977.25**	**551.85**	**38.72**
Draft Animals	4.09	4.18	0.09	2.30	10.19	**8.39**	**-1.79**	**-17.59**	0.41	**0.30**	**-0.12**	**-28.35**	14.69	**12.87**	**-1.82**	**-12.36**
Machinery	312.98	317.75	4.77	1.52	560.01	587.57	27.56	4.92	24.05	25.02	0.97	4.01	897.04	930.33	33.29	3.71
Seeds Cost	142.57	146.00	3.43	2.40	230.19	240.33	10.14	4.40	16.21	**16.17**	**-0.04**	**-0.25**	388.98	402.50	13.52	3.48
Manure	47.46	48.05	0.59	1.25	111.17	**109.17**	**-2.00**	**-1.80**	9.32	12.18	2.85	30.59	167.96	169.40	1.45	0.86
Fertilizers	261.95	266.37	4.42	1.69	602.30	**576.26**	**-26.03**	**-4.32**	28.28	**27.67**	**-0.62**	**-2.18**	892.53	**870.30**	**-22.23**	**-2.49**
Insecticides	50.70	50.70	0.00	0.00	28.90	38.41	9.51	32.90	5.46	**5.23**	**-0.23**	**-4.22**	85.06	94.34	9.28	10.91
Laser land leveling cost	18.24	**18.24**	**0.00**	**0.00**	23.76	**23.48**	**-0.27**	**-1.16**	0.75	1.13	0.38	50.86	42.75	42.85	0.10	0.25
Other Expenses	124.40	125.53	1.13	0.91	323.05	335.04	11.99	3.71	13.07	14.15	1.08	8.24	460.51	474.71	14.20	3.08
Rent	990.04	1019.77	29.7	3.00	1375.05	1396.20	21.15	1.54	66.58	67.52	0.94	1.41	2431.67	2483.50	51.8	2.13
Crop revenue	19876.41	**27255.50**	**7379.10**	**37.12**	8251.47	**10591.09**	**2339.62**	**28.35**	367.62	**760.18**	**392.56**	**106.78**	28495.50	**38606.77**	**10111.3**	**35.48**
Crop profit	17445.97	**22621.35**	**5175.39**	**29.67**	4630.84	**6027.53**	**1396.69**	**30.16**	249.70	**347.56**	**97.86**	**39.19**	22326.51	**28996.44**	**6669.9**	**29.87**
Crop water consumption	2306.31	**1342.60**	**-963.70**	**-41.79**	6338.33	**3817.88**	**-2520.45**	**-39.77**	243.28	**156.20**	**-87.07**	**-35.79**	8887.91	**5316.68**	**-3571.2**	**-40.18**
Kerosene fuel Liter	334.93	**286.83**	**-48.10**	**-14.36**	331.24	**312.39**	**-18.85**	**-5.69**	21.41	**18.28**	**-3.13**	**-14.60**	687.57	**617.50**	**-70.1**	**-10.19**
Energy consumption in cultivation MJ	10224.81	**8724.37**	**-1500.45**	**-14.67**	19645.26	**17134.74**	**-2510.53**	**-12.78**	1147.22	**975.48**	**-171.74**	**-14.97**	31017.30	**26834.58**	**-4182.7**	**-13.49**

Data source: (1) MALR (2016); (2) RMS model (2016)(3) MWRI (2016)

Table 4. Changes in new land area, unit values and aggregate zones flow values from mean (2011/2012-2013/2014) to RMS (boold is values that have increased, gray is values that have decreased)

Lower Egypt	Winter Cultivation				Summer Cultivation				Nili Cultivation				Cultivation in Lower Egypt			
	Mean	AWSCS	Change	%	Mean	AWSCS	Change	%	Mean	AWSCS	Change	%	Mean	AWSCS	Change	%
Irrigated area of crop in old land	237.69	**237.69**	**0.00**	**0.00**	207.94	**207.94**	**0.00**	**0.00**	15.93	**15.93**	**0.00**	**0.00**	461.56	**461.56**	**0.00**	**0.00**
Soil type	0.92	1.18	0.26	27.71	0.47	0.65	0.19	39.80	0.08	**0.07**	**-0.01**	**-10.34**	1.47	1.90	0.43	29.54
Main crop yield	7.27	**11.65**	**4.38**	**60.30**	4.53	**7.54**	**3.01**	**66.57**	0.23	**0.60**	**0.37**	**164.05**	12.02	**19.79**	**7.77**	**64.61**
Secondry crop yield	2.06	**2.90**	**0.84**	**40.80**	1.77	**2.22**	**0.45**	**25.49**	0.07	**0.15**	**0.07**	**95.38**	3.90	**5.26**	**1.36**	**34.91**
Main crop price	654.64	**752.77**	**98.13**	**14.99**	591.80	**922.25**	**330.45**	**55.84**	10.78	**54.49**	**43.72**	**405.66**	1257.22	**1729.52**	**472.30**	**37.57**
Secondry crop price	40.64	**41.64**	**1.00**	**2.47**	6.43	**7.29**	**0.86**	**13.33**	0.20	**0.43**	**0.23**	**112.58**	47.27	**49.36**	**2.09**	**4.42**
Total crop production cost	1906.26	1972.70	66.44	3.49	1764.92	2364.65	599.73	33.98	158.58	171.05	12.47	7.87	3829.76	4508.40	678.64	17.72
*Labor Wages	340.14	**586.66**	**246.52**	**72.48**	412.84	**647.24**	**234.40**	**56.78**	16.44	**48.25**	**31.81**	**193.54**	769.41	**1282.14**	**512.73**	**66.64**
Draft Animals	3.52	**2.79**	**-0.72**	**-20.55**	17.55	**6.12**	**-11.43**	**-65.11**	0.16	**0.13**	**-0.03**	**-15.96**	21.22	**9.05**	**-12.18**	**-57.36**
Machinery	245.12	**238.65**	**-6.47**	**-2.64**	234.30	**221.29**	**-13.00**	**-5.55**	15.23	16.79	1.56	10.27	494.65	**476.74**	**-17.91**	**-3.62**
Seeds Cost	108.53	110.46	1.93	1.78	150.05	151.04	0.99	0.66	14.48	**14.21**	**-0.27**	**-1.88**	273.06	275.72	2.65	0.97
Manure	42.71	**40.06**	**-2.65**	**-6.20**	87.33	**78.46**	**-8.87**	**-10.16**	6.64	7.42	0.78	11.70	136.68	**125.94**	**-10.74**	**-7.86**
Fertilizers	221.04	229.51	8.48	3.84	241.23	**223.71**	**-17.51**	**-7.26**	20.05	20.07	0.02	0.12	482.31	**473.30**	**-9.01**	**-1.87**
Insecticides	56.84	59.91	3.07	5.40	64.87	66.47	1.60	2.47	5.38	5.49	0.11	2.03	127.09	131.87	4.78	3.76
Laser land leveling cost	13.85	**13.85**	**0.00**	**0.00**	8.00	**8.00**	**0.00**	**0.00**	0.63	0.63	0.00	0.00	22.47	**22.47**	**0.00**	**0.00**
Other Expenses	100.65	**99.87**	**-0.79**	**-0.78**	135.53	**118.13**	**-17.40**	**-12.84**	9.58	10.36	0.77	8.06	245.76	**228.35**	**-17.41**	**-7.08**
Rent	915.67	**771.88**	**-143.8**	**-15.70**	721.17	**657.05**	**-64.12**	**-8.89**	45.38	47.70	2.32	5.11	1682.22	**1476.64**	**-205.6**	**-12.22**
Crop revenue	8836.68	**13828.61**	**4991.93**	**56.49**	5732.57	**7604.38**	**1871.80**	**32.65**	148.47	**632.07**	**483.60**	**325.72**	14717.72	**22065.05**	**7347.3**	**49.92**
Crop profit	7287.01	**10318.93**	**3031.92**	**41.61**	4240.11	**5320.07**	**1079.95**	**25.47**	90.06	**302.21**	**212.15**	**235.55**	11617.19	**15941.20**	**4324.0**	**37.22**
Crop water consumption	1391.65	**791.62**	**-600.03**	**-43.12**	2309.53	**1262.90**	**-1046.62**	**-45.32**	122.49	**81.71**	**-40.79**	**-33.30**	3823.67	**2136.23**	**-1687.4**	**-44.13**
Kerosene fuel Liter	368.10	**320.04**	**-48.06**	**-13.06**	193.08	**187.55**	**-5.53**	**-2.87**	12.95	**10.06**	**-2.89**	**-22.35**	574.14	**517.65**	**-56.5**	**-9.84**
Energy consumption in cultivation MJ	8009.23	**7017.60**	**-991.64**	**-12.38**	10246.11	**9413.11**	**-833.01**	**-8.13**	743.59	**661.84**	**-81.75**	**-10.99**	18998.93	**17092.54**	**-1906.4**	**-10.03**

continued on following page

Table 4. Continued

Middle Egypt	Winter Cultivation				Summer Cultivation				Nili Cultivation				Cultivation in Middle Egypt			
	Mean	AWSCS	Change	%	Mean	AWSCS	Change	%	Mean	AWSCS	Change	%	Mean	AWSCS	Change	%
Irrigated area of crop in old land	47.81	47.81	0.00	0.00	31.49	31.49	0.00	0.00	4.17	4.17	0.00	0.00	83.47	83.47	0.00	0.00
Soil type	0.20	0.19	-0.01	-5.67	0.12	0.20	0.07	58.32	0.01	0.01	0.01	82.13	0.34	0.40	0.07	20.03
Main crop yield	1.78	2.16	0.38	21.08	0.88	1.30	0.42	47.62	0.11	0.18	0.07	63.90	2.77	3.64	0.87	31.22
Secondry crop yield	0.52	0.60	0.08	14.34	0.32	0.48	0.16	49.51	0.01	0.01	0.00	4.39	0.86	1.09	0.23	27.28
Main crop price	121.94	174.28	52.35	42.93	55.80	96.45	40.64	72.83	6.14	14.06	7.92	128.97	183.88	284.79	100.91	54.88
Secondry crop price	6.31	7.47	1.16	18.37	1.17	1.51	0.34	29.05	0.04	0.04	0.01	14.18	7.52	9.03	1.51	20.01
Total crop production cost	467.19	473.21	6.03	1.29	336.93	285.78	-51.15	-15.18	41.74	45.62	3.88	9.29	845.86	804.62	-41.24	-4.88
*Labor Wages	99.11	126.87	27.77	28.02	62.68	95.73	33.05	52.74	7.08	12.47	5.39	76.06	168.86	235.07	66.21	39.21
Draft Animals	0.59	0.43	-0.15	-26.29	0.60	0.36	-0.24	-39.59	0.02	0.02	0.00	3.65	1.20	0.81	-0.39	-32.44
Machinery	63.57	51.17	-12.40	-19.50	38.92	29.39	-9.53	-24.49	2.68	4.35	1.67	62.16	105.17	84.91	-20.26	-19.26
Seeds Cost	37.63	32.28	-5.35	-14.22	22.86	18.16	-4.70	-20.55	3.26	3.58	0.32	9.71	63.76	54.02	-9.73	-15.27
Manure	13.37	10.31	-3.06	-22.87	13.63	12.30	-1.33	-9.75	0.95	1.85	0.91	95.30	27.95	24.47	-3.48	-12.46
Fertilizers	52.48	49.71	-2.77	-5.27	36.52	33.26	-3.26	-8.93	3.62	5.08	1.46	40.35	92.62	88.05	-4.57	-4.93
Insecticides	13.59	12.40	-1.19	-8.76	8.18	7.37	-0.81	-9.94	1.16	1.89	0.73	62.49	22.94	21.66	-1.28	-5.57
Laser land leveling cost	2.60	2.60	0.00	0.00	1.31	1.31	0.00	0.00	0.12	0.15	0.03	27.11	4.02	4.06	0.03	0.81
Other Expenses	29.05	23.27	-5.78	-19.88	20.21	16.95	-3.26	-16.13	1.73	2.73	1.00	57.61	50.99	42.95	-8.04	-15.76
Rent	155.90	151.14	-4.8	-3.05	101.54	90.64	-10.91	-10.74	7.94	13.49	5.55	69.88	265.38	255.27	-10.1	-3.81
Crop revenue	2356.13	2843.79	487.66	20.70	622.10	1052.36	430.26	69.16	105.62	183.34	77.72	73.58	3083.85	4079.49	995.6	32.29
Crop profit	1923.71	2331.85	408.14	21.22	430.46	726.82	296.36	68.85	67.67	137.86	70.18	103.71	2421.84	3196.53	774.7	31.99
Crop water consumption	304.04	174.23	-129.81	-42.69	308.63	200.96	-107.67	-34.89	36.09	25.06	-11.03	-30.56	648.76	400.25	-248.5	-38.30
Kerosene fuel Liter	86.99	75.79	-11.19	-12.87	27.85	26.48	-1.36	-4.89	3.82	3.48	-0.34	-8.87	118.65	105.76	-12.9	-10.87
Energy consumption in cultivation MJ	1900.88	1636.29	-264.60	-13.92	1467.82	1329.59	-138.24	-9.42	214.17	204.38	-9.79	-4.57	3582.88	3170.25	-412.6	-11.52

continued on following page

Table 4. Continued

Upper Egypt	Winter Cultivation				Summer Cultivation				Nili Cultivation				Cultivation in Upper Egypt			
	Mean	AWSCS	Change	%	Mean	AWSCS	Change	%	Mean	AWSCS	Change	%	Mean	AWSCS	Change	%
Irrigated area of crop in old land	87.60	87.60	0.00	0.00	64.37	64.37	0.00	0.00	6.58	6.58	0.00	0.00	158.56	158.56	0.00	0.00
Soil type	0.43	0.51	0.08	18.68	0.32	0.31	-0.02	-4.75	0.02	0.04	0.02	69.41	0.78	0.86	0.08	10.37
Main crop yield	2.71	4.10	1.38	50.99	2.57	3.88	1.31	50.81	0.09	0.18	0.09	99.94	5.38	8.16	2.78	51.72
Secondry crop yield	1.25	1.70	0.45	36.18	0.37	0.67	0.30	81.27	0.03	0.11	0.07	204.99	1.65	2.48	0.82	49.87
Main crop price	190.91	200.13	9.22	4.83	149.13	235.57	86.44	57.96	8.63	11.28	2.65	30.66	348.67	446.97	98.30	28.19
Secondry crop price	16.90	20.67	3.77	22.32	1.18	2.00	0.82	69.45	0.23	0.32	0.09	40.35	18.30	22.99	4.68	25.59
Total crop production cost	898.91	969.66	70.75	7.87	742.52	803.37	60.85	8.20	37.57	61.26	23.69	63.05	1679.00	1834.29	155.29	9.25
*Labor Wages	151.12	237.87	86.75	57.41	141.78	207.37	65.59	46.26	7.80	17.35	9.55	122.48	300.69	462.58	161.89	53.84
Draft Animals	1.45	1.04	-0.41	-28.46	1.13	0.72	-0.41	-36.13	0.07	0.06	0.00	-6.71	2.65	1.82	-0.83	-31.19
Machinery	105.58	103.48	-2.10	-1.99	86.41	81.51	-4.91	-5.68	6.28	6.38	0.10	1.56	198.28	191.36	-6.91	-3.49
Seeds Cost	54.79	55.99	1.20	2.19	70.81	66.55	-4.26	-6.02	4.59	4.23	-0.36	-7.92	130.19	126.77	-3.43	-2.63
Manure	20.15	18.36	-1.79	-8.89	13.88	15.65	1.77	12.72	2.65	2.98	0.33	12.39	36.69	36.99	0.30	0.83
Fertilizers	93.85	95.97	2.12	2.26	94.55	103.26	8.72	9.22	6.65	6.93	0.28	4.20	195.04	206.16	11.12	5.70
Insecticides	23.86	22.40	-1.46	-6.12	7.98	8.60	0.63	7.84	1.27	1.58	0.31	24.31	33.11	32.58	-0.53	-1.59
Laser land leveling cost	5.13	5.13	0.00	0.00	3.93	3.93	0.00	0.00	0.20	0.30	0.10	48.97	9.26	9.36	0.10	1.06
Other Expenses	46.45	44.77	-1.67	-3.60	55.66	53.41	-2.25	-4.03	2.08	3.61	1.52	73.14	104.19	101.79	-2.40	-2.30
Rent	332.97	299.69	-33.3	-9.99	205.82	217.82	11.99	5.83	10.53	17.35	7.31	69.43	549.32	535.36	-14.0	-2.54
Crop revenue	2714.06	3605.48	891.42	32.84	1470.72	2565.11	1094.39	74.41	86.49	126.88	40.39	46.70	4271.26	6297.47	2026.2	47.44
Crop profit	2030.77	2635.81	605.05	29.79	958.47	1823.47	865.00	90.25	56.67	65.93	9.26	16.34	3045.91	4525.22	1479.3	48.57
Crop water consumption	656.19	414.94	-241.25	-36.76	1154.48	725.30	-429.18	-37.18	67.94	44.89	-23.05	-33.93	1878.61	1185.13	-693.5	-36.91
Kerosene fuel Liter	131.85	112.79	-19.06	-14.45	63.60	58.77	-4.83	-7.59	5.47	4.42	-1.05	-19.26	200.92	175.98	-24.9	-12.41
Energy consumption in cultivation MJ	3538.66	3046.23	-492.43	-13.92	3545.46	3194.39	-351.07	-9.90	297.11	266.94	-30.17	-10.15	7381.23	6507.56	-873.7	-11.84

Data source: (1) MALR (2016); (2) RMS model (2016); (3) MWRI (2016)

the considered scenario as a homogeneous character is higher than their heterogeneous one. In addition, the water requirement rates of the homogeneous consideration were less than the heterogeneous one. Tables 3 and 4 show economic evaluations for optimum cultivation based on production and technical risk management, laser land leveling in old and new land of the Nile valley, water and it is comparable with the existing condition of Egypt. Figures 3 and 4 show changes in crops area aggregates in the three zones and seasons from mean 2011/2012-2013/2014 to RMS in old and new land of Egypt. The results show that area of crops would be 5419.251 and 703.581 hectares cultivated in old and new land of Egypt, respectively, and the proposed model provided higher net benefit than the existing model in all cases. The sum of net benefit for heterogeneous case (342614.709 and 32442.008 Million E.P.) was higher than the sum of homogeneous case (220095.868 and 22072.840 Million E.P.) in old and new land of Egypt and the sum of crop water consumption for heterogeneous case (28538.383 and 3721.615 Million cub. m.) was lower than the sum of homogeneous case (46267.812 and 6351.040 Million cub. m.). It indicates that the variation of heterogeneous character had a large impact on the optimal solution. For this reason, the RMS model with heterogeneous character of land area was appropriate for finding cultivation based on production and technical risk management after laser land leveling in old and new land of the Nile valley in Egypt have done. An agro-climatic adaptability classification (for each crop) should be

Figure 3. Changes in crops area aggregates in the three zones and seasons flow values from mean (2009/2010-2011/2012) to RMS. Data source: (1) MALR (2016); (2) RMS model (2016).

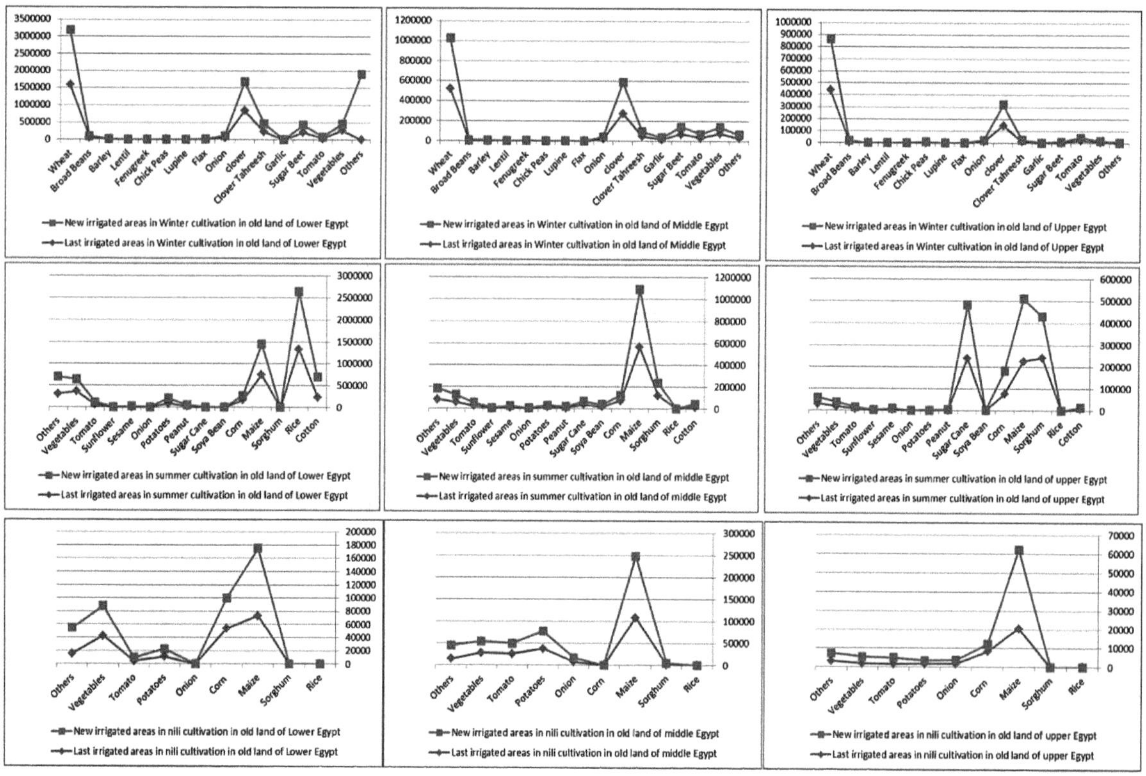

Table 5. Changes of economic and financial values in old land zones from mean (2011/2012-2013/2014) to RMS (boold is values that have increased, gray is values that have decreased

Lower Egypt	Winter Cultivation				Summer Cultivation				Nili Cultivation				Cultivation in Lower Egypt			
	Mean	AWSCS	Change	%	Mean	AWSCS	Change	%	Mean	AWSCS	Change	%	Mean	AWSCS	Change	%
Irrigated area of crop in old land	2922.5	**2922.5**	**0.0**	**0.0**	2253.2	**2253.2**	**0.0**	**0.0**	243.5	**243.5**	**0.0**	**0.0**	5419.3	**5419.3**	**0.0**	**0.0**
Total crop production cost	-23410.9	-26607.4	-3196.4	13.7	-24969.4	-26684.4	-1715.0	6.9	-2381.4	-2484.7	-103.4	4.3	-50761.8	**-55776.5**	**-5014.8**	**9.9**
Crop revenue	157470.7	**257548.6**	**100078.0**	**63.6**	57960.8	**76210.0**	**18249.2**	**31.5**	4664.4	**8856.0**	**4191.7**	**89.9**	220095.9	**342614.7**	**122518.8**	**55.7**
Crop profit	139919.9	**226132.7**	**86212.9**	**61.6**	36205.9	**49525.6**	**13319.8**	**36.8**	3142.7	**4370.2**	**1227.4**	**39.1**	179268.5	**280028.5**	**100760.1**	**56.2**
Rate of return (IRR)	5.7	**8.7**	**3.0**	**51.6**	1.3	**1.9**	**0.5**	**40.5**	1.0	**2.6**	**1.6**	**167.5**	3.3	**5.1**	**1.8**	**54.2**
Absolute Risk	0.4	**0.3**	**-0.2**	**-38.9**	0.2	**0.2**	**-0.1**	**-23.9**	0.6	**0.3**	**-0.3**	**-47.3**	0.4	**0.3**	**-0.1**	**-35.8**
Middle Egypt	**Winter Cultivation**				**Summer Cultivation**				**Nili Cultivation**				**Cultivation in Lower Egypt**			
	Mean	**AWSCS**	**Change**	**%**	**Mean**	**AWSCS**	**Change**	**%**	**Mean**	**AWSCS**	**Change**	**%**	**Mean**	**AWSCS**	**Change**	**%**
Irrigated area of crop in old land	473.5	**473.5**	**0.0**	**0.0**	423.8	**423.8**	**0.0**	**0.0**	113.8	**113.8**	**0.0**	**0.0**	1011.1	**1011.1**	**0.0**	**0.0**
Total crop production cost	-3866.1	-4731.8	-865.7	22.4	-4536.6	-4584.1	-47.5	1.0	-1163.7	-1234.1	-70.5	6.1	-9566.4	**-10550.1**	**-983.7**	**10.3**
Crop revenue	30891.9	**38839.1**	**7947.2**	**25.7**	10071.2	**13621.4**	**3550.2**	**35.3**	2545.3	**4337.2**	**1792.0**	**70.4**	43508.3	**56797.7**	**13289.3**	**30.5**
Crop profit	27025.1	**33748.8**	**6723.7**	**24.9**	6582.9	**9037.3**	**2454.4**	**37.3**	1903.6	**2346.4**	**442.7**	**23.3**	35511.6	**45132.4**	**9620.8**	**27.1**
Rate of return (IRR)	7.0	**7.2**	**0.2**	**3.1**	1.2	**2.0**	**0.8**	**61.6**	1.2	**2.5**	**1.3**	**111.8**	3.5	**4.4**	**0.8**	**23.6**
Absolute Risk	0.2	**0.1**	**0.0**	**-20.5**	0.2	**0.2**	**-0.1**	**-26.1**	0.5	**0.3**	**-0.2**	**-41.3**	0.2	**0.2**	**-0.1**	**-23.4**
Upper Egypt	**Winter Cultivation**				**Summer Cultivation**				**Nili Cultivation**				**Cultivation in Lower Egypt**			
	Mean	**AWSCS**	**Change**	**%**	**Mean**	**AWSCS**	**Change**	**%**	**Mean**	**AWSCS**	**Change**	**%**	**Mean**	**AWSCS**	**Change**	**%**
Irrigated area of crop in old land	290.0	**290.0**	**0.0**	**0.0**	372.6	**372.6**	**0.0**	**0.0**	25.3	**25.3**	**0.0**	**0.0**	688.0	**688.0**	**0.0**	**0.0**
Total crop production cost	-2477.1	-2992.2	-515.2	20.8	-4388.5	-4563.6	-175.1	4.0	-227.2	-238.1	-11.0	4.8	-7092.7	**-7793.9**	**-701.2**	**9.9**
Crop revenue	19876.4	**27255.5**	**7379.1**	**37.1**	8251.5	**10591.1**	**2339.6**	**28.4**	367.6	**760.2**	**392.6**	**106.8**	28495.5	**38606.8**	**10111.3**	**35.5**
Crop profit	17446.0	**22621.4**	**5175.4**	**29.7**	4630.8	**6027.5**	**1395.7**	**30.2**	249.7	**347.6**	**97.9**	**39.2**	22326.5	**28996.4**	**6669.9**	**29.9**
Rate of return (IRR)	7.0	**8.1**	**1.1**	**15.4**	0.9	**1.3**	**0.4**	**50.0**	0.6	**2.2**	**1.6**	**254.5**	3.0	**4.0**	**0.9**	**31.0**
Absolute Risk	0.3	**0.2**	**-0.1**	**-27.1**	0.2	**0.2**	**0.0**	**-22.1**	0.8	**0.4**	**-0.4**	**-51.6**	0.3	**0.2**	**-0.1**	**-26.2**

Data source: (1) MALR (2016); (2) RMS model (2016); (3) MWRI (2016)

Table 6. Changes of economic and financial values in new land zones from mean (2011/2012-2013/2014) to RMS (boold is values that have increased, gray is values that have decreased)

Lower Egypt	**Winter Cultivation**				**Summer Cultivation**				**Nili Cultivation**				**Cultivation in Lower Egypt**			
	Mean	**AWSCS**	**Change**	%	**Mean**	**AWSCS**	**Change**	%	**Mean**	**AWSCS**	**Change**	%	**Mean**	**AWSCS**	**Change**	%
Irrigated area of crop in old land	237.687	**237.687**	**0.000**	**0.000**	207.94	**207.94**	**0.00**	**0.00**	15.93	**15.93**	**0.00**	**0.00**	461.56	**461.56**	**0.00**	**0.00**
Total crop production cost	-1906.256	-1972.698	-66.442	3.485	-1764.92	-2364.65	-599.73	33.98	-158.58	-171.05	-12.47	7.87	-3829.76	-4508.40	-678.64	17.72
Crop revenue	8836.677	**13828.607**	**4991.929**	**56.491**	5732.57	**7604.38**	**1871.80**	**32.65**	148.47	**632.07**	**483.60**	**325.72**	14717.72	**22065.05**	**7347.33**	**49.92**
Crop profit	7287.009	**10318.926**	**3031.916**	**41.607**	4240.11	**5320.07**	**1079.95**	**25.47**	90.06	**302.21**	**212.15**	**235.55**	11617.19	**15941.20**	**4324.02**	**37.22**
Rate of return (IRR)	3.636	**6.010**	**2.374**	**65.309**	2.25	2.22	-0.03	-1.43	-0.06	**2.70**	**2.76**	**-4329.64**	2.84	**3.89**	**1.05**	**36.98**
Absolute Risk	0.399	**0.255**	**-0.144**	**-36.099**	23.09%	**17.41%**	**-5.68%**	**-24.61**	230.32%	**54.10%**	**-176.22%**	**-76.51**	35.30%	**23.55%**	**-11.75%**	**-33.30**
Middle Egypt	**Winter Cultivation**				**Summer Cultivation**				**Nili Cultivation**				**Cultivation in Lower Egypt**			
	Mean	**AWSCS**	**Change**	%	**Mean**	**AWSCS**	**Change**	%	**Mean**	**AWSCS**	**Change**	%	**Mean**	**AWSCS**	**Change**	%
Irrigated area of crop in old land	47.81	**47.81**	**0.00**	**0.00**	31.49	**31.49**	**0.00**	**0.00**	4.17	**4.17**	**0.00**	**0.00**	83.47	**83.47**	**0.00**	**0.00**
Total crop production cost	-467.19	-473.21	-6.03	1.29	-336.93	-285.78	51.15	-15.18	-41.74	-45.62	-3.88	9.29	-845.86	**-804.62**	**41.24**	**-4.88**
Crop revenue	2356.13	**2843.79**	**487.66**	**20.70**	622.10	**1052.36**	**430.26**	**69.16**	105.62	**183.34**	**77.72**	**73.58**	3083.85	**4079.49**	**995.64**	**32.29**
Crop profit	1923.71	**2331.85**	**408.14**	**21.22**	430.46	**726.82**	**296.36**	**68.85**	67.67	**137.86**	**70.18**	**103.71**	2421.84	**3196.53**	**774.69**	**31.99**
Rate of return (IRR)	4.04	**5.01**	**0.97**	**23.90**	0.85	**2.68**	**1.84**	**216.92**	1.53	**3.02**	**1.49**	**97.28**	2.65	**4.07**	**1.42**	**53.83**
Absolute Risk	14.64%	**12.13%**	**-2.51%**	**-17.15**	48.90%	**28.91%**	**-19.99%**	**-40.89**	52.03%	**29.98%**	**-22.06%**	**-42.39**	22.83%	**17.26%**	**-5.57%**	**-24.41**
Upper Egypt	**Winter Cultivation**				**Summer Cultivation**				**Nili Cultivation**				**Cultivation in Lower Egypt**			
	Mean	**AWSCS**	**Change**	%	**Mean**	**AWSCS**	**Change**	%	**Mean**	**AWSCS**	**Change**	%	**Mean**	**AWSCS**	**Change**	%
Irrigated area of crop in old land	87.60	**87.60**	**0.00**	**0.00**	64.37	**64.37**	**0.00**	**0.00**	6.58	**6.58**	**0.00**	**0.00**	158.56	**158.56**	**0.00**	**0.00**
Total crop production cost	-898.91	-969.66	-70.75	7.87	-742.52	-803.37	-60.85	8.20	-37.57	-61.26	-23.69	63.05	-1679.00	-1834.29	-155.29	9.25
Crop revenue	2714.06	**3605.48**	**891.42**	**32.84**	1470.72	**2565.11**	**1094.39**	**74.41**	86.49	**126.88**	**40.39**	**46.70**	4271.26	**6297.47**	**2026.20**	**47.44**
Crop profit	2030.77	**2635.81**	**605.05**	**29.79**	958.47	**1823.47**	**865.00**	**90.25**	56.67	**65.93**	**9.26**	**16.34**	3045.91	**4525.22**	**1479.31**	**48.57**
Rate of return (IRR)	2.02	**2.72**	**0.70**	**34.62**	0.98	**2.19**	**1.21**	**123.61**	1.30	**1.07**	**-0.23**	**-17.73**	1.54	**2.43**	**0.89**	**57.60**
Absolute Risk	23.22%	**17.48%**	**-5.74%**	**-24.72**	52.62%	**30.17%**	**-22.45%**	**-42.66**	33.02%	**22.51%**	**-10.51%**	**-31.83**	33.54%	**22.75%**	**-10.79%**	**-32.17**

Data source: (1) MALR (2016); (2) RMS model (2016); (3) MWRI (2016)

Table 7. Changes crop emission in cultivation in old land zones flow values from mean (2011/2012-2013/2014) to RMS (boold is values that have increased, gray is values that have decreased)

Lower Egypt	Winter Cultivation				Summer Cultivation				Nili Cultivation				Cultivation in Lower Egypt			
	Mean	AWSCS	Change	%	Mean	AWSCS	Change	%	Mean	AWSCS	Change	%	Mean	AWSCS	Change	%
NOx	1.711	**1.412**	**-0.299**	**-17.491**	0.643	**0.559**	**-0.084**	-13.101	0.041	**0.035**	**-0.005**	**-13.393**	2.395	**2.006**	**-0.389**	**-16.241**
SO2	8.255	**6.811**	**-1.444**	**-17.491**	3.105	**2.698**	**-0.407**	-13.101	0.198	**0.171**	**-0.026**	**-13.393**	11.557	**9.680**	**-1.877**	**-16.241**
CO2	8298.159	**6846.747**	**-1451.41**	**-17.491**	3120.392	**2712.030**	**-408.86**	-13.101	198.618	**172.018**	**-26.60**	**-13.393**	#######	**9730.795**	**-1886.87**	**-16.241**
SO3	nugatory	**nugatory**	**nugatory**	**nugatory**	nugatory	**nugatory**	**nugatory**	**nugatory**	nugatory	**nugatory**	**nugatory**	**nugatory**	nugatory	**nugatory**	**nugatory**	**nugatory**
CO	2.637	**2.176**	**-0.461**	**-17.491**	0.992	**0.862**	**-0.130**	-13.101	0.063	**0.055**	**-0.008**	**-13.393**	3.692	**3.092**	**-0.600**	**-16.241**
CH	nugatory	**nugatory**	**nugatory**	**nugatory**	nugatory	**nugatory**	**nugatory**	**nugatory**	nugatory	**nugatory**	**nugatory**	**nugatory**	nugatory	**nugatory**	**nugatory**	**nugatory**
SPM	nugatory	**nugatory**	**nugatory**	**nugatory**	nugatory	**nugatory**	**nugatory**	**nugatory**	nugatory	**nugatory**	**nugatory**	**nugatory**	nugatory	**nugatory**	**nugatory**	**nugatory**
Middle Egypt	**Winter Cultivation**				**Summer Cultivation**				**Nili Cultivation**				**Cultivation in Middle Egypt**			
	Mean	**AWSCS**	**Change**	**%**	**Mean**	**AWSCS**	**Change**	**%**	**Mean**	**AWSCS**	**Change**	**%**	**Mean**	**AWSCS**	**Change**	**%**
NOx	0.325	**0.280**	**-0.044**	**-13.686**	0.189	**0.152**	**-0.038**	**-19.874**	0.047	**0.042**	**-0.005**	**-11.320**	0.561	**0.474**	**-0.087**	**-15.574**
SO2	1.566	**1.352**	**-0.214**	**-13.686**	0.914	**0.732**	**-0.182**	**-19.874**	0.229	**0.203**	**-0.026**	**-11.320**	2.709	**2.287**	**-0.422**	**-15.574**
CO2	1574.494	**1359.015**	**-215.48**	**-13.686**	918.878	**736.261**	**-182.62**	**-19.874**	229.837	**203.820**	**-26.02**	**-11.320**	2723.209	**2299.097**	**-424.11**	**-15.574**
SO3	nugatory	**nugatory**	**nugatory**	**nugatory**	nugatory	**nugatory**	**nugatory**	**nugatory**	nugatory	**nugatory**	**nugatory**	**nugatory**	nugatory	**nugatory**	**nugatory**	**nugatory**
CO	0.500	**0.432**	**-0.068**	**-13.686**	0.292	**0.234**	**-0.058**	**-19.874**	0.073	**0.065**	**-0.008**	**-11.320**	0.865	**0.731**	**-0.135**	**-15.574**
CH	nugatory	**nugatory**	**nugatory**	**nugatory**	nugatory	**nugatory**	**nugatory**	**nugatory**	nugatory	**nugatory**	**nugatory**	**nugatory**	nugatory	**nugatory**	**nugatory**	**nugatory**
SPM	nugatory	**nugatory**	**nugatory**	**nugatory**	nugatory	**nugatory**	**nugatory**	**nugatory**	nugatory	**nugatory**	**nugatory**	**nugatory**	nugatory	**nugatory**	**nugatory**	**nugatory**
Upper Egypt	**Winter Cultivation**				**Summer Cultivation**				**Nili Cultivation**				**Cultivation in Upper Egypt**			
	Mean	**AWSCS**	**Change**	**%**	**Mean**	**AWSCS**	**Change**	**%**	**Mean**	**AWSCS**	**Change**	**%**	**Mean**	**AWSCS**	**Change**	**%**
NOx	0.167	**0.143**	**-0.024**	**-14.360**	0.165	**0.156**	**-0.009**	**-5.690**	0.011	**0.009**	**-0.002**	**-14.601**	0.342	**0.308**	**-0.035**	**-10.191**
SO2	0.805	**0.689**	**-0.116**	**-14.360**	0.796	**0.751**	**-0.045**	**-5.690**	0.051	**0.044**	**-0.008**	**-14.601**	1.652	**1.484**	**-0.168**	**-10.191**
CO2	809.050	**692.869**	**-116.18**	**-14.360**	800.139	**754.612**	**-45.53**	**-5.690**	51.713	**44.163**	-7.55	**-14.601**	1660.902	**1491.643**	**-169.26**	**-10.191**
SO3	nugatory	**nugatory**	**nugatory**	**nugatory**	nugatory	**nugatory**	**nugatory**	**nugatory**	nugatory	**nugatory**	**nugatory**	**nugatory**	nugatory	**nugatory**	**nugatory**	**nugatory**
CO	0.257	**0.220**	**-0.037**	**-14.360**	0.254	**0.240**	**-0.014**	**-5.690**	0.016	**0.014**	**-0.002**	**-14.601**	0.528	**0.474**	**-0.054**	**-10.191**
CH	nugatory	**nugatory**	**nugatory**	**nugatory**	nugatory	**nugatory**	**nugatory**	**nugatory**	nugatory	**nugatory**	**nugatory**	**nugatory**	nugatory	**nugatory**	**nugatory**	**nugatory**
SPM	nugatory	**nugatory**	**nugatory**	**nugatory**	nugatory	**nugatory**	**nugatory**	**nugatory**	nugatory	**nugatory**	**nugatory**	**nugatory**	nugatory	**nugatory**	**nugatory**	**nugatory**

Data source: (1) MALR (2016); (2) RMS model (2016)

Table 8. Changes crop emission in cultivation in new land zones flow values from mean (2011/2012-2013/2014) to RMS (boold is values that have increased, gray is values that have decreased)

Lower Egypt	Winter Cultivation				Summer Cultivation				Nili Cultivation				Cultivation in Lower Egypt			
	Mean	RMS	Change	%	Mean	RMS	Change	%	Mean	RMS	Change	%	Mean	RMS	Change	%
NOx	0.183	0.159	-0.024	-13.057	0.096	0.093	-0.003	-2.866	0.006	0.005	-0.001	-22.345	0.286	0.258	-0.028	-9.840
SO2	0.885	0.769	-0.115	-13.057	0.464	0.451	-0.013	-2.866	0.031	0.024	-0.007	-22.345	1.380	1.244	-0.136	-9.840
CO2	889.194	773.089	-116.10	-13.057	466.416	453.047	-13.37	-2.866	31.292	24.300	-6.99	-22.345	1386.902	1250.436	-136.47	-9.840
SO3	nugatory	nugatory	nugatory	nugatory	nugatory	nugatory	nugatory	nugatory	nugatory	nugatory	nugatory	nugatory	nugatory	nugatory	nugatory	nugatory
CO	0.283	0.246	-0.037	-13.057	0.148	0.144	-0.004	-2.866	0.010	0.008	-0.002	-22.345	0.441	0.397	-0.043	-9.840
CH	nugatory	nugatory	nugatory	nugatory	nugatory	nugatory	nugatory	nugatory	nugatory	nugatory	nugatory	nugatory	nugatory	nugatory	nugatory	nugatory
SPM	nugatory	nugatory	nugatory	nugatory	nugatory	nugatory	nugatory	nugatory	nugatory	nugatory	nugatory	nugatory	nugatory	nugatory	nugatory	nugatory
Middle Egypt	Winter Cultivation				Summer Cultivation				Nili Cultivation				Cultivation in Middle Egypt			
	Mean	RMS	Change	%	Mean	RMS	Change	%	Mean	RMS	Change	%	Mean	RMS	Change	%
NOx	0.043	0.038	-0.006	-12.868	0.014	0.013	-0.001	-4.894	0.002	0.002	0.000	-8.867	0.059	0.053	-0.006	-10.868
SO2	0.209	0.182	-0.027	-12.868	0.067	0.064	-0.003	-4.894	0.009	0.008	-0.001	-8.867	0.285	0.254	-0.031	-10.868
CO2	210.129	183.090	-27.04	-12.868	67.263	63.971	-3.29	-4.894	9.221	8.404	-0.82	-8.867	286.613	255.465	-31.15	-10.868
SO3	nugatory	nugatory	nugatory	nugatory	nugatory	nugatory	nugatory	nugatory	nugatory	nugatory	nugatory	nugatory	nugatory	nugatory	nugatory	nugatory
CO	0.067	0.058	-0.009	-12.868	0.021	0.020	-0.001	-4.894	0.003	0.003	0.000	-8.867	0.091	0.081	-0.010	-10.868
CH	nugatory	nugatory	nugatory	nugatory	nugatory	nugatory	nugatory	nugatory	nugatory	nugatory	nugatory	nugatory	nugatory	nugatory	nugatory	nugatory
SPM	nugatory	nugatory	nugatory	nugatory	nugatory	nugatory	nugatory	nugatory	nugatory	nugatory	nugatory	nugatory	nugatory	nugatory	nugatory	nugatory
Upper Egypt	Winter Cultivation				Summer Cultivation				Nili Cultivation				Cultivation in Upper Egypt			
	Mean	RMS	Change	%	Mean	RMS	Change	%	Mean	RMS	Change	%	Mean	RMS	Change	%
NOx	0.066	0.056	-0.009	-14.454	0.032	0.029	-0.002	-7.594	0.003	0.002	-0.001	-19.260	0.100	0.088	-0.012	-12.414
SO2	0.317	0.271	-0.046	-14.454	0.153	0.141	-0.012	-7.594	0.013	0.011	-0.003	-19.260	0.483	0.423	-0.060	-12.414
CO2	318.491	272.455	-46.04	-14.454	153.644	141.976	-11.67	-7.594	13.215	10.670	-2.55	-19.260	485.350	425.101	-60.25	-12.414
SO3	nugatory	nugatory	nugatory	nugatory	nugatory	nugatory	nugatory	nugatory	nugatory	nugatory	nugatory	nugatory	nugatory	nugatory	nugatory	nugatory
CO	0.101	0.087	-0.015	-14.454	0.049	0.045	-0.004	-7.594	0.004	0.003	-0.001	-19.260	0.154	0.135	-0.019	-12.414
CH	nugatory	nugatory	nugatory	nugatory	nugatory	nugatory	nugatory	nugatory	nugatory	nugatory	nugatory	nugatory	nugatory	nugatory	nugatory	nugatory
SPM	nugatory	nugatory	nugatory	nugatory	nugatory	nugatory	nugatory	nugatory	nugatory	nugatory	nugatory	nugatory	nugatory	nugatory	nugatory	nugatory

Data source: (1) MALR (2016); (2) RMS model (2016)

Figure 4. Changes in crops area aggregates in the three zones and seasons flow values from mean (2009/2010-2011/2012) to RMS. Data source: (1) MALR (2016); (2) RMS model (2016).

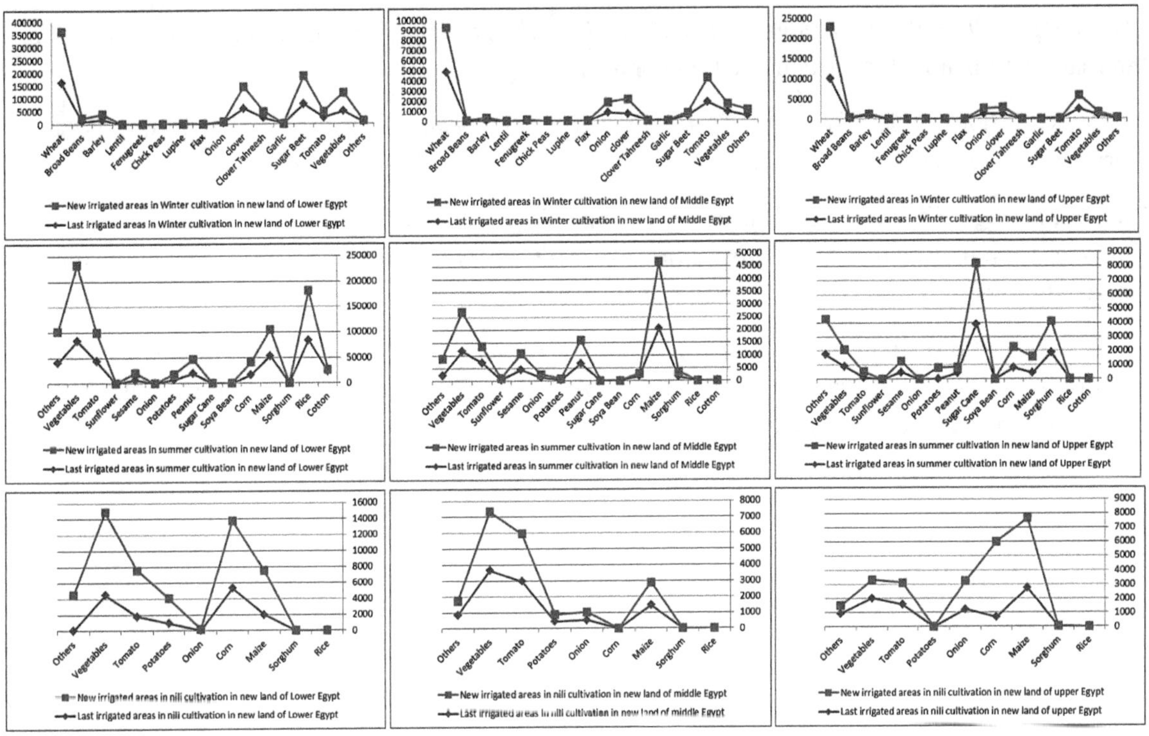

established in a form suitable for matching crops with climate and soil resources and crop production cost established according to soil and climatic zone, sufficient to judge whether yields exceed costs. According the financial and economic analyses, the annual internal rate of return (IRR) became higher than the existing model for the three zones and increased by 1.807, 0.836 and 0.936% in Lower, Middle and Upper of old land of Egypt and increased by 1.051, 1.424 and 0.889% in Lower, Middle and Upper of new land of Egypt, respectively. Absolute risk of optimum cultivation reduced by -35.760, -23.398, -26.191% in Lower, Middle and Upper of old land of Egypt and reduced by -33.299, -24.406, -32.175% in Lower, Middle and Upper of new land of Egypt, respectively (Table 5 and 6). The proposed model provided low greenhouse gas emission than the existing model for all agricultural operations. Pollutant causes destruction of ecosystem, damage to structures and people's health. The social cost of each ton emission of greenhouse gases and air pollutants was accounted for data on optimal use of energy in old and new land of Egypt in Tables 7 and 8. Finally, laser land leveling should make by the farmers because it is the best solution for the Egyptian question, it is a low cost (261.904 Egyptian pounds per hectare in northern Egypt, 333.333 Egyptian pounds per hectare in southern Egypt), high benefits. (61255.438 E.P./hectare) and save water consumption by 38.691%.

REFERENCES

AGRICORA. (2016). *Climate Risk Management for agriculture: Research and Tools*. AGRICORA.

APF. (2003). *Building a Business Risk Management, System for Agriculture in the 21st Century*. Retrieved from www.agr.gc.ca/puttingcanadafirst

IISD. (2013). *Climate risk management for agriculture in Peru: Focus on the regions of Junín and Piura*. International Institute for Sustainable Development.

Kahan, D. (2008). *Managing risk in farming, Farm management extension guide*. Rome: Food and Agriculture Organization of the United Nations.

Lim, B., Spanger-Siegfried, E., Burton, I., Malone, E., & Huq, S. (2005). *Adaptation Policy Frameworks for Climate Change: Developing Strategies, Policies and Measures*. Cambridge, UK: Cambridge University Press.

RMS. (2016). *Production and technical risk management, prepared by Youssef m. Hamada*. Department of Water and Land Economics Researches, Agricultural Economics Research Institute, Ministry of Agriculture & Land Reclamation.

Chapter 15
Social-Based Product Innovation and Governance in The Milk Sector:
The Case of Carciocacio and Innonatura

Giovanni Quaranta
University of Basilicata, Italy

Rosanna Salvia
University of Basilicata, Italy

ABSTRACT

The chapter reports and discusses a case study on an economic and scientific partnership which has developed a successful innovation, a cheese produced with vegetable (artichoke) rennet, using measure 124 of Campania region's Rural Development Plan, in a sector which is currently in crisis. The case study shows how the initiative's key to success is not only in product innovation but, more importantly, in the innovation of governance in the production chain and in the composition of the partnership. The latter includes not only the actors traditionally involved in processes of innovation in the agricultural sector (producers/adopters of innovation) but, also, new figures capable of producing organisational models to increase competitiveness in this struggling sector.

INTRODUCTION

In Europe, rural areas cover nearly 90% of its territory and are home to more than 50% of its population (Knickel et al., 2009). The effects of globalization, coupled with social, political and environmental changes, have left rural socio-ecological systems at high risk (Bardsley & Bardsley, 2014). Rural regions are, in fact, facing numerous and complex challenges to their economic viability and sustainability. Common features of rural areas, such as population ageing, out migration, low-waged and low-educated workers and cuts to public services, are all contributing factors to the social and economic decline of these regions (Slight et al., 2016; Ward & Brown, 2009).

DOI: 10.4018/978-1-5225-2107-5.ch015

The effects of external shocks are most evident in marginal rural areas, where low population density, a lack of economic diversification and a lack of control over economic policies imposed by national governments (Freshwater, 2015), undermine their resilience, defined as the ability of rural economies to resist disturbances and/or return to a pre-shock state (Pike et al., 2010; Perrings, 2006; Wilson, 2012). As a result, current rural system analysis in developed countries is focused on creating the conditions to support and increase local resilience in the face of external changes (Fielke & Bardsley, 2013; Milestad &Darnhofer, 2003) and the crucial role that the public and private sector can play in these transformations (Lemos &Agrawal, 2006). In particular, decision makers are increasingly asked for innovative policies (Bardsley &Bardsley, 2016).

State support for agriculture has also undergone a progressive reduction in the last thirty years, in line with the increasing predominance of neoliberal policies. Price support mechanisms for agricultural commodities have been significantly reduced in OECD countries and there has been a tendency to move away from a productivist model of agriculture to a model based on rural multi-functionality, without sufficient investment in innovation and economic growth (Marsden & Sonnino, 2008).

In order to face these challenges and relaunch their economies, rural communities are looking for new opportunities for growth and innovation. Promoting economic growth is, therefore, a priority in rural regions, and the tools needed for this objective are largely considered to be better access to financial capital and, most importantly, innovations (Sarkkinen & Kässi, 2013).

European policies have increasingly tried to promote interaction between farmers, researchers and rural businesses in an attempt to produce an interactive model of innovation, which is inclusive of all the various rural actors (EU, 2009). The European Union's interest in innovation arises from its positive contribution to growth, but especially from the realization that production systems in Europe are not sufficiently innovative and have not yet started moving down the desired path towards a knowledge-based economy.

Innovation is considered important to tackle issues such as response to climate change, conservation of biodiversity, maintaining water quality, exploiting renewable energy sources and has an important role to play in restructuring the European dairy sector. Innovation is also considered essential to meeting objectives for competitiveness, quality of life, diversification and territorial cohesion, all of which are key issues in the EU agenda.Despite working with a reduced budget, the EU is still highly committed to supporting innovation and research across all sectors, including agriculture (Bonfiglio et al., 2015). In fact, research and development is one of the EU's five priority targets in its ten year strategy launched in 2010 for sustainable and inclusive economic growth (the Europe 2020 Strategy).

This chapter aims to add to the policy debate surrounding innovation in agriculture through a case study within the NOVOROD Project, which has successfully developed innovation in the struggling Italian dairy sector by building an economic and scientific partnership. The case study analysis shows how the success of the project was more to do with innovations in governance in the production chain and the makeup of the project's partnership rather than innovations in the product. The project added new and innovative figures to the traditional actors involved in introducing innovations (producers/adopters of innovation) who were able to produce organizational models capable of increasing the competitiveness of this sector, which is currently in crisis.

BACKGROUND

The progressive recognition of the multi-functionality of agriculture and rural areas, now freed from the mono-function of food production, is changing the traditional idea of agriculture and, with this, the role of farmers. To respond to these changes in the European Union's new Common Agricultural Policy (CAP) has redefined its objectives, both in Pillar I, aimed at market measures, and in Pillar II, policies on Rural Development (RD). In this revised approach to rural development, economic diversification and environmental sustainability have taken an ever more central role in defining processes for territorial development.

As Knickel et al. (2009) outline, in order to be coherent with the objectives of the new agricultural agenda, policies on rural innovation must abandon the old models and adopt "second order" innovations, meaning innovations that involve a systemic approach based on new objectives and new frameworks. If first order innovations are developed from existing results and pursue well-trodden paths, "second order" innovations involve adopting new paradigms and a new set of rules. Innovation policies should nurture "second order" innovation so that they may better adapt to system shocks by implementing already experimented alternatives (Knickel et al., 2009).

Farmers will also have to adapt to the new rules and redefine their roles. However, in many instances there is a gap between the changes farmers need to make and the capacity of the institutional framework to support such changes though innovations (Knickel et al., 2009). The predominant model that governs the transfer of innovations tends, in fact, to follow the simplistic view of a linear "from creation to adoption" model. According to this system, innovation is the result of knowledge gained through scientific research, which is applied to a production process and, if economically viable, spread via imitation or via initiatives to promote knowledge transfer (Godin, 2006). This model is unsustainable because innovation is not linear, nor exclusively the result of formal scientific research but rather a social process which involves a multitude of social, economic and institutional actors and their formal and informal relationships (Camagni, 1991). An interesting approach to non-linear innovation processes is found from the review of the triple helix and learning region model (Wellbrock et al., 2012) which identifies three large structures (rural territory, system of knowledge and innovation support, public sector), each of which are home to important mechanisms and processes. However, rural innovation cannot come about solely via the actions in one of the three systems, but requires interaction among the various actors operating in each system in order to promote knowledge transfer, funding, market studies, valorisation of products, etc., which then result in processes of real rural innovation (Esparcia, 2014).

This territorial and systemic view of development and, therefore, of innovation overtakes the concept of "agricultural knowledge systems" (AKIS), developed in the 1990s and based on an interventionist agricultural policy which believed that in order to accelerate processes of modernization in agriculture, the transfer of innovations had to be highly coordinated and implemented by four main actors: research, extension services, education and training. However, as highlighted by Van der Ploeg (2003), if agricultural knowledge were simply the product of the work done by a team of experts, it would be far removed from farm level realities and could never provide a realistic representation of the correct path to follow.

For current systems of innovation to meet the real needs of farmers, they must distance themselves from the objectives of the so-called "productivist era" (Wilson & Rigg, 2003; Van der Ploeg et al., 2000; Knickel et al., 2005) and instead stick to the principals of economic, environmental and social sustainability, as formulated by European agricultural policies. In recent years, innovation transfer in the agricultural sector has been increasingly led by user-centered research systems (Klerkx & Leeuwis,

2009; Neef & Neubert, 2011). There has also been increasing acceptance that innovation in agriculture has been just as much about changing organizational and institutional models as it is about introducing new technologies (Hounkonnou et al., 2012; Klerkx & Nettle, 2013).According to the new approach, innovation does not involve technology alone, but is instead the combined result of technological, social, economic and institutional change (Kilelu et al., 2013). In terms of agricultural innovation, this change could be the result of top-down interventions or bottom-up farmer's grassroots activities (Smith et al., 2014).

Whatever method is chosen to support innovation, research has clearly shown that successful innovation requires end users to be included in decision-making processes (Douthwaite, 2002; Klerkx & Leeuwis, 2008; Klerkx et al., 2006; Neef & Neubert, 2011; Poulton et al.,2010). Building on this comes the concept of social innovation, which promotes the active participation of end users in every part of the innovation process via built local connections and a common learning culture (Dargan & Shucksmith, 2008).

A good example of how the principles of social innovation can be applied to rural areas is the EU funded project C@R ("Collaboration@Rural: a collaborative platform for working and living in rural areas") (Schaffers et al., 2010). C@R applies the research concept of 'living labs', user-centered innovation ecosystems capable of tailoring research and validation activities to the real needs of local stakeholders and users, to struggling rural regions with an aim to catalyze sustainable rural development. Over its ten year implementation, the C@R project found that building strong stakeholder networks was key to exploiting the full benefits of living labs and ensuring their long-term sustainability.

THE CHALLENGE OF INNOVATING A SECTOR IN CRISIS IN A FRAGILE RURAL AREA

The initiative under study, the NOVOROD project (Validation of new dairy products and dairy cattle feed to improve the overall quality of the Dairy Cow Milk System), was implemented in Campania, a region in South Italy, and financed by the Campania region's Rural Development Plan 2007-2013 under measure 124 "Cooperation for the development of new products, processes and technologies in the agriculture and food sectors". This measure was set up to promote initiatives aimed at increasing the competitiveness of the agricultural sector and improving product quality, environmental performance and safety in the workplace through testing, adopting and disseminating technological, processing, product and organisational innovations in the pre-competitive phase, as well as supporting actions which promote cooperation between producers, farmers and processing industry representatives, research organisations and other economic operators.

The analysis focuses on the mechanisms that explain how the introduction of innovation in the dairy cow milk sector can prove crucial to its survival, especially in light of the competitive landscape of the sector and the new demands made of agriculture, namely the need to be environmentally friendly whilst simultaneous increasing productivity and efficiency. The introduction of elements of innovation in the sector has also required a reassessment of the best instruments to use for their dissemination. The choice of which innovations disseminate and support becomes, therefore, an increasingly complex task that goes beyond the technical aspects to also consider a range of variables relating to economic, social and environmental sustainability.

THE MILK SECTOR: REFERENCE CONTEXT AND PROBLEMS ON THE GROUND

Competitive Landscape

The global cheese market, although mature, is expanding rapidly and the sector is showing signs of significant innovation, diversification, concentration and investment.At a European level, medium term prospects for the sector are also good as increased global demand should boost exports and help maintain price levels, (EU-DG Agri, 2012; EU-DG Agri 2013).

The international cheese market will, however, be strongly affected in the coming years by the 2015 abolition of milk quotas. Since the beginning of April 2015 EU milk production has been liberalized and European dairy farmers are therefore no longer constrained by production caps. Although effects should be contained in Europe as a whole, the end of the milk quota system could further expose the domestic market to fluctuations in world milk prices and put less efficient dairy farmers at risk. To soften the anticipated effects of the milk quota abolition, the EU introduced policy instruments such as "The Milk Package" and the "The Quality Package". These new measures deal with contractual relations in the sector and strengthen the market power of dairy farmers by giving them the possibility to negotiate contract terms and prices collectively through Producer Organizations (POs). The packages also provide possibilities for the supply management of cheese covered by PDO or PGI (Inea, 2013) and seek to simplify the process for certification under geographic indication schemes for food and dairy products, as well as strengthen legal safeguards (Inea, 2013).

The Situation in Italy and in the Campania Region

The dairy sector is a very important part of Italy's agri-food sector. In terms of value, dairy farming represents over 9% of Italy's agricultural output and Italy's dairy processing industries contribute to around 12% of the total value of the Italian food industry (Inea, 2013). The dairy sector also plays a crucial role in Italian food export markets.The Italian cheese sector alone represents 88.3% of the total value of products destined for foreign markets (Pieri, 2014), with products with Protected Designation of Origin certification proving especially desirable abroad.

The sector has been subject to a deep structural re-organization over the last few decades, largely focused on concentration and re-organization of production. In recent decades the dairy sector in the Campania region has largely mirrored national and European trends. On the plains, production has seen more concentration and specialization thanks to the larger financial returns obtained through economies of scale and investments in innovation. The re-structuring of the sector in inland hilly and mountainous areas, however, has been greatly hindered by geographic constraints, linked to the nature of the terrain, climate and environment, and socioeconomic barriers, linked to the fragility of the local production system.

In Campania, the number of dairy cows is going down whilst the number of buffalo, sheep and goat in the dairy sector is going up. The reverse is true on a national level (ISTAT, 2016).

The dairy processing industry in Campania counts 1,210 plants and is heavily localized with most dairies concentrated in the provinces of Naples, Salerno and Caserta (CCIAA, 2016).

The region mainly produces fresh cheeses (15% of national production comes from the region) and semi-hard cheeses (8% of the national production). In terms of import and exports, Campania has seen an inverse trend than that of Italy with net exports generally higher than imports.

Profile of Demand for Dairy Products

Consumer demand and consumer behavior are key variables for evaluating agri-food systems, and are decisive in successfully orientating research and development and innovations. Understanding trends in consumption and consumer demand for innovative foods is much more complex today than it was in even the recent past because postmodern consumers often display contradictory behaviors (Cicia et al., 2012). On the one hand, there is a preference for "all natural" foods, and, on the other,demand for enriched products that promise health benefits. Demand for convenience foods has also risen dramatically, in line with the huge societal changes seen in recent decades (population ageing, dual career families, urbanization etc.).

The features that shape demand for dairy products are the same of those that drive demand for food products generally. Over time, patterns of cheese consumption have changed as well as consumer preferences. There has been a significant increase in innovative products, aimed at meeting demand for a wider range of consumption options. The Italian cheese market is characterized by a high level of per capita consumption. There are a vast array of products; although the majority of sales are for a select few types of cheese, (aged cheeses are particularly in decline). Noteworthy is the impact the financial crisis of 2008 had on average food expenditure in Italy, which was long thought to be immune to squeezes. Despite the negative picture that emerges from official data, food products with certified geographical indication, and certified cheeses, especially, have held their ground (Ismea-Qualivita, 2013). A significant portion (35%) of total expenditure on cheese went to cheeses with denomination of origin certification (Ismea-Qualivita, 2013).

Institutional Framework and Instruments

The EU has committed to increasing the spending on research and development to 3% of GDP in the European countries by 2020 under the renewed Lisbon Agenda of 2004, which sought to boost Europe's innovation and competitiveness (EC, 2010). The Lisbon Agenda helped promote innovation as a socio-economic process rather than just a technological one, although critics are quick to point out that the policy framework adopted an unrealistic linear model of innovation (formal knowledge passed on to industry to adopt) which did not leave space for informal knowledge sharing or bottom-up approaches (Kronjee and Nooteboom, 2008).

More broadly speaking, the EU provides an array of measures to promote knowledge transfer and the creation and dissemination of innovation under its Rural Development Policy. The objectives for rural development set out by EU policy are achieved by individual Member States under national Rural Development Plans (RDPs), which are in turn defined and implemented at a regional level. In Italy, the Campania region's Rural Development Plan 2007-2013 provided funding under Measure 124 to promote initiatives aimed at increasing the competitiveness of the agricultural sector through experimenting, adapting and disseminating technological, processing, product and organizational innovations in the pre-competitive phase, i.e. before the product is ready for market. The measure also intended to support actions, which promote cooperation between producers, farmers and processing industry representatives, research organizations and other economic operators.

The measure came about from the realization that one of the biggest obstacles to innovation in the agricultural sector is the lack of integration between operators in the sector, both horizontally, within supply and production chains, and vertically, between the different sectors of production, processing and sales and, especially, between research, consultancy and training bodies.

The implementation of the Measure 124 in Campania varied from that in other Italian regions, extended its scope, and reach so much so that this measure unintentionally became a precursor to Measure 16, the new measure aimed at tackling competitiveness in the primary sector in the next RDP planning period 2014/2020.

The main features that made Measure 124's implementation so successful in the Campania region were:

- The inclusion of a research body, public or private, in the partnership to work alongside farm, processing and retail sector representatives was compulsory;
- Research bodies were not excluded from taking a leadership role in future Temporary Associations of Companies (ATS-*Associazione temporanea di scopo*) that may form, thereby giving associations greater freedom of self-organization;
- The Measure could be implemented directly, with projects promoted by individual ATS, and also through so-called Integrated Supply Chain Projects (PIF- *Programmi Integrati di filiera*), and a territorial approach (LEADER projects).

Measure 124 was designed as a model to foster the creation of networks. At its heart is the recognition that policy intervention must start from the needs of local stakeholders and that the uptake of innovations should be seen as a shared learning experience, which involves farms, agri-food businesses, scientific and research bodies, institutions and technicians. Unlike previous dissemination models, which were based on a top-down approach, the current model for Measure 124 in the Campania region forces innovations and innovators to adapt to different local contexts. Consumers too are given an ever greater role in this process.

The NOVOROD Project: The Concept Idea and the Planned Activities

The idea for the NOVOROD project was inspired by a mix of social, economic and environmental needs. The current crisis in the bovine milk and processing sector in the region has had serious economic effects, in terms of attention to cost structures (e.g. farmers looking for reductions in the cost of dairy waste disposal methods or cattle feed), environmental effects, especially on the degree of extensive grazing in cattle farming and on use of pastures, and social effects, in that the dismantlement of such a labour intensive sector exacerbates the problem of depopulation. The NOVOROD project set about to address each of these issues, as well as address the needs of new health-conscious consumers looking for natural products and ethical production (e.g. demand for cheese products made with non-animal derivatives or reconstituted elements (GMO).

The project, therefore, foresees actions intended to introduce elements of innovation along the entire production chain of the cattle, dairy sector, thereby helping to increase the sector's competitiveness with new technologies and innovations to both products and processes. The project involved the following phases of the production chain.

Breeding

Farming systems have been implemented that view livestock as sources of "cheese" and no longer just "milk". This was done through: valorisation of dairy cattle breeds which are better suited to cheese production (*Bruna* and *Pezzata Rossa*) and less intensive farming models; the trial of cattle feed based on forage crops which can modify the nutritional content of milk; the trial of cattle feed which is high in polyunsaturated fatty acids; the trial of protocols for the production of forage crops.

Processing

Innovative production lines have been tested regarding: vegetable rennet cheeses, produced using the white artichoke of Pertosa (Slow Food Presidium); the innovative (re)use of a milk processing waste by-product, whey, into creamy whey-cheese spreads / whey-desserts with berries from the Alburni area (found within the project area); naturally enriched cheese, thanks to cattle feed based on forage with a high potential to modify the nutritional content in milk.

The final product of the innovative project, *Carciocacio* valorises the milk produced in more extensive farming systems oriented towards the production of milk for cheese, making it quite different, in terms of microbes and nutritional content, from milk produced for general consumption. The production of this naturally enriched milk has significant organizational and managerial implications for cattle farms because it requires cattle feeding systems that favor fodder which is able to naturally enrich the milk to bring human health benefits. These types of forage crops are particularly suited to the Mediterranean productive context and could play an important economic and ecological role in the local and regional production system. In addition, some feeding protocols tested in the project involve recycling waste products left over from the processing of olives, which would contribute significantly to reducing the environmental impact associated with the production of olive oils. The use of artichoke rennet, and in particular the white artichoke of Pertosa, as a coagulant is another important element of connection within the production chain. The choice to use this type of vegetable based rennet helps valorise artichokes on their second or third crop of flowers, which have a lower commercial value. Their commercialization could represent an important source of extra income for artichoke producers in the territory.

In addition to Carciocacio, the project was also involved in the production of innovative whey based products. New processing machinery for producing whey concentrate was piloted in one of the partner dairies. The whey concentrate produced was used to make creamy whey concentrate based products (whey-dessert). The introduction of the machinery to produce the whey concentrate brought immediate benefits to the dairy that piloted the technology: a 30% reduction in disposal costs of whey, a by-product of cheese production, which is known to be problematic for disposal (Sepe et al., 2014). The products obtained from the concentrated whey were further processed by adding a seedless puree of berries (blackberries, raspberries, strawberries). The development of newer and high-quality whey-based products, made with recipes to appeal to a wider audience, will be subject to further evaluation.

Valorisation and Commercialisation of Innovative Cheeses

Implementation of measures that aim to contribute to the strategic repositioning of the bovine dairy sector through: the transfer of technological and product innovations to businesses; the strengthening of collaboration between firms and research organizations; the provision of training and the promotion of new products.

Dissemination of the Project's Results

A summary of the results and the technological standardization of the innovative cheeses produced under the project was presented in a final booklet, prepared by the MEDES Foundation, entitled "Production of innovative cheeses: Carciocacio. Cheese with vegetable rennet, a result of a validation initiative supported by the RDP 2007-2013 Campania, Measure 124".

Training seminars and conferences were also organized as part of the dissemination, valorisation and commercialization of the innovations piloted under Project NOVOROD, aimed at transferring the techniques and results of the project to operators in the bovine dairy sector. Participants in the training initiatives included public bodies, local authorities, consumers, the media, primary producers, processing/marketing firms and economic and social partners, all in cooperation with the relevant regional authorities.

The Project as a Rural Living Lab and the Active Role of Consumers

The partnership that started the NOVOROD project and the partner roles are reported in Table 1. The partners already had good relationships from previous projects they had worked on together and so proved very effective in defining and implementing the project's activities. This allowed the research partners to respond quickly to producers' requests. The exchange was in no means one way, however, farmers and processors actively participated in adapting the innovations to their specific needs.

Consumers were involved at every stage of the project: their support helped shape the initial idea for the project and their contribution throughout helped inform the choices the partnership made in product development.

Table 2 summarises all the meetings where the product was tested by consumers.

Table 1. NOVOROD partnership

Members of Partnership		Role in the Project
Lead Partner	MEDES Foundation	In charge of dissemination activities
Primary Producers, Farms	Azienda Agricola Pucciarelli Paolo, Azienda Agricola Alburni Natura di Turco Anna; Azienda Agricola Valitutto Antonio	Producing and supplying the artichokes and mixed berries for the project
Primary Producers, Cattle Farms	Azienda Agricola Sant'Antonio, Azienda Agricola Formentin Angelo, Azienda Agricola Catale Gerardo, Azienda Agricola Mario D'angelo, Azienda Agricola Tonino D'Iorio.	Producing and supplying the milk used in the project
Processing Industry/Sales	Caseificio Campolongo srl, Caseificio P. & P. srl, Caseificio F.lli Starace srl, Caseificio Senatore srl, Caseificio Mediterraneo snc	Processing the milk (dairies)
Research Bodies	CREA-ZOE, CREA-ORT, University of Basilicata	Producing the initial innovation (artichoke rennet) and adapting the innovation to the local context of the project
Territorial Actors	MIdA Foundation	Management body for natural and cultural patrimony

Table 2. Meetings with consumers

Date	Location	Participants	Objective
28 June 2011	MEDES Foundation Offices–Sicignano degli Alburni (SA)	18 partners from the project	Product tasting (fresh and semi-soft, caciotte cheeses made from vegetable rennet) to gather opinions/feedback and invite suggestions from partners
21 December 2011	MIdA Foundation offices - Pertosa (SA)	Mainly local consumers and local authority and administration representatives(around 50 people)	Product tasting and raising awareness of objectives of the NOVOROD project. Cheeses served were produced by cattle fed flax enriched diets. Participants were given a questionnaire to rate the products and were invitedto leave comments.
25 February 2012	MIdA Foundation – Pertosa (SA)	Mainly "extra local" consumers and spectators to the production "Dante's Inferno" held by the MIdA foundation (around 200 people)	Product tasting. Selection of cheeses served, produced by different cattle feed diets which had been piloted on the project's cattle farms
March 2012	Vinitaly – Verona	Exhibition visitors and vendors	Many different consumers were invited to leave feedback on the cheese and also the name and logo (choice of the name Carciocacio) via a questionnaire
May 2012	Forum PA (2012) Public Administration Forum – Rome	Forum visitors and staff	Tasting of Carciocacio cheese paired with wines from the Campania region
October 2012	Salone del Gusto- International Food Exhibition Turin	Exhibition visitors and vendors	Tasting of Carciocacio cheese and whey-dessert
March 2013	Agrosud Exhibition –Naples	Exhibition visitors and vendors	Tasting of Carciocacio cheese and whey-dessert
May 2013	Forum PA (2013)–Public Administration Forum Rome	Forum visitors and staff	Tasting of Carciocacio cheese and whey-dessert
June 2013	Vitignoitalia Wine Tasting–Naples	Exhibition visitors and vendors	Tasting of Carciocacio cheese and whey-dessert
September 2013	"Back to Cheese" Trade fair–City of Bra	Exhibition visitors and vendors	Tasting of Carciocacio cheese and whey-dessert
April 2014	Agricultural Fair Pastorano	Exhibition visitors and vendors	Tasting of Carciocacio cheese and whey-dessert
May 2014	Cibus International Food Exhibition–Parma	Exhibition visitors and vendors	Tasting of Carciocacio cheese and whey-dessert

At all the meetings there was a high level of consumers participation, which allowed the project to gather a significant amount of feedback, which was subsequently used to refine and improve the product. On a scale of one to ten, Carciocacio showed an increasing trend in consumer acceptance from December 2011 (5.7) to July 2014 (8.1), thanks in part also to the standardization of the cheese production technique. The product's name was chosen directly by consumers from a range of different options.

Training activities also helped make the project a real living lab. These sessions were designed to transmit the theoretical and also practical insights gained from the project and were held at dairies and on cattle farms with retails outlets that were located in areas of production considered to be representative of the wider regional context. The training covered all aspects of the project, from the production of the forage crops used for cattle feed, to the cultivation of the artichokes for the rennet and, finally, the processing of the milk into cheese.

The INNONATURA Consortium: The Sustainability of the Project in the Future

The Consortium INNONATURA was set up in February 2014. Its purpose is to foster greater cooperation between the world of production and research organizations. It is made up of six partners (five actors in the Agri-livestock productive chain and one research organization specializing in sustainable development and innovations to product and processes for environmentally and health conscious products). All the INNONATURA actors had previously been involved in the NOVOROD project. INNONATURA began as a natural evolution to the NOVOROD project and continues its good work by safeguarding the production of Carciocacio and overseeing the correct application of the production protocols conceived during the initial project. The Consortium's mission, however, also includes promoting other innovative cheeses and local typical products, as well as represent/aggregating specific needs of regional/local supply chains for innovation, facilitating the dialogue with innovation centres and informing potential beneficiaries on the opportunity to invest in innovation. The success in establishing the INNONATURA consortium is proof of the very high levels of trust and cooperation that formed between the different partners during the NOVOROD project.

SOLUTIONS AND RECOMMENDATIONS

The key to the NOVOROD Project's success was the combination of innovative governance structures and the choice to develop the different phases of the project as a "living lab" with the active involvement of end users. The MEDES Foundation, the lead partner, played a crucial role in governance. As has already been highlighted by other studies (Cristiano et al., 2013), the partner who acts as the mediator of innovation is usually the lead partner, who also plays another key function in coordinating and disseminating information regarding the project, at each stage of its execution. The management of the projects' activities, and the planning and implementation of awareness raising activities and dissemination of results, were two essential elements in consolidating and strengthening the integrated and systemic approach that was promoted by the NOVOROD project.

The Campania Region has, in fact, been strongly committed to disseminating of the project's results and promoting knowledge transfer so that the innovative production techniques piloted in the project could be made available to the widest possible audience. Having been financed under Measure 124 of the Campania Region's RDP, the innovation developed within the NOVOROD project was not protected nor designed to give only the participating dairies a market edge. Instead, the project was aimed at transferring the knowledge gained through the project's activities so that not only individual firms could benefit, but that they may then also lead the way to an overall increased market competitiveness of the local production chain. A successful dissemination of results also helps increase the uptake of the innovation and protects the investment made in its development. For this reason, choosing the right approach to dissemination is vital. The NOVOROD project favored a learning by doing approach, with all training and activities designed for maximum interaction between individual farmers and between farmers and other project actors. This approach proved very successful and helped foster trust between the project's actors and increased farmers' awareness of the benefits of introducing elements of innovation into their business.

Measure 124 was also set up to establish greater collaboration between researchers and farmers with an aim to persuade farmers of the importance of innovation in the current competitive landscape.The living-lab that sprang up inside the project became the space in which the scientific know-how applied in the product development phase was married with the need for clear marginal benefits for participating firms in terms of competitive advantage, cost reduction and improvements to organizational efficiency and production processes.

A large part of the project's success can be attributed to the fact that the partners had a good working knowledge of the specific local challenges faced by the agricultural sector and had previously built good working relationships. This prevented any barriers to communication and made it much easier to reach a consensus on the initial idea for the project and to begin to build the partnership. Good relationships and communication are particularly important in rural areas where the fragmentation between different components of the social-ecological system can present real barriers to progress. Given the particular rural context of the agricultural sector, spatial clustering, which is often the key to success for the creation and diffusion of innovations, is not an option. Instead, there needs to be support for building and consolidating territorial networks, formal or informal, which bring together different economic, social and institutional actors. In the case of the NOVOROD project, the pre-existing relationship between the Research Centre that first developed the innovation and the MEDES Foundation, and their rooting in the local community, played a key role in every stage of the project, from evaluating the market feasibility and economic sustainability of the innovation to the implementation and coordination of the dissemination of results.

All of this was fundamental to the effective adaptation of the innovation to the real needs of farmers and helped ensure the aims of the project were met. The methodology adopted under Measure 124, the choice to put the results of research already carried out to the test on the farm floor, placed the needs of the partner farmers at the center of the project. It is also an example of how research centres translate the results of their research to farm scale. The case of Carciocacio, in fact, is emblematic of how support under measure 124 has been used to adapt techniques and practices to the individual requirements of local farms. There was good interaction between the project's researchers and cattle farmers during the experimental development phase of the product. The researchers provided made-to-measure technical support to partner farms in order to support the correct implementation of the innovation and frequent on-site visits also allowed researchers to troubleshoot and provide fast solutions to problems. Another aspect to highlight is the holistic and systemic approach the project adopted. The link the project created between milk/cheese production and processors, involving also artichoke farmers for the production of the rennet used, was designed to be both economically sustainable (reduce costs) and environmentally sustainable (recycling of whey waste product, pasture productivity, reuse of crop residues and pomace: encouraging more extensive livestock farming). In addition, increasing the nutritional properties of the cheese manufactured increased its value and opened up alternative markets.

Another key point of success was the project's ability to grow and strengthen networks of actors. Inclusive mechanisms that promote the transfer of knowledge and assist in managing change are essential. In other words, the link between local and extra-local actors, and the acquisition of skills *in itinere* prove decisive factors in adopting innovations. Generating and transferring innovation in rural territories increasingly takes place through actions to increase farmers' capabilities, promote change in business models and improve the rural economy's propensity for innovation/adaptation. The NOVOROD project brought about real change in farmers' attitudes to innovation. Interaction inside a space where knowledge is co-generated, in fact, creates the right conditions for long-term forms of collaboration, which may continue knowledge exchange long after the current project has ended and promote copycat initiatives.

The establishment of the INNONATURA Consortium, which was strongly supported by farms, will channel the needs of farmers to the scientific community so that solutions may be jointly found and will help access the resources necessary to continue the process of innovation already started. The NOVOROD project helped consolidate relationships between partners and build trust by proving that collaboration promotes proactive investment in innovation and creates a dynamic business environment.

The more points of contact (potential sources of information) available to farmers, the greater the opportunities are for learning and consequently, the implementation of innovations. However, in this type of project, the full participation of all the various actors right from the very start of the project, and the clear definition of their respective roles in the trial/implementation phases is essential to achieving an effective application of the innovation, as well as a good dissemination of results. Furthermore, farmers who have already successfully implemented an innovation in their business prove to be the best advertisement to persuade more reluctant farmers to adopt innovation in their own businesses.

The final important aspect of the project was the role played by policy-makers and the financial support. Politicians still prove crucial to securing the funding needed to transfer, innovation to rural areas and setting up a framework of reference for greater communication between actors.

FUTURE RESEARCH DIRECTIONS

The project went far beyond a mere technical intervention in the dairy sector to improve farmer's general attitude to change, encourage collaboration between rural actors, build strong territorial networks, boost innovation in the sector and increase overall sectorial and territorial competitiveness. Naturally, the full effect of the project will have to be seen in the long term. It will be interesting to monitor the project's influence on similar and complementary actions, for example, investments in the certification/qualification of productions, in innovative solutions to gain better access to markets, and in generally fostering entrepreneurship in the rural territory, as well as investments to tackle the long-standing problems of rural under-employment and depopulation.

CONCLUSION

It is well known that in terms of experimentation and research the agricultural sector is characterized by a limited circulation of innovation and limited cooperation between different actors. This is particularly true in disadvantaged areas. There are also cases where the key strategic role of Universities and Research institutions in promoting innovation and long-term partnerships along and across agri-food chains is evident. Given the trend towards social innovation, which relies on stakeholder involvement to produce collaborative actions, the EU highlights the fact that evaluation methods should be more geared to measuring outcomes in terms of sustainable benefits to local actors rather than empirical outputs in terms of technical efficiencies.

The European Union's Rural Development Plan 2007-2013 included a specific measure, Measure 124 "Cooperation for the development of new products, processes and technologies in the agricultural and food sector", to promote the greater diffusion of innovation in the agricultural sector. Through this measure, a group of cow milk producers in the Campania Region have promoted an initiative of cooperation to introduce a true innovation in the dairy sector with a product, which is completely different from traditional local products yet in line with new market demands. The process involved the active participation of consumers and of research bodies, which developed innovations never before introduced into production, processing industries, and also a third party in the form of an organization whose role was to produce and create a tailor-made model of management and governance of the whole chain of production. The result of this cooperation was a real *Living Lab,* which has not only been able to transfer, innovation from the laboratory to industry to create a brand new product, *Carciocacio* (cheese made with artichoke rennet), but has also brought the chain of production full circle and carries new and positive implications for the local economy, environment and social growth (through the creation of the Consortium *INNONATURA*).

This product innovation has also fostered the introduction of new sustainable technologies orientated towards energy production from the waste material left over after production and the constitution of a new organizational model, which has permanently altered the whole chain of production in the area. This initiative, which began in a highly economical and environmentally fragile area, has thus brought a new wave of innovation to an entire region eventually setting the stage for an increase in entrepreneurship in the rural territory. Focusing on the process of building a network, which is able to connect all the various social and economic actors in a territory, the initiative has showed how the difficulties linked to very poor levels of collaboration and exchange between local actors could be reversed. This is particularly relevant in such as context where economic fragility results from various driving forces, mainly related to the depopulation and geographical and psychological marginalization. The remoteness of many rural businesses creates a competitive disadvantage (Hall & Williams, 2008) and a peripheralization of the local market compared to densely populated urban areas where demand is concentrated. This fuels a vicious circle of out-migration and lack of entrepreneurial opportunities (Veeck et al, 2006, Meccheri & Pelloni, 2006) and, also, the increased fragmentation and lack of coordination of rural activities (Hjalager & Johansen, 2013). The creation of an innovative and inclusive environment can help in mitigate the impact that the lack of infrastructure and those related to the high dispersion of the firms determine in rural areas, i.e. an exclusion of many rural enterprises from the benefits of the entrepreneurial embeddedness (Granovetter, 1985). These benefits include increased productivity and competitiveness and better access to information, a favorable environment for start-ups, virtuous links between the business sectors and the local context able to guide firms towards virtuous path of sustainable exploitation of available resources. These characteristics are found in many rural areas around the world, which makes the results of this study generally applicable to a wider context.

The management model adopted in the NOVOROD project could also prove a useful tool to develop future innovation projects in the next rural development programming period. In fact the co-generation of the innovation through a full participatory lab that was the keystone of the project, is going to pervade the new innovation measure of the RDP programme implemented in the Region. NOVOROD project has been included among the best practices of the Campania Region both for development of the innovation and its implementation and follow up. This has been particularly effective in a context where most of the innovation are either industry-sector driven or designed to meet general and local needs, turning in a quite poor diffusion and effectiveness.

REFERENCES

EU - DG Agri. (2012). *Prospects for Agricultural Markets and Income in the EU 2012-2022*. Commission's Directorate-General for Agriculture and Rural development, European Commission, Agriculture and Rural Development.

EU - DG Agri. (2013). *Prospects for Agricultural Markets and Income in the EU 2013-2023*. Commission's Directorate-General for Agriculture and Rural development, European Commission, Agriculture and Rural Development.

Bardsley, D. K., & Bardsley, A. M. (2014). Organising for socio-ecological resilience: The roles of the mountain farmer cooperative Genossenschaft Gran Alpin in Graubünden, Switzerland. *Ecological Economics*, *98*, 11–21. doi:10.1016/j.ecolecon.2013.12.004

Bonfiglio, A., Camaioni, B., Coderoni, S., Esposti, R., Pagliacci, F., & Sotte, F. (2015). *Innovation and education within rural development policy. Which role for rural and peripheral EU regions?* Paper presented at the 4th AIEAA Conference "Innovation, productivity and growth: towards sustainable agri-food production", Ancona, Italy.

Camagni, R. (Ed.). (1991). *Innovation networks: Spatial perspectives*. London, UK: Belhaven Press.

CCIAA. (2016). Retrieved from http://www.infocamere.it/

Cicia, G., Cembalo, L., Del Giudice, T., &Verneau, F. (2012). Il sistema agroalimentare ed il consumatore post-moderno: Nuove sfide per la ricerca e per il mercato. *Rivista di Economia Agroalimentare*, (1), 117-142.

Cristiano, S., & Proietti, P. (2013). Farm innovation through rural development programmes: Experiences and pathways of innovation in Italy. *Proceedings of the 21st ESEE European Seminar on Extension Services - Extension Education Worldwide trends, challenges and cases*. Antalya, Turkey: Akdeniz University, Faculty of Agriculture Department of Agricultural Economics Extension Science.

Dargan, L., & Shucksmith, M. (2008). LEADER and innovation. *Sociologia Ruralis*, *48*(3), 274–291. doi:10.1111/j.1467-9523.2008.00463.x

Douthwaite, B. (2002). *Enabling Innovation: A Practical guide to understanding and fostering technological change*. London, UK: Zed Books.

EC 2010. (2010). *Europe 2020: A European Strategy for smart, sustainable growth*. COM(2010) 2020.

Esparcia, J. (2014). Innovation and networks in rural areas. An analysis from European innovative projects. *Journal of Rural Studies*, *34*, 1–14. doi:10.1016/j.jrurstud.2013.12.004

Europe 2020. (n.d.). Retrieved from http://ec.europa.eu/

Fielke, S. J., & Bardsley, D. K. (2013). South Australian farmers markets: Tools for enhancing the multifunctionality of Australian agriculture. *GeoJournal*, *78*(5), 759–776. doi:10.1007/s10708-012-9464-8

Freshwater, D. (2015). Vulnerability and resilience: Two dimensions of rurality. *Sociologia Ruralis*, *55*(4), 497–515. doi:10.1111/soru.12090

Godin, B. (2006). The linear model of innovation: The historical construction of an analytical framework. *Science, Technology & Human Values, 31*(6), 639–667. doi:10.1177/0162243906291865

Granovetter, M. (1985). Economic action and social structure: The problem of embeddedness. *American Journal of Sociology, 91*(3), 481–510. doi:10.1086/228311

Hall, C. M., & Williams, A. (2008). *Tourism and Innovation*. London, UK: Routledge.

Hjalager, A. M., & Johansen, P. H. (2013). Food tourism in protected areas—sustainability for producers, the environment and tourism? *Journal of Sustainable Tourism, 21*(3), 417–433. doi:10.1080/09669582.2012.708041

Hounkonnou, D., Kossou, D., Kuyper, T. W., Leeuwis, C., Nederlof, E. S., Röling, N., & Van Huis, A. et al. (2012). An innovation systems approach to institutional change: Smallholder development in West Africa. *Agricultural Systems, 108*, 74–83. doi:10.1016/j.agsy.2012.01.007

Inea. (2013). *Annuario dell'Agricoltura Italiana– 2012* (vol. 66). Author.

Ismea-Qualivita. (2013). *Rapporto 2013 – Qualivita*. ISMEA.

Istat. (2016). Retrieved from www.istat.it

Kilelu, C. W., Klerkx, L., & Leeuwis, C. (2013). Unravelling the role of innovation platforms in supporting co-evolution of innovation: Contributions and tensions in a smallholder dairy development programme. *Agricultural Systems, 118*, 65–77. doi:10.1016/j.agsy.2013.03.003

Klerkx, L., De Grip, K., & Leeuwis, C. (2006). Hands off but strings attached: The contradictions of policy-induced demand-driven agricultural extension. *Agriculture and Human Values, 23*(2), 189–204. doi:10.1007/s10460-005-6106-5

Klerkx, L., & Leeuwis, C. (2008). Institutionalizing end-user demand steering in agricultural R&D: Farmer levy funding of R&D in The Netherlands. *Research Policy, 37*(3), 460–472. doi:10.1016/j.respol.2007.11.007

Klerkx, L., & Leeuwis, C. (2009). Operationalizing demand-driven agricultural research: Institutional influences in a public and private system of research planning in The Netherlands. *Journal of Agricultural Education and Extension, 15*(2), 161–175. doi:10.1080/13892240902909080

Klerkx, L., & Nettle, R. (2013). Achievements and challenges of innovation coproduction support initiatives in the Australian and Dutch dairy sectors: A comparative study. *Food Policy, 40*, 74–89. doi:10.1016/j.foodpol.2013.02.004

Knickel, K., Brunori, G., Rand, S., & Proost, J. (2009). Towards a better conceptual framework for innovation processes in agriculture and rural development: From linear models to systemic approaches. *Journal of Agricultural Education and Extension, 15*(2), 131–146. doi:10.1080/13892240902909064

Knickel, K., & Peter, S. (2005). Amenity-led development of rural areas: The example of the Regional Action pilot programme in Germany. In G. P. Green, D. Marcouiller & S. Deller (Eds.), Amenities and rural development: Theory, methods and public policy (pp. 302-321). Northampton, UK: Edward Elgar Publishing.

Kronjee, G., & Nooteboom, B. (2008). Research, higher education, and innovation. In B. Nooteboom & E. Stam (Eds.), *Micro-Foundations for innovation policy*. Amsterdam, NL: Amsterdam University Press.

Lemos, M. C., & Agrawal, A. (2006). Environmental governance. *Annual Review of Environment and Resources*, *31*(1), 297–325. doi:10.1146/annurev.energy.31.042605.135621

Marsden, T., & Sonnino, R. (2008). Rural development and the regional state: Denying multifunctional agriculture in the UK. *Journal of Rural Studies*, *24*(4), 422–431. doi:10.1016/j.jrurstud.2008.04.001

Meccheri, N., & Pelloni, G. (2006). Rural Entrepreneurs and Institutional Assistance: An Empirical Study from Mountainous Italy. *Entrepreneurship & Regional Development*, *18*(5), 371–392. doi:10.1080/08985620600842113

Milestad, R., & Darnhofer, I. (2003). Building farm resilience: The prospects and challenges of organic farming. *Journal of Sustainable Agriculture*, *22*(3), 81–97. doi:10.1300/J064v22n03_09

Neef, A., & Neubert, D. (2011). Stakeholder participation in agricultural research projects: A conceptual framework for reflection and decision-making. *Agriculture and Human Values*, *28*(2), 179–194. doi:10.1007/s10460-010-9272-z

Perrings, C. (2006). Resilience and sustainable development. *Environment and Development Economics*, *11*(4), 417–427. doi:10.1017/S1355770X06003020

Pieri, R. (Ed.). (2014). Il mercato del latte, Rapporto 2013. Milano, Italy: Franco Angeli.

Pike, A., Dawley, S., & Tomaney, J. (2010). Resilience, adaptation and adaptability. *Cambridge Journal of Regions Economy and Society*, *3*(1), 59–70. doi:10.1093/cjres/rsq001

Poulton, C., Dorward, A., & Kydd, J. (2010). The future of small farms: New directions for services, institutions, and intermediation. *World Development*, *38*(10), 1413–1428. doi:10.1016/j.worlddev.2009.06.009

Sarkkinen, M., & Kässi, T. (2013). Characterization of innovation situation in a remote rural region. *International Journal of Engineering Business Management*, *5*(1), 1.

Schaffers, H., García Guzman, J., Merz, C., & Navarro, M. (Eds.). (2010). *Living labs and rural development; Results from the C@R Project*. Madrid: TRAGSA.

Sepe, L., & Paladino, F. (2014). I formaggi validati durante il progetto NOVOROD. In G. Quaranta (Ed.), *Produzione di formaggi innovativi: CarcioCacio. Formaggi a caglio vegetale frutto di una validazione sostenuta dal PSR Campania 2007-13, misura 124. Ascea Print Service srl, Ascea Marina (Sa).*

Slight, P., Adams, M., & Sherren, K. (2016). Policy support for rural economic development based on Hollings ecological concept of panarchy. *International Journal of Sustainable Development and World Ecology*, *23*(1), 1–14. doi:10.1080/13504509.2015.1103801

Smith, A., Fressoli, M., & Thomas, H. (2014). Grassroots innovation movements: Challenges and contributions. *Journal of Cleaner Production*, *63*, 114–124. doi:10.1016/j.jclepro.2012.12.025

Van der Ploeg, J. D. (2003). *The virtual farmer. Past, present and future of the Dutch peasantry*. Assen, The Netherlands: Van Gorcum.

Van der Ploeg, J. D., Renting, H., Brunori, G., Knickel, K., Mannion, J., Marsden, T., & Ventura, F. et al. (2000). Rural development: From practices and policies towards theory. *Sociologia Ruralis*, *40*(4), 391–408. doi:10.1111/1467-9523.00156

Veeck, G., Che, D., & Veeck, A. (2006). Americas changing farmscape: A study of agricultural tourism in Michigan. *The Professional Geographer*, *58*(3), 235–248. doi:10.1111/j.1467-9272.2006.00565.x

Ward, N., & Brown, D. (2009). Placing the rural in regional development. *Regional Studies*, *43*(10), 1237–1244. doi:10.1080/00343400903234696

Wellbrock, W., Roep, D., & Wiskerke, J. S. C. (2012). An integrated perspective on rural regional learning. *European Countryside*, *4*(1), 1–16. doi:10.2478/v10091-012-0010-y

Wilson, G., & Rigg, J. (2003). Post-productivist agricultural regimes and the south: Discordant concepts? *Progress in Human Geography*, *27*(6), 681–707. doi:10.1191/0309132503ph450oa

Wilson, G. A. (2012). Community resilience, globalization, and transitional pathways of decision-making. *Geoforum*, *43*(6), 1218–1231. doi:10.1016/j.geoforum.2012.03.008

Chapter 16
A System Dynamics Model and Interface for the Simulation and Analysis of Milk Supply Chains

Georgios Tsaples
University of Macedonia, Greece

Theodore Tarnanidis
University of Macedonia, Greece

ABSTRACT

The objective of this chapter is the development of a System Dynamics model for the study of the milk supply chain and how an extreme event can affect its behavior. A simple interface is developed that can be used to increase the ease of communication and provide an interactive approach to the decision-making process. The model contains three echelons: farmers, processors and retailers. The main results show that under normal circumstances, the behavior of the system reaches equilibrium after a few oscillations. However, these oscillations can be smoothed out if the adjustment time of the order placement is increased. Under an extreme event that reduces the demand for milk, behavior changes and the system remains in dis-equilibrium for the entire simulation. Once again, adjustment times remain the leverages that can influence and mitigate those negative effects. Finally, a more robust and collaborative decision-making process among the actors of the chain could be beneficial for all not only under normal circumstances, but also in the presence of extreme uncertainty.

INTRODUCTION

The supply chain is defined as a network of flows and processes, in which companies cooperate along the chain from the initial raw materials all the way to the delivery of the final product to the end-user (Li, Zhang, & Jiang, 2008; Ramanathan, 2013; Schimith et al., 2015). A supply chain is a highly dynamic system that is subject to supply availability and demand uncertainty. Along with supply chain comes the term of supply chain management, which focuses on planning, coordinating and integrating

DOI: 10.4018/978-1-5225-2107-5.ch016

the material, information and capital flows along the chain (Li, Zhang, & Jiang, 2008; Seuring, 2013), with the objective of providing maximum customer satisfaction at the lowest cost (Chu, 2003). Hence, supply chain management incorporates both strategic and tactical objectives and the management of the chain in a sustainable manner has become an increasing concern for industries across a wide range of areas (Seuring, 2013).

The dairy supply chain shares the same structure as any other chain, it is considered especially complex and it entails many links that pass from the producer to retail and the final consumer (Shepherd & Flanders, 2008; Robinson, 2009; Simonson, 2009; Kumar & Nigmatullin, 2011). However, the dairy supply chain, being part of the vast food industry, is characterized by several attributes that make it unique and require a different strategy for the management to be successful (Ruteri, 2009). Firstly, at the supplier's/producer's side there is a great differentiation of products in an increasingly competitive market (Georgiadis, Vlachos, & Iakovou, 2005). Furthermore, the end products are considered highly perishable and fragile (Ayağ, Samanlioglu, & Büyüközkan, 2013), which furthermore are constrained by specificity in terms of deadline of consumption while there is limited storage capacity (Minegishi & Thiel, 2000). Thus, there is the need to transfer the end-products in a cost-effective way (Georgiadis, Vlachos, & Iakovou, 2005), while at the same time the supplier must consider the low value to size ratio (Ayağ, Samanlioglu, & Büyüközkan, 2013) and the appropriate packaging which must comply with the legislation (Minegishi & Thiel, 2000). Finally, there are concerns about the quality and safety of the end-products (Gereffi & Lee, 2009).

From the consumer's side, there is the demand for homogeneity of batches and a long duration of presence on the shelf (Minegishi & Thiel, 2000). Furthermore, there is a fixation with the price of milk (Ayağ, Samanlioglu, & Büyüközkan, 2013), because milk is considered one of the most important elements of nutrition, accompanied by sensitivity on place of origin (Georgiadis, Vlachos, & Iakovou, 2005) integrity of sources (Kumar & Nigmatullin, 2011) and means of production. Furthermore, seasonality and varying consumer tastes put an extra burden originating from consumers (Kumar & Nigmatullin, 2011). Finally, consumers are extremely serious when it comes to quality and safety issues (Georgiadis, Vlachos, & Iakovou, 2005; Enderwick, 2009). Hence, it seems that the dairy supply chain management is not limited to the various facets of production, but entails a wider range of management objectives (Holzworth, et al., 2014; Moore, et al., 2014).

The key level of the whole chain is, nonetheless, the farm (Andersen, Elbersen, Godeschalk, & Verhood, 2007), where profit maximization, efficiency of utilization of resources and competition due to globalization (O'Hara & Stagl, 2001); (Marletta & Biere, 2009; Castelán-Ortega, et al., 2016). However, managing even small aspects of this whole process is a complex task, and the farm manager must into account both quantitative and qualitative aspects of the operation (and the whole chain), properly evaluate the current state of the entire system and timely anticipate potential evolutions in the future with bounded knowledge (Snow & Lovatt, 2008).

Yet, there is a scarcity of research concerning the management of a farm in relation to the entire dairy supply chain; models and applications are either focused on the macroscopic view of the entire chain or are concerned only with the automation of the production process (Minegishi & Thiel, 2000). As a result, there is the need of development of decision aid tools that take into account the management of the farm, how the decisions made at that level affect the entire chain and how changes in the chain affect the management and of the farm and finally be user friendly. Moreover, there is a gap on how unexpected events (such as milk adulteration) affect the behavior of the entire chain.

The purpose of this chapter is to propose such a tool with the deployment of the System Dynamics methodology, investigate the interdependencies in the dairy supply chain with a special focus on milk and gain insights into supply chains in uncertain and high-risk environments, especially when an extreme event influences the demand for milk. The rest of the chapter is structured as follows: in section 2 the methodology is presented, with a description of the complexity that is inherent in all forms of supply chain, the method that will be used and the structure of the developed model. In section 3 scenarios, simulations and results are presented and analyzed and finally in section 4 there is a discussion on the merits of the proposed methodology and model and directions of future research.

METHODOLOGY

The problems of the dairy supply chain that were mentioned in the introduction, demonstrate that the situation is not only complex, but also dynamic; more than one decisions may be necessary to achieve the goal, the decisions are interrelated and there is feedback from the decision that may be the result of the decision or not (Sterman, 1989). As a result, many attempts to apply a policy in such an environment often fail, despite being well intended. The reasons for those failures can be found is the misunderstanding on the part of the manager of the relation and the nature of the structure of the system-under-study and in particular its connection with the environment (Otto, 2008). Moreover, the environment does not stand in isolation; other stakeholders may have interests and react to the applied decision. Finally, the reasons that policies/decisions fail is that they are applied based on the notion that cause and effect are closely linked not only in time but also in space. As a result, traditional models are committed to linearities that lead to unrealistic assumptions and are especially concerned with equilibrium, a state that is hardly ever met in real-world systems (Forrester, 2003). Moreover, they fail to include and account for feedback mechanisms, time delays and a general misunderstanding of how the phenomena arise (Tsaples & Armenia, 2016).

This inability to solve problems with the traditional tools led to the rise of Systems Thinking, which supports that the world-view must be one of a complex, whole system, in which every element is connected to the other and can have an effect on the performance of the whole (Ackoff, 1994). Thus, Systems Thinking promoted a more holistic approach to problem solving, trying to identify where those points in the system's structure that generate the most problematic behavior could increase the potential of a policy (Sterman, 2000). To achieve this type of holistic approach Senge (2006) proposed the use of mental models. Mental models are assumptions, generalizations or even pictures of how a manager understands the world and consequently they affect the decision-making process that he/she is using simply because mental models are easier to work with than their real-life counterparts (Ford, 1999). However, mental models are not adequate enough to help explain the behavior of complex systems.

SYSTEM DYNAMICS

System Dynamics (Forrester, 1997) is a computer-based methodology that quantifies Systems Thinking in understanding the dynamic behavior of complex systems over time (Sterman, 2000). Contrary to traditional tools, System Dynamics avoids trying to "forecast" future behavior based on past data and assumptions (Lyneis, 2000); its main goal is to provide insights into how a system's behavior emerges and use this understanding for a more efficient decision-making (Santos, Belton, & Howick, 2002). The building block of the methodology are stocks, flows, feedback loops and time delays, all of which are inherent in socioeconomic systems (Sterman, 1989). Stocks represent the state of the system and are considered its "memory".

They are increased by inflows and decreased by outflows and their mathematical representation is provided by the equation:

$$Stock(t) = Stock(t0) + \int_{t0}^{t} \left(Flow1(s) - Flow2(s) \right) ds$$

A feedback loop is a closed chain of relationships among various elements of the system under study (García, 2006). For example, if there is a system with three variables (A, B and C) and a change in a variable A causes a direct positive change in a variable B, variable B causes a direct positive change in a variable C and variable C causes again (after a time delay) an opposite change in A, maybe even cancelling the initial change of A (negative feedback loop). Feedback loops generate non-linear behavior, whether the relations among the variables are linear or not (Pruyt, 2013) (see Figure 1).

In complex systems, feedback loops are rarely this simple. However, the combinations of those simple and primary loops generate interesting behaviors that can be met in various real-life systems (Senge, 2006). Thus, System Dynamics is focused on understanding those behaviors and how the system's internal structure can generate them (Schwaninger & Rios, 2008). Moreover, the holistic, top-down approach used by the methodology allows the generic representation of different classes of systems in one model that can be adapted to account for the individual attributes of any system in the class (Forrester J. W., 2003). Furthermore, it allows for the inclusion of aspects of human behavior, even if these aspects are not easily defined and/or quantified (Tsaples & Armenia, 2016). Finally, System Dynamics is easy to communicate, which favors the participation of citizens and non-experts to the modeling process.

System Dynamics has been used extensively in various aspects of the supply chain management, such as inventory decision and policy development, demand amplification, capacity planning of remanufacturing networks, closed-loop supply chains with recycling, bullwhip effect and oscillations, international supply chain management, demand forecasting, etc. (Coppini, Rossignoli, Rossi, & Strozzi, 2010; Georgiadis & Athanasiou, 2010; Özbayrak, Theopisti, & Akgun, 2007; Smith, 2005; Saeed, 2008). For the food supply chain in general and the dairy chain in particular Nicholson & Kaiser (2008) studied the effects

Figure 1. Example of a negative feedback loop

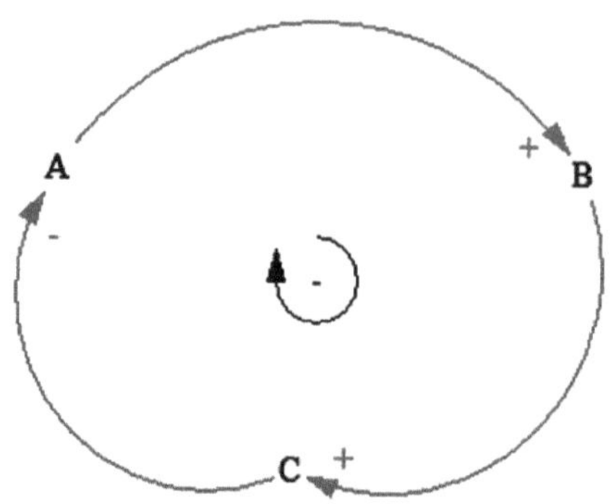

of generic advertising to the dynamics of the whole chain. Minegishi & Thiel (2000) demonstrated how System Dynamics can be applied to a particular food supply chain. Schimith, et al. (2015) studied how genetic improvement can affect the milk supply chain. Non-perishable products were studied by Kumar & Nigmatullin (2011), while Georgiadis, Vlachos, & Iakovou (2005) defined the general framework for using System Dynamics for the strategic management of food supply chains. Finally, an extensive overview of the simulation methods that have been in the dairy supply chain can be found in Kleijnen (2005) and Li, Zhang, & Jiang (2008).

MODEL STRUCTURE

The following paragraphs provide a top-level description of the system of the milk supply chain with a description of the main variables. The simulation model includes only milk produced by cows and does not include other animals. Furthermore, the data and the structure of the model correspond to a generic supply chain in Greece. As such, no imports and exports are included in the model. The dairy supply chain model consists of three echelons: the farms where the milk is produced, the processors where the milk is tested and processed and the retailers. At the level of the farm, the number of productive cows determines how many new animals will be born; the higher number of existing cows the higher the number of new births. Similarly, the higher the number of newborns the larger the potential number of productive cows, after a maturity period. No imports or exports of animals were considered for the model. Similarly, the higher number of productive cows the larger the quantity of the milk that will be produced under normal circumstances. After the milk is produced, it spends a number of hours to a cooling facility, after which it is ready to be transferred/sold to the processors. The first echelon of the milk supply chain, the main variables and the loops that are formed are illustrated in the red rectangle in Figure 2.

The second echelon starts with the testing of the milk. If the quality of the milk is found to be at the appropriate levels, then it proceeds to the processing phase where it is prepared to be in a shape suitable for selling (for example bottles or cartons). In the next echelon, the final products are transferred to the retail centers, where they are placed on the shelves (after a delay) and finally sold to the consumers based on the demand (blue rectangle of Figure 2).

Figure 2. General causal loop diagram of the milk supply chain model

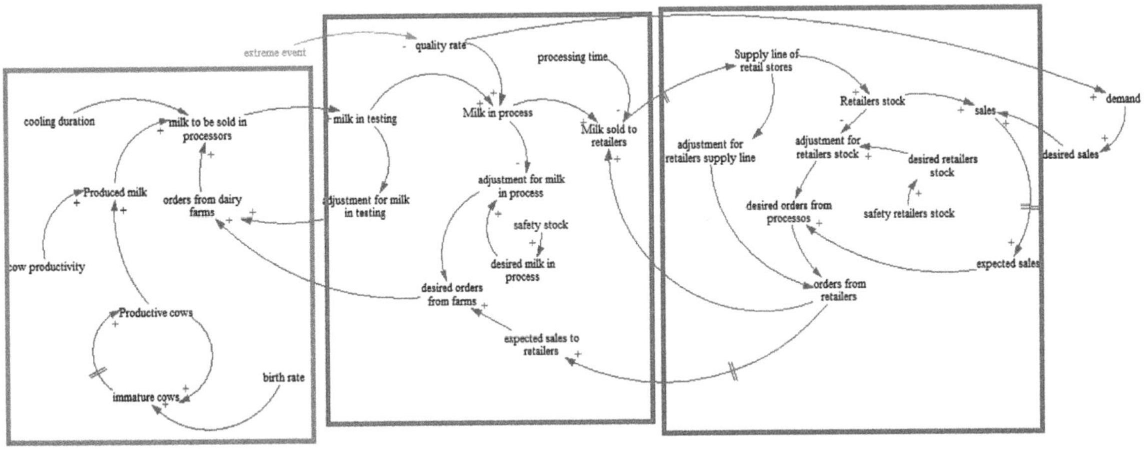

The feedback loops in the system and the origin of the oscillation in the chain are formed when the information flows the opposite way than the product flow. Demand and the actual sales determine the expected sales that the retailers use to place their orders (after an adjustment for their current stokes and a time delay which they use to "forecast" their needs) to the processors. Similarly, the processors sell to the retailers based on the orders they receive and their actual stock of milk. The information on the sales to the retailers are used to calculate/"forecast" their demand for milk from the farmers. Thus, farmers sell milk to the processors based on what is their productivity and the actual orders from the processors. Farmers are the ones that have the least control of this information flow; the milk that is produced depends only on the number of productive cows and what is the productivity of the animals. They could act to boost the productivity or decide on the number of animals on their farms. However, the abstraction of the model that is being developed would not represent accurately at this bottom level of the chain. Thus, imports and decision on increasing the animal productivity are omitted from the model.

Finally, to investigate the behavior of the entire supply chain in the face of an extreme and critical event, a variable is introduced named "extreme event". This variable affects straight the quality rate at the level of processors. As soon as it is discovered that the quality of milk is not at an appropriate level, the milk is discarded. After a time delay, the variable also has an effect on the demand at the other end of the chain (i.e. the consumers). The extreme event is modeled in the form of a graphical function that changes during the simulation time.

Figure 3 illustrates the form of the graphical function. On the *x-axis* there is the simulation time (150 days) and on the *y-axis* there is the value of the extreme event. In the particular example, as the simulation begins, there is no extreme event happening in the chain (value of 0). When the time reaches the point of 20 the value begins to rise, reaches its maximum value almost at the half of the simulation time and then descends again towards 0. The value of the extreme event can be thought of as a percentage of the effect that it will have on the demand for milk. Thus, even though it is completely qualitative, it can be included in the model to investigate the effect of the demand to the entire chain. The variable will be used to generate different scenarios and study the behavior of the entire supply chain. The data that was used for the calibration of the model was from the case of Greece (see Table 1).

Figure 3. Graphical function of the extreme event

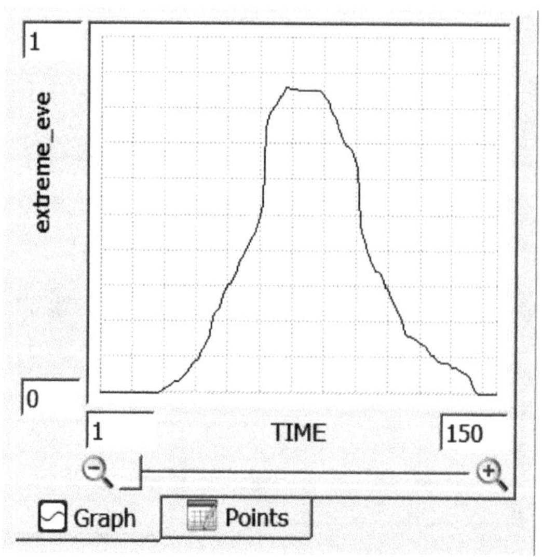

Table 1. Data for the milk supply chain in Greece

Number of Dairy Cows	135000 [animals]	(EUROSTAT, 2015)
Milk Production	769000 [tones/year]	
Yield	4553 [kg/(cow*year)]	
Total Domestic Consumption	2147000 [tones/year]	
Imports	0	

User Interface

The model was developed and simulated with the software ithink[1], which offers the possibility to create an interface that increases the usability of the underlying model in a friendly manner without losing any of the insights that would be gained by the actual model and analysis. Such an interface can be used by all stakeholders in the chain to estimate future states of the system and test different scenarios and different policies. Thus, the decision-making process can become more effective, the stakeholders can have an overview of how every decision at every level can affect the whole supply chain and finally the interface allows for a better communication and justification of the decision that was taken.

For the present study, the focus of the interface will be on the information flow that moves contrary to the milk flow; the time of the decision to place the orders to the previous echelon of the chain and the safety stock that is considered appropriate. Furthermore, the shape of the extreme event will be one of the elements of the interface, so that the decisions can be tested against an array of different scenarios. Finally, another decision is added which takes the form of a switch: during a scenario, if the switch is activated, all the stocks that contain milk in every echelon will be emptied. The rationale behind the decision is to investigate how the entire chain will be affected if it decided that the entire milk will be discarded due to low quality or a disease (a possible interpretation of the extreme event).

Figure 4 depicts the user interface and the variables that can be manipulated by the end-user.

SIMULATIONS AND SCENARIOS

The following paragraphs describe the scenarios that were performed with the model and a sensitivity analysis and a sensitivity analysis with the variables that are in the interface to investigate their importance in an array of possible future states. The model was simulated for a time of 150 days with the data of Greece and a calibration of the variables for which values could not be found.

No Critical Event

Firstly, the model is tested for the default values and without any critical event. Thus, the variable "extreme" event is a straight line with a value of 0 for the entire simulation time. For the processors, the adjustment time is set at 4 days; every 4 days there is a calculation of the current stock of milk added to the safety stock and an estimation of future demand by the retailers and the orders are placed to the farmers. Similarly, the time is set to 3 days for the retailers; due to an immediate transaction with the consumers, the retailers must place orders faster than the processors and with a smaller safety stock.

Figure 4. User interface of the model

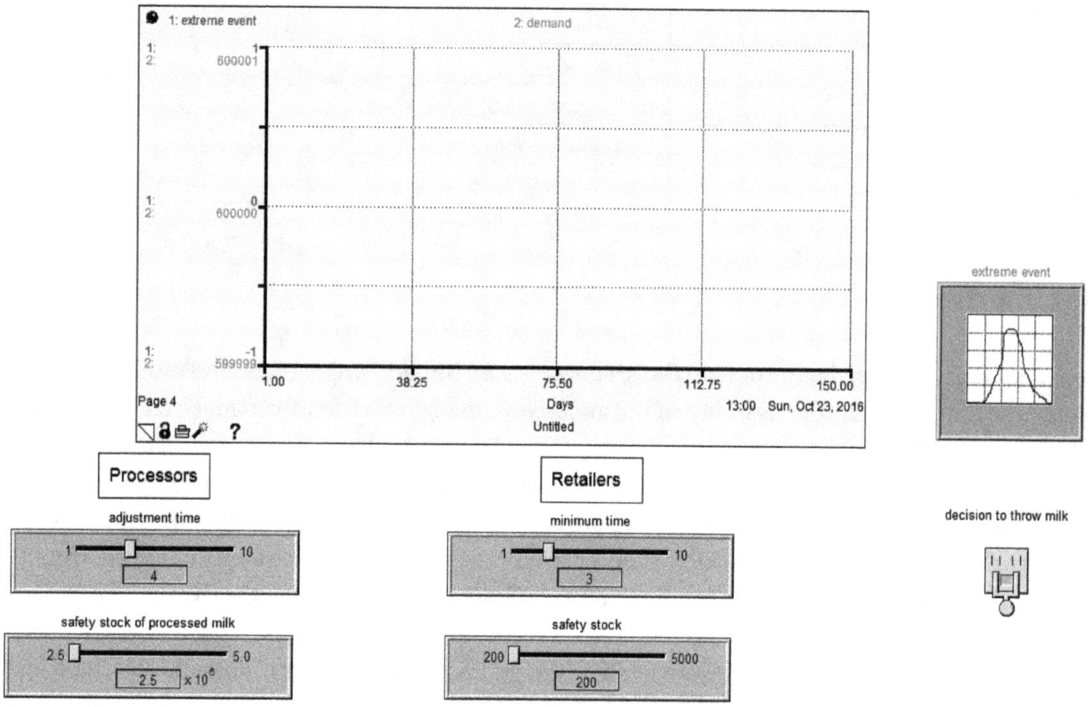

Since there is no extreme event, the demand remains steady for the entire simulation time. Thus, the milk that is sold to the processors by the farmers is almost steady, because the productivity of the cows cannot be controlled. Thus, farmers remain with a large stock of milk that is not sold (see Figure 5).

For the model, other uses of milk were not included (for example, milk sold to companies for cheese production) and the time duration of the milk is not important because of the abstraction of the model and its continuous nature. As a result, the milk that the processors buy from the farmers increases and

Figure 5. Produced milk and milk sold to the processors by the farmers

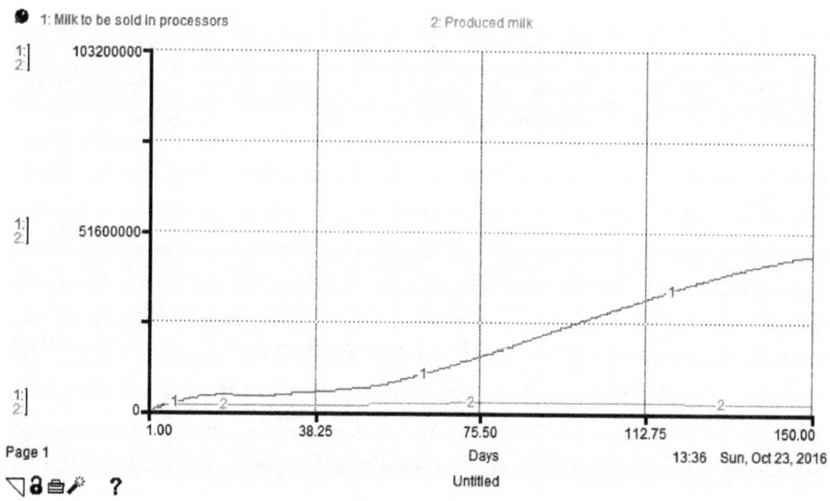

is transferred for processing where it is sold based on the orders from the retailers. In this part of the chain, small oscillations are observed in the first days of the simulation time. This occurs because of the calibration of the model, until it reaches an equilibrium which lasts until the end of the simulation time (Figure 6 left). The constant difference between the stock of milk in process and that in testing is due to the large amount of the safety stock that was used for the initial values of the model. Similarly for the retailers, the stocks that describe the behavior of their part of the system oscillate at the initial stages of the simulation and reach equilibrium at the end of the simulation time (Figure 6 right).

Thus, in the scenario without an extreme event and the default values of the model, the behavior of the entire system is stable and predictable. Apart from small oscillations in the beginning of the simulation time, the model stabilizes around the values that are observed in the case of Greece; with a steady demand the milk that is produced locally covers the needs of the consumers. The differences between the stocks of the various sub-sectors (processors and retailers) are attributed to the height of the safety stock that has been attributed to each partner in the chain. The first change in the models will be performed

Figure 6. Milk in testing and in processing in the processors echelon

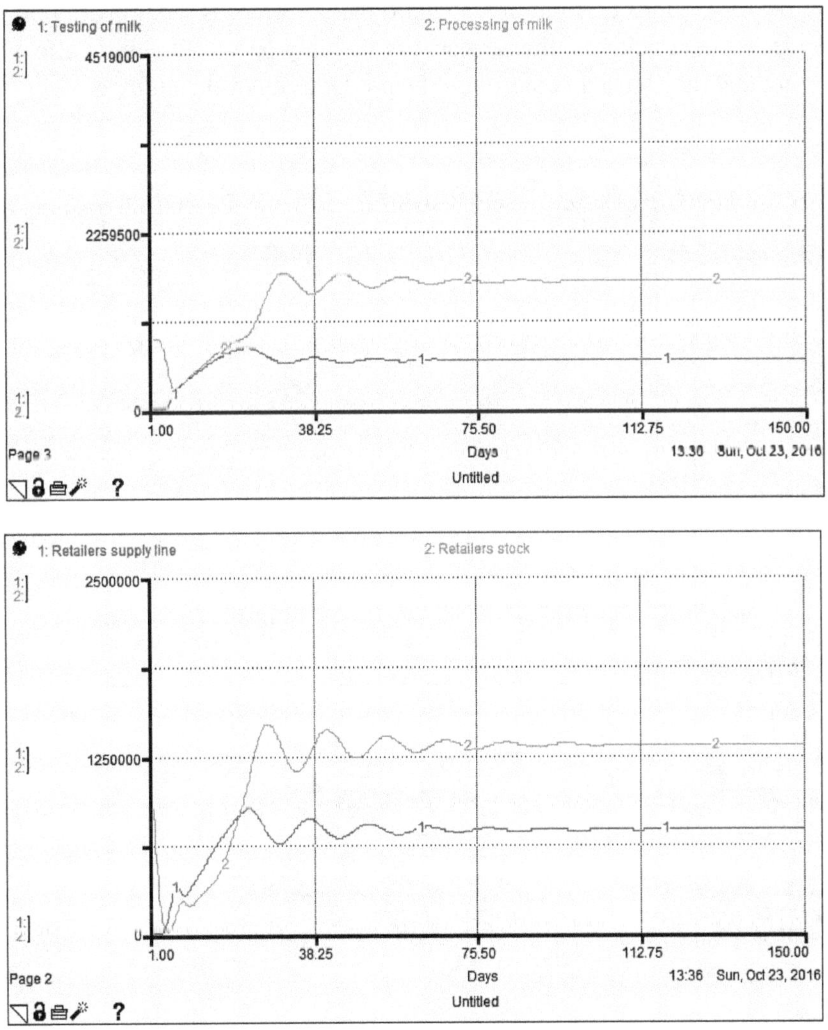

by reducing the safety stocks for the processors and the retailers and simulate the model again without an extreme event. Figure 7 demonstrates the changes in the interface.

By reducing the safety stocks to their minimum values, the system does not change behavior, neither for the processors nor for the retailers (see Figure 8). The only difference is the normal, reduced difference observed between the stocks of each sector because of the reduced safety stocks.

The next change in the model is with the times that the orders are placed; they are increased; hence processors and retailers take larger times to calculate how many orders they will need from the other parts of the chain. With these changes the model alters its behavior. The oscillations are almost eliminated and the model/chain reaches the equilibrium state faster than before. Thus, the information flow can greatly influence the behavior of the system, especially since the milk flow is more rigid (see Figure 9). What is also surprising is that for the three simulations the stocks that describe the farmers' structure remain unchanged. Hence, the echelon of the farmers is influenced only by the final demand and not by the next echelon in the chain, especially if there is no fluctuation in demand.

Conclusively, without an extreme event that will affect the final demand, the model is stable. It reaches an equilibrium state in the early stages of the simulation. The important aspects of the model are the times that determine the information flow of the chain; the larger they are (the larger time to calculate and place orders to the previous echelon in the chain) the smoother is the transition to the equilibrium state. Finally, under these conditions, the farmers are the only parts that cannot affect the behavior of the chain, assuming that they have no control on the productivity of the animals.

Figure 7. The interface and the changes that are made for the new simulation

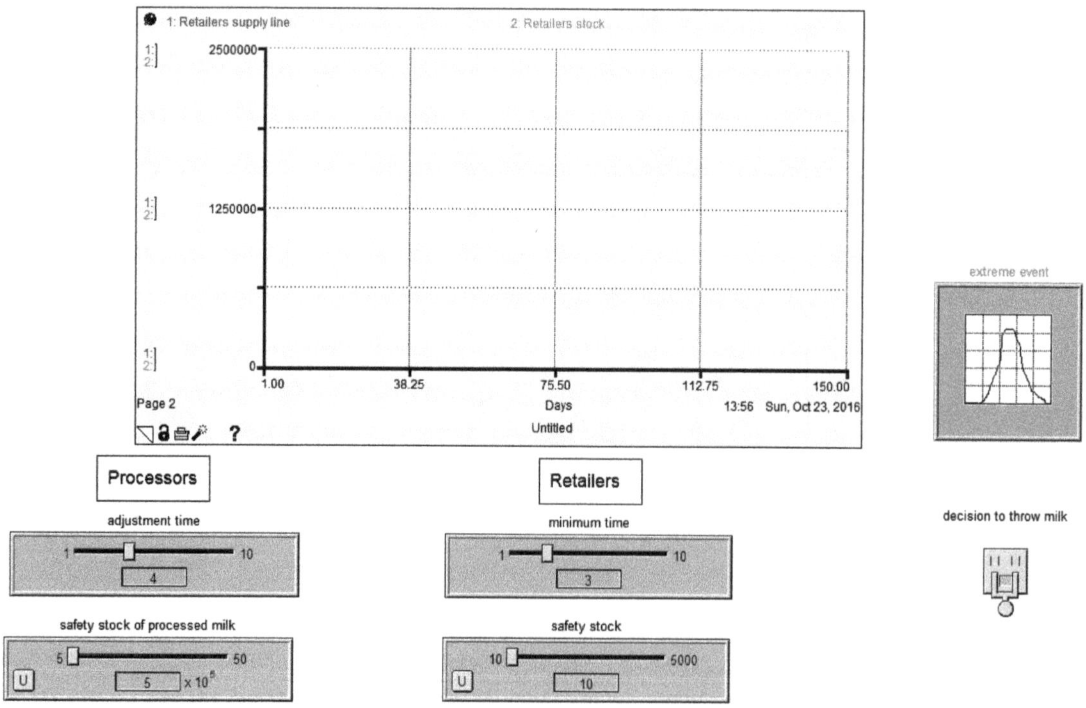

Figure 8. The main stocks of the processors (left) and retailers (right)

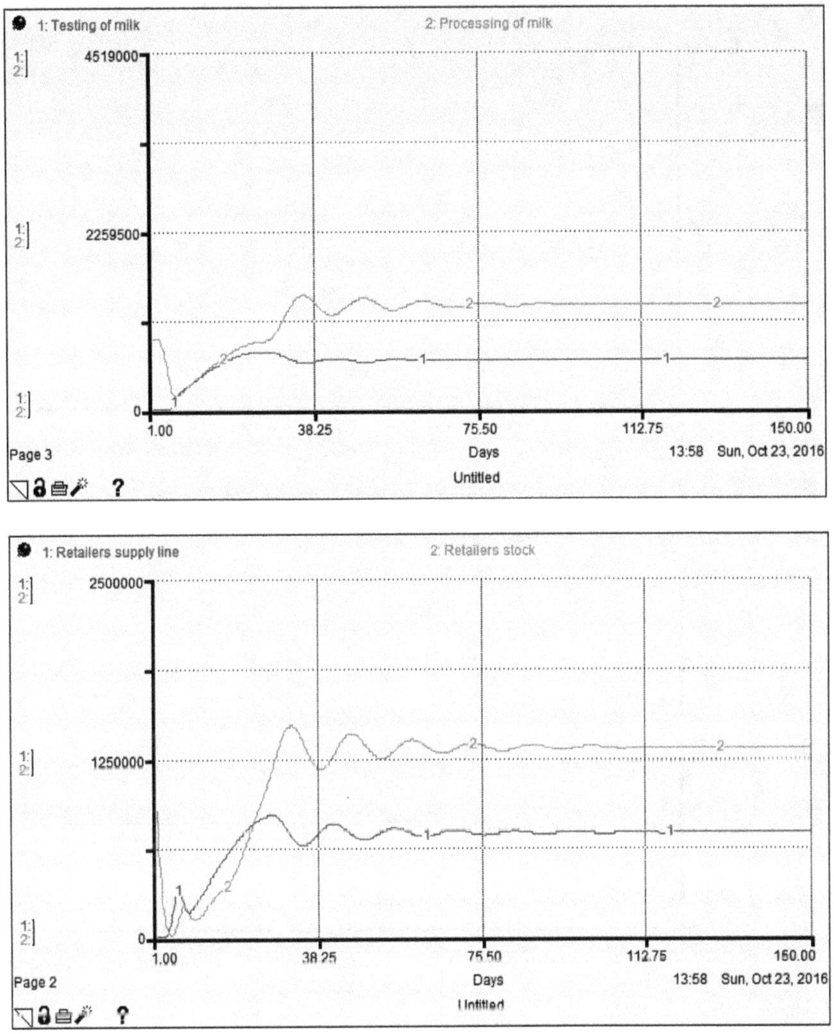

SENSITIVITY ANALYSIS OF THE SCENARIO WITHOUT AN EXTREME EVENT

To test how the variables in the interface affect the behavior of the entire chain under different sets of values, a sensitivity analysis with 50 simulations was performed in various combinations of the four variables.

For the processors, it can be seen from Figure 10 that the behavior depends heavily from the particular variables because the system generates different types of behavior. There are cases/simulations where the stock of milk in process reaches equilibrium without oscillations (similar to the case where there were large adjustment times for orders), there are simulations were small oscillations are observed before the equilibrium state and finally there are simulations where the system never reaches an equilibrium. The magnitude of those oscillations depends not only on the adjustment times but also on the volume of the safety stock. Similar behavior is observed for the Retailers stock.

Figure 9. The main stocks for processors (left) and retailers (right) for increased times of orders' calculation

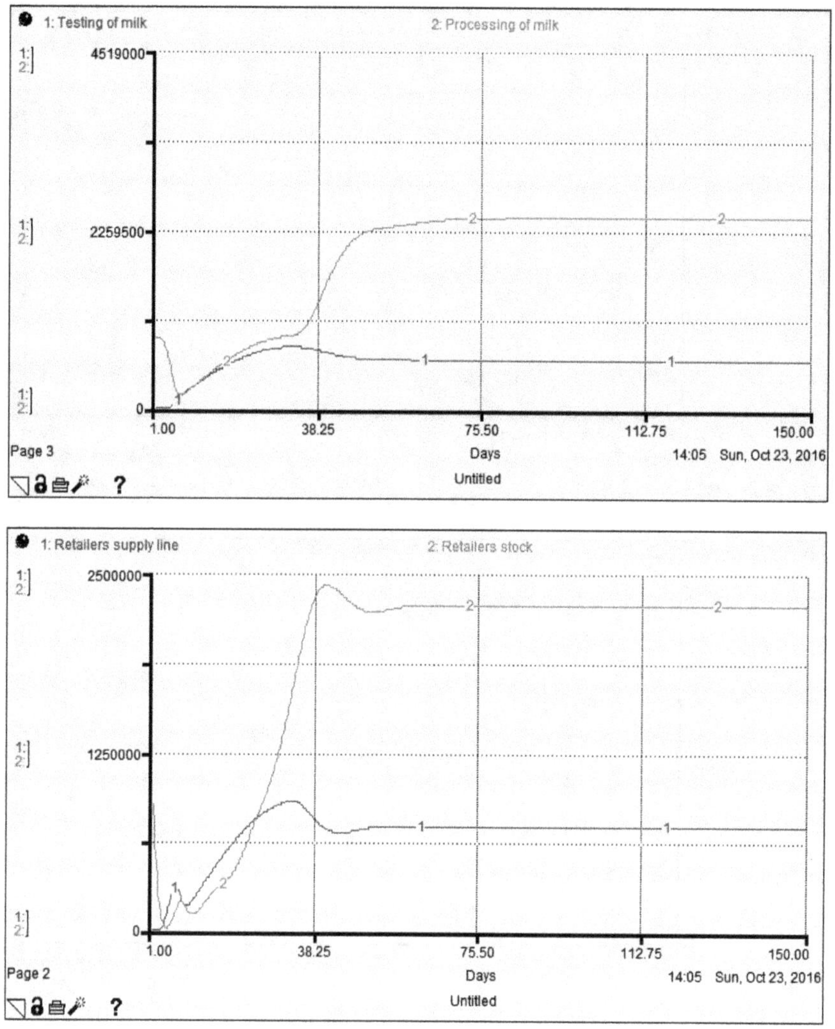

Figure 10. Results of the sensitivity analysis of the processing of milk stock

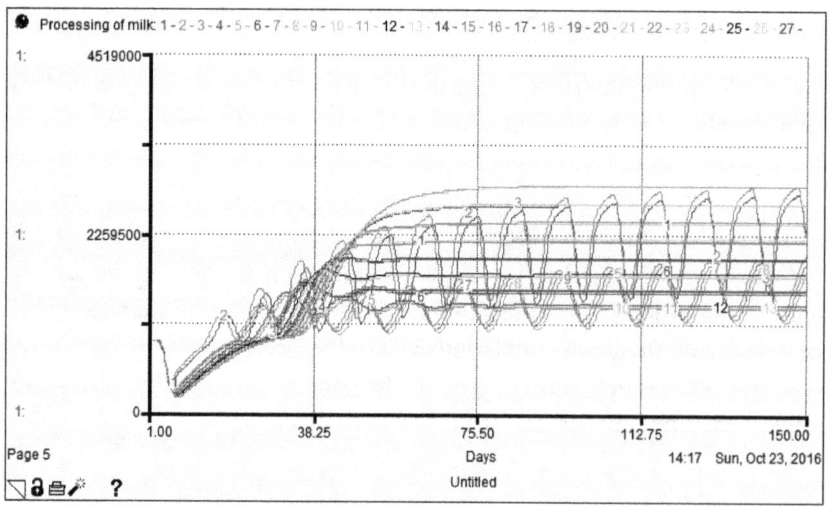

These oscillations influence the finances of the sub-sectors, which are modeled implicitly with the sales to the next step in the chain. For example, the sales of the retailers to the consumers escape the stability of the constant demand and can be reduced up to 50% of the sales under default values (Figure 11 left). These oscillations have also an effect on the farmers and their finances. Since the behavior of the other parts of the chain oscillate, the sales of milk from farmers to processors are also affected (Figure 11 right). The only difference is that in the model have no power to control their decisions and the behavior of their sector.

In conclusion, the sensitivity analysis demonstrated that the variables that determine the information flow of the chain can have a great effect on the system's behavior even when the demand for milk is constant. This effect is not limited to their respective sectors, but propagates through the entire chain, affecting even the farmers, who have no power in this process.

Figure 11. Sales from the retailers to the end consumers (left) and sales from farmers to the processors (right)

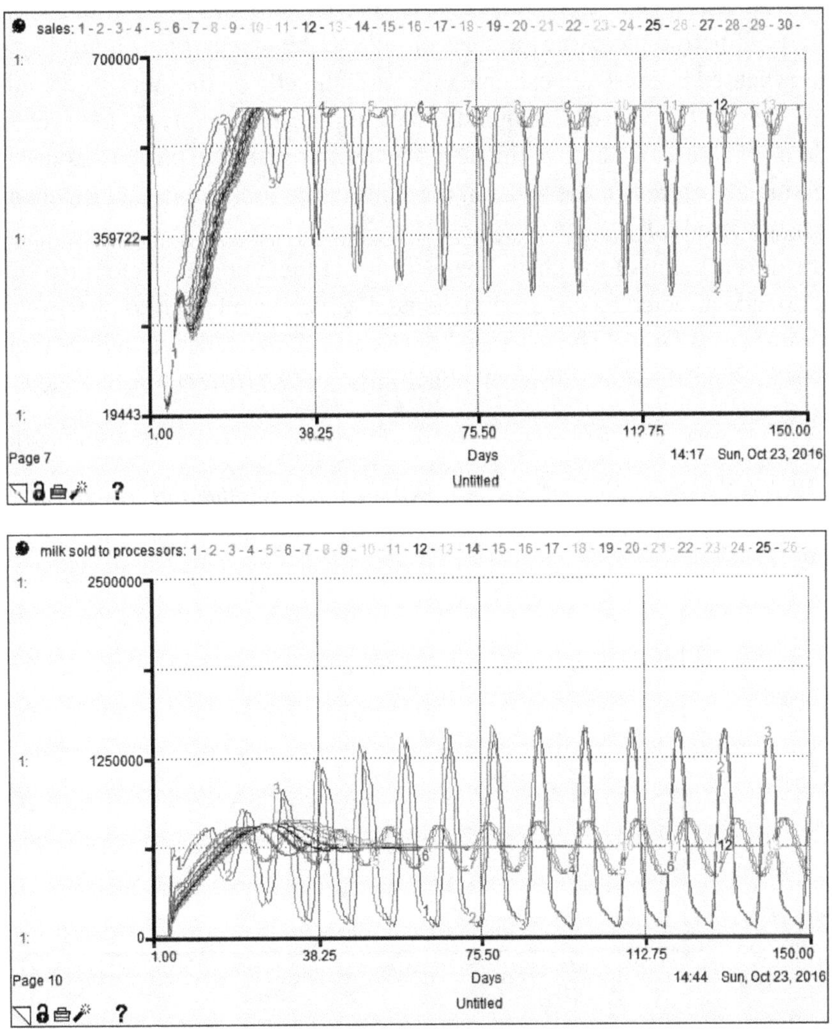

EXTREME EVENT

The shape of the graph of the extreme event was presented in Figure 3. The extreme event begins around 20 days into the simulation day, reaches its peak 60 into the simulation after which it slowly starts to descend.

Figure 12 illustrates the effect that the extreme event has on the demand. Despite the initial increase, the fall in demand does not occur instantly and at a great rate. However, as the extreme event unfolds with greater intensity, so does the decrease in demand. The steep decrease in demand, though, begins only when the magnitude of the event approaches its peak value (blue circle). This is because of the delay in the information transmission; the consumers reduce the demand as the news about the event unfold and the greater the magnitude of the event the bigger the volume of the news about it. This delay is present also when the event begins to withdraw: the ascent to the demand begins only when the magnitude of the event is half way to the end (red circle). After that, the demand gradually increases until it reaches its original levels at the end of the simulation time.

With the change in demand, the stocks that determines the volume of milk that can be sold to the processors by the farmers, changes behavior. There is a drop in the stock because processors ask for more milk, which for a brief period increases the sales of milk to the processors (Figure 13 right).

The behavior is caused because the extreme event acts directly on the stock of the milk in testing; as the magnitude of the event increases, the quality to the mil decreases forcing the processors to discard milk (blue circle on Figure 14). Thus, in the next batches of orders the processors need to compensate for the discarded milk. As a result, the spike in the sales of the milk from the farmers occurs when the demand has already started falling.

Moving to the next echelon of the chain, the behavior of the retailers' sector also changes behavior. Firstly, they are directly affected by the drop in demand, which results in a smaller volume of sales and an increased stock of unsold milk. The increased stock causes the retailers to reduce their buys from the processors in the next cycle, thus the transfer of milk from processors to retailers is reduced (see Figure 15).

Figure 12. Shape of the extreme event (blue line) and the effect on demand (red line)

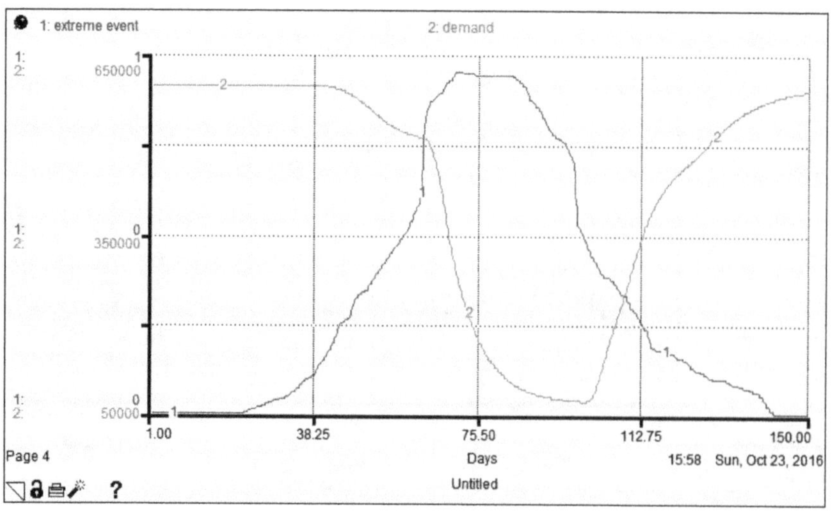

Figure 13. Produced milk and milk to be sold in processors under an extreme event (left) and milk sold to the processors (right)

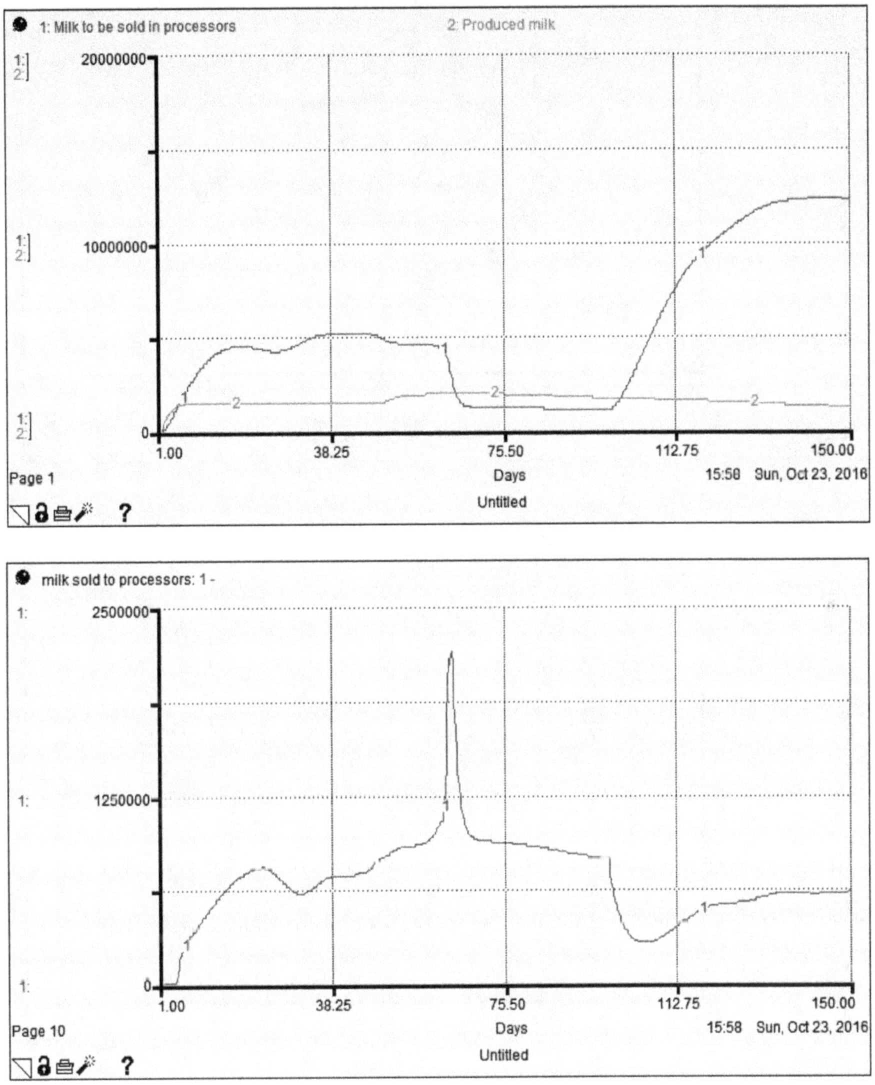

If the times to place the orders for processors and retailers increased, then the milk to be sold by farmers to processors generates almost the same behavior with the default values, but the oscillations are minimized (see Figure 16).

The oscillations are also smoothed for the processors, however the larger adjustment times result in bigger orders, especially when the demand starts to increase again. As a result, the processors end up with a higher stock of processed milk that is not sold to retailers (Figure 17 left). Similar behavior is observed in the retailer's stock, only the oscillations are not smoothed out completely, but the most intense ones remain (Figure 17 right).

Hence, the increased times that the processors and retailers calculate the volume of their orders plays a crucial role also when an extreme event occurs in the chain. Nonetheless, contrary to the scenario with

Figure 14. Milk in testing and process in the scenario with the extreme event

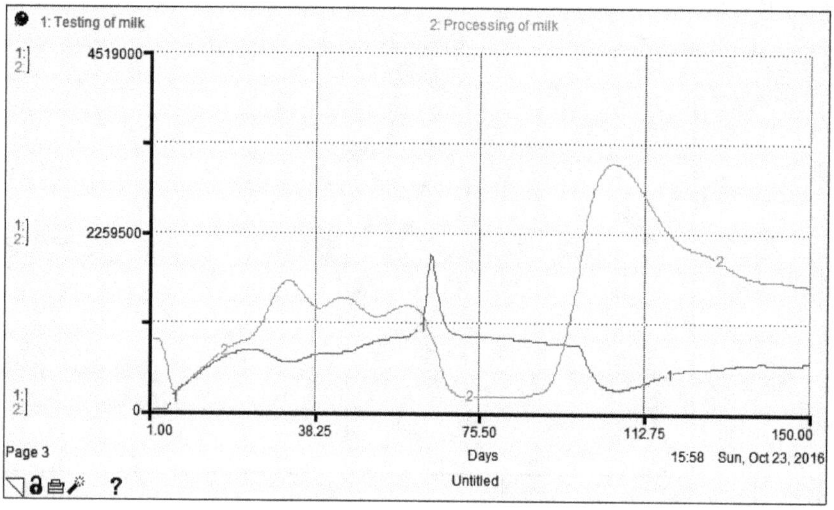

Figure 15. Retailers' supply line and stock of milk for the extreme event scenario

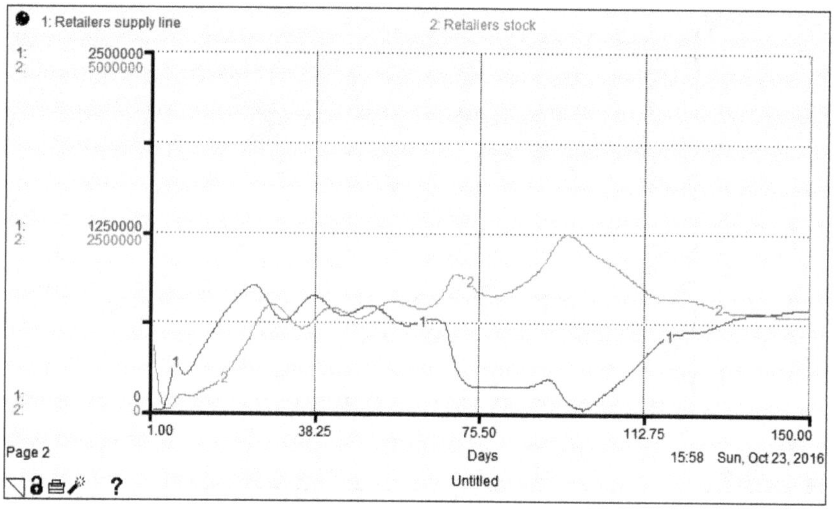

no event, there is also an unwanted consequence in the form of increased stock of (unsold) milk during the period when the demand starts to increase again (end of the simulation time). This fact provides a valuable insight into the function of the entire chain. Under normal circumstances big delays are beneficial to processors and retailers but in uncertain periods adaptiveness seems to be the most correct state of mind. The adjustment times that are necessary for the calculation of orders, should constantly be evaluated because demand is constantly changing. By having constant adjustment times for the entire period of uncertainty (for example, during an extreme event), fluctuations appear in the entire system, causing even more difficulties than already present due to the extreme event.

One decision that is always in the minds of decision-makers during such periods is the withdrawal/discard of products. Thusly, in a new simulation this decision is represented by the switch in the interface

Figure 16. Comparison of milk to be sold in processors by farmers in the extreme event scenario

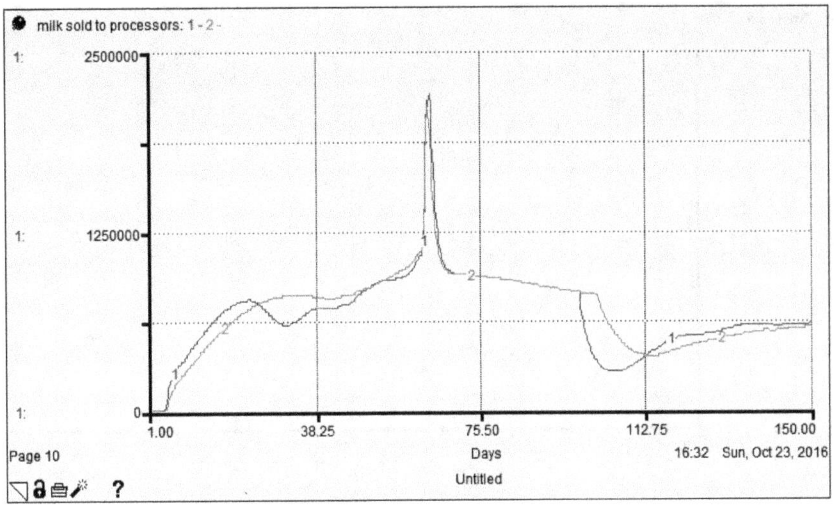

(Figure 4), which is activated for a period of time, when the magnitude of the event is at around its peak and deactivated when the situation begins to return to normal levels.

Thus, the stock of milk that can be sold to processors by the farmers is very low compared to the previous simulations of the extreme events scenario (Figure 18 left). This results, of course, in reduced sales by the farmers, although they increase above the levels of the previous simulations once demand begins to increase (red circle Figure 18 right). Consequently, the decision to discard milk is beneficial for the farmers in the long run.

The situation is similar for the processors, however, the sales to the retailers are more oscillatory during the last days of the simulation time. As a result, the decision to discard milk might create several problems to the processors, although the stocks that describe their sector do not present as many fluctuations as in the other simulations (see Figure 19). Consequently, the decision-makers of the sector must find the appropriate trade-offs between the sales and the smooth operation of the sector; thus, the extreme event in combination with the decision to discard/withdraw products, creates an extra level of complexity in the decision-making process of the processing sector.

The stock of the retailers reaches its lowest levels with the decision to discard milk. However, the timing of the decision results in the absence of fluctuations for the stock (Figure 20 left).The worst effect occurs in sales to the consumers (Figure 20 right). During the period when demand falls to its lowest levels, the sales remain even lower than the previous simulations in the extreme event scenario, although they return to their normal values at the end of the simulation time.

In conclusion, the decision to discard milk loses its attractiveness as we move upwards in the supply chain; it is beneficial to the farmers, beneficial under certain conditions for the processors and has the worst impacts on the retailers. In general, the extreme event is negative for the entire chain. The measures that stabilized the chain under normal circumstances are not a panacea in such a situation. The system is characterized by uncertain behavior and oscillations and the solution to minimize the negative effects seems to be adaptability in the parameters that determine the decision-making process (adjustment times, safety stocks etc.).

Figure 17. Comparison of the milk in processing (left) and retailers stock (right)

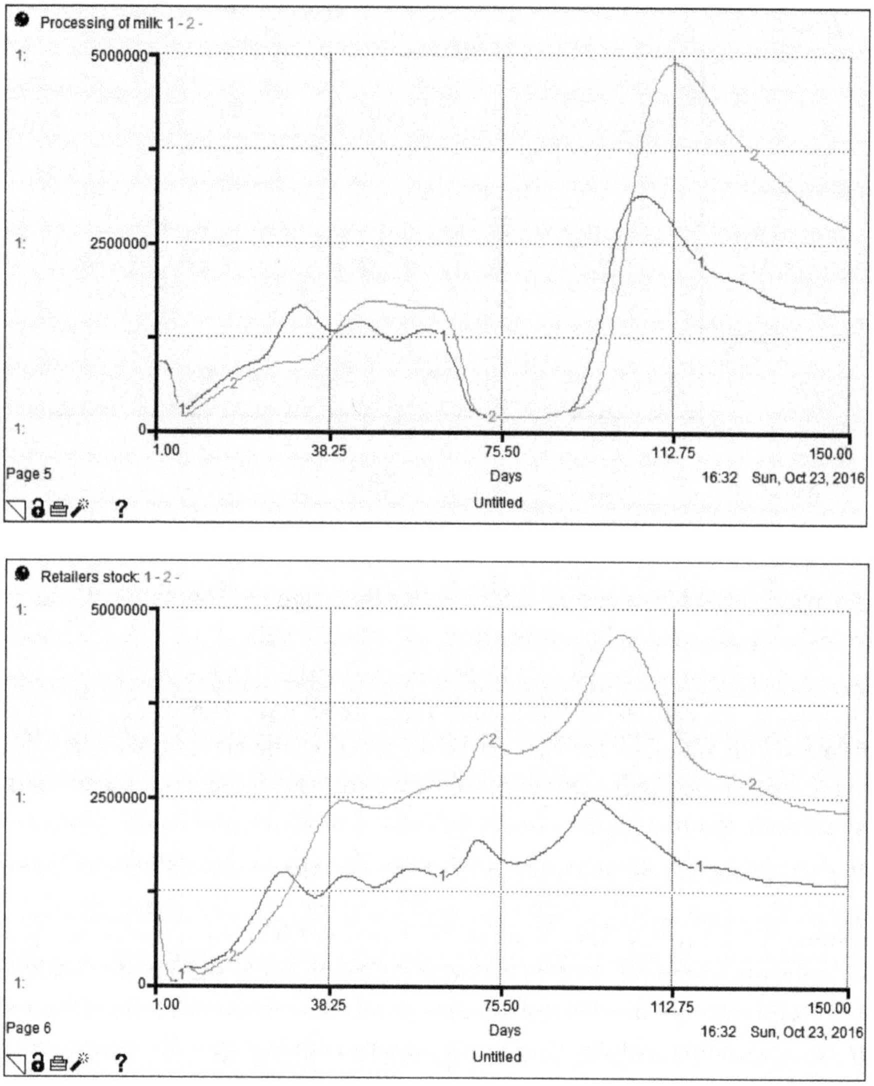

CONCLUSION AND FUTURE RESEARCH

The purpose of this chapter was the development of a System Dynamics model of the milk supply chain, the study of the interdependencies among the various echelons and the investigation of the system's behavior when an extreme event affects the demand and the quality of the milk. The developed model consisted of three echelons: farmers, processors and retailers. Without an extreme event that will affect the final demand, the model is stable. It reaches an equilibrium state in the early stages of the simulation. The important aspects of the model are the times that determine the rate of orders by the processors and the retailers; the larger they are the smoother is the transition to the equilibrium state. Under these conditions, the farmers are the only parts that cannot affect the behavior of the chain, assuming that they have no control on the productivity of the animals.

Figure 18. Milk to be sold in processors (left) and actual sales (right) with the decision to discard milk and comparison with previous simulations

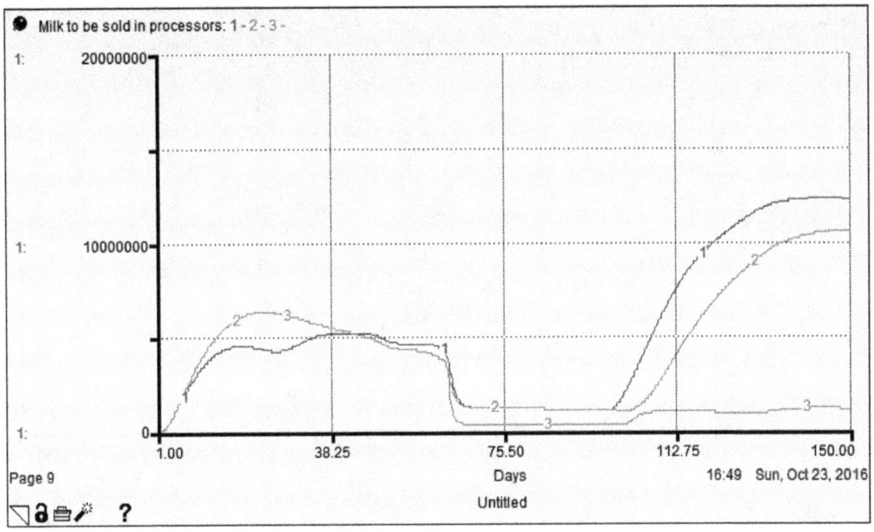

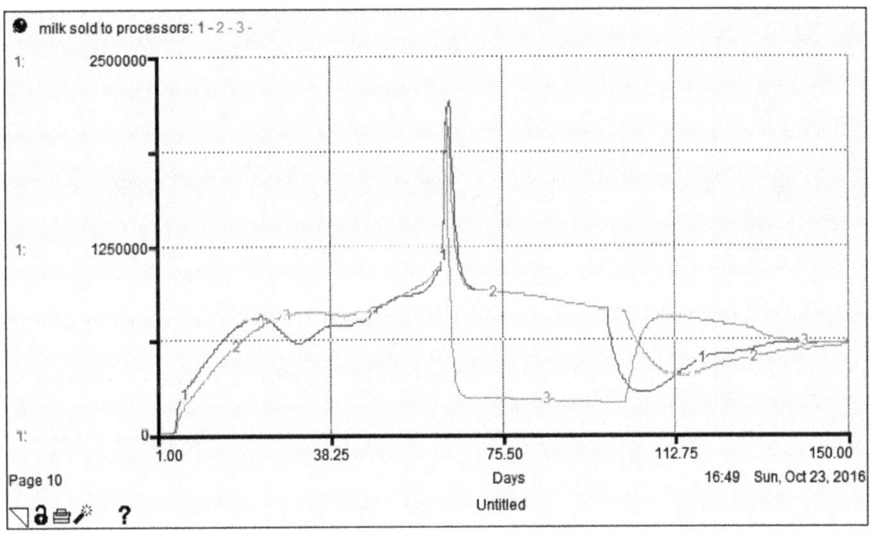

With an extreme event, which reduces the demand for milk, big delays to calculate the orders for processors and retailers create more problems in the chain. The adjustment times that are necessary for the calculation of orders, should constantly be evaluated because demand is constantly changing. But adaptiveness does not come without a side-effect; By having constant adjustment times for the entire period of uncertainty (for example, during an extreme event), fluctuations appear in the entire system, causing even more difficulties than already present due to the extreme event.

To gain further insights, a simple interface was created that can be used for an enhanced learning experience of the decision-makers of every echelon of the milk supply chain. The particular interface can be also used as a tool for communication among the different sectors in order to achieve a more robust decision-making process and collectively face uncertain situations and risks with the minimum

Figure 19. Milk in testing (left) and sales to the retailers (right) for the various simulations of the extreme event scenario

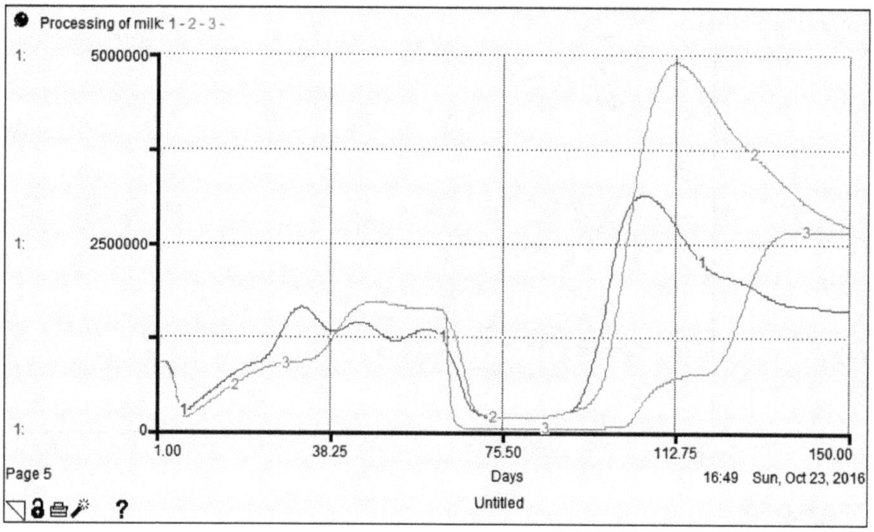

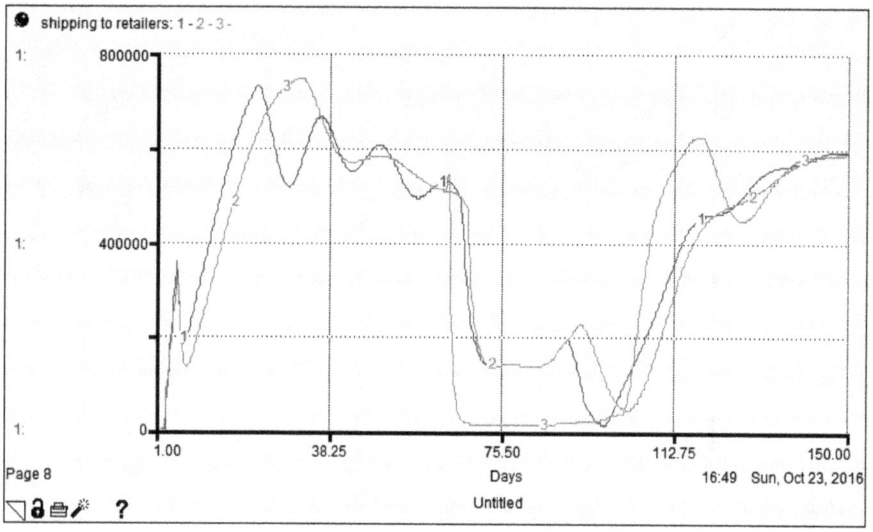

loss of gains. Several scenarios were simulated that demonstrated that the information flow affects the behavior of the entire system. This flow is determined by the individual decision-making processes of each sector and the simulations showed that a more robust process/flow results in decreased oscillations and more stability for the whole supply chain. Thus, a more collaborative approach by the various partners could be beneficial for all. In conclusion, the model-along with its interface- can be used for the strategic management of every echelon of the milk supply chain. Using it in combination with other programs/pieces of software that handle the day-to-day operations, a manager can gain a more thorough view not only for the firm he represents, but also for the whole chain.

Figure 20. Retailers stock (left) and sales to the consumers (right) for the various simulations of the extreme event scenario

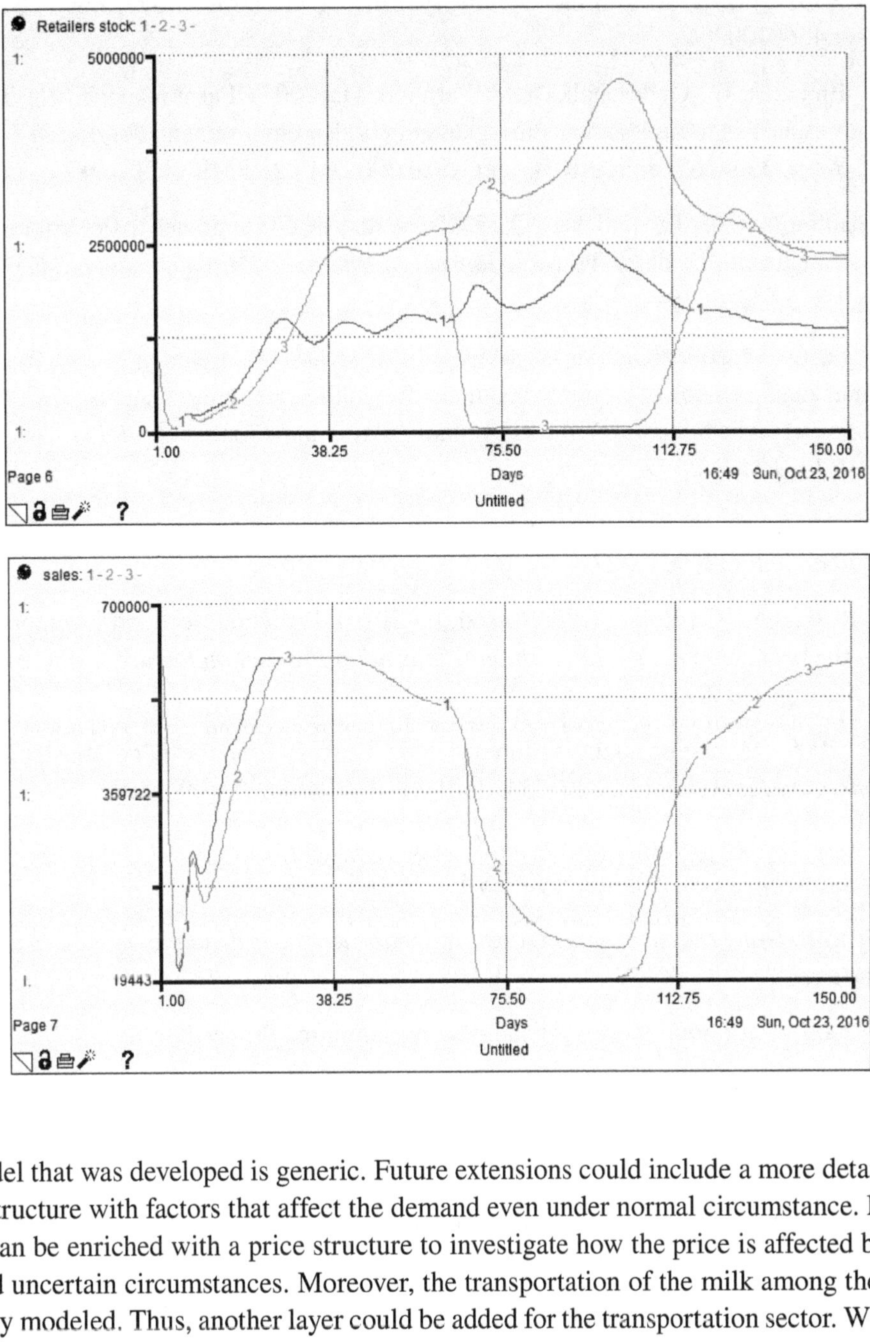

The model that was developed is generic. Future extensions could include a more detailed demand/ consumer structure with factors that affect the demand even under normal circumstance. Furthermore, the model can be enriched with a price structure to investigate how the price is affected by changes in demand and uncertain circumstances. Moreover, the transportation of the milk among the sectors was not explicitly modeled. Thus, another layer could be added for the transportation sector. With regards to the analysis of the simulations, System Dynamics can be integrated with statistical or machine learning techniques to investigate which factors affect the variables of interest, in what way and which patterns remain unchanged under a wide array of scenarios and simulations. Thus, a robust decision-making process could be designed that would incorporate the elements that maximize profits under normal circumstances and minimize the negative effects of an uncertain event for all partners in the supply chain.

REFERENCES

Ackoff, R. L. (1994). Systems thinking and thinking systems. *System Dynamics Review*, *10*(2-3), 175–188. doi:10.1002/sdr.4260100206

Andersen, E., Elbersen, B., Godeschalk, F., & Verhood, D. (2007). Farm management indicators and farm typologies as a basis for assessments in a changing policy environment. *Journal of Environmental Management*, *82*(3), 353–362. doi:10.1016/j.jenvman.2006.04.021 PMID:17126987

Ayağ, Z., Samanlioglu, F., & Büyüközkan, G. (2013). A fuzzy QFD approach to determine supply chain management strategies in the dairy industry. *Journal of Intelligent Manufacturing*, *24*(6), 1111–1122. doi:10.1007/s10845-012-0639-4

Castelán-Ortega, O., Martinez-Garcia, C., Mould, F., Dprward, P., Rehman, T., & Rayas-Amor, A. (2016). Optimal management of on-farm resources in small-scale dairy systems of Central Mexico: Model development and evaluation. *Tropical Animal Health and Production*, *48*(5), 1–8. doi:10.1007/s11250-016-1041-6 PMID:26992734

Chu, E. (2003). A review of simulation studies on supply chain management. *Journal of Academy of Business and Economics, 2*(1), 22-27.

Coppini, M., Rossignoli, C., Rossi, T., & Strozzi, F. (2010). Bullwip effect and inventory oscillations analysis using the beer game model. *International Journal of Production Research*, *48*(13), 3943–3956. doi:10.1080/00207540902896204

Enderwick, P. (2009). Managing quality failure in China: Lessons from the dairy industry. *International Journal of Emerging Markets*, *4*(3), 220–234. doi:10.1108/17468800910968391

EUROSTAT. (2015). *Agriculture, Forestry and Fishery Statistics.* Retrieved from http://ec.europa.eu/eurostat/en/web/products-statistical-books/-/KS-FK-15-001

Ford, A. (1999). *Modeling the environment: an introduction to system dynamics models of environmental systems*. Island Press.

Forrester, J. (2003). Economic theory for the new millennium. *Proceeding of the 21st International Conference of the System Dynamics Society.*

Forrester, J. W. (1997). Industrial Dynamics. *The Journal of the Operational Research Society*, *48*(10), 1037–1041. doi:10.1057/palgrave.jors.2600946

Forrester, J. W. (2003). Dynamic models of economic systems and industrial organizations. *System Dynamics Review*, *19*(4), 329–345. doi:10.1002/sdr.284

García, J. M. (2006). *Theory and practical exercises of system dynamics*. MIT Sloan School of Management.

Georgiadis, P., & Athanasiou, E. (2010). The impact of two-product joint lifecycles on capacity planning of remanufacturing networks. *European Journal of Operational Research, 202*(2), 420–433. doi:10.1016/j.ejor.2009.05.022

Georgiadis, P., Vlachos, D., & Iakovou, E. (2005). A system dynamics modeling framework for the strategic supply chain management of food chains. *Journal of Food Engineering, 70*(3), 351–364. doi:10.1016/j.jfoodeng.2004.06.030

Gereffi, G., & Lee, J. (2009). *A global value chain approach to food safety and quality standards*. Global Health Diplomacy for Chronic Disease Prevention, Working Paper Series. Duke University.

Holzworth, D. P., Huth, N. I., de Voil, P. G., Zurcher, E. J., Herrmann, N. I., McLean, G., & Keating, B. A. et al. (2014). APSIM–evolution towards a new generation of agricultural systems simulation. *Environmental Modelling & Software, 62*, 327–350. doi:10.1016/j.envsoft.2014.07.009

Kleijnen, J. P. (2005). Supply chain simulation tools and techniques: A survey. *International Journal of Simulation and Process Modelling, 1*(1-2), 82–89. doi:10.1504/IJSPM.2005.007116

Kumar, S., & Nigmatullin, A. (2011). A system dynamics analysis of food supply chains- Case studies with non-perishable products. *Simulation Modelling Practice and Theory, 19*(10), 2151–2168. doi:10.1016/j.simpat.2011.06.006

Li, W., Zhang, F., & Jiang, M. (2008). A simulation- and optimisation-based decision support system for an uncertain supply chain in a dairy firm. *International Journal of Business Information Systems, 3*(2), 183–200. doi:10.1504/IJBIS.2008.016585

Lyneis, J. (2000). System Dynamics for market forecasting and structural analysis. *System Dynamics Review, 16*(1), 3–25. doi:10.1002/(SICI)1099-1727(200021)16:1<3::AID-SDR183>3.0.CO;2-5

Marletta, V. R., & Biere, A. (2009). Entepreneurial supply chains and strategic collaboration: the case of Bagòss cheese in Bagolino, Italy. *International Food and Agribusiness Management Review, 12*(3), 49-68.

Minegishi, S., & Thiel, D. (2000). System dynamics modeling and simulation of a particular food supply chain. *Simulation Practice and Theory, 8*(5), 321–339. doi:10.1016/S0928-4869(00)00026-4

Moore, A., Holzworth, D., Herrmann, N., Brown, H., de Voil, P., Snow, V., & Huth, N. et al. (2014). Modelling the manager: Representing rule-based management in farming systems simulation models. *Environmental Modelling & Software, 62*, 399–410. doi:10.1016/j.envsoft.2014.09.001

Nicholson, C. F., & Kaiser, H. M. (2008). Dynamic market impacts of generic dairy advertising. *Journal of Business Research, 61*(11), 1125–1135. doi:10.1016/j.jbusres.2007.11.011

OHara, S. U., & Stagl, S. (2001). Global food markets and their alternatives: A socio-ecological economic perspective. *Population and Environment, 22*(6), 533–554. doi:10.1023/A:1010795305097

Otto, P. (2008). A system dynamics model as a decision aid in evaluating and communicating complex market entry strategies. *Journal of Business Research, 61*(11), 1173–1181. doi:10.1016/j.jbusres.2007.11.013

Özbayrak, M., Theopisti, P., & Akgun, M. (2007). System dynamics modelling of a manufacturing supply chain system. *Simulation Modelling Practice and Theory*, *15*(10), 1338–1355. doi:10.1016/j.simpat.2007.09.007

Pruyt, E. (2013). *Small System dynamics models for big issues: Triple jump towards real-world complexity*. Delft: TU Delft Library.

Ramanathan, U. (2013). Aligining supply chain collaboration using Analytic Hierarchy Process. *Omega*, *41*(2), 431–440. doi:10.1016/j.omega.2012.03.001

Robinson, A. (2009). Driving efficiencies in the cold supply chains. *Food Logistics*, 10-14.

Ruteri, J. M. (2009). Supply chain management and challenges facing the food industry sector in Tanzania. *International Journal of Business and Management*, *4*(12), 70. doi:10.5539/ijbm.v4n12p70

Saeed, K. (2008). Trend forecasting for stability in supply chains. *Journal of Business Research*, *61*(11), 1113–1124. doi:10.1016/j.jbusres.2007.11.005

Santos, S., Belton, V., & Howick, S. (2002). Adding value to performance measurement by using system dynamics and multicriteria analysis. *International Journal of Operations & Production Management*, *22*(11), 1246–1272. doi:10.1108/01443570210450284

Schimith, C., Scavarda, A., Bittencourt, S., Santos, M., Vacarro, G., Gabi, A., & Weise, A. (2015). The System Dynamics Use for Measurement of the Results of Technological Applications for Genetic Improvement in Milk Supply Chain. *Chemical Engineering (Albany, N.Y.)*, 45.

Schwaninger, M., & Rios, J. (2008). System dynamics and cybernetics: A synergetic pair. *System Dynamics Review*, *24*(2), 145–174. doi:10.1002/sdr.400

Senge, P. (2006). *The fifth discipline: The art and practice of the learning organization*. Broadway Business.

Seuring, S. (2013). A review of modeling approaches for sustainable supply chain management. *Decision Support Systems*, *54*(4), 1513–1520. doi:10.1016/j.dss.2012.05.053

Shepherd, T., & Flanders, A. (2008). Do Shifting Milk-Production Patterns Weaken the Dairy Supply Chain? The Case for the Southeastern US. *Journal of Food Distribution Research*, *39*(1), 130.

Simonson, S. (2009). Transforming the fresh food supply-chain. *Food Logist*, 14-15.

Smith, J. N. (2005). Specialized logistics for a longer perishable supply chain. *World Trade*, *18*(11), 46.

Snow, V., & Lovatt, S. (2008). A general planner for agro-ecosystem models. *Computers and Electronics in Agriculture*, *60*(2), 201–211. doi:10.1016/j.compag.2007.08.001

Sterman, J. D. (1989). Modeling managerial behavior: Misperceptions of feedback in a dynamic decision making experiment. *Management Science*, *35*(3), 321–339. doi:10.1287/mnsc.35.3.321

Sterman, J. D. (2000). *Business dynamics: Systems Thinking and Modeling for a Complex World*. Boston, MA: Irwin MacGraw-Hill.

Tsaples, G., & Armenia, S. (2016). Studying pension systems and retirement age: A simple system dynamics model for a complex issue. *International Journal of Applied Systemic Studies*, *6*(3), 258–270. doi:10.1504/IJASS.2016.079095

ENDNOTE

1 http://www.iseesystems.com/

Compilation of References

AAFC. (2013). *Agriculture and Agri-Food Canada, Global Analysis Division, Ethical Food Trends in Japan*. Retrieved August 17, 2016, fromhttp://www5.agr.gc.ca/resources/prod/Internet-Internet/MISB-DGSIM/ATS-SEA/PDF/6473-eng.pdf

Abate, F., Di Leo, G., Paolillo, A., & Paciello, V. (2014) An advanced traceability system based on semi-active RFID devices. *20th IMEKO TC4 International Symposium and 18th International Workshop on ADC Modelling and Testing Research on Electric and Electronic Measurement for the Economic Upturn*, 1106-1109.

Abatekassa, G., & Peterson, H. C. (2011). Market access for local food through the conventional food supply chain. *International Food and Agribusiness Management Review, 14*(1), 63–82.

Abatekassa, G., & Peterson, H. C. (2011). Market Access for Local Food through the Conventional Food Supply Chain. *International Food and Agribusiness Management Review, 14*(1), 63–82.

Abdul Kadir, E., Shamsuddin, S. M., Supriyanto, E., Sutopo, W., & Rosa, S. L. (2015). Food Traceability in Supply Chain Based on EPCIS Standard and RFID Technology. *TELKOMNIKA Indonesian Journal of Electrical Engineering, 13*(1), 187–194. doi:10.11591/telkomnika.v13i1.6919

Abdullah, N. H., Shamsuddin, A., Wahab, E., & Abdul Hamid, N. (2012). Preliminary qualitative findings on technology adoption of Malaysian SMEs. *2012 IEEE Colloquium on Humanities, Science and Engineering Research (CHUSER 2012)*.

Abello, J., Pardalos, P. M., & Resende, M. (2002). *Handbook of massive data sets*. Kluwer Academic Publishers Norwell. doi:10.1007/978-1-4615-0005-6

Acerbi, C., & Tasche, D. (2002). On the coherence of expected shortfall. *Journal of Banking & Finance, 26*(1487).

Ackoff, R. L. (1994). Systems thinking and thinking systems. *System Dynamics Review, 10*(2-3), 175–188. doi:10.1002/sdr.4260100206

Adams, D. C., & Salois, M. J. (2010). Local versus organic: A turn in consumer preferences and willingness-to-pay. *Renewable Agriculture and Food Systems, 25*(04), 331–341. doi:10.1017/S1742170510000219

Adnan, N., Nordin, S. M., & Noor, A. N. (2017). *A Comparative Study on GFT Adoption Behaviour Among Malaysian Paddy Farmers. In Green Marketing and Environmental Responsibility in Modern Corporations* (pp. 239–263). IGI Global. doi:10.4018/978-1-5225-2331-4.ch012

Adnan, N., Nordin, S. M., & Redza, A. (2017). *Benefit of One Baja Fertilizer for Attaining Agricultural Sustainability among Malaysian Paddy Farmers: Agricultural Sustainability among Malaysian Paddy Farmers. In Business Infrastructure for Sustainability in Developing Economies* (pp. 143–160). IGI Global. doi:10.4018/978-1-5225-2041-2.ch008

Advisory, F. (1993). Development and Information Network for Asia and the Pacific (FADINAP). *Proceedings of the regional FADINAP seminar on fertilisation and the environment*.

Aertsens, J., Mondelaers, K., Verbeke, W., Buysse, J., & Van Huylenbroeck, G. (2011). The influence of subjective and objective knowledge on attitude, motivations and consumption of organic food. *British Food Journal, 113*(11), 1353–1378. doi:10.1108/00070701111179988

Aertsens, J., Verbeke, W., Mondelaers, K., & Van Huylenbroeck, G. (2009). Personal determinants of organic food consumption: A review. *British Food Journal, 111*(10), 1140–1167. doi:10.1108/00070700910992961

Aftalion, F., & Poncet, P. (1991). Les mesures de performance des OPCVM: Problemes et solutions. *Revue Banque, 517*, 582–588.

AGRICORA. (2016). *Climate Risk Management for agriculture: Research and Tools*. AGRICORA.

Ahi, P., & Searcy, C. (2013). A comparative literature analysis of definitions for green and sustainable supply chain management. *Journal of Cleaner Production, 52*, 329–341. doi:10.1016/j.jclepro.2013.02.018

Ahmad, N. N., Tarmidi, M., Ridzwan, I. U., Masdiah Abdul Hamid, M., & Rusli Abdul Roni, R. (2014). The application of the Unified Theory of Acceptance and Use of Technology (UTAUT) in for predicting the usage of e-Zakat online system. *International Journal of Science and Research, 3*(4), 63–70.

Ahmad, S., & Tahar, R. M. (2014). Selection of renewable energy sources for sustainable development of electricity generation system using analytic hierarchy process: A case of Malaysia. *Renewable Energy, 63*, 458–466. doi:10.1016/j.renene.2013.10.001

Ahmed, O. H., Hanif, M., Husni, A., Abd Rahim, A., & Musa, M. H. (2001). Pineapple residue management practices and fertilizer regimes: Effects on P and K uptake, yield and some economic implications. *Pertanika. Journal of Tropical Agricultural Science, 24*(2), 123–129.

Akama, J. S., & Kieti, D. M. (2003). Measuring Tourist Satisfaction with Kenyas Wildlife Safari: A Case Study of Tasavo West National Park. *Tourism Management, 24*(1), 73–81. doi:10.1016/S0261-5177(02)00044-4

Alawadhi, A., & Morris, A. (2008). The use of the UTAUT model in the adoption of *e*-government services in Kuwait. *Proceedings of the 41st Hawaii International Conference on System Sciences*. doi:10.1109/HICSS.2008.452

Amilien, V., Schjøll, A. &Vramo, L.M. (2008). *Forbrukernes forståelse av lokal mat.* Statens Institutt for Forbruksforskning. Oslo, Fagrapport nr. 1-2008.

Amores, A., & Contreras, I. (2009). New approach for the assignment of new European agricultural subsidies using scores from data envelopment analysis: Application to olive-growing farms in Andalusia (Spain). *European Journal of Operational Research, 193*(3), 718–729. doi:10.1016/j.ejor.2007.06.059

Andersen, E., Elbersen, B., Godeschalk, F., & Verhood, D. (2007). Farm management indicators and farm typologies as a basis for assessments in a changing policy environment. *Journal of Environmental Management, 82*(3), 353–362. doi:10.1016/j.jenvman.2006.04.021 PMID:17126987

Andersson, F., Mausser, H., & Uryasev, S. (2000). Credit Risk Optimization With Conditional Value at Risk Criterion. *Journal of Mathematical Programming*.

Anikina, E. V., & Djordjevic, M. (2015). International Experience of winemaking production traceability system use. *Journal of Hygienic Engineering and Design, 12*, 85–89.

Antonaras, A., & Memtsa, C. (2009). *Business Ethics and Sustainable Excellence*. Presented and published in the proceedings of the 3rd Annual Quality Congress of the Middle East, organised by e-University in partnership with MEQA, Berkley University of California and the International Academy for Quality, Dubai, UAE.

Antonaras, A., Iacovidou, M., & Memtsa, C. (2011). Measuring social return on investment using the EBEN GR Business Ethics Excellence Model. *Current Issues of Business and Law*, *6*(1), 69–89. doi:10.5200/1822-9530.2011.04

APF. (2003). *Building a Business Risk Management, System for Agriculture in the 21st Century.* Retrieved from www.agr.gc.ca/puttingcanadafirst

Appelhanz, S., Osburg, V. S., Toporowski, W., & Schumann, M. (2016). Traceability system for capturing, processing and providing consumer relevant information about wood products: System solution and its economic feasibility. *Journal of Cleaner Production*, *110*, 132–148. doi:10.1016/j.jclepro.2015.02.034

Arabi, B., Munisamy, S., Emrouznejad, A., & Shadman, F. (2014). Powerindustryrestructuringandeco-efficiency changes: A new slacks-basedmodelinMalmquist–LuenbergerIndexmeasurement. *Energy Policy*, *68*, 132–145. doi:10.1016/j.enpol.2014.01.016

Ariza, C., Rugeles, L., Saavedra, D., & Guaitero, B. (2013). Measuring Innovation in Agricultural Firms: A Methodological Approach. *Electronic Journal of Knowledge Management*, *11*(3), 185–198.

Armstrong, G., Adam, S., Denize, S., & Kotler, P. (2014). *Principles of marketing*. Pearson Australia.

Artzner, P., Delbaen, F., Eber, J.-M., & Heath, D. (1999). Coherent measures of risk. *Journal of Mathematical Finance*, *9*(3), 203–228. doi:10.1111/1467-9965.00068

Asfaw, S., Mithöfer, D., & Waibel, H. (2010a). Agrifood supply chain, private-sector standards, and farmers health: Evidence from Kenya. *Agricultural Economics*, *41*(3–4), 251–263. doi:10.1111/j.1574-0862.2010.00443.x

Asfaw, S., Mithöfer, D., & Waibel, H. (2010b). What impact are EU supermarket standards having on developing countries export of high-value horticultural products? Evidence from Kenya. *Journal of International Food & Agribusiness Marketing*, *22*(3–4), 252–276. doi:10.1080/08974431003641398

Ashok, K. R., & Balasubramanian, R. (2006). *Role of Infrastructure in Productivity and Diversification of Agriculture, A Research Report*. Islamabad, Pakistan: SANEI.

Atănăsoaie, G. (2011). Distribution channels on the organic foods market. *Journal of Horticulture. Forestry and Biotechnology*, *15*(3), 19–25.

Athanassopoulos, A., & Curram, S. (1996). A Comparison of Data Envelopment Analysis and Artificial Neural Networks as Tools for Assessing the Efficiency of Decision Making Units. *The Journal of the Operational Research Society*, *47*(8), 1000–1016. doi:10.1057/jors.1996.127

ATTRA. (2003). *Applying the Principles of Sustainable Farming - Fundamentals of Sustainable Agriculture*. Retrieved August 22, 2016, fromhttp://ipm.ifas.ufl.edu/pdfs/Applying_the_Principles_of_Sustainable_Farming.pdf?pub=295%5D

Ayağ, Z., Samanlioglu, F., & Büyüközkan, G. (2013). A fuzzy QFD approach to determine supply chain management strategies in the dairy industry. *Journal of Intelligent Manufacturing*, *24*(6), 1111–1122. doi:10.1007/s10845-012-0639-4

Azadi, H., Ho, P., & Hasfiati, L. (2011). Agricultural land conversion drivers: A comparison between less developed, developing and developed countries. *Land Degradation & Development*, *22*(6), 596–604. doi:10.1002/ldr.1037

Bach, C. F., Frandsen, S. E., & Jensen, H. G. (2000). Agricultural and Economy-Wide Effects of European Enlargement: Modelling the Common Agricultural Policy. *Journal of Agricultural Economics*, *51*(2), 162–180. doi:10.1111/j.1477-9552.2000.tb01221.x

Bacon, C. R. (2008). *Practical Portfolio Performance Measurement and Attribution* (Vol. 546). Wiley.

Badia-Melis, R., Mishra, P., & Ruiz-García, L. (2015). Food traceability: New trends and recent advances. A review. *Food Control*, *57*, 393–401. doi:10.1016/j.foodcont.2015.05.005

Bakar, B. B. (2009). *The Malaysian agricultural industry in the new millennium: Issues and challenges*. Academic Press.

Baker, S., Thompson, K. E., Engelken, J., & Huntley, K. (2004). Mapping the values driving organic food choice: Germany vs the UK. *European Journal of Marketing*, *38*(8), 995–1012. doi:10.1108/03090560410539131

Balbus, J. M., Boxall, A., Fenske, R. A., McKone, T. E., & Zeise, L. (2013). Implications of global climate change for the assessment and management of human health risks of chemicals in the natural environment. *Environmental Toxicology and Chemistry*, *32*(1), 62–78. doi:10.1002/etc.2046 PMID:23147420

Ballantyne, B., Packer, J., & Hughes, K. (2008). Environmental Awareness, Interests, and Motives of Botanic Gardens Visitors: Implications for Interpretive Practice. *Tourism Management*, *29*(3), 439–444. doi:10.1016/j.tourman.2007.05.006

Balogun, R. B., Ogbu, J. U., Umeokechukwu, E. C., & Kalejaiye-Matti, R. B. (2016). *Effective Micro-organisms (EM) as Sustainable Components in Organic Farming: Principles, Applications and Validity. In Organic Farming for Sustainable Agriculture* (pp. 259–291). Springer.

Bang, J. M. (2005). *Ecovillages: A Practical Guide to Sustainable Communities*. New Society Publishers.

Bardsley, D. K., & Bardsley, A. M. (2014). Organising for socio-ecological resilience: The roles of the mountain farmer cooperative Genossenschaft Gran Alpin in Graubünden, Switzerland. *Ecological Economics*, *98*, 11–21. doi:10.1016/j.ecolecon.2013.12.004

Barett, C. B., & Reardon, T. (2001). *Asset, activity, and income diversification among African agriculturists: Some practical issues*. UASID Basis CRSP.

Basu, A. K., Chau, N. H., & Grote, U. (2003). Eco-labeling and stages of development. *Review of Development Economics*, *7*(2), 228–247. doi:10.1111/1467-9361.00188

Batte, M. T., Hooker, N. H., Haab, T. C., & Beaverson, J. (2007). Putting their money where their mouths are: Consumer willingness to pay for multi-ingredient, processed organic food products. *Food Policy*, *32*(2), 145–159. doi:10.1016/j.foodpol.2006.05.003

Bayo-Moriones, A., Billon, M., & Lera-Lopez, F. (2013). Perceived performance effects of ICT in manufacturing SMEs. *Industrial Management & Data Systems*, *113*(1), 117–135. doi:10.1108/02635571311289700

Bayo-Moriones, A., & Lera-Lopez, F. (2007). A firm-level analysis of determinants of ICT adoption in Spain. *Technovation*, *27*(6-7), 352–366. doi:10.1016/j.technovation.2007.01.003

Bean, J., & Hussey, L. (2012). *Marketing Public Sector Services: essential skills for the public sector*. HB Publications.

Benzecri, J. P. (1992). *Correspondence Analysis Handbook*. New York: Taylor & Francis.

Bera, A. K., & Park, S. Y. (2004). *Optimal portfolio diversification using maximum entropy*. Working Paper.

Bernaert, M., Poels, G., Snoeck, M., & De Barker, M. (2014). Enterprise architecture for small and medium-sized enterprises: A starting point for bringing EA to SMs, based on adoption models. In *Information Systems for Small and Medium-Sized Enterprises* (pp. 67–96). Berlin: Springer. doi:10.1007/978-3-642-38244-4_4

Bertsimas, D., Lauprete, G., & Samarov, A. (2004). Shortfall as a risk measure: Properties and optimization. *Journal of Economic Dynamics & Control*, *28*(7), 1353–1381. doi:10.1016/S0165-1889(03)00109-X

Bewley, J. M., & Russell, R. A. (2010). Reasons for slow adoption rates of precision dairy farming technologies: evidence from a producer survey. *Proceedings* of *the First North American Conference on Precision Dairy Management.*

Biglova, A., Ortobelli, S., Rachev, S. T., & Stoyanov, S. (2004). Different Approaches to Risk Estimation in Portfolio Theory. *Journal of Portfolio Management*, *31*, 103–112.

Black, F., & Litterman, R. (1990). *Asset Allocation: Combining Investors Views with Market Equilibrium.* Fixed Income Research. Goldman, Sachs & Company.

Blewitt, J. (2012). *Understanding Sustainable Development.* London: Earthscan.

Bodini, A., Richter, T., & Felder, R. (2009). Quality related communication approaches for organic food. *Journal of Food Products Marketing*, *15*(3), 364–377. doi:10.1080/10454440902942117

Bollerslev, T. (1986). Generalized Autoregressive Conditional Heteroskedasticity. *Journal of Econometrics*, *31*(3), 307–327. doi:10.1016/0304-4076(86)90063-1

Bonfiglio, A., Camaioni, B., Coderoni, S., Esposti, R., Pagliacci, F., & Sotte, F. (2015). *Innovation and education within rural development policy. Which role for rural and peripheral EU regions?* Paper presented at the 4th AIEAA Conference "Innovation, productivity and growth: towards sustainable agri-food production", Ancona, Italy.

Bougherara, D., Grolleau, G., & Thiébaut, L. (2005). Can labelling policies do more harm than good? An analysis applied to environmental labelling schemes. *European Journal of Law and Economics*, *19*(1), 5–16. doi:10.1007/s10657-005-5273-6

Boussofiane, A., Dyson, R. G., & Thanassoulis, E. (1991). Applied data envelopment analysis. *European Journal of Operational Research*, *52*(1), 1–15. doi:10.1016/0377-2217(91)90331-O

Brashear Alejandro, T., Javalgi, R. G., Gross, A. C., Benoy Joseph, W., & Granot, E. (2011). Assessing competitive advantage of emerging markets in knowledge intensive business services. *Journal of Business and Industrial Marketing*, *26*(3), 171–180. doi:10.1108/08858621111115895

Brooker, D. (2012). Build it and they will come? A critical examination of utopian planning practices and their socio-spatial impacts in Malaysias intelligent city. *Asian Geographer*, *29*(1), 39–56. doi:10.1080/10225706.2012.659192

Brown, C., & Miller, S. (2008). The impacts of local markets: A review of research on farmers markets and community supported agriculture (CSA). *American Journal of Agricultural Economics*, *90*(5), 1298–1302. doi:10.1111/j.1467-8276.2008.01220.x

Bukeviciute, L., Dierx, A., & Ilzkovitz, F. (2009) The functioning of the food supply chain and its effect on food prices in the European Union. *European Economy, Occasional Papers*, 47.

Burke, A. E., FitzRoy, R. L., & Nolan, M. A. (2002). Self-employment, wealth and job creation: The roles of gender, non-pecuniary motivation and entrepreneurial ability. *Small Business Economics*, *19*(3), 255–270. doi:10.1023/A:1019698607772

Burton, M., Rigby, D., Young, T., & James, S. (2001). Consumer attitudes to genetically modified organisms in food in the UK. *European Review of Agriculture Economics*, *28*(4), 479–498. doi:10.1093/erae/28.4.479

Butler, R. W. (1999). Sustainable Tourism: A state-of-the Art Review. *Tourism Geographies*, *1*(1), 7–25. doi:10.1080/14616689908721291

Byun, J., & Kim, D. (2015). Enhanced Object Traceability on Unified and Linked EPCIS Events. *Lecture Notes in Computer Science*, *9418*, 232–246. doi:10.1007/978-3-319-26190-4_16

Camagni, R. (Ed.). (1991). *Innovation networks: Spatial perspectives.* London, UK: Belhaven Press.

Canavari, C., Fritz, M., Hofstede, G. J., Matopoulos, A., & Vlachopoulou, M. (2010). The role of trust in the transition from traditional to electronic B2B relationships in agri-food chains. *Computers and Electronics in Agriculture*, *70*(2), 321–327. doi:10.1016/j.compag.2009.08.014

Canavari, M., Deiters, J., Fritz, M., Hofstede, G. J., Sprundel, G. V., Pignatti, E., & Fernández, M. C. et al. (2016). A typology of trust when we introduce information technology. In M. Canavari, M. Fritz, & G. Schiefer (Eds.), *Food supply networks: trust and e-business* (pp. 142–159). CABI Publishing. doi:10.1079/9781845936396.0142

Caporin, M., Lisi, F., & Janin, M. (2012). *A survey on Four Families of Performance Measures*. Working Papers Series.

Caporin, M., & Lisi, F. (2011). Comparing and Selecting Performance Measures using Rank Correlations *Economics. The Open-Acces, Open-Assessment E-journal.*, *5*, 31.

Cappelli, L., Guglielmetti, R., Mattia, G., Merli, R., & Renzi, M. F. (2011). Peer evaluation to develop benchmarking in the public sector. *Benchmarking an International Journal*, *4*(4), 490–509. doi:10.1108/14635771111147605

Castelán-Ortega, O., Martinez-Garcia, C., Mould, F., Dprward, P., Rehman, T., & Rayas-Amor, A. (2016). Optimal management of on-farm resources in small-scale dairy systems of Central Mexico: Model development and evaluation. *Tropical Animal Health and Production*, *48*(5), 1–8. doi:10.1007/s11250-016-1041-6 PMID:26992734

Cater, E. (1993). Ecotourism in the Third World: Problems for Sustainable Tourism Development. *Tourism Management*, *14*(2), 85–90. doi:10.1016/0261-5177(93)90040-R

CCIAA. (2016). Retrieved from http://www.infocamere.it/

CFS. (2014). *Principles for Responsible Investment in Agriculture and Food Systems, World Food Security (CFS)*. Retrieved August 19, 2016, fromhttp://www.fao.org/fileadmin/templates/cfs/Docs1314/rai/CFS_Principles_Oct_2014_EN.pdf

CGAP. (2004). *Key Principles of Microfinance - Building financial systems for the poor*. Retrieved August 19, 2016, fromhttps://www.cgap.org/sites/default/files/CGAP-Consensus-Guidelines-Key-Principles-of-Microfinance-Jan-2004.pdf

Chakrabarti, S. (2010). Factors influencing organic food purchase in India-expert survey insights. *British Food Journal*, *112*(8), 902–915. doi:10.1108/00070701011067497

Charnes, A., Cooper, W. W., & Rhodes, E. L. (1978). Measuring the efficiency of decision making units. *European Journal of Operational Research*, *2*(6), 429–444. doi:10.1016/0377-2217(78)90138-8

Chatzipetrou, E., & Moschidis, O. (2016). Quality costing: A survey in Greek supermarkets using multiple correspondence analysis. *International Journal of Quality & Reliability Management*, *33*(5), 615–632. doi:10.1108/IJQRM-01-2014-0004

Chaves, D., Hsu, J., Li, F., & Shakernia, O. (2012). *Efficient Algorithms for Computing Risk Parity Portfolio Weights*. Working Paper.

Cheng, D., Liu, G. & Qian, C. (2008). On determinants of user acceptance of Internet banking: A theoretical framework and empirical study. *Advanced Management of Information for Globalized Enterprises.*

Cheng, M. L., & Simmons, J. E. L. (1994). Traceability in manufacturing systems. *International Journal of Operations & Production Management*, *14*(10), 4–16. doi:10.1108/01443579410067199

Chen, J., Lobo, A., & Rajendran, N. (2014). Drivers of organic food purchase intentions in mainland China–evaluating potential customers attitudes, demographics and segmentation. *International Journal of Consumer Studies*, *38*(4), 346–356. doi:10.1111/ijcs.12095

Chen, M. F. (2007). Consumer attitudes and purchase intentions in relation to organic foods in Taiwan: Moderating effects of food-related personality traits. *Food Quality and Preference*, *18*(7), 1008–1021. doi:10.1016/j.foodqual.2007.04.004

Chen, R.-Y. (2017). An intelligent value stream-based approach to collaboration of food traceability cyber physical system by fog computing. *Food Control*, *71*, 124–136. doi:10.1016/j.foodcont.2016.06.042

Chinchuluun, R., Lee, W. S., Bhorania, J., & Pardalos, P. M. (2009). Clustering and Classification Algorithms in Food and Agricultural Applications:A Survey. In Advances in Modeling Agricultural Systems. Springer Science + Business Media, LLC.

Chitura, T., Mupemhi, S., Dube, T., & Bolongkikit, J. (2008b). Barriers to electronic commerce adoption in small and medium enterprises: A critical literature review. *Journal of Internet Banking and Commerce*, *13*(2), 1–13.

Chodorow, K. (2013). *MongoDB - The definitive guide*. O'Reilly.

Choi, J. B., Rogers, J., & Jones, E. C. (2015) The impact of a shared pharmaceutical supply chain model on counterfeit drugs, diverted drugs, and drug shortages. *Portland International Conference on Management of Engineering and Technology*, 1879-1889. doi:10.1109/PICMET.2015.7273165

Christian, D. L. (2003). *Creating a Life Together: Practical Tools to Grow Ecovillages and Intentional Communities.* New Society Publishers.

Chu, E. (2003). A review of simulation studies on supply chain management. *Journal of Academy of Business and Economics, 2*(1), 22-27.

Cicia, G., Cembalo, L., Del Giudice, T., &Verneau, F. (2012). Il sistema agroalimentare ed il consumatore post-moderno: Nuove sfide per la ricerca e per il mercato. *Rivista di Economia Agroalimentare*, (1), 117-142.

Cimino, M. G. C. A., & Marcelloni, F. (2012). Enabling Traceability in the Wine Supply Chain. *Lecture Notes in Computer Science*, *7200*, 397–412. doi:10.1007/978-3-642-31739-2_20

Closs, D. J., Swink, M., & Nair, A. (2005). The role of information connectivity in making logistics programs successful. *International Journal of Physical Distribution & Logistics Management*, *35*(4), 258–277. doi:10.1108/09600030510599922

Colucci, S. (2013). *A quick introduction to quantitative models that discard estimation of expected returns for portfolio construction*. Working Paper.

Cook, W., & Seiford, L. (2009). Data envelopment analysis (DEA) – Thirty years on. *European Journal of Operational Research*, *192*(1), 1–17. doi:10.1016/j.ejor.2008.01.032

Coppini, M., Rossignoli, C., Rossi, T., & Strozzi, F. (2010). Bullwip effect and inventory oscillations analysis using the beer game model. *International Journal of Production Research*, *48*(13), 3943–3956. doi:10.1080/00207540902896204

Cornell University, INSEAD, & WIPO. (2015). *The Global Innovation Index 2015: Effective Innovation Policies for Development*. Authors.

Cristiano, S., & Proietti, P. (2013). Farm innovation through rural development programmes: Experiences and pathways of innovation in Italy. *Proceedings of the 21st ESEE European Seminar on Extension Services - Extension Education Worldwide trends, challenges and cases*. Antalya, Turkey: Akdeniz University, Faculty of Agriculture Department of Agricultural Economics Extension Science.

Crockford, D. (2006). *RFC 4627, Network Working Group*. Available online: http://www.ietf.org/rfc/rfc4627.txt

Croft, M. J. (1994). *Market segmentation: A step-by-step guide to profitable new business*. Cengage Learning Emea.

CSR Europe. (2008). *ProactiveStakeholderEngagement, Practical Guide for companies and stakeholders, CSR Europe Toolbox*. Retrieved August 19, 2016, from http://www.csreurope.org/csr-europes-toolbox-user-manual#.V8VoVyiLSUm

Cudec, R. (2000). Exploratory factor analysis. In Handbook of Applied Multivariate Statistics and Mathematical Modeling. New York: Elsevier Inc. doi:10.1016/B978-012691360-6/50011-2

Cullinane, K., Wang, T.-F., Song, D.-W., & Ji, P. (2006). The technical efficiency of container ports: Comparingdata envelopment analysis and stochastic frontier analysis. *Transportation Research Part A, Policy and Practice*, *40*(4), 354–374. doi:10.1016/j.tra.2005.07.003

D'Angelo, G., De Luca, G., & Rampone, S. (2014). Food safety and traceability system based on RFID technologies and services. *First International Workshop on Global Sustainability Inside and Outside the Territory.*

Danielsson, J., & de Vries, C. (n.d.). *Beyond the Sample: Extreme Quantile and Probability Estimation.* London School of Economics, Discussion Paper 298.

Danielsson, J., & de Vries, C. (1997). Tail Index and Quantile Estimation with Very High Frequence Data. *Journal of Empirical Finance*, *4*(2-3), 241–257. doi:10.1016/S0927-5398(97)00008-X

Darby, K., Batte, M. T., Ernst, S., & Roe, B. (2008). Decomposing Local: A Conjoint Analysis of Locally Produced Foods. *American Journal of Agricultural Economics*, *90*(2), 476–486. doi:10.1111/j.1467-8276.2007.01111.x

Dargan, L., & Shucksmith, M. (2008). LEADER and innovation. *Sociologia Ruralis*, *48*(3), 274–291. doi:10.1111/j.1467-9523.2008.00463.x

Darnall, N., Ji, H., & Vázquez-Brust, D. A. (2016). Third-party certification, sponsorship, and consumers' ecolabel use. *Journal of Business Ethics*, 1–17. doi:10.1007/s10551-016-3138-2

Dasmann, R. F., Milton, J. P., & Freeman, P. H. (1973). *Ecological Principles for Economic Development.* London: John Wiley.

Davies, A., Titterington, A. J., & Cochrane, C. (1995). Who buys organic food? A profile of the purchasers of organic food in Northern Ireland. *British Food Journal*, *97*(10), 17–23. doi:10.1108/00070709510104303

Davis, B., Winters, P., Carletto, C., Covarrubias, K., Quinones, E., & Zezza, A. (2007). Rural income generating activities: Across country comparison. *World Development*, 119–123.

Dawson, J. (2006). *Ecovillages: New Frontiers for Sustainability.* Totnes: Green Books.

deJanvry, A., & Sadoulet, E. (2003). *Progress in the modeling of rural households' behavior under market failures. Chapters in Honor of Erik Thorbecke.* New York: Kluwer Publishing.

Delmas, M. A., & Grant, L. E. (2014). Eco-labeling strategies and price-premium: The wine industry puzzle. *Business & Society*, *53*(1), 6–44. doi:10.1177/0007650310362254

deMagistris, T., & Gracia, A. (2008). The decision to buy organic food products in Southern Italy. *British Food Journal*, *110*(9), 929–947. doi:10.1108/00070700810900620

Dholakia, R. R., & Kshetri, N. (2004). Factors impacting the adoption of the internet among SMEs. *Small Business Economics*, *23*(4), 311–322. doi:10.1023/B:SBEJ.0000032036.90353.1f

Didier, T., & Lucie, S. (2008). Measuring consumers willingness to pay for organic and Fair Trade products. *International Journal of Consumer Studies*, *32*(5), 479–490. doi:10.1111/j.1470-6431.2008.00714.x

Dimitri, C., & Dettmann, R. L. (2012). Organic food consumers: What do we really know about them? *British Food Journal*, *114*(8), 1157–1183. doi:10.1108/00070701211252101

Dine, M., & Haynes, K. E. (1999). Sources of regional inefficiency. An integrated shift-share, data envelopment analysis and input-output approach. *The Annals of Regional Science*, (33): 469–489.

Djellal, F., Francoz, D., Gallouj, C., Gallouj, F., & Jacquin, Y. (2003). Revising the definition of research and development in the light of the specificities of services. *Science & Public Policy, 30*(6), 415–429. doi:10.3152/147154303781780227

Dlamini, B., Kirsten, J., & Masuku, M. (2014). Factors Affecting the Competitiveness of the Agribusiness Sector in Swaziland. *Journal of Agricultural Studies, 2*(1).

Do, H., Mazzarol, T., Volery, T., Geoff, S., & Reboud, S. (2015). SME owner-manager's anticipated returns to commercialisation: The influence of organisational factors. *60th International Council for Small Business (ICSB) World Conference 2015: Entrepreneurship at a Global Crossroads.*

Dooley, J. A., Jones, S. C., & Iverson, D. (2012). Web 2.0: An assessment of social marketing principles. *Journal of Social Marketing, 2*(3), 207–221. doi:10.1108/20426761211265195

DOSM. (2014a). *SMEs Census 2011*. Available at www.statistics.my.gov/

DOSM. (2014b). *Statistics Yearbook Malaysia 2012*. Retrieved from http://www.statistics.gov.my/portal/download_Buku_Tahunan/files/BKKP/2012/pdf/TAB10.pdf

Douthwaite, B. (2002). *Enabling Innovation: A Practical guide to understanding and fostering technological change*. London, UK: Zed Books.

Dowd, K. (2000). Adjusting for Risk: An Improved Sharpe Ratio. *International Review of Economics & Finance, 9*(3), 209–222. doi:10.1016/S1059-0560(00)00063-0

EC 2010. (2010). *Europe 2020: A European Strategy for smart, sustainable growth.* COM(2010) 2020.

Eco Connexion. (2014). *Rangpur Ecovillage*. Retrieved from: bit.ly/1nbUsgl

Eggert, A., & Helm, S. (2003). Exploring the impact of relationship transparency on business relationships: A cross-sectional study among purchasing managers in Germany. *Industrial Marketing Management, 32*(2), 101–108. doi:10.1016/S0019-8501(02)00224-9

Ellis, F. (2000). *Rural Livelihoods and Diversity in Developing Countries*. Oxford, UK: Oxford University Press.

Embrechts, P., Kluppelberg, C., & Mikosch, T. (1997). *Modelling External Events for Insurance and Finance*. Springer.

Enderwick, P. (2009). Managing quality failure in China: Lessons from the dairy industry. *International Journal of Emerging Markets, 4*(3), 220–234. doi:10.1108/17468800910968391

Engelseth, P., & Hogset, H. (2016). Adapting Supply Chain Management for Local Foods Logistics. *Proceedings in Food System Dynamics 2016.*

Engelseth, P. (2015). Customer-Responsive Supply of Local Foods. *Journal of Operations and Supply Chain Management, 8*(3), 111–119.

Engelseth, P., Wongthatsanekorn, W., & Charoensiriwath, C. (2014). Food Product Traceability and Customer Value. *Global Business Review, 15*(4suppl.), 87S–105S. doi:10.1177/0972150914550549

Entrepreneurial Learning Initiative. (2016). *Entrepreneurial Mindset Education and Training Programs*. Retrieved September 1, 2016 from https://elimindset.com/about/

EPCIS. (2014). *EPC Information Services (EPCIS) Version 1.1 Specification*. GS1 Standard Version 1.1, May 2014.

Ergönül, B., & Ergönül, P. G. (2015). Consumer motivation for organic food consumption. *Emirates Journal of Food and Agriculture*, *27*(5), 416.

Esparcia, J. (2014). Innovation and networks in rural areas. An analysis from European innovative projects. *Journal of Rural Studies*, *34*, 1–14. doi:10.1016/j.jrurstud.2013.12.004

Esterhuizen, D. (2006). *An Evaluation of the Competitiveness of the South African Agribusiness Sector*. University of Pretoria. Available from UPeTD: Retrieved June 12, 2016, from http://upetd.up.ac.za/ thesis/ available/ etd-12082006144349/ unrestricted/ 00front.pdf,9/3/2010

EU - DG Agri. (2012). *Prospects for Agricultural Markets and Income in the EU 2012-2022*. Commission's Directorate-General for Agriculture and Rural development, European Commission, Agriculture and Rural Development.

EU - DG Agri. (2013). *Prospects for Agricultural Markets and Income in the EU 2013-2023*. Commission's Directorate-General for Agriculture and Rural development, European Commission, Agriculture and Rural Development.

EurekAlert. (2015). *The Global Source for Science News, World crop diversity survives in small farms from peri-urban to remote rural locations, Article by Penn State*. Retrieved August 26, 2016, from http://www.eurekalert.org/pub_releases/2015-02/ps-wcd021115.php

Europe 2020. (n.d.). Retrieved from http://ec.europa.eu/

European Commission. (2005). e-Business Watch. *ICT and Electronic Business in the Food and Beverages Industry. Report 2005 edition. ICT adoption and e-business activity in 2005*. Retrieved July, 2016, from http://www.e-business-watch.org/

European Commission. (2007). e-Business Watch. *The European E-Business Report2006/07edition. A Portrait Of E-Business In 10 Sectors Of The EU Economy, 5th Synthesis Report of the e-Business*. Retrieved July, 2016, from http://www.e-business-watch.org/

European Commission. (2013). *Overview of CAP reform 2014-2020*. Retrieved May 19, 2016, fromhttp://ec.europa.eu/agriculture/policy-perspectives/index_en.htm

European Commission. (2014). *Rural development programme*. Retrieved June 10, 2016, from http://ec.europa.eu/agriculture/rural-development-2014-2020/

European Commission. (2016). *Factsheet on 2014-2020 Rural development programme for Greece*. Retrieved July 12, 2016, from http://ec.europa.eu/agriculture/rural-development-2014-2020/country-files/el/factsheet-greece_en.pdf

EUROSTAT. (2015). *Agriculture, Forestry and Fishery Statistics*. Retrieved from http://ec.europa.eu/eurostat/en/web/products-statistical-books/-/KS-FK-15-001

Eurostat. (2015). *Eurostat News Release 81/2015*. Retrieved May 7, 2015, from http://ec.europa.eu/eurostat/documents/2995521/6823708/3-07052015-AP-EN.pdf/7e507ea0-43c7-452f-8e6a-b479c89d2bd6

Evans, H. E., & Ngau, P. (1991). Rural-urban relations, household income diversification and agricultural productivity. *Development and Change*, *22*(3), 519–545. doi:10.1111/j.1467-7660.1991.tb00424.x

Exposito, I. J. A., Gay-Fernandez, I., & Cuinas, A. (2013). Complete Traceability System for a Wine Supply Chain Using Radio-Frequency Identification and Wireless Sensor Network. *IEEE Antennas and Propagation Magazine*, *55*(2), 255–267. doi:10.1109/MAP.2013.6529365

FADINAP. (1984). Report. Marketing, distribution and use of fertilizer in Malaysia. FADINAP.

Fahmi, Z., Samah, B. A., & Abdullah, H. (2013). Paddy Industry and Paddy Farmers Well-being: A Success Recipe for Agriculture Industry in Malaysia. *Asian Social Science*, *9*(3), 177. doi:10.5539/ass.v9n3p177

Fairtrade. (2014). *FairtradeInternational - Sustainable Development Goals and Fairtrade: The case for partnership.* Retrieved August 17, 2016, from http://www.fairtrade.net/fileadmin/user_upload/content/2009/resources/15-10_Sustainable_Development_Report.pdf

Fama, E. (1965). The behavior of Stock Market Prices. *The Journal of Business*, *38*(1), 34–105. doi:10.1086/294743

Fang, W. C., Li, M. W., & Liu, C. W. (2008). Measurement of the knowledge-sharing efficacy of Web2.0 site constructed on the basis of knowledge-based systems by applying the model of UTAUT: Evidence of the early adopters. *Proceeds of the 3rd International Conference on Innovative Computing Information and Control (ICICIC'08).*

FAO. (2009). *How to Feed the World by 2050, High-Level Expert Forum.* Retrieved August 27, 2016, from http://www.fao.org/fileadmin/templates/wsfs/docs/Issues_papers/HLEF2050_Global_Agriculture.pdf

FAO. (2010). *World deforestation decreases, but remains alarming in many countries.* Retrieved August 27, 2016, from http://www.fao.org/news/story/en/item/40893/icode/

FAO. (2011). *The State of Food Insecurity in the World, How does international price volatility affect domestic economies and food security?* Retrieved August 27, 2016, from http://www.fao.org/docrep/014/i2330e/i2330e.pdf

FAO. (2012). Retrieved from http://www.fao.org/docrep/x0262e/x0262e06.htm

FAO. (2013). *Enabling environments for agribusiness and agro-industries development: Regional and country perspectives. Agribusiness and Food Industries Series 1* (pp. 1–61). Rome: FAO, United Nations.

FAO. (2013). *Post-harvest food losses estimation - Development of consistent methodology.* Retrieved August 27, 2016, from http://www.fao.org/fileadmin/templates/ess/documents/meetings_and_workshops/GS_SAC_2013/Improving_methods_for_estimating_post_harvest_losses/Final_PHLs_Estimation_6-13-13.pdf

FAO. (2014a). *Family Farmers: Feeding the world, caring for the earth.* Retrieved August 27, 2016, from http://www.fao.org/docrep/019/mj760e/mj760e.pdf

FAO. (2014b). *CFS-Committee on World Food Security, CFS Chair's Statement for the FAO Regional Conferences 2014.* Retrieved August 27, 2016, from http://www.fao.org/fileadmin/templates/cfs/Docs1314/RC/CFS_Chairs_RC_Statement_EN.pdf

FAO. (2014c). *DEEP ROOTS.* Retrieved August 27, 2016, from http://www.fao.org/3/a-i3976e.pdf

FAO. (2014d). *State of Food and Agriculture in the Africa Region and Comprehensive Africa Agriculture Development Programme Implementation with a Specific Focus on Smallholder Farmers and Family Farming.* Africa Regional Conference, March 2014. ARC/14/3. Retrieved August 27, 2016, from http://www.fao.org/docrep/meeting/030/mj566e.pdf

FAO. (2014e). *Food Losses and Waste in Europe and Central Asia.* Document ERC/14/3. Retrieved August 17, 2016, fromhttp://www.fao.org/docrep/meeting/030/mj621e.pdf)

FAO. (2014f). *Building a common vision for sustainable food and agriculture - Principles and Approaches.* Retrieved August 22, 2016, fromhttp://www.fao.org/3/a-i3940e.pdf

Farming First. (2015). *The story of Agriculture and the Sustainable Development Goals.* Retrieved August 29, 2016, from http://www.farmingfirst.org/sdg-toolkit#home

Farrell, M.J., (1957). The measurement of productive efficiency. *Journal of the Royal Statistical Society, Series A, General, 120*(3), 253–281.

Farrell, B. H. (1999). Conventional or Sustainable Tourism? No Room for Choice. *Tourism Management*, *20*(2), 189–191.

Fellowship for Intentional Community. (2005). *Communities Directory: A Comprehensive Guide to Intentional Communities and Cooperative Living*. Rutledge, MO: Fellowship for Intentional Community.

Feng, J., Fu, Z., Wang, Z., Xu, M., & Zhang, X. (2013). Development and evaluation on a RFID-based traceability system for cattle/beef quality safety in China. *Food Control*, *31*(2), 314–325. doi:10.1016/j.foodcont.2012.10.016

FiBL-IFOAM Survey. (2016). *The World of Organic Agriculture: Statistics & Emerging Trends 2016*. Retrieved November 05, 2016 from https://shop.fibl.org/fileadmin/documents/shop/1698-organic-world-2016.pdf

Fielke, S. J., & Bardsley, D. K. (2013). South Australian farmers markets: Tools for enhancing the multifunctionality of Australian agriculture. *GeoJournal*, *78*(5), 759–776. doi:10.1007/s10708-012-9464-8

Fink, D., & Disterer, G. (2006). International case studies: To what extent is ICT infused into the operations of SMEs? *Journal of Enterprise Information Management*, *19*(6), 608–624. doi:10.1108/17410390610708490

Fishbein, M., & Ajzen, I. (1985). Belief, attitudes, intention and behaviour. An introduction to theory and research. Reading, MA: Addison-Wesley.

Folinas, D., Manikas, I., & Manos, B. (2006). Traceability data management for food chains. *British Food Journal*, *108*(8), 622–633. doi:10.1108/00070700610682319

Ford, A. (1999). *Modeling the environment: an introduction to system dynamics models of environmental systems*. Island Press.

Forrester, J. (2003). Economic theory for the new millennium. *Proceeding of the 21st International Conference of the System Dynamics Society*.

Forrester, J. W. (1997). Industrial Dynamics. *The Journal of the Operational Research Society*, *48*(10), 1037–1041. doi:10.1057/palgrave.jors.2600946

Forrester, J. W. (2003). Dynamic models of economic systems and industrial organizations. *System Dynamics Review*, *19*(4), 329–345. doi:10.1002/sdr.284

Fotopoulos, C., & Krystallis, A. (2002). Purchasing motives and profile of the Greek organic consumer: A countrywide survey. *British Food Journal*, *104*(9), 730–765. doi:10.1108/00070700210443110

Främling, K., Parmar, S., Hinkka, V., Tätilä, J., & Rodgers, D. (2013). Assessment of EPCIS Standard for Interoperable Tracking in the Supply Chain. *Studies in Computational Intelligence*, *472*, 119–134.

Franklin, S. (1995). Science as Culture, Cultures of Science. *Annual Review of Anthropology*, *24*(1), 163–184. doi:10.1146/annurev.an.24.100195.001115

Fred van Raaij, W., & Verhallen, T. M. (1994). Domain-specific market segmentation. *European Journal of Marketing*, *28*(10), 49–66. doi:10.1108/03090569410075786

Freshwater, D. (2015). Vulnerability and resilience: Two dimensions of rurality. *Sociologia Ruralis*, *55*(4), 497–515. doi:10.1111/soru.12090

Fritz, M., & Canavari, M. (2008). Management of Perceived e-Business Risks in Food-Supply Networks: e-Trust as Prerequisite for Supply-Chain System Innovation. *Agribusiness*, *24*(3), 355–368. doi:10.1002/agr.20163

Furnham, A., & Gunter, B. (2015). *Corporate Assessment (Routledge Revivals): Auditing a Company's Personality*. Routledge.

Gandino, F., Rebaudengo, M., & Sanchez, E. R. (2009). On Improving Automation by Integrating RFID in the Traceability Management of the Agri-Food Sector. *IEEE Transactions on Industrial Electronics*, *56*(7), 2357–2365. doi:10.1109/TIE.2009.2019569

García, J. M. (2006). *Theory and practical exercises of system dynamics*. MIT Sloan School of Management.

Garnett, T. (2011). Where are the best opportunities for reducing greenhouse gas emissions in the food system (including the food chain)? *Food Policy*, *36*(1), S23–S32. doi:10.1016/j.foodpol.2010.10.010

General Secretariat of Research and Technology-GSRT. (2013). *Research and Innovation in the Agrobioalimentary cluster for the 2014-2020 period* (in Greek). Retrieved July 12, 2016, from http://www.gsrt.gr/Financing/Files/ProPeFiles48/ΤΟ%20ΑΓΡΟΒΙΟΔΙΑΤΡΟΦΙΚΟ%20ΣΥΜΠΛΕΓΜΑ%202014-2020_ΕΣΠΕΚ_10-7-2013_A_FINAL.pdf

Georgiadis, P., & Athanasiou, E. (2010). The impact of two-product joint lifecycles on capacity planning of remanufacturing networks. *European Journal of Operational Research*, *202*(2), 420–433. doi:10.1016/j.ejor.2009.05.022

Georgiadis, P., Vlachos, D., & Iakovou, E. (2005). A system dynamics modeling framework for the strategic supply chain management of food chains. *Journal of Food Engineering*, *70*(3), 351–364. doi:10.1016/j.jfoodeng.2004.06.030

Gereffi, G., & Lee, J. (2009). *A global value chain approach to food safety and quality standards*. Global Health Diplomacy for Chronic Disease Prevention, Working Paper Series. Duke University.

Giddens, A. (1991). *Modernity and Self-Identity*. Cambridge, UK: Polity Press.

Gil, J. M., Gracia, A., & Sanchez, M. (2000). Market segmentation and willingness to pay for organic products in Spain. *The International Food and Agribusiness Management Review*, *3*(2), 207–226. doi:10.1016/S1096-7508(01)00040-4

Gilman, R. (1991). The Ecovillage Challenge - The Challenge of Developing a Community Living in Balanced Harmony - With Itself as Well as Nature - is Tough, but Attainable. *In Context, IC#29*. Retrieved from: http://bit.ly/19AH38j

Global Ecovillage Network. (2014). *What is an Ecovillage?* Retrieved from: http://bit.ly/1b4GgTC

Godin, B. (2006). The linear model of innovation: The historical construction of an analytical framework. *Science, Technology & Human Values*, *31*(6), 639–667. doi:10.1177/0162243906291865

Goodwin, H. (1995). Tourism and the Environment. *Biologist (Columbus, Ohio)*, *42*(3), 129–133.

GoP. (2011). *Economic survey of Pakistan. Economic Advisory Wing. Finance division*. Islamabad: MINFAL Pakistan.

Gorton, M., Duarin, E., Davidova, S., & Latruffe, L. (2008). Attitudes to agricultural policy and farming futures in the contextof the 2003 CAP reform: A comparison of farmers in selected established and new Member States. *Journal of Rural Studies*, *24*(3), 322–336. doi:10.1016/j.jrurstud.2007.10.001

Govindasamy, R., & Italia, J. (1999). Predicting willingness-to-pay a premium for organically grown fresh produce. *Journal of Food Distribution Research*, *30*, 44–53.

Gracia, A., & de Magistris, T. (2013). Organic food product purchase behaviour: A pilot study for urban consumers in the South of Italy. *Spanish Journal of Agricultural Research*, *5*(4), 439–451. doi:10.5424/sjar/2007054-5356

Granovetter, M. (1985). Economic action and social structure: The problem of embeddedness. *American Journal of Sociology*, *91*(3), 481–510. doi:10.1086/228311

Grassini, P., Eskridge, K. M., & Cassman, K. G. (2013). Distinguishing between yield advances and yield plateaus in historical crop production trends. *Nature Communications*, 4. PMID:24346131

Greenacre, M. (2007). *Correspondence Analysis in Practice*. Boca Raton, FL: Chapman and Hall/CRC. doi:10.1201/9781420011234

Grimm, J. H., Hofstetter, J. S., & Sarkis, J. (2016). Exploring sub-suppliers compliance with corporate sustainability standards. *Journal of Cleaner Production*, *112*, 1971–1984. doi:10.1016/j.jclepro.2014.11.036

Grunert, K. G., & Ramus, K. (2005). Consumers Willingness To Buy Food Through The Internet: A Review Of The Literature And A Model For Future Research. *British Food Journal*, *107*(6), 381–403. doi:10.1108/00070700510602174

GS1. (2008). *GS1 Working Group, Wine Supply Chain Traceability*. GS1 Application Guideline, Brussels, Belgium.

GS1. (2016). *GS1 General Specifications*. Available online at http://www.gs1.org/docs/barcodes/GS1_General_Specifications.pdf

Gunnlaugsson, V., Thakur, M., Forås, E., Ringsberg, H., Gran-Larsen, Ö., & Margeirsson, S. (2011) EPCIS standard used for improved traceability in the redfish value chain. *International Conference on Modern Information Technology in the Innovation Processes of the Industrial Enterprises*, 182-192.

Haggblade, S., Hazell, P., & Reardon, T. (2010). The Rural Non-farm Economy: Prospects for Growth and Poverty Reduction. *World Development*, *38*(10), 1429–1441. doi:10.1016/j.worlddev.2009.06.008

Hair, J. F. J., Black, W. C., Babin, B. J., Anderson, R. E., & Tatham, R. L. (2010). *Multivariate data analysis a global perspective. Pearson Education Inc.*

Hall, C. M., & Williams, A. (2008). *Tourism and Innovation*. London, UK: Routledge.

Hampton, R., Fromm, I., & Nyhodo, B. (2007). *International Trade, Consumer Behaviour and Trust: Factors Affecting Agribusinesses in Developing Countries*. Marketing Department Faculty Publications. Retrieved June 20, 2016), from http://digitalcommons.unl.edu/marketingfacpub/6

Hamzaoui Essoussi, L., & Zahaf, M. (2009). Exploring the decision-making process of Canadian organic food consumers: Motivations and trust issues. *Qualitative Market Research: An International Journal*, *12*(4), 443–459. doi:10.1108/13522750910993347

Hamzaoui-Essoussi, L., Sirieix, L., & Zahaf, M. (2013). Trust orientations in the organic food distribution channels: A comparative study of the Canadian and French markets. *Journal of Retailing and Consumer Services*, *20*(3), 292–301. doi:10.1016/j.jretconser.2013.02.002

Hamzaoui-Essoussi, L., & Zahaf, M. (2012). Canadian organic food consumers profile and their willingness to pay premium prices. *Journal of International Food & Agribusiness Marketing*, *24*(1), 1–21. doi:10.1080/08974438.2011.621834

Handschuch, C., Wollni, M., & Villalobos, P. (2013). Adoption of food safety and quality standards among Chilean raspberry producers – Do smallholders benefit? *Food Policy*, *40*, 64–73. doi:10.1016/j.foodpol.2013.02.002

Hansson, H. (2007). Strategy factors as drivers and restraints on dairy farmperformance: Evidence from Sweden. *Agricultural Systems*, *94*(3), 726–737. doi:10.1016/j.agsy.2007.03.002

Hansson, H., & Ohlmer, B. (2008). The effect of operational managerial practices on economic, technical and allocative efficiency at Swedish dairy farms. *Livestock Science*, *118*(1-2), 34–43. doi:10.1016/j.livsci.2008.01.013

Haq, Z., & Meilke, K. (2010). Do the BRICs and Emerging Markets Differ in their Agrifood Trade? *Journal of Agricultural Economics*, *61*(1), 1–14. doi:10.1111/j.1477-9552.2009.00229.x

Hardy, D. (2000). *Utopian England: Community Experiments 1900-1945*. London: E. and F.W. Spon.

Haris, N. B. M., Hamzah, A., Krauss, S. E., & Ismail, I. A. (2013). Relationship between Decision-Making Inputs and Productivity among Paddy Farmers" in Integrated Agriculture Development Areas (IADAs), in Malaysia. *International Journal on Advanced Science, Engineering and Information Technology*, *3*(1), 64–70.

Harper, G. C., &Makatouni, A. (2002). Consumer perception of organic food production and farm animal welfare. *British Food Journal, 104*(3/4/5), 287-299.

Hartmann, J., & Moeller, S. (2014). Chain liability in multitier supply chains? Responsibility attributions for unsustainable supplier behaviour. *Journal of Operations Management*, *32*(5), 281–294. doi:10.1016/j.jom.2014.01.005

Hashim, H., & Ho, W. S. (2011). Renewable energy policies and initiatives for a sustainable energy future in Malaysia. *Renewable & Sustainable Energy Reviews*, *15*(9), 4780–4787. doi:10.1016/j.rser.2011.07.073

Hassan, A., & Forhad, A. (2013). The Role of NGOs in the Sustainable Development of Bangladesh. *Present Environment and Sustainable Development*, *7*(2), 59–72.

Hassini, E., Surti, C., & Searcy, C. (2012). A literature review and a case study of sustainable supply chains with a focus on metrics. *International Journal of Production Economics*, *140*(1), 69–82. doi:10.1016/j.ijpe.2012.01.042

Hatanaka, M., Bain, C., & Busch, L. (2005). Third-party certification in the global agri-food system. *Food Policy*, *30*(3), 354–369. doi:10.1016/j.foodpol.2005.05.006

Hattam, C. E., Lacombe, D. J., & Holloway, G. J. (2012). Organic certification, export market access and the impacts of policy: Bayesian estimation of avocado smallholder times-to-organic certification in Michoacán Mexico. *Agricultural Economics*, *43*(4), 441–457. doi:10.1111/j.1574-0862.2012.00595.x

Hausen, T., Fritz, M., & Schiefer, G. (2006). Potential Of Electronic Trading In Complex Supply Chains: An Experimental Study. *International Journal of Production Economics*, *104*(2), 580–597. doi:10.1016/j.ijpe.2005.04.010

Hawkins, D. E. (1993). Global Assessment of Tourism Policy: A Process Model. In D. G. Pearce & R. W. Butler (Eds.), *Tourism Research: Critiques and Challenges* (pp. 175–200). London: Routledge.

Heidari, M. D., & Omid, M. (2011). Energy use patterns and econometric models of major greenhouse vegetable productions in Iran. *Energy*, *36*(1), 220–225. doi:10.1016/j.energy.2010.10.048

Heinrichs, A. J., Jones, C. M., Gray, S. M., Heinrichs, P. A., Cornelisse, S. A., & Goodling, R. C. (2013). Identifying efficient dairy heifer producersusingproduction costs and data envelopment analysis. *Journal of Dairy Science*, *90*(11), 7355–7362. doi:10.3168/jds.2012-6488 PMID:24054291

Hellenic Statistical Authority (ELSTAT). (2009). *Results of Agricultural-Livestock Census 2009*. Retrieved from http://www.statistics.gr/el/statistics/-/publication/SPG31/-

Hellenic Statistical Authority (ELSTAT). (2016). *Gross value added by industry (A10) - NACE REV.2, the sector of Agriculture, Forestry*. Retrieved June 10, 2016, from http://www.statistics.gr/el/statistics/-/publication/SEL45/-

Helms, M. M., & Nixon, J. (2010). Exploring SWOT analysis - where are we now? A review of academic researchfrom the last decade. *Journal of Strategy and Management*, *3*(3), 215–251.

Hennik, M. M. (2013). *International Focus Group Research*. Cambridge, UK: Cambridge University Press.

Henson, S., & Reardon, T. (2005). Private agri-food standards: Implications for food policy and the agri-food system. *Food Policy*, *30*(3), 241–253. doi:10.1016/j.foodpol.2005.05.002

Hertel, T. (1997). *Global Trade Analysis: Modeling and Applications*. Published by Cambridge University Press.

Hertz, T. (2009). The effect of nonfarm income on investment in Bulgarian. *Agricultural Economics*, *2*(40), 161–176. doi:10.1111/j.1574-0862.2009.00367.x

Herzfeld, T., Drescher, L. S., & Grebitus, C. (2011). Cross-national adoption of private food quality standards. *Food Policy*, *36*(3), 401–411. doi:10.1016/j.foodpol.2011.03.006

Higon, D. A. (2011). The impact of ICT on innovation activities: Evidence for UK SMEs. *International Small Business Journal*, *30*(6), 684–699. doi:10.1177/0266242610374484

Hjalager, A. M., & Johansen, P. H. (2013). Food tourism in protected areas—sustainability for producers, the environment and tourism? *Journal of Sustainable Tourism*, *21*(3), 417–433. doi:10.1080/09669582.2012.708041

Holloway, M. (1951). *Heavens on Earth: Utopian Communities in America*. New York: Library Publishers.

Holzworth, D. P., Huth, N. I., de Voil, P. G., Zurcher, E. J., Herrmann, N. I., McLean, G., & Keating, B. A. et al. (2014). APSIM–evolution towards a new generation of agricultural systems simulation. *Environmental Modelling & Software*, *62*, 327–350. doi:10.1016/j.envsoft.2014.07.009

Homburg, C., Kuester, S., & Krohmer, H. (2013). *Marketing management: A contemporary perspective*. McGraw-Hill Higher Education.

Honey, M. (1999). *Who Owns Paradise? Ecotourism and Sustainable Development*. Washington, DC: Island Press.

Honkanen, P., Verplanken, B., & Olsen, S. O. (2006). Ethical values and motives driving organic food choice. *Journal of Consumer Behaviour*, *5*(5), 420–430. doi:10.1002/cb.190

Horne, , Fien, , Beza, , & Nelson., (Eds.). (2016). *Sustainability, citizenship and cities: theory and practice*. New York: Routledge.

Hosseini, S. E., & Wahid, M. A. (2013). Feasibility study of biogas production and utilization as a source of renewable energy in Malaysia. *Renewable & Sustainable Energy Reviews*, *19*, 454–462. doi:10.1016/j.rser.2012.11.008

Hounkonnou, D., Kossou, D., Kuyper, T. W., Leeuwis, C., Nederlof, E. S., Röling, N., & Van Huis, A. et al. (2012). An innovation systems approach to institutional change: Smallholder development in West Africa. *Agricultural Systems*, *108*, 74–83. doi:10.1016/j.agsy.2012.01.007

Hsu, Y.-C., Chen, A. P., & Wang, C.-H. (2008) A RFID-Enabled Traceability System for the Supply Chain of Live Fish. *IEEE International Conference on Automation and Logistics*, 81-86.

Huanga, J., Wu, Y., & Rozelle, S. (2009). Moving off the farm intensifying agricultural production in Shandong: A case study of rural labor market linkages in China. *Agricultural Economics*, *2*(40), 203–218. doi:10.1111/j.1574-0862.2009.00370.x

Hughner, R. S., McDonagh, P., Prothero, A., Shultz, C. J., & Stanton, J. (2007). Who are organic food consumers? A compilation and review of why people purchase organic food. *Journal of Consumer Behaviour*, *6*(2-3), 94–110. doi:10.1002/cb.210

Hunter, W., Hutschemaekers, G. J., Unger, J.-P., de Paepe, P., Ghilbert, P., Soors, W., & Rissanen, P. et al. (2003). AP Nauta, J. von Grumbkow. *Policy*, *3*, 4.

Hutt, M. D., & Speh, T. W. (1984). The marketing strategy center: Diagnosing the industrial marketers interdisciplinary role. *Journal of Marketing*, *48*(4), 53–61. doi:10.2307/1251510

Hutt, M., & Speh, T. (2012). *Business marketing management: B2B*. Cengage Learning.

Hwang, Y.-M., Moon, J., & Yoo, S. (2015). Developing A RFID-based Food Traceability System in Korea Ginseng Industry: Focused on the Business Process Reengineering. *International Journal of Control and Automation*, *8*(4), 397–406. doi:10.14257/ijca.2015.8.4.36

ICSU. (2016). *International Council for Science, A draft framework for understanding SDG interactions by Måns Nilsson, Dave Griggs, Martin Visbeck and Claudia Ringler*. Retrieved August 29, 2016, from http://www.icsu.org/publications/reports-and-reviews/working-paper-framework-for-understanding-sdg-interactions-2016/SDG-interactions-working-paper.pdf

Idota, H., Bunno, T., & Tsuji, M. (2011). Empirical analysis of internal social media and product innovation: Focusing on SNS and social capital. *Proceedings of 22nd European Regional ITS Conference in Budapest.*

Idota, H., Ueki, Y., Bunno, T., Shinohara, S., & Tsuji, M. (2014). Role of ICT in the innovation process based on firm-level evidence from four ASEAN economies: An SEM approach. *25th European Regional Conference on the International Telecommunications Society (ITS).*

IFAD. (2010). *International Fund for Agricultural Development - Rural Poverty Report 2011: New realities, new challenges: new opportunities for tomorrow's generation, IFAD, Rome*. Retrieved August 20, 2016, from https://www.ifad.org/documents/10180/c47f2607-3fb9-4736-8e6a-a7ccf3dc7c5b

IISD. (2013). *Climate risk management for agriculture in Peru: Focus on the regions of Junín and Piura*. International Institute for Sustainable Development.

ILO. (2013). *International Labour Organisation. Global Employment Trends 2013*. Retrieved August 29, 2016, from http://www.ilo.org/wcmsp5/groups/public/---dgreports/---dcomm/---publ/documents/publication/wcms_202326.pdf

ILO. (2016). *World Employment Social Outlook - Trends 2016*. Retrieved August 29, 2016, from http://www.ilo.org/wcmsp5/groups/public/---dgreports/---dcomm/---publ/documents/publication/wcms_443480.pdf

INE. (2013). *Industrial Companies Survey*. CNAE 09. Retrieved July, 2016, from http://www.ine.es/jaxiT3/Tabla.htm?t=2539

Inea. (2013). *Annuario dell'Agricoltura Italiana– 2012* (vol. 66). Author.

Iribarren, D., Hospido, A., Moreira, M. T., & Feijoo, G. (2011). Benchmarking environmental and operational parameters through eco-efficiency criteria for dairy farms. *The Science of the Total Environment*, *409*(10), 1786–1798. doi:10.1016/j.scitotenv.2011.02.013 PMID:21371738

Iribarren, D., Vasquez-Rowe, I., Moreira, M. T., & Feijoo, G. (2010). Further potentials in the joint implementation of life cycle assessment and data envelopment analysis. *The Science of the Total Environment*, *408*(22), 5265–5272. doi:10.1016/j.scitotenv.2010.07.078 PMID:20797768

Ismaila, S. S. (2008). *Off farm employment among paddy farmers in the Muda and Kemasin Semarak granary areas*. Universiti Putra Malaysia.

Ismail, R., Jeffery, R., & Van Belle, J.-P. (2011). Using ICT as a value adding tool in South African SMEs. *Journal of African Research in Business & Technology*, *2011*, 1–12. doi:10.5171/2011.470652

Ismea-Qualivita. (2013). *Rapporto 2013 – Qualivita*. ISMEA.

Istat. (2016). Retrieved from www.istat.it

Jackson, H., & Svensson, K. (2002). *Ecovillage Living: Restoring the Earth and Her People*. Totnes: Green Books.

Jackson, R. (2004). The Ecovillage Movement. *Permaculture Magazine*, *40*, 25–30.

Jahanshahloo, G. R., Lotfi, F. H., Shoja, N., Tohidi, G., & Razavyan, S. (2005). Undesirable inputs and outputs in DEA models. *Applied Mathematics and Computation*, *169*(2), 917–925. doi:10.1016/j.amc.2004.09.069

Jakkhupan, W., Arch-int, S., & Li, Y. (2015). An RFID-based traceability system A case study of rice supply chain. *Telecommunication Systems*, *58*(3), 243–258. doi:10.1007/s11235-014-9866-7

Janssen, M., & Hamm, U. (2012). Product labelling in the market for organic food: Consumer preferences and willingness-to-pay for different organic certification logos. *Food Quality and Preference*, *25*(1), 9–22. doi:10.1016/j.foodqual.2011.12.004

Jarossova, M. A., &Mind'asova, K. (2015). Comparison of the application of marketing communication tools used by organic food producers in Slovakia and Austria. *NaukiInżynierskieiTechnologie*, *4*(19).

Jenkins, M., & McDonald, M. (1997). Market segmentation: Organizational archetypes and research agendas. *European Journal of Marketing*, *31*(1), 17–32. doi:10.1108/03090569710157016

Johnson, A., Dibrell, C., & Hansen, E. (2009, Spring/Fall). Market Orientation, Innovativeness, and Performance of Food Companies. *Journal of Agribusiness*, *27*(1/2), 85–106.

Joines, J. L., Scherer, C. W., & Scheufele, D. A. (2003). Exploring Motivations For Consumer Web Use And Their Implications For E-Commerce. *Journal of Consumer Marketing*, *20*(2), 90–108. doi:10.1108/07363760310464578

Jurin, R. (2012). *Principles of Sustainable Living: A New Vision for Health, Happiness, and Prosperity*. Champaign, IL: Human Kinetics.

Kahan, D. (2008). *Managing risk in farming, Farm management extension guide*. Rome: Food and Agriculture Organization of the United Nations.

Kang, Y.-S., Park, I.-H., Rhee, J., & Lee, Y.-H. (2016). MongoDB-Based Repository Design for IoT-Generated RFID/Sensor Big Data. *IEEE Sensors Journal*, *16*(2), 485–497. doi:10.1109/JSEN.2015.2483499

Karipidis, P., & Sartzetakis, E. (2013). Firm's decisions based on consumers' choices in eco-certified food markets. *Economics Research International*, 1–10. doi:10.1155/2013/920164

Karipidis, P., Chrysochou, P., & Karypidou, I. (2016, June). *Does the Greek food supply respond to the world challenges?* Paper presented at the second International Conference of Development and Economy, Thessaloniki, Greece.

Karipidis, P., Tsakiridou, E., Aggelopoulos, S., & Belidis, A. (2010). Consumers' purchasing and store switching intentions in respect of eco-marked products. *International Journal of E-Business Research*, *2*(6), 511–524. doi:10.1504/IJEBR.2010.035702

Karipidis, P., Tselempis, D., Karypidou, I., & Aggelopoulos, S. (in press). Market-driven or policy-directed quality certification? *Sinergie: Italian Journal of Management*.

Karipidis, P., & Tselepmis, D. (2014). Farmers intention to maintain quality certification. *EuroMed Journal of Business*, *9*(1), 93–110. doi:10.1108/EMJB-04-2013-0015

Karkazis, J., & Thanassoulis, E. (1998). Assessing the Effectiveness of Regional Development Policies in Northern Greece Using Data Envelopment Analysis. *Socio-Economic Planning Sciences*, *32*(2), 123–137. doi:10.1016/S0038-0121(97)00023-2

Kassahun, A., Hartog, R. J. M., Sadowski, T., Scholten, H., Bartram, T., Wolfert, S., & Beulens, A. J. M. (2014). Enabling chain-wide transparency in meat supply chains based on the EPCIS global standard and cloud-based services. *Computers and Electronics in Agriculture*, *109*, 179–190. doi:10.1016/j.compag.2014.10.002

Kassahun, A., Hartog, R. J. M., & Tekinerdogana, B. (2016). Realizing chain-wide transparency in meat supply chains based on global standards and a reference architecture. *Computers and Electronics in Agriculture*, *123*, 275–291. doi:10.1016/j.compag.2016.03.004

Kastberg, P. (2015). Promoting communication, participation, and learning with regard to organic food products: A communication theoretical approach. *Ecology and Society*, *20*(1), 3. doi:10.5751/ES-07139-200103

Kaynak, E., Tatoglu, E., & Kula, V. (2005). An analysis of the factors affecting the adoption of electronic commerce by SMEs: Evidence from an emerging market. *International Marketing Review*, *22*(6), 623–640. doi:10.1108/02651330510630258

Kearney, D. (2013). Food Deprivations as Crimes against Humanity. *NYUJ Int'l L. & Pol.*, *46*, 253.

Kelepouris, T., Pramatari, K., & Doukidis, G. (2007). RFID-enabled traceability in the food supply chain. *Industrial Management & Data Systems*, *107*(2), 183–200. doi:10.1108/02635570710723804

Kenny, D. A. (2014). *Measuring model fit*. Available online at http://davidakenny.net/cm/fit.htm#RMSEA

Kenny, D. A., Kaniskan, B., & McCoach, D. B. (2014). The performance of RMSEA in models with small degrees of freedom. *Sociological Methods & Research*.

Kersting, S., & Wollni, M. (2012). New institutional arrangements and standard adoption: Evidence from small-scale fruit and vegetable farmers in Thailand. *Food Policy*, *37*(4), 452–462. doi:10.1016/j.foodpol.2012.04.005

Kesse-Guyot, E., Peneau, S., Mejean, C., de Edelenyi, F. S., Galan, P., Hercberg, S., & Lairon, D. (2013). Profiles of organic food consumers in a large sample of French adults: Results from the Nutrinet-Sante cohort study. *PLoS ONE*, *8*(10), e76998. doi:10.1371/journal.pone.0076998 PMID:24204721

Khoshnevisan, B., Rafiee, S., Omid, M., & Mousazadeh, H. (2013). Applying data envelopment analysis approach to improve energy efficiency and reduce GHG (greenhouse gas) emission of wheat production. *Energy*, *58*, 588–593. doi:10.1016/j.energy.2013.06.030

Khoshnevisan, B., Rafiee, S., Omid, M., & Mousazadeh, H. (2013). Reduction of CO_2 emission by improving energy use efficiency of greenhouse cucumber production using DEA approach. *Energy*, *55*, 676–682. doi:10.1016/j.energy.2013.04.021

Khoshnevisan, B., Rafiee, S., Omid, M., Mousazadeh, H., & Sefeedpari, P. (2013). Prognostication of environmental indices in potato production using artificial neural networks. *Journal of Cleaner Production*, *52*, 402–409. doi:10.1016/j.jclepro.2013.03.028

Khoshnevisan, B., Rafiee, S., Omid, M., Yousefi, M., & Movahedi, M. (2013). Modeling of energy consumption and GHG (greenhouse gas) emissions in wheat production in Esfahan province of Iran using artificial neural networks. *Energy*, *52*, 333–338. doi:10.1016/j.energy.2013.01.028

Khoshroo, A., Mulwa, R., Emrouznejad, A., & Arabi, B. (2013). A non-parametric Data Envelopment Analysis approach for improving energy efficiency of grape production. *Energy*, *63*, 189–194. doi:10.1016/j.energy.2013.09.021

Kijima, Y., Matsumoto, T., & Yamano, T. (2006). Nonfarm employment, agricultural shocks, poverty dynamics: Evidence from rural Uganda. *Agricultural Economics*, 459-467.

Kilelu, C. W., Klerkx, L., & Leeuwis, C. (2013). Unravelling the role of innovation platforms in supporting co-evolution of innovation: Contributions and tensions in a smallholder dairy development programme. *Agricultural Systems*, *118*, 65–77. doi:10.1016/j.agsy.2013.03.003

Kilic, T., Carletto, C., Miluka, J., & Savastano, S. (2009). Rural nonfarm income its effect on agriculture: Evidence from Albania. *Agricultural Economics*, *2*(40), 139–160. doi:10.1111/j.1574-0862.2009.00366.x

Kim, Y. G., & Woo, E. (2016). Consumer acceptance of a quick response (QR) code for the food traceability system: Application of an extended technology acceptance model (TAM). *Food Research International*, *85*, 266–272. doi:10.1016/j.foodres.2016.05.002

King, R. P., Hand, M. S., Di Giacomo, G., Clancy, K., Gomez, M. I., Hardesty, S. D., . . . McLaughlin, E. W. (2010). *Comparing the Structure, Size and Performance of Local and Mainstream Food Supply Chains.* USDA ERS Report Number 99.

Kleijnen, J. P. (2005). Supply chain simulation tools and techniques: A survey. *International Journal of Simulation and Process Modelling*, *1*(1-2), 82–89. doi:10.1504/IJSPM.2005.007116

Klerkx, L., De Grip, K., & Leeuwis, C. (2006). Hands off but strings attached: The contradictions of policy-induced demand-driven agricultural extension. *Agriculture and Human Values*, *23*(2), 189–204. doi:10.1007/s10460-005-6106-5

Klerkx, L., & Leeuwis, C. (2008). Institutionalizing end-user demand steering in agricultural R&D: Farmer levy funding of R&D in The Netherlands. *Research Policy*, *37*(3), 460–472. doi:10.1016/j.respol.2007.11.007

Klerkx, L., & Leeuwis, C. (2009). Operationalizing demand-driven agricultural research: Institutional influences in a public and private system of research planning in The Netherlands. *Journal of Agricultural Education and Extension*, *15*(2), 161–175. doi:10.1080/13892240902909080

Klerkx, L., & Nettle, R. (2013). Achievements and challenges of innovation coproduction support initiatives in the Australian and Dutch dairy sectors: A comparative study. *Food Policy*, *40*, 74–89. doi:10.1016/j.foodpol.2013.02.004

Knickel, K., & Peter, S. (2005). Amenity-led development of rural areas: The example of the Regional Action pilot programme in Germany. In G. P. Green, D. Marcouiller & S. Deller (Eds.), Amenities and rural development: Theory, methods and public policy (pp. 302-321). Northampton, UK: Edward Elgar Publishing.

Knickel, K., Brunori, G., Rand, S., & Proost, J. (2009). Towards a better conceptual framework for innovation processes in agriculture and rural development: From linear models to systemic approaches. *Journal of Agricultural Education and Extension*, *15*(2), 131–146. doi:10.1080/13892240902909064

Konno, H., & Yamazaki, H. (1991). Mean-Absolute Deviation Portfolio Optimization Model and Its Applications to Tokyo Stock Market. *Management Science*, *37*(5), 519–531. doi:10.1287/mnsc.37.5.519

Kotler, P., & Keller, K. L. (2009). *Dirección de marketing*. Pearson educación.

Kotler, P. (2012). *Kotler on marketing*. Simon and Schuster.

Kotler, P., & Armstrong, G. (2014). *Principles of Marketing* (15th ed.). Pearson.

Kotler, P., Armstrong, G., Saunders, J., & Wong, V. (2003). *Principi di Marketing (Italian Translation of Principles of Marketing)*. ISEDE and Prentice Hall International.

Kotler, P., & Caslione, J. A. (2009). *Chaotics: the business of managing and marketing in the age of turbulence*. AMACOM Div American Mgmt Assn.

Kottila, M., & Rönni, P. (2010). Communication between actors of food chains: Case studies of two organic food chains in Finland. In A. Lindgreen, M. K. Hingley, D. Harness, & P. Custance (Eds.), *Market orientation: Transforming food and agribusiness around the customer* (pp. 171–186). Surrey, UK: Gower.

Kotz, S, Banks, D.L.,& Read, C.B.(1999). *Wiley Series in Probability and Statistics*. Wiley-Interscience.

Krippendorff, K. (2004). Reliability in content analysis: Some common misconceptions and recommendations. *Human Communication Research*, *30*(3), 411–433.

Kristoferson, L. A., & Bokalders, V. (2013). *Renewable energy technologies: their applications in developing countries*. Elsevier.

Kronjee, G., & Nooteboom, B. (2008). Research, higher education, and innovation. In B. Nooteboom & E. Stam (Eds.), *Micro-Foundations for innovation policy*. Amsterdam, NL: Amsterdam University Press.

Krystallis, A., & Chryssohoidis, G. (2005). Consumers willingness to pay for organic food: Factors that affect it and variation per organic product type. *British Food Journal*, *107*(5), 320–343. doi:10.1108/00070700510596901

Krystallis, A., Fotopoulos, C., & Zotos, Y. (2006). Organic consumers profile and their willingness to pay (WTP) for selected organic food products in Greece. *Journal of International Consumer Marketing*, *19*(1), 81–106. doi:10.1300/J046v19n01_05

Kumar, S., & Nigmatullin, A. (2011). A system dynamics analysis of food supply chains- Case studies with non-perishable products. *Simulation Modelling Practice and Theory*, *19*(10), 2151–2168. doi:10.1016/j.simpat.2011.06.006

Kuosmanen, T., & Kortelainen, M. (2005). Measuring Eco-efficiency of Production with DataEnvelopment Analysis. *Journal of Industrial Ecology*, *9*(4), 59–72. doi:10.1162/108819805775247846

Laari, S., Töyli, J., Solakivi, T., & Ojala, L. (2016). Firm performance and customer-driven green supply chain management. *Journal of Cleaner Production*, *112*, 1960–1970. doi:10.1016/j.jclepro.2015.06.150

Laarman, J. C., & Perdue, R. R. (1989). Tropical Science and Tourism: The Case of OTS in Costa Rica. *Annals of Tourism Management*, *10*(1), 29–38. doi:10.1016/0261-5177(89)90032-0

Lakes, T., Muller, D., & Kruger, C. (2009). Cropland change in southern Romania: A comparison of logistic regressions and artificial neural networks. *Landscape Ecology*, *24*(9), 1195–1206. doi:10.1007/s10980-009-9404-2

Lambert, D. M., Stock, J. R., & Pagh, J. D. (1998). Supply Chain Management: Implementation Issues and Research Opportunities. *The International Journal of Logistics Management*, *9*(2), 1–19. doi:10.1108/09574099810805807

Lea, E., & Worsley, T. (2005). Australians organic food beliefs, demographics and values. *British Food Journal*, *107*(11), 855–869. doi:10.1108/00070700510629797

Leavitt, N. (2010). Will NoSQL databases live up to their promise? *Computer*, *43*(2), 12–14. doi:10.1109/MC.2010.58

Lee, H. J., & Yun, Z. S. (2015). Consumers perceptions of organic food attributes and cognitive and affective attitudes as determinants of their purchase intentions toward organic food. *Food Quality and Preference*, *39*, 259–267. doi:10.1016/j.foodqual.2014.06.002

Lee, K., Kim, H., & Vohs, K. D. (2011). Stereotype threat in the marketplace: Consumer anxiety and purchase intentions. *The Journal of Consumer Research*, *38*(2), 343–357. doi:10.1086/659315

Lehmann, R. J., Reiche, R., & Schiefer, G. (2012). Future Internet And The Agri-Food Sector: State-Of-The-Art In Literature And Research. *Computers and Electronics in Agriculture*, *89*, 158–174. doi:10.1016/j.compag.2012.09.005

Lemos, M. C., & Agrawal, A. (2006). Environmental governance. *Annual Review of Environment and Resources*, *31*(1), 297–325. doi:10.1146/annurev.energy.31.042605.135621

Levina, O. (2010). *Business market segmentation: Segmenting market for AmCham Finland*. Academic Press.

Liang, R. D. (2014). Enthusiastically consuming organic food:: an analysis of the online organic food purchasing behaviors of consumers with different food-related lifestyles. *Internet Research: Electronic Networking Applications and Policy, 24*(5), 3-4.

Lim, B., Spanger-Siegfried, E., Burton, I., Malone, E., & Huq, S. (2005). *Adaptation Policy Frameworks for Climate Change: Developing Strategies, Policies and Measures*. Cambridge, UK: Cambridge University Press.

Lim, C., & McAleer, M. (2005). Ecologically Sustainable Tourism Management. *Environmental Modelling & Software*, *20*(11), 1431–1438. doi:10.1016/j.envsoft.2004.09.023

Lindh, H., & Olsson, A. (2010). Communicating imperceptible product attributes through traceability: A case study in an organic food supply chain. *Renewable Agriculture and Food Systems*, *25*(04), 263–271. doi:10.1017/S1742170510000281

Litfin, K. T. (2014). *Ecovillages: Lessons for Sustainable Community*. Cambridge, UK: Polity Press.

Little, P., Stone, P., Mogues, T., Castro, P., & Negatu, W. (2006). Moving in place drought and poverty dynamics in South Wollo, Ethiopia. *Development Studies*, (42), 200-225.

Li, W., Zhang, F., & Jiang, M. (2008). A simulation- and optimisation-based decision support system for an uncertain supply chain in a dairy firm. *International Journal of Business Information Systems*, *3*(2), 183–200. doi:10.1504/IJBIS.2008.016585

Llanto, G. (2012). *The impact of infrastructure on agricultural productivity*. Philippine Institute for Development Studies, Series: Discussion paper series (Philippine Institute for Development Studies), no. 2012-12.

Lobo, A., Mascitelli, B., & Chen, J. (2014). Opportunities for small and medium enterprises in the innovation and marketing of organic food: Investigating consumers purchase behaviour of organic food products in Victoria, Australia. *AI & Society*, *29*(3), 311–322. doi:10.1007/s00146-013-0457-z

Lockie, S., Lyons, K., Lawrence, G., & Grice, J. (2004). Choosing organics: A path analysis of factors underlying the selection of organic food among Australian consumers. *Appetite*, *43*(2), 135–146. doi:10.1016/j.appet.2004.02.004 PMID:15458800

Lockie, S., Lyons, K., Lawrence, G., & Mummery, K. (2002). Eating green: Motivations behind organic food consumption in Australia. *Sociologia Ruralis*, *42*(1), 23–40. doi:10.1111/1467-9523.00200

Loebnitz, N., & Aschemann-Witzel, J. (2016). Communicating organic food quality in China: Consumer perceptions of organic products and the effect of environmental value priming. *Food Quality and Preference*, *50*, 102–108. doi:10.1016/j.foodqual.2016.02.003

Loureiro, M. L., & Hine, S. (2002). Discovering niche markets: A comparison of consumer willingness to pay for local (Colorado grown), organic, and GMO-free products. *Journal of Agricultural and Applied Economics*, *34*(03), 477–487. doi:10.1017/S1074070800009251

Lozano, S., Iribarren, D., Moreira, M. T., & Feijoo, G. (2009). The link between operational efficiency environmental impacts. A joint application of Life Cycle Assessment and Data Envelopment Analysis. *The Science of the Total Environment*, *407*(5), 1744–1754. doi:10.1016/j.scitotenv.2008.10.062 PMID:19056104

Lozano, S., Villa, G., & Brannlund, R. (2009). Centralized reallocation of emission permits using DEA. *European Journal of Operational Research*, *193*(3), 752–760. doi:10.1016/j.ejor.2007.07.029

Lyneis, J. (2000). System Dynamics for market forecasting and structural analysis. *System Dynamics Review*, *16*(1), 3–25. doi:10.1002/(SICI)1099-1727(200021)16:1<3::AID-SDR183>3.0.CO;2-5

MacCallum, R. C., Browne, M. W., & Sugawara, H. M. (1996). Power analysis and determination of sample size for covariance structure modelling. *Psychological Methods*, *1*(2), 130–149. doi:10.1037/1082-989X.1.2.130

Magnusson, M. K., Arvola, A., Koivisto Hursti, U.-K., Åberg, L., & Sjödén, P.-O. (2001). Attitudes towards organic foods among Swedish consumers. *British Food Journal, 103*(3), 209–227. doi:10.1108/00070700110386755

MAGRAMA. (2008). *La Distribución Agroalimentaria Y Transformaciones Estratégicas En La Cadena De Valor*. Retrieved July, 2016, from https://www.google.es/search?q=La+Distribuci%C3%B3n+Agroalimentaria+Y+Transformaciones+Estrat%C3%A9gicas+En+La+Cadena+De+Valor&ie=utf-8&oe=utf-8&gws_rd=cr&ei=p3YSVvFFxLZ43KeIsA0

MAGRAMA. (2013). *Canales Cortos De Comercialización En El Sector Agroalimentario*. Retrieved July, 2016, from http://www.magrama.gob.es/es/alimentacion/servicios/observatorio-de-precios-de-los-alimentos/estudioccc_tcm7-272187.pdf

MAGRAMA. (2014). *Industria Alimentaria En 2013-2014*. Retrieved July, 2016, from http://www.magrama.gob.es/es/alimentacion/temas/industria-agroalimentaria/_informeindustriaalimentaria2013-2014_tcm7-156934.pdf

Maillard, S., Roncalli, T., & Teiletche, J. (2012). *Managing Risk Exposures using Risk Budgeting Approach*. Working Paper.

Maillard, S., Roncalli, T., & Teiletche, J. (2009). On the properties of equally weighted risk contributions portfolios. *Journal of Portfolio Management*.

Mainetti, L., Mele, F., Patrono, L., Simone, F., Stefanizzi, M. L., & Vergallo, R. (2013). An RFID-Based Tracing and Tracking System for the Fresh Vegetables Supply Chain. *International Journal of Antennas and Propagation. Article ID, 531364*, 1–15.

Makatouni, A. (2002). What motivates consumers to buy organic food in the UK? Results from a qualitative study. *British Food Journal, 104*(3/4/5), 345-352.

Maldonado, U. P. T., Khan, G. F., Moon, J., & Rho, J. J. (2009). E-learning motivation, students' acceptance/use of educational portal in developing countries. *Proceeds of the 4th International Conference on Computer Sciences and Convergence Information Technology*.

Malhan, S. (2015). Impact of demographic and entrepreneur's profile on level of supply management chain practices in selected small scale and medium enterprises (SMEs) in retail sector. *International Journal of Business Management and Scientific Research, 2*.

Malhotra, N. K. (2004). *Marketing research: An applied orientation*. Upper Saddle River, NJ: Prentice-Hall.

Mandilas, A., Karasavvoglou, A., Nikolaidis, M., & Tsourgiannis, L. (2013). Predicting Consumers Perceptions in On-line Shopping. *Procedia Technology, 8*, 435–444. doi:10.1016/j.protcy.2013.11.056

Manganelli, S., & Engle, R. F. (2001). *Value at Risk Models in Finance*. European Central Bank (ECB) Research Paper Series Working Papers.

Manikas, I., Hamann, K., & Sentic, A. (2016). Quality management schemes and connections to the concept of sustainability in the food chain. In E. Iakovou, D. Bochtis, D. Vlachos, & D. Aidonis (Eds.), *Supply chain management for sustainable food networks* (pp. 233–254). Chichester, UK: Wiley. doi:10.1002/9781118937495.ch9

Marhieson, A., & Wall, G. (1982). *Tourism: Economic, Physical and Social Impacts*. Harlow: Pearson.

Marian, L., Chrysochou, P., Krystallis, A., & Thøgersen, J. (2014). The role of price as a product attribute in the organic food context: An exploration based on actual purchase data. *Food Quality and Preference, 37*, 52–60. doi:10.1016/j.foodqual.2014.05.001

Marketing4Ecommerce. (2013). El comercio electrónico alcanzacifras record en España. *Marketing4Ecommerce*. Retrieved July, 2016, from http://marketing4ecommerce.net/el-comercio-electronico-alcanza-cifras-record-en-espana/)

Markowitz, H.M. (1952). Portfolio selection. *Journal of Finance, 7*, 77-91.

Markowitz, H. (1987). *Mean-Variance analysis in portfolio choice and capital markets*. Basil Blackwell.

Marletta, V. R., & Biere, A. (2009). Entepreneurial supply chains and strategic collaboration: the case of Bagòss cheese in Bagolino, Italy. *International Food and Agribusiness Management Review, 12*(3), 49-68.

Marsden, T., Banks, J., & Bristow, G. (2000). Food supply chain approaches: Exploring their role in rural development. *Sociologia Ruralis, 40*(4), 424–438. doi:10.1111/1467-9523.00158

Marsden, T., & Sonnino, R. (2008). Rural development and the regional state: Denying multifunctional agriculture in the UK. *Journal of Rural Studies, 24*(4), 422–431. doi:10.1016/j.jrurstud.2008.04.001

Martinez, S., Hand, M., Da Pra, M., Pollack, S., Ralston, K., Smith, T., . . . Newman, C. (2010). *Local Food Systems Concepts, Impacts and Issues*. Economic Research Report No. 97, United State Department of Agriculture.

Martin, L. M., & Matlay, H. (2001). Blanket approach to promoting ICT in small firms: Some lessons from the DTI ladder adoption model in the UK. *Internet Research, 11*(5), 399–410. doi:10.1108/EUM0000000006118

Masakure, O., Cranfield, J., & Henson, S. (2011). Factors affecting the incidence and intensity of standards certification evidence from exporting firms in Pakistan. *Applied Economics, 43*(8), 901–915. doi:10.1080/00036840802600103

Massaroni, E., Cozzolino, A., & Wankowicz, E. (2015). Sustainability in supply chain management – A literature review. *Sinergie: Italian Journal of Management, 33*(98), 331–355. doi:10.7433/1115

Mata, F. J., & Quesada, A. (2014). Web 2.0, Social Networks And E-Commerce As Marketing Tools. *Journal of Theoretical and Applied Electronic Commerce Research, 9*(1), 56–69. doi:10.4067/S0718-18762014000100006

McCain. (2016). *KarposFrontidas*. Retrieved September 1, 2016 from http://www.karposfrontidas.gr/

McCain. (2016a). *Responsible Business*. Retrieved September 1, 2016 from http://www.mccain.com/good-business/responsible-business

McDonald, M. (2012). *Market Segmentation*. Wiley Online Library. doi:10.1002/9781119207863

McDonald, M., & Wilson, H. (2011). *Marketing plans: How to prepare them, how to use them*. John Wiley & Sons.

McDougall, R. A., Elbehri, A., & Truong, T. P. (1998). Global Trade Assistance and Protection: The GTAP 4 Data Base. Center for Global Trade Analysis, Purdue University.

McEachern, M. G., & Mcclean, P. (2002). Organic purchasing motivations and attitudes: Are they ethical? *International Journal of Consumer Studies, 26*(2), 85–92. doi:10.1046/j.1470-6431.2002.00199.x

McLaughlin, C., & Davidson, G. (1984). *Leadership and Power in Intentional Communities. In Context, 7*, 23.

McNeil, A. J., & Frey, R. (2000). Estimation of Tail-Related Risk Measures for Heteroscedastic Financial Time Series: An Extreme Value Approach. *Journal of Empirical Finance, 7*(3-4), 271–300. doi:10.1016/S0927-5398(00)00012-8

Meccheri, N., & Pelloni, G. (2006). Rural Entrepreneurs and Institutional Assistance: An Empirical Study from Mountainous Italy. *Entrepreneurship & Regional Development, 18*(5), 371–392. doi:10.1080/08985620600842113

Meijl, H. V., Rheenen, T. V., Tabeau, A., & Eickhout, B. (2006). The impact of different policy environments on agricultural land use in Europe. *Agriculture, Ecosystems & Environment, 114*(1), 21–38. doi:10.1016/j.agee.2005.11.006

Melchior, M. R. (2011). From Design Nations to Fashion Nations? Unpacking Contemporary Scandinavian Fashion Dreams. *Fashion Theory, 15*(2), 177–200. doi:10.2752/175174111X12954359478681

MERCASA. (2013). *Informe Sobre Producción, Industria, Distribución Y Consumo de Alimentación en España*. Retrieved July, 2016, from http://www.mercasa.es/noticias/post/informe_2013_sobre_produccion__industria__distribucion_y_consumo_de_alimentacion_en_espana

Merton, R. C. (1972). An Analytic derivation of the e¢ cient portfolio frontier. *Journal of Financial and Quantitative Analysis*, *7*(4), 1851–1872. doi:10.2307/2329621

Merton, R. C. (1980). On estimating the expected return on the market. *Journal of Financial Economics*, *8*(4), 323–361.

Meyer, L. R. (2011). *Microcredit and Agriculture: Challenges, Successes, and Prospects*. Retrieved August 20, 2016, from https://www.incofin.com/sites/default/files/attachments/publications/Microfinance%20and%20agriculture.pdf

Michaelidou, N., & Hassan, L. M. (2008). The role of health consciousness, food safety concern and ethical identity on attitudes and intentions towards organic food. *International Journal of Consumer Studies*, *32*(2), 163–170. doi:10.1111/j.1470-6431.2007.00619.x

Migliore, G., Forno, F., Guccione, G.D., & Schifani, G. (2014). Food Community Networks as sustainable self-organized collective action: a case study of a solidarity purchasing group. *New medit: Mediterranean journal of economics, agriculture and environment- Revue méditerranéenne d'economie, agriculture et environment*, *13*(4), 54-62.

Milestad, R., & Darnhofer, I. (2003). Building farm resilience: The prospects and challenges of organic farming. *Journal of Sustainable Agriculture*, *22*(3), 81–97. doi:10.1300/J064v22n03_09

Millennium Ecosystem Assessment. (2005). *Ecosystems and Human Well-being: Synthesis*. Retrieved August 20, 2016, from http://www.millenniumassessment.org/documents/document.356.aspx.pdf

Miller, G., & Spoolman, S. (2012). *Living in the Environment: Principles, Connections, and Solutions*. Belmont: CENGAGE Learning.

Minegishi, S., & Thiel, D. (2000). System dynamics modeling and simulation of a particular food supply chain. *Simulation Practice and Theory*, *8*(5), 321–339. doi:10.1016/S0928-4869(00)00026-4

Miorandi, D., Sicari, S., De Pellegrini, F., & Chlamtac, I. (2012). Internet of things: Vision, applications and research challenges. *Ad Hoc Networks*, *10*(7), 1497–1516. doi:10.1016/j.adhoc.2012.02.016

Mobtaker, H. G., Akram, A., Keyhani, A., & Mohammadi, A. (2012). Optimization of energy required for alfalfa production using data envelopment analysis approach. *Energy for Sustainable Development*, *16*(2), 242–248. doi:10.1016/j.esd.2012.02.001

Moeskops, B., Buchan, D., Sleutel, S., Herawaty, L., Husen, E., Saraswati, R., & De Neve, S. et al. (2010). Soil microbial communities and activities under intensive organic and conventional vegetable farming in West Java, Indonesia. *Applied Soil Ecology*, *45*(2), 112–120. doi:10.1016/j.apsoil.2010.03.005

Mohammadi, A., Shahin, R., Jafari, A., Dalgaard, T., Trydeman Knudsen, M., Keyhani, A., & Hermansen, J. E. et al. (2013). Potential greenhouse gas emission reductions in soybean farming: A combined use of Life Cycle Assessment and Data Envelopment Analysis. *Journal of Cleaner Production*, *54*, 89–100. doi:10.1016/j.jclepro.2013.05.019

Mohapatra, S. (2013). Sustainability in e-commerce adoption in small and medium enterprises (SMEs): A case study in Odisha, India. *International Journal of Green Computing*, *4*(2), 12–23. doi:10.4018/jgc.2013070102

Moore, A., Holzworth, D., Herrmann, N., Brown, H., de Voil, P., Snow, V., & Huth, N. et al. (2014). Modelling the manager: Representing rule-based management in farming systems simulation models. *Environmental Modelling & Software*, *62*, 399–410. doi:10.1016/j.envsoft.2014.09.001

Morey, M. R., & Vinod, H. D. (2001). A Double Sharpe Ratio. *Advances in Investments Analysis and Portfolio Management, 8*, 57–61.

Moschidis, O. (2006). A proposal of comparative evaluation with the correspondence analysis [in Greek]. *Journal SPOUDAI, 56*(3), 95–113.

Moschidis, O. (2015). Unified coding of qualitative and quantitative variables and their analysis with ascendant hierarchical classification. *International Journal of Data Analysis Techniques and Strategies, 7*(2), 114–128. doi:10.1504/IJDATS.2015.068745

Moschidis, O. E. (2009). A different approach to multiple correspondence analysis (MCA) than that of specific MCA. *Mathématiques et sciences humaines. Mathematical Social Sciences*, (186), 77–88.

Moschidis, O., & Arabatzis, G. (2013). SHP stations and integrated rural development: A multivariate statistical approach. *International Journal of Green Economics, 7*(4), 333–347. doi:10.1504/IJGE.2013.058561

Moschidis, O., Spathis, C., & Floropoulos, I. (2009). Methodological approach to Multidimensional exploratory evaluation of Taxis (Taxation information system) usefulness:Greek economy in Perspective. *Journal of Financial Management and Analysis:International Review of Finance, 22*(2), 1–12.

Mundubat (2012). *Circuitos Cortos De Comercialización En Euskal Herria*. Retrieved July, 2016, from http://www.baserribizia.info/index.php/euskalherria/ekimenak/4134 dossiersobrecircuitosc ortosdecomercializaccioneneuskalherria

Muriithi, B. W., Mburu, J., & Ngigi, M. (2011). Constraints and determinants of compliance with EurepGap standards: A case of smallholder French bean exporters in Kirinyaga district, Kenya. *Agribusiness, 27*(2), 193–204. doi:10.1002/agr.20261

Musa, A., & Abba Dabo, A.-A. (2016). A Review of RFID in Supply Chain Management: 2000–2015. *Global Journal of Flexible Systems Management, 17*(2), 189–228. doi:10.1007/s40171-016-0136-2

Mylan, J., Geels, F. W., Gee, S., McMeekin, A., & Foster, C. (2014). Eco-innovation and retailers in milk, beef and bread chains: Enriching environmental supply chain management with insights from innovation studies. *Journal of Cleaner Production, 107*, 20–30. doi:10.1016/j.jclepro.2014.09.065

Nadeem, N., Mushtaq, K., & Javed, M. I. (2011). Impact of social and physical infrastructure on agricultural productivity in Punjab, Pakistan - A production function approach. *Pakistan Journal of Life and Social Sciences, 9*(2).

Nagarajan, S. (2008). Indians in Malaysia towards vision 2020. *Rising India and Indian communities in East Asia*, 375-398.

NamGung, K., Choi, Y., Park, S., & Jun, C. (2012). The Development of e-Pedigree Model for Securing Transparent Pharmaceutical Distribution Channel in Korea. *Convergence and Hybrid Information Technology*, 226-234.

Nasir, V. A., & Karakaya, F. (2014). Underlying motivations of organic food purchase intentions. *Agribusiness, 30*(3), 290–308. doi:10.1002/agr.21363

Neef, A., & Neubert, D. (2011). Stakeholder participation in agricultural research projects: A conceptual framework for reflection and decision-making. *Agriculture and Human Values, 28*(2), 179–194. doi:10.1007/s10460-010-9272-z

Nhamo, N., Rodenburg, J., Zenna, N., Makombe, G., & Luzi-Kihupi, A. (2014). Narrowing the rice yield gap in East and Southern Africa: Using and adapting existing technologies. *Agricultural Systems, 131*, 45–55. doi:10.1016/j.agsy.2014.08.003

Nicholson, C. F., & Kaiser, H. M. (2008). Dynamic market impacts of generic dairy advertising. *Journal of Business Research, 61*(11), 1125–1135. doi:10.1016/j.jbusres.2007.11.011

Nie, C., & Zepeda, L. (2011). Lifestyle segmentation of US food shoppers to examine organic and local food consumption. *Appetite*, *57*(1), 28–37. doi:10.1016/j.appet.2011.03.012 PMID:21477631

Njoh, A. J. (2012). Implications of spatial and physical structures for ICT as a tool of urban management and development in Cameroon. *Habitat International*, *36*(3), 343–351. doi:10.1016/j.habitatint.2011.06.006

Noppers, E. H., Keizer, K., Bolderdijk, J. W., & Steg, L. (2014). The adoption of sustainable innovations: Driven by symbolic and environmental motives. *Global Environmental Change*, *25*, 52–62. doi:10.1016/j.gloenvcha.2014.01.012

Nordin, S. M., Noor, S. M., & Saad, M. S. M. (2014). Innovation Diffusion of New Technologies in the Malaysian Paddy Fertilizer Industry. *Procedia: Social and Behavioral Sciences*, *109*, 768–778. doi:10.1016/j.sbspro.2013.12.542

NRI. (2001). *Natural Resources Institute - University of Greenwich, Ethical Consumers & Ethical Trade: A Review Of Current Literature*. Retrieved August 17, 2016, from http://www.nri.org/projects/publications/policyseries/PolicySeriesNo12.pdf

Nunnally, J. C. (1978). *Psychometric theory*. New York, NY: McGraw-Hill.

OECD. (2008). *Promoting Sustainable Consumption, Good Practices In OECD Countries*. Retrieved August 17, 2016, from http://www.oecd.org/greengrowth/40317373.pdf

OECD. (2015). *International Standards Considered In The Oecd-FAO Guidance For Responsible Agricultural Supply Chains*. Retrieved August 19, 2016, from https://mneguidelines.oecd.org/OECD-FAO-Guidance_International-Standards.pdf

OHara, S. U., & Stagl, S. (2001). Global food markets and their alternatives: A socio-ecological economic perspective. *Population and Environment*, *22*(6), 533–554. doi:10.1023/A:1010795305097

Olech, E., & Kuboń, M. (2016). Clients preferences and development of organic food distribution channels. *Agricultural Engineering*, *20*(1), 119–125. doi:10.1515/agriceng-2016-0012

Olowa O.W., & Olowa, O.A. (2015). Factors Affecting Entrepreneurship Development in Agribusiness Enterprises in Lagos State. *Global Journal of Management and Business, 15*(7).

Onyango, B. M., Hallman, W. K., & Bellows, A. C. (2007). Purchasing organic food in US food systems: A study of attitudes and practice. *British Food Journal*, *109*(5), 399–411. doi:10.1108/00070700710746803

Organic Trade Association. (2015). *2015 Organic Industry Survey*. Retrieved November 05, 2016 from https://ota.com/sites/default/files/indexed_files/PolicyConference2015_Infographic_8.5x11_1a_0.pdf

Organic Trade Association. (2016). *State of the Organic Industry 2016*. Retrieved November 05, 2016 from http://ota.com/sites/default/files/indexed_files/OTA_StateofIndustry_2016.pdf

Ortobelli, S., Rachev, S. T., Biglova, A., & Stoyanov, S. (2010). Portfolio Selection based on a Simulated Copula. *Journal of Applied Functional Analysis*, *5*, 177–194.

Orton, J. D., & Weick, K. E. (1990). Loosely coupled systems: A reconceptualization. *Academy of Management Review*, *15*(2), 203–223.

Oseni, G., & Winters, P. (2009). Rural nonfarm activities and agricultural crop. *Agricultural Economics*, *2*(40), 189–201. doi:10.1111/j.1574-0862.2009.00369.x

Othman, Z. (2012). *Information and communication technology innovation as a tool for promoting sustainable agriculture: a case study of paddy farming in west Malaysia*. University of Malaya.

Otto, P. (2008). A system dynamics model as a decision aid in evaluating and communicating complex market entry strategies. *Journal of Business Research*, *61*(11), 1173–1181. doi:10.1016/j.jbusres.2007.11.013

Oubiña, J. (2000). *Las Relaciones De Poder Y Conflicto En Los Canales De Distribucion De Productos De Granconsumo: Un Analisis De La Satisfaccion En Las Relaciones Fabricante-Distribuidor Desde La Perspectiva Del Fabricante* (Unpublished doctoral dissertation). Universidad Autonoma de Madrid, Spain.

Oxfam. (2014). *Even it Up: Time to End Extreme Inequality*. Retrieved August 19, 2016, from https://www.oxfam.org/sites/www.oxfam.org/files/file_attachments/cr-even-it-upextreme-inequality-291014-en.pdf

Özbayrak, M., Theopisti, P., & Akgun, M. (2007). System dynamics modelling of a manufacturing supply chain system. *Simulation Modelling Practice and Theory*, *15*(10), 1338–1355. doi:10.1016/j.simpat.2007.09.007

Padel, S., & Foster, C. (2005). Exploring the gap between attitudes and behaviour: Understanding why consumers buy or do not buy organic food. *British Food Journal*, *107*(8), 606–625. doi:10.1108/00070700510611002

Pahlavan, R., Omid, M., & Akram, A. (2011). Energy use efficiency in greenhouse tomato production in Iran. *Energy*, *36*(12), 6714–6719. doi:10.1016/j.energy.2011.10.038

Pahlavan, R., Omid, M., & Akram, A. (2012). Energy input-output analysis and application of artificial neural networks for predicting greenhouse basil production. *Energy*, *37*(1), 171–176. doi:10.1016/j.energy.2011.11.055

Pahlavan, R., Omid, M., Rafiee, S., & Mousavi-Avval, S. H. (2012). Optimization of energy consumption for rose production in Iran. *Energy for Sustainable Development*, *16*(2), 236–241. doi:10.1016/j.esd.2011.12.001

Papaioannou, E., Georgiadis, C. K., Moschidis, O., & Manitsaris, A. (2015). Factors Affecting Customers Perceptions and Firms Decisions Concerning Online Fast Food Ordering. *International Journal of Agricultural and Environmental Information Systems*, *6*(1), 48–78. doi:10.4018/ijaeis.2015010104

Papajorgji, P. J., & Pardalos, P. M. (2005). *Software Engineering Techniques Applied to Agricultural Systems: An Object-Oriented and UML Approach (Applied Optimization)*. Secaucus, NJ: Springer-Verlag New York, Inc.

Parreño-Marchante, A., Alvarez-Melcon, A., Trebar, M., & Filippin, P. (2014). Advanced traceability system in aquaculture supply chain. *Journal of Food Engineering*, *122*, 99–109. doi:10.1016/j.jfoodeng.2013.09.007

Patel, R. (2013). The long green revolution. *The Journal of Peasant Studies*, *40*(1), 1–63. doi:10.1080/03066150.2012.719224

Paul, J., & Rana, J. (2012). Consumer behavior and purchase intention for organic food. *Journal of Consumer Marketing*, *29*(6), 412–422. doi:10.1108/07363761211259223

Pérez Mesa, J. C., & Galdeano-Gómez, E. (2015). Collaborative firms managing perishable products in a complex supply network: An empirical analysis of performance. *Supply Chain Management: An International Journal*, *20*(2), 128–138. doi:10.1108/SCM-06-2014-0185

Perrings, C. (2006). Resilience and sustainable development. *Environment and Development Economics*, *11*(4), 417–427. doi:10.1017/S1355770X06003020

Petroni, A., & Rizzi, A. (2001). Antecedents of MRP Adoption in small and medium-sized firms. *Benchmarking: An Integrated Journal*, *8*(2), 144–156. doi:10.1108/14635770110389852

Petrovic, D. (2010). *What is e-Marketing*. Retrieved July, 2016, from http://analogik. com/articles/231/e-marketing

Pfeiffer, L., Feldman, L. A., & Taylor, J. (2009). Is off-farm income reforming the farm? *Evidence from Maxico.*, 2(40), 125–138.

Picazo-Tadeo, A., Beltran-Esteve, M., & Gomez-Limon, J. A. (2012). Assessing eco-efficiency with directional distance functions. *European Journal of Operational Research*, *220*(3), 298–309. doi:10.1016/j.ejor.2012.02.025

Picazo-Tadeo, A., Gomez-Limon, J. A., & Martinez, E. R. (2011). Assessing farming eco-efficiency: A Data Envelopment Analysis approach. *Journal of Environmental Management*, *92*(4), 1154–1164. doi:10.1016/j.jenvman.2010.11.025 PMID:21193265

Piercy, N., & Giles, W. (1989). Making SWOT Analysis Work. *Marketing Intelligence & Planning*, *7*(5/6), 5–7. doi:10.1108/EUM0000000001042

Pieri, R. (Ed.). (2014). Il mercato del latte, Rapporto 2013. Milano, Italy: Franco Angeli.

Pike, A., Dawley, S., & Tomaney, J. (2010). Resilience, adaptation and adaptability. *Cambridge Journal of Regions Economy and Society*, *3*(1), 59–70. doi:10.1093/cjres/rsq001

Pinstrup-Andersen, P., & Shimokawa, S. (2007). Rural Infrastructure and Agricultural Development. In F. Bourguignon & B. Pleskovic (Eds.), *Annual World Bank Conference on Development Economics – Global 2007: Rethinking Infrastructure for Development* (pp. 175-203). Washington, DC: The World Bank.

Piramuthu, S., Farahani, P., & Grunow, M. (2013). RFID-generated traceability for contaminated product recall in perishable food supply networks. *European Journal of Operational Research, 225*(2), 253–262. doi:org/10.1016/j.ijpe.2013.04.035

Pjevčević, D., Radonjić, A., Hrle, Z., Čolić, V., (2012). DEA Window Analysis for Measuring Port Efficiencies in Serbia. *Traffic Management Review, 24*(1), 63-72.

Polyzos, S., & Arabatzis, G. (2005). *Labour Productivity of the Agricultural Sector in Greece: Determinant Factors and Interregional Differences Analysis.* Department of Planning and Regional Development, School of Engineering, University of Thessaly, Discussion Paper Series.

Porter, M. E., & Kramer, M. R. (2011, January–February). Creating shared value: How to reinvent capitalism—and unleash a wave of innovation and growth. *Harvard Business Review*, 1–17.

Poulton, C., Dorward, A., & Kydd, J. (2010). The future of small farms: New directions for services, institutions, and intermediation. *World Development*, *38*(10), 1413–1428. doi:10.1016/j.worlddev.2009.06.009

Prabovo, N. E., Foster, H. L., Nelson, S., Sitepu, B., & Nelson, P. (2012). *Practical use of oil palm nutrient physiological efficiency with regard to nutrient recovery and agronomic efficiencies at different Sumatran sites*. Academic Press.

Pramatari, K. (2016). Information technology for food supply chains. In E. Iakovou, D. Bochtis, D. Vlachos, & D. Aidonis (Eds.), *Supply chain management for sustainable food networks* (pp. 183–204). Chichester, UK: Wiley. doi:10.1002/9781118937495.ch7

Pruyt, E. (2013). *Small System dynamics models for big issues: Triple jump towards real-world complexity*. Delft: TU Delft Library.

PwC. (2016). *Global entertainment and media outlook 2016-2020*. PwC.

Qian, E. (2005). *Risk Parity Portfolios*. PanAgora Asset Management.

Rahman, H., & Ramos, I. (2014). Open innovation in SMEs: Prospects and Challenges. In *Information Systems for Small and Medium-sized Enterprises* (pp. 313–335). Berlin, Germany: Springer. doi:10.1007/978-3-642-38244-4_16

Ramanathan, U. (2013). Aligining supply chain collaboration using Analytic Hierarchy Process. *Omega*, *41*(2), 431–440. doi:10.1016/j.omega.2012.03.001

Ramayah, T., Ling, N. S., Taghizadeh, S. K., & Rahman, S. A. (2016). Factors influencing SMEs website continuance intention in Malaysia. *Telematics and Informatics*, *33*(1), 150–164. doi:10.1016/j.tele.2015.06.007

Ramli, N. S., Hassan, M. S., Samah, B. A., Ali, M. S. S., Shaffril, H. A. M., & Azaharian, Z. S. (2013). Problems in using agricultural information from television among farmers in Malaysia. *Asian Social Science*, *9*(9), 70. doi:10.5539/ass.v9n9p70

Ramus, K., & Nielsen, N. A. (2005). Online grocery retailing: What do consumers think? *Internet Research*, *15*(3), 335–352. doi:10.1108/10662240510602726

Renting, H., Marsden, T. K., & Banks, J. (2003). Understanding alternative food networks: Exploring the role of short food supply chains in rural development. *Environment & Planning A*, *35*(3), 393–411. doi:10.1068/a3510

Reuters, U. S. (2015). *Build better roads in developing world to bolster food supplies: study, Article by Alister Doyle*. Retrieved August 25, 2016, from http://www.reuters.com/article/us-environment-food-idUSKBN0L200520150129

Rios, V. R., & Campo, E. P. (2013). *Business Research Methods: Theory and Practice*. Madrid: ESIC Editorial.

Risk Metrics. (1996). *Technical Document*. Morgan Guaranty Trust Company of NewYork.

Rizal, A. R. A., Nordin, S. M., Saad, M. S., & Ismail, K. (2012). Engagement Strategies for Stakeholder Management in New Technology Development in the Fertilizer Industry-A Conceptual Framework. *Proceedings of World Academy of Science, Engineering and Technology*.

Rizal, A. R. A., Nordin, S. M., Saad, S. M., & Wahab, H. (2013). Communication Management between Organizations in a Subsidized Market: The Case of Paddy Fertilizer in Malaysia. *Asian Journal of Research in Social Sciences and Humanities*, *3*(12), 1.

Rizal, A., Redza, A., Md Nordin, S., Saad, M., Shamsuri, M., & Wahab, H. (2014). Inter-Organization Communication Management between Organizations in a Subsidized Fertilizer Market in Malaysia. *Global Business and Management Research: An International Journal*, *6*(3), 229–236.

RMS. (2016). *Production and technical risk management, prepared by Youssef m. Hamada*. Department of Water and Land Economics Researches, Agricultural Economics Research Institute, Ministry of Agriculture & Land Reclamation.

Robinson, A. (2009). Driving efficiencies in the cold supply chains. *Food Logistics*, 10-14.

Rockfellar, R.T., & Uryasev, S. (2000). Optimization of Conditional Value at Risk. *The Journal of Risk*.

Rödiger, M., & Hamm, U. (2015). How are organic food prices affecting consumer behaviour? A review. *Food Quality and Preference*, *43*, 10–20. doi:10.1016/j.foodqual.2015.02.002

Rogers, E. M. (2003). *Diffusion of Innovations* (5th ed.). New York: Free Press.

Roopan, S. M., Madhumitha, G., Rahuman, A. A., Kamaraj, C., Bharathi, A., & Surendra, T. (2013). Low-cost and eco-friendly phyto-synthesis of silver nanoparticles using Cocos nucifera coir extract and its larvicidal activity. *Industrial Crops and Products*, *43*, 631–635. doi:10.1016/j.indcrop.2012.08.013

Ross, S., & Wall, G. (1999). Ecotourism: Towards Congruence between Theory and Practice. *Tourism Management*, *20*(1), 123–132. doi:10.1016/S0261-5177(98)00098-3

Rota, C., Reynolds, N., & Zanasi, C. (2013). Sustainable Food Supply Chains: The Role of Collaboration and Sustainable Relationships. *International Journal of Business and Social Science*, *4*(4), 45–53.

Rousseau, S., & Vranken, L. (2013). Green market expansion by reducing information asymmetries: Evidence for labeled organic food products. *Food Policy*, *40*, 31–43. doi:10.1016/j.foodpol.2013.01.006

Ruben, R., & Van Den Berg, P. (2001). Nonfarm employment and poverty alleviation of rural farm households in Honduras. *World Development*, *3*(29), 549–560. doi:10.1016/S0305-750X(00)00107-8

Ruteri, J. M. (2009). Supply chain management and challenges facing the food industry sector in Tanzania. *International Journal of Business and Management*, *4*(12), 70. doi:10.5539/ijbm.v4n12p70

Rutsaert, P., Regan, Á., Pieniak, Z., McConnon, Á., Moss, A., Wall, P., & Verbeke, W. (2013). The use of social media in food risk and benefit communication. *Trends in Food Science & Technology*, *30*(1), 84–91. doi:10.1016/j.tifs.2012.10.006

Sacks, F. M., Pfeffer, M. A., Moye, L. A., Rouleau, J. L., Rutherford, J. D., Cole, T. G., & Wun, C.-C. et al. (1996). The effect of pravastatin on coronary events after myocardial infarction in patients with average cholesterol levels. *The New England Journal of Medicine*, *335*(14), 1001–1009. doi:10.1056/NEJM199610033351401 PMID:8801446

Sadalage, P. J., & Fowler, M. (2013). *NoSQL Distilled*. Addison-Wesley.

Saeed, K. (2008). Trend forecasting for stability in supply chains. *Journal of Business Research*, *61*(11), 1113–1124. doi:10.1016/j.jbusres.2007.11.005

Safitri, R., & Andriyani, A. (2011). Keefektifan Pemberian Posisi Semi Fowler Terhadap Penurunan Sesak Nafas Pada Pasien Asma di Ruang Rawat Inap Kelas III RSUD Dr. Moewardi Surakarta. *Gaster Jurnal Ilmu Kesehatan*, *8*(2), 783-792.

Sampson, S. E. (2000). Customer–supplier duality and bidirectional supply chains in service organizations. *International Journal of Service Industry Management*, *11*(4), 348–364. doi:10.1108/09564230010355377

Sangkumchaliang, P., & Huang, W. C. (2012). Consumers' perceptions and attitudes of organic food products in Northern Thailand. *International Food and Agribusiness Management Review*, *15*(1), 87–102.

Santini, F., & Gomez y Paloma, S. G. (Eds.). (2013). *Short Food Supply Chains and Local Food Systems in the EU: a state of play of their socio-economic characteristics*. Luxembourg: Publications Office of the European Union.

Santos, S., Belton, V., & Howick, S. (2002). Adding value to performance measurement by using system dynamics and multicriteria analysis. *International Journal of Operations & Production Management*, *22*(11), 1246–1272. doi:10.1108/01443570210450284

Sarkkinen, M., & Kässi, T. (2013). Characterization of innovation situation in a remote rural region. *International Journal of Engineering Business Management*, *5*(1), 1.

Schaffers, H., García Guzman, J., Merz, C., & Navarro, M. (Eds.). (2010). *Living labs and rural development; Results from the C@R Project*. Madrid: TRAGSA.

Schimith, C., Scavarda, A., Bittencourt, S., Santos, M., Vacarro, G., Gabi, A., & Weise, A. (2015). The System Dynamics Use for Measurement of the Results of Technological Applications for Genetic Improvement in Milk Supply Chain. *Chemical Engineering (Albany, N.Y.)*, 45.

Schmid, E., & Sinabell, F. (2007). On the choice of farm management practices after the reform of the Common Agricultural Policy in 2003. *Journal of Environmental Management*, *82*(3), 332–340. doi:10.1016/j.jenvman.2005.12.027 PMID:17095138

Schmid, E., Sinabell, F., & Hofreither, M. F. (2007). Phasing out of environmentally harmful subsidies:Consequences of the 2003 CAP reform. *Ecological Economics*, *60*(3), 596–604. doi:10.1016/j.ecolecon.2005.12.017

Schwaninger, M., & Rios, J. (2008). System dynamics and cybernetics: A synergetic pair. *System Dynamics Review, 24*(2), 145–174. doi:10.1002/sdr.400

Schwarz, J., Schuster, M., Annaert, B., Maertens, M., & Mathijs, E. (2016). Sustainability of global and local food value chains: An empirical comparison of Peruvian and Belgian asparagus. *Sustainability, 8*(4), 344. doi:10.3390/su8040344

Seiford, L. M., & Zhu, J. (2002). Modeling undesirable factors in efficiency evaluation. *European Journal of Operational Research, 142*(1), 16–20. doi:10.1016/S0377-2217(01)00293-4

Sekaran, U. (2003). *Research methods for business: A skill building approach*. New York, NY: Wiley.

Senge, P. (2006). *The fifth discipline: The art and practice of the learning organization*. Broadway Business.

Sepe, L., & Paladino, F. (2014). I formaggi validati durante il progetto NOVOROD. In G. Quaranta (Ed.), *Produzione di formaggi innovativi: CarcioCacio. Formaggi a caglio vegetale frutto di una validazione sostenuta dal PSR Campania 2007-13, misura 124. Ascea Print Service srl, Ascea Marina (Sa).*

Seuring, S. (2013). A review of modeling approaches for sustainable supply chain management. *Decision Support Systems, 54*(4), 1513–1520. doi:10.1016/j.dss.2012.05.053

Seyfang, G. (2008). Avoiding Asda? Exploring consumer motivations in local organic food networks. *Local Environment, 13*(3), 187–201. doi:10.1080/13549830701669112

Shaffril, H. A. M., Asmuni, A., & Ismail, A. (2010). Menu. *Journal of International Social Research, 3*, 450–458.

Shaffril, M., Azril, H., Uli, J., Asiah, M., Ali, N. A., Abu Samah, B., & Thomas, K. et al. (2010). Can quality of work life affect work performance among government agriculture extension officers? A case from Malaysia. *Journal of Social Sciences, 6*(1), 64–73. doi:10.3844/jssp.2010.64.73

Shafie, F. A., & Rennie, D. (2012). Consumer perceptions towards organic food. *Procedia: Social and Behavioral Sciences, 49*, 360–367. doi:10.1016/j.sbspro.2012.07.034

Shakya, M. (2009). *Risk, vulnerability and tourism in developing countries: The case of Nepal*. Berlin: Logos Verlag Berlin GMBH.

Shang, R. A., Chen, Y. C., & Shen, L. (2005). Extrinsic Versus Intrinsic Motivations For Consumers To Shop On-Line. *Information & Management, 42*(3), 401–413. doi:10.1016/j.im.2004.01.009

Sharma, P. (2013). Rural entrepreneurship, science and technology and innovations in farm-based entrepreneurship venture: A case study of a farmer inventing accustomed machine for tillage farming. *International Journal of Innovative Research and Development, 2*(4), 6-14. Available at www.ijird.com

Sharpe, W. (1964). Capital Asset Prices: A Theory of Market Equilibrium under Conditions of Risk. *The Journal of Finance, 19*, 425–442.

Sharpe, W. (1966). Mutual Fund Performance. *The Journal of Business, 39*(S1), 119–138. doi:10.1086/294846

Shepherd, T., & Flanders, A. (2008). Do Shifting Milk-Production Patterns Weaken the Dairy Supply Chain? The Case for the Southeastern US. *Journal of Food Distribution Research, 39*(1), 130.

Silva, E., & Stefanou, S. E. (2003). Nonparametric Dynamic Production Analysis and the Theory of Cost. *Journal of Productivity Analysis, 19*(1), 5–32. doi:10.1023/A:1021865018717

Simaan, Y. (1997). Estimation Risk in Portfolio Selection: The Mean Variance Model Versus the Mean Absolute Deviation Model. *Management Science, 43*(10), 1437–1446. doi:10.1287/mnsc.43.10.1437

Simonson, S. (2009). Transforming the fresh food supply-chain. *Food Logist*, 14-15.

Śledzik, K. (2013). Schumpeter's view on innovation and entrepreneurship. In Management Trends in Theory and Practice. University of Zilina. doi:10.2139/ssrn.2257783

Slight, P., Adams, M., & Sherren, K. (2016). Policy support for rural economic development based on Hollings ecological concept of panarchy. *International Journal of Sustainable Development and World Ecology*, *23*(1), 1–14. doi:10.1080/13504509.2015.1103801

SMECORP. (2014). *Definition of SMEs (Appendix 4)*. Available at www.smecorp.gov.my

SMECORP. (2015). SME development framework: The Malaysian case. *Proceedings of the Cairo AMC-CBE-WBG Conference*. Available at http://www.cbe.org.eg/NR/rdonlyres/95EA10FA-62EE-4469-A40E-A952FDE2B434/2737/CairoSMEConferenceJan2015Session3DatoHafsahMalaysi.pdf Retrieved 6 March 2016.

Smith, A., Fressoli, M., & Thomas, H. (2014). Grassroots innovation movements: Challenges and contributions. *Journal of Cleaner Production*, *63*, 114–124. doi:10.1016/j.jclepro.2012.12.025

Smith, J. N. (2005). Specialized logistics for a longer perishable supply chain. *World Trade*, *18*(11), 46.

Smith, P., & Mayston, D. J. (1987). Measuringefficiency in the public sector. *Omega*, *15*(3), 181–189. doi:10.1016/0305-0483(87)90068-5

Snow, V., & Lovatt, S. (2008). A general planner for agro-ecosystem models. *Computers and Electronics in Agriculture*, *60*(2), 201–211. doi:10.1016/j.compag.2007.08.001

Solanki, M., Brewster, C. (2015). OntoPedigree: Modelling pedigrees for traceability in supply chains. *Semantic Web.*

Solanki, M., & Brewster, C. (2014). Modelling and Linking Transformations in EPCIS Governing Supply Chain Business Processes. *Lecture Notes in Business Information Processing*, *188*, 46–57. doi:10.1007/978-3-319-10491-1_5

Soltani, S., Azadi, H., Mahmoudi, H., & Witlox, F. (2014). Organic agriculture in Iran: Farmers barriers to and factors influencing adoption. *Renewable Agriculture and Food Systems*, *29*(2), 126–134. doi:10.1017/S1742170513000069

Sonawane, M. B. (2014). Entrepreneurship (entrepreneurial mind) in emerging economies. *Journal of Business Management and Social Sciences Researches, 3*(2), 24-26. Available at www.borjounals.com

Song, M., An, Q., Zhang, W., Wang, Z., & Wu, J. (2012). Environmental efficiency evaluation based on data envelopment analysis: A review. *Renewable & Sustainable Energy Reviews*, *16*(7), 4465–4469. doi:10.1016/j.rser.2012.04.052

Sozen, A., Alp, I., & Ozdemir, A. (2010). Assessment of operational and environmental performance of the thermal power plants in Turkey by using data envelopment analysis. *Energy Policy*, *38*, 6194–6203.

Stabell, C. B., & Fjeldstad, Ø. D. (1998). Configuring value for competitive advantage: On chains, shops, and networks. *Strategic Management Journal*, *19*(5), 413–437. doi:10.1002/(SICI)1097-0266(199805)19:5<413::AID-SMJ946>3.0.CO;2-C

Statista. (2016a). *Organic packaged food and beverage sales in the United States in 2014, by distribution channel.* Retrieved November 05, 2016 from https://www.statista.com/statistics/249231/share-of-organic-packaged-food-and-beverage-sales-in-the-us-by-distribution-channel/

Stefanovits, D. (2010). *Equal Contributions to Risk and Portfolio Construction* (Master Thesis). ETH Zurich.

Sterman, J. D. (1989). Modeling managerial behavior: Misperceptions of feedback in a dynamic decision making experiment. *Management Science*, *35*(3), 321–339. doi:10.1287/mnsc.35.3.321

Sterman, J. D. (2000). *Business dynamics: Systems Thinking and Modeling for a Complex World.* Boston, MA: Irwin MacGraw-Hill.

Stewart, D. W., & Shamdasani, P. N. (1990). *Focus groups – Theory and practice.* Newbury Park, CA: Sage.

Stokes, J. R., Tozer, P. R., & Hyde, J. (2007). Identifying Efficient Dairy Producers Using Data Envelopment Analysis. *Journal of Dairy Science, 90*(5), 2555–2562. doi:10.3168/jds.2006-596 PMID:17430960

Sturiale, L., & Scuderi, A. (2013). Evaluation Of Social Media Actions For The agrifood System. *Procedia Technology, 8*, 200–208. doi:10.1016/j.protcy.2013.11.028

Taalika, S. (2004). Assessing the managerial influence as a facilitating factor in innovation adoption process. *Proceedings of the IAMOT Conference.*

Taiwo, A. D., & Downe, A. G. (2013). The Theory of User Acceptance and Use of Technology (UTAUT): A meta-analytic review of empirical findings. *Journal of Theoretical and Applied Information Technology, 49*(1).

Taiwo, A. A., Downe, A. G., & Mahmood, A. K. (2012). User acceptance of E-government: Integrating risk and trust dimensions with UTAUT Model. *Proceedings at the International Conference on Computer and Information sciences.*

Tarkiainen, A., & Sundqvist, S. (2005). Subjective norms, attitudes and intentions of Finnish consumers in buying organic food. *British Food Journal, 107*(11), 808–822. doi:10.1108/00070700510629760

Tasche, D. (2000). *Conditional expectation as quantile derivative. Risk Contributions and Performance Measurement.* Department of Mathematics, TU-Munchen.

Taulbert, C., & Schoeniger, G. (2010). *Who own the ice house? Eight life lessons from an unlikely entrepreneur.* ELI Press.

Taylor, J. E., & Martin, P. (2001). Human capital: Migration and rural population change. In B. Gardener & G. Rausser (Eds.), *Handbook of Agricultural Economics* (Vol. 1, pp. 457–511). Amsterdam: Elsevier.

Taylor, S., & Todd, P. (1995). Decomposition and crossover effects in the theory of planned behavior: A study of consumer adoption intentions. *International Journal of Research in Marketing, 12*(2), 137–155. doi:10.1016/0167-8116(94)00019-K

Teng, C. C., & Wang, Y. M. (2015). Decisional factors driving organic food consumption: Generation of consumer purchase intentions. *British Food Journal, 117*(3), 1066–1081. doi:10.1108/BFJ-12-2013-0361

Teo, T., Su Luan, W., & Sing, C. C. (2008). A cross-cultural examination of the intention to use technology between Singaporean and Malaysian pre-service teachers: An application of the Technology Acceptance Model (TAM). *Journal of Educational Technology & Society, 11*(4), 265–280.

Tey, Y. S. (2013). *The Adoption of Sustainable Agricultural Practices: An Integrative Approach for Malaysian Vegetable Farmers.* The University of Adelaide.

Thakur, M., & Forås, E. (2015). EPCIS based online temperature monitoring and traceability in a cold meat chain. *Computers and Electronics in Agriculture, 117*, 22–30. doi:10.1016/j.compag.2015.07.006

Thanassoulis, E. (1999). Data Envelopment Analysis and Its Use in Banking. *Interfaces, 29*(3), 1–13. doi:10.1287/inte.29.3.1

Thanassoulis, E., & Dunstan, P. (1994). Guidingschools to improved performance usingdata envelopment analysis: An illustrationwith data from a local education authority. *The Journal of the Operational Research Society, 45*(11), 1247–1262. doi:10.1057/jors.1994.198

The Babui Ecoresort. (2014). *About Us.* Retrieved from: http://bit.ly/1kySE15

The Daily Star. (2014). *Ecovillage Raises Hope for Aila Victims.* Retrieved from: http://bit.ly/199aimH

The International Union for Conservation of Nature (IUCN). (1980). *World Conservation Strategy*. Gland, Switzerland: International Union for the Conservation of Nature and Natural Resources. Retrieved from: http://bit.ly/1bWsFdJ

The Panigram Resort. (2014). *Home*. Retrieved from: http://www.panigram.com

Thompson, G. D. (1998). Consumer demand for organic foods: What we know and what we need to know. *American Journal of Agricultural Economics*, *80*(5), 1113–1118. doi:10.2307/1244214

Thompson, J. D. (1967). *Organizations in Action*. New York: McGraw Hill.

Tian, F. (2016). A Quality and Safety Control System for China's Dairy Supply Chain Based on RACCP & GS1. *13th International Conference on Service Systems and Service Management*, 1-6.

Touzard, J.-M., & Maffezzoli, C. (2015). *Glamur project multidimensional comparison of a local and global wine supply chain*. INRA.

Trainer, F. E. (1998). Towards a Checklist for Ecovillage Development. *Local Government*, *3*(1), 79–84.

Tranter, R. B., Swinbank, A., Wooldridge, M. J., Costa, L., Knapp, T., Little, G. P. J., & Sottomayor, M. L. (2007). Implications for food production, land use and rural development of the European Unions Single FarmPayment: Indications from a survey of farmersintentions in Germany, Portugal and the UK. *Food Policy*, *32*(5-6), 656–671. doi:10.1016/j.foodpol.2007.04.001

Trienekens, J. H., Wognum, P. M., Beulens, A. J., & van der Vorst, J. G. (2012). Transparency in complex dynamic food supply chains. *Advanced Engineering Informatics*, *26*(1), 55–65. doi:10.1016/j.aei.2011.07.007

Tsaples, G., & Armenia, S. (2016). Studying pension systems and retirement age: A simple system dynamics model for a complex issue. *International Journal of Applied Systemic Studies*, *6*(3), 258–270. doi:10.1504/IJASS.2016.079095

Tsekouropoulos, G., Andreopoulou, Z. S., Koliouska, C., Stavroula lefa, Koutroumanidis, T., & Batzios, C. (2011). e-Marketing and Internet Functions of Agricultural Products In SME In Greece. *Proceedings of the International Conference on Information and Communication Technologies*, 213-224.

Tselempis, D., Karipidis, P., Pavloudi, A., & Semos, A. (2015). Is the quality certification in crop and vegetable production a market driven choice in Greece? *Agricultural and Food Economics*, *3*(13), 1–12. doi:10.1186/s40100-015-0032-7

Tubiello, F. N., Salvatore, M., Cóndor Golec, R. D., Ferrara, A., Rossi, S., Biancalani, R., & Flammini, A. (2014). *Agriculture, forestry and other land use emissions by sources and removals by sinks*. Rome, Italy: FAO.

UN Global Compact. (2014). *Food and Agriculture Business Principles*. Retrieved August 19, 2016, fromhttps://www.unglobalcompact.org/take-action/action/food

UN Global Compact. (2016). *Food and Agriculture*. Retrieved August 29, 2016, from https://www.unglobalcompact.org/what-is-gc/our-work/environment/food-agriculture

UNDP. (2009). Impact of UNDP microfinance programme on poverty alleviation among farmers in selected local government areas of Kaduna State, Nigeria. *International Journal of Sociology and Anthropology*, *1*(6), 99-103. Retrieved August 20, 2016, from http://www.academicjournals.org/article/article1379409974_Kudi%20et%20al.pdf

United Nations Global Compact and Business for Social Responsibility. (2014). *A guide to traceability: A practical approach to advance sustainability in global supply chains*. New York, NY: United Nations Global Compact Office.

United Nations. (2015). *Transforming our world: the 2030 Agenda for Sustainable Development*. Retrieved August 29, 2016 from https://sustainabledevelopment.un.org/post2015/transformingourworld

UNSG. (2014). *Agriculture development, food security and nutrition*. Report of the UN Secretary-General, A/69/279, 7 August 2014.

Ureña, F., Bernabéu, R., & Olmeda, M. (2008). Women, men and organic food: Differences in their attitudes and willingness to pay. A Spanish case study.*international Journal of consumer. Studies*, *32*(1), 18–26.

USAID. (2014). *Land tenure in urban environments*. Retrieved August 27, 2016, from http://www.usaidlandtenure.net/sites/default/files/USAID_Land_Tenure_Urban_Brief_061214.pdf

Van der Ploeg, J. D. (2003). *The virtual farmer. Past, present and future of the Dutch peasantry*. Assen, The Netherlands: Van Gorcum.

Van der Ploeg, J. D., Renting, H., Brunori, G., Knickel, K., Mannion, J., Marsden, T., & Ventura, F. et al. (2000). Rural development: From practices and policies towards theory. *Sociologia Ruralis*, *40*(4), 391–408. doi:10.1111/1467-9523.00156

Van Doorn, J., & Verhoef, P. C. (2011). Willingness to pay for organic products: Differences between virtue and vice foods. *International Journal of Research in Marketing*, *28*(3), 167–180. doi:10.1016/j.ijresmar.2011.02.005

van Duren, E., Sparling, D., Turvey, C., & Lake, L. (2003). An assessment of the strategies and strengths of medium-sized food processors. *Agribusiness*: *International Journal (Toronto, Ont.)*, *19*(1), 115–132.

Van Schyndel Kasper, D. (2008). Redefining Community in the Ecovillage. *Human Ecology Review*, *15*, 12–24.

Vasquez-Rowe, I., Iribarren, D., Moreira, M. T., & Feijoo, G. (2010). Combined application of life cycle assessment and data envelopment analysis as a methodological approach for the assessment of fisheries. *International Journal of Life Cycle Assessment*, *15*(3), 272–283. doi:10.1007/s11367-010-0154-9

Vasquez-Rowe, I., Villanueva-Rey, P., Iribarren, D., Moreira, M. T., & Feijoo, G. (2012). Joint life cycle assessment and data envelopment analysis of grape production for vinification in the RíasBaixas appellation (NW Spain). *Journal of Cleaner Production*, *27*, 92–102. doi:10.1016/j.jclepro.2011.12.039

Veeck, G., Che, D., & Veeck, A. (2006). Americas changing farmscape: A study of agricultural tourism in Michigan. *The Professional Geographer*, *58*(3), 235–248. doi:10.1111/j.1467-9272.2006.00565.x

Veldstra, M. D., Alexander, C. E., & Marshall, M. I. (2014). To certify or not to certify? Separating the organic production and certification decisions. *Food Policy*, *49*(2), 429–436. doi:10.1016/j.foodpol.2014.05.010

Venkatesh, V., Morris, M. G., & Ackerman, P. L. (2000). A longitudinal field investigation of gender difference in individual technology adoption decision-making processes. *Organizational Behavior and Human Decision Processes*, *83*(1), 36–60. doi:10.1006/obhd.2000.2896 PMID:10973782

Venkatesh, V., Morris, M. G., Davis, G. B., & Davis, F. D. (2003). Use acceptance of information technology: Toward a unified view. *Management Information Systems Quarterly*, *27*(3), 425–478.

Vennesland, B. (2005). Measuring rural economic development in Norway using data envelopment analysis. *Forest Policy and Economics*, *7*(1), 109–119. doi:10.1016/S1389-9341(03)00025-X

Verdouw, C. N., Wolfert, J., Beulens, A. J. M., & Rialland, A. (2016). Virtualization of food supply chains with the internet of things. *Journal of Food Engineering*, *176*, 128–136. doi:10.1016/j.jfoodeng.2015.11.009

Vlontzos, G., Arabatzis, G., & Manos, B. (2014). Investigation of the relative efficiency of LEADER+ in rural areas of Northern Greece. *International Journal of Green Economics*, *8*(1), 37–48. doi:10.1504/IJGE.2014.064470

Vukatana, K., Sevrani, K., & Hoxha, E. (2016). Wine Traceability: A Data Model and Prototype in Albanian Context. *Foods*, *5*(11), 1–11.

Wang, K., Wei, Y. M., & Zhang, X. (2012). A comparative analysis of Chinas regional energy and emission performance: Which is the better way to deal with undesirable outputs? *Energy Policy*, *46*, 574–584. doi:10.1016/j.enpol.2012.04.038

Wang, Y., & Hajli, M. N. (2014). Co-creation in branding through social commerce: The role of social support, relationship quality and privacy concerns. *Proceedings of Twentieth Americas Conference on Information Systems.*

Ward, N., & Brown, D. (2009). Placing the rural in regional development. *Regional Studies*, *43*(10), 1237–1244. doi:10.1080/00343400903234696

Weber, C. L., & Matthews, H. S. (2008). Food-miles and the relative climate impacts of food choices in the United States. *Environmental Science & Technology*, *42*(10), 3508–3513. doi:10.1021/es702969f PMID:18546681

Weick, K. E. (1976). Educational organizations as loosely couples systems. *Administrative Science Quarterly*, *21*(1), 1–19. doi:10.2307/2391875

Weitz, N., Nilsson, M., & Davis, M. (2014). A Nexus Approach to the Post-2015 Agenda: Formulating Integrated Water, Energy, and Food SDGs. *SAIS Review of International Affairs*, *34*(2), 37–50.

Wellbrock, W., Roep, D., & Wiskerke, J. S. C. (2012). An integrated perspective on rural regional learning. *European Countryside*, *4*(1), 1–16. doi:10.2478/v10091-012-0010-y

West, P. (2008). Tourism as Science and Science as Tourism Environment, Society, Self, and Other in Papua New Guinea. *Current Anthropology*, *49*(4), 597–626. PMID:19230266

West, P., & Daniel, B. (2006). Some Unexpected Consequences of Protected Areas: An Anthropological Perspective. *Conservation Biology*, *20*, 609–616. doi:10.1111/j.1523-1739.2006.00432.x PMID:16909541

WFP. (2015). *World Food Programme, P4P Purchase for Progress - The Story: Connecting farmers to markets*. Retrieved August 27, 2016, from https://www.wfp.org/sites/default/files/the%20P4P%20Story%20-%20Connecting%20Farmers%20to%20Markets,%20Feb%202015.pdf

Wilson, G. A. (2012). Community resilience, globalization, and transitional pathways of decision-making. *Geoforum*, *43*(6), 1218–1231. doi:10.1016/j.geoforum.2012.03.008

Wilson, G., & Rigg, J. (2003). Post-productivist agricultural regimes and the south: Discordant concepts? *Progress in Human Geography*, *27*(6), 681–707. doi:10.1191/0309132503ph450oa

Windeshausen, H. N. (1973). Market Segmentation: Concepts and Applications. *Journal of Marketing (pre-1986)*, *37*(4), 124.

Wittman, H., Beckie, M., & Hergesheimer, C. (2012). Linking local food systems and the social economy? Future roles for farmers markets in Alberta and British Columbia*. *Rural Sociology*, *77*(1), 36–61. doi:10.1111/j.1549-0831.2011.00068.x

Wognum, P. N., Bremmers, H., Trienekens, J. H., van der Vorst, J. G., & Bloemhof, J. M. (2011). Systems for sustainability and transparency of food supply chains – Current status and challenges. *Advanced Engineering Informatics*, *25*(1), 65–76. doi:10.1016/j.aei.2010.06.001

Womack, J., & Jones, D. (2003). *Lean thinking: Banish waste and create wealth in your corporation*. New York, NY: Simon & Schuster.

Wong, C. W. (2013). Leveraging environmental information integration to enable environmental management capability and performance. *Journal of Supply Chain Management*, *49*(2), 114–136. doi:10.1111/jscm.12005

World Bank. (2014). *Employment in agriculture*. Retrieved August 28, 2016, from http://data.worldbank.org/indicator/SL.AGR.EMPL.ZS?page=2159Donor

World Bank. (2016). *Agriculture*. Retrieved August 28, 2016, from http://www.worldbank.org/en/topic/agriculture/overview

World Commission on Environment and Development (WCED). (1987). *Our Common Future*. Oxford, UK: Oxford University Press.

World Economic Forum. (2013). *The Global Competitiveness Report 2013–2014*. Geneva: World Economic Forum.

World Economic Forum. (2015). Retrieved from http://www.weforum.org/reports/global-competitiveness-report-2014-2015

WRAP. (2015). *Waste and Resources Action Programme. Strategies to achieve economic and environmental gains by reducing food waste – Final Report*. Retrieved August 25, 2016, from http://static.newclimateeconomy.report/wp-content/uploads/2015/02/WRAP-NCE_Economic-environmental-gains-food-waste.pdf

Wu, Y. L., Tao, Y. H., & Yang, P. (2007). Using UTAUT to explore the behaviour of 3G mobile communication users. In *Industrial Engineering and Engineering management, 2007 IEEE International Conference on* (pp. 199-203). IEEE.

Wu, H., Du, S., Liang, L., & Zhou, Y. (2013). DEA-based approach for fair reduction and reallocation of emission permits. *Mathematical and Computer Modelling*, *58*(5-6), 1095–1101. doi:10.1016/j.mcm.2012.03.008

Xie, B., Wang, L., Yang, H., Wang, Y., & Zhang, M. (2015). Consumer perceptions and attitudes of organic food products in Eastern China. *British Food Journal*, *117*(3), 1105–1121. doi:10.1108/BFJ-09-2013-0255

Yan, B., Hu, D., & Shi, P. (2012). A traceable platform of aquatic foods supply chain based on RFID and EPC Internet of Things. *International Journal of RF Technologies*, *4*(1), 55–70.

Yan, B., Hu, D., & Shi, P. (2012). A traceable platform of aquatic foods supply chain based on RFID and EPC Internet of things. *International Journal of RF Technologies: Research and Applications*, *4*(1), 55–70. doi:10.3233/RFT-2012-0035

Yang, H.-H., & Chang, Ch. (2009). Using DEAwindowanalysistomeasureefficienciesofTaiwansIntegratedtelecommunicationfirms. *Telecommunications Policy*, *33*(1-2), 98–108. doi:10.1016/j.telpol.2008.11.001

Yitzhaki, S. (1982). Stochastic Dominance, Mean Variance and Gini's Mean Difference. *The American Economic Review*, *72*, 178–185.

Young, M. (1998). A Minimax Portfolio Selection Rule with Linear Programming Solution. *Management Science*, *44*(5).

Zagata, L. (2012). Consumers beliefs and behavioural intentions towards organic food. Evidence from the Czech Republic. *Appetite*, *59*(1), 81–89. doi:10.1016/j.appet.2012.03.023 PMID:22504401

Zainal, Z., Shamsudin, M. N., Mohamed, Z. A., Adam, S. U., & Kaffashi, S. (2014). Assessing the Impacts of Climate Change on Paddy Production in Malaysia. *Research Journal of Environmental Sciences*, *8*(6), 33.

Zakowska-Biemans, S. (2011). Polish consumer food choices and beliefs about organic food. *British Food Journal*, *113*(1), 122–137. doi:10.1108/00070701111097385

Zander, K., & Hamm, U. (2012). Information search behaviour and its determinants: The case of ethical attributes of organic food. *International Journal of Consumer Studies*, *36*(3), 307–316. doi:10.1111/j.1470-6431.2011.00998.x

Zanoli, R., & Naspetti, S. (2002). Consumer motivations in the purchase of organic food: A means-end approach. *British Food Journal*, *104*(8), 643–653. doi:10.1108/00070700210425930

Zepeda, L., & Li, J. (2007). Characteristics of organic food shoppers. *Journal of Agricultural and Applied Economics*, *39*(01), 17–28. doi:10.1017/S1074070800022720

Zhu, X., Zhang, Q., Zhang, L., & Yang, J. (2014). Consumers' Demand And Online Promotion Of The Food Retailing Through The E-Commerce Websites In China. *Advance Journal of Food Science and Technology*, *6*(1), 102–111.

Ziesemer, J. (2007). *Energy use in organic food systems*. Rome, Italy: FAO.

Zopounidis, C., & Pardalos, P. M. (2010). *Handbook of Multicriteria Analysis*. Applied Optimization, Springer-Verlag Berlin Heidelberg. doi:10.1007/978-3-540-92828-7

About the Contributors

Theodore Tarnanidis is a researcher in the Business Administration Department at the University of Macedonia. He finished his postdoctoral research at the University of Macedonia, in November 2015, in the area of sustainable entrepreneurship. He obtained a Ph.D. from the University of London Met., UK. He received his M.B.A. from Liverpool University, UK and is a graduate from the University of Macedonia (Business Administration) and Alexander Technological Educational Institute (Marketing). His research focuses on Decision-Making Processes in Sustainable Innovations and Entrepreneurship, Conjoint Models and preference measurement techniques, modelling of purchases and consumer behavior. His work has been published in various internationally renowned scientific conferences and journals mostly from the fields of Business Administration.

Maro Vlachopoulou is a Professor at the Department of Applied Informatics, Dean of the School of Information Sciences, University of Macedonia, Greece. Degree in Business Administration, Aristotle University of Thessaloniki, Greece (1978). Degree in Law, Aristotle University of Thessaloniki, Greece (1989). Postgraduate studies in Marketing, University of Mannheim, Germany (1981), and MBS in Business Administration, Aristotle University of Thessaloniki (1987). PhD in Marketing Information Systems, University of Macedonia, Department of Applied Informatics, Greece (1992). Visiting research professor at the University of Sunderland, UK (1995). Her professional expertise, research and teaching interests include: Internet Marketing, Marketing Information Systems, E-Business/ e-marketing models, emarketplaces, E-Business plan, new technologies and marketing, mobile Business / marketing, SEO & SEM, social media / networks and marketing, ERP (Enterprise Resource Planning), CRM (Customer Relationship Management) systems, Supply Chain Management (SCM) Systems, Knowledge management systems, e-supply chain management / logistics, with particular emphasis on specific applications (agrifood, e-trust, e-health, government, tourism, environment). She has published over 100 papers in scientific journals and organized numerous of conferences.

Jason Papathanasiou is an Assistant Professor at the Department of Business Administration, University of Macedonia, Greece. He holds a PhD in Operational Research and Informatics and a degree in Physics from the Aristotle University of Thessaloniki. He has worked for a number of years at various institutes and has organised and participated in a number of international scientific conferences and workshops. He has published papers in international scientific peer referred journals like the Environmental Monitoring and Assessment, Regional Studies and European Journal of Operational Research and has participated in many research projects in FP6, FP7, Interreg and COST; he is also a member of the TDP Panel of COST.

* * *

Nadia Adnan received her EMBA (Marketing) from Institution of Business Administration (IBA), Karachi, Pakistan and MBA(Marketing)from Jinnah University for Women, Karachi, Pakistan and Bachelor in Business Administration (Honors)from Jinnah University for Women, Karachi, Pakistan. Currently, she is pursuing Doctor of philosophy from the Department of Management and Humanities,Universiti Teknologi PETRONAS, Malaysia.

Alexandros Antonaras holds a BSc (Hons) in Mathematics & Statistics with Management from Brunel University of West London [UK], an MSc in Quality Management and a PhD in Quality & Process Improvement from Cranfield University [UK]. Alexandros is the Director of Institutional Excellence & Business Services and an Assistant Professor of Management at the University of Nicosia. He is a Certified Entrepreneurship Facilitator by the Entrepreneurship Learning Initiative (USA) and Member of the Board of Directors of the Centre for Innovation, Entrepreneurship and New Technologies. Alexandros is an active European Excellence Assessor and Validator for the EFQM Excellence Model and the President of the Cyprus Network for CSR (CSR Cyprus). He is a member of the Project Management Institute (PMI) and a Project Management Professional (PMP). He has considerable graduate and undergraduate teaching experience and has published and presented many research papers in national and international journals and conferences. His research interests include business excellence, business ethics, entrepreneurship, corporate social responsibility (CSR) and quality management.

Giulia Bruno holds a doctorate in Information and System Engineering and she is currently a post-doc researcher at the department of Management and Production Engineering of the Politecnico di Torino, Italy. Her research activity is focused on knowledge management, data mining and production system analysis. Furthermore, she is working in the fields of human-robot collaboration and product traceability along the whole supply chain. She was involved in several European projects in the context of product lifecycle management, healthcare systems and SME networks. Previous research interests also include the analysis of gene expression data and the development of data mining algorithms for clinical analysis.

Ahmed Elghannam is an Egyptian researcher who was born in 1987. He works as Assistant lecturer of Agricultural Economics at the Faculty of Agriculture, Damanhour University (Egypt). He studied in Spain at the Mediterranean Agronomic Institute of Zaragoza where he successfully completed his master degree in Agro-Food Marketing (with a Cum laude distinction). As a researcher he continued for another degree at Extremadura University (Spain) pursuing his PhD studies in Electronic and Social Marketing in the food sector.

Per Engelseth has been an associate professor at Molde University College since 2012 and received his Dr. Oecon degree in logistics and SCM from BI Norwegian School of Management in 2007. His main research interests include engineering management, case studies of end-to-end supply chains, food product traceability, supply chain collaboration including value co-creation, local foods chains, and complex adaptive systems. He has published numerous articles in academic journals as well as in anthologies including publications in International Journal of Production Economics, International Journal of Physical Distribution and Logistics Management and Supply Chain Forum: An International Journal.

Youssef Hamada has completed his PhD from Al-Azhar University. He is a Professor in Agricultural Economy, Agricultural Economics Research Institute, Agricultural Research Center, Egypt. He has published more than 24 papers in local and international journals and conferences.

Azizul Hassan is a PhD candidate at the Cardiff Metropolitan University, UK and a member of the Tourism Consultants Network of the Tourism Society, UK. His main areas of research interest are technology supported marketing in tourism, innovative marketing dynamics, destination branding in tourism, cultural heritage tourism, heritage interpretation and sustainable management/marketing alternatives for cultural heritage industries.

Md. Salleh Hj. Hassan, PhD, Prof., is a professor at Communication Department, Universiti Putra Malaysia since 2004. He conducts research, writes and teaches in the fields of agricultural studies, development communication and information and communication technology (ICT). He has published book chapters and cited in numerous journals.

Adamkolo Mohammed Ibrahim did his Master's degree programme in Development Communication at Universiti Putra Malaysia in 2017 and received his BA Mass Communication certificate from University of Maiduguri, Nigeria in 2007. He is a lecturer at his Alma Mata department and university where he teaches Mass Communication studies. He conducts research and writes in ICT adoption/e-adoption for socio-economic development and social media. He has published book chapters and journal articles.

Muhammad Ishaq is Director, Social Sciences Division, Pakistan Agricultural Research Council, Islamabad, Pakistan.

Vasileios Ismyrlis works in the Hellenic Statistical Authority as a statistician since two years. He was an administrative employee, for eleven years in the social insurance institute IKA-ETAM. He has also a two-year experience in a bank and an experience of teaching financial lessons in vocational post-secondary education and in tertiary education. He graduated with a degree in Statistics and Insurance sciences from the University of Piraeus. His postgraduate studies are in the field of Quality Assurance. He has obtained his Ph.D. in the field of ''Quality management and Multidimensional statistics'', at the Department of Business Administration of the University of Macedonia, Greece. He has published eight articles in scientific journals and conferences' proceedings. He speaks well English and French.

Philippos I. Karipidis is a Professor at the Alexander Technological Educational Institute of Thessaloniki, department of Agricultural Technology –Agricultural Economics. He has a 37 year academic experience as well as working experience in the industry. He has published a large number of peer-reviewed articles in many journals and conference proceedings in English as well as a large number of articles, books, and book chapters in Greek. He participated in many research projects as a coordinator or a member of scientific committees in the fields of Agricultural Marketing, Quality Management, Green Marketing, Social Responsibility, and Environmental Management.

Zia Khan is a Lecturer, Department of Economics, University of Swabi, Khyber Pakhtunkhwa, Pakistan.

Alexandros Kostopoulos is a CSR and Sustainability Expert, currently working as the Head of Business Development at CSR Hellas (Hellenic Network for Corporate Social Responsibility). He joined CSR Hellas in 2006 as Project Manager and over the years his responsibilities included among others: coordination of several European and national projects, surveys and researches; organisation of more than 10 conferences on CSR in Greece, as well as the active participation in the preparatory group for the Greek National CSR Strategy Plan in 2014. Through his role he also cooperates closely with the UN Global Compact for the monitoring of the integration process of the EU Directive on the disclosure of non-financial information in the national legislation of EU member-states. Prior to this he has work in textile industry, and as business consultant. In 2003 he completed his MBA studies in Coventry University (UK) with a focus on Multinational Corporations while his thesis focused on CSR. He writes regularly on issues related to CSR, Sustainability and Business Responsibility.

Francisco J. Mesías is a Professor at the Department of Economics of the University of Extremadura, Spain. He has a background in agricultural economics, with a PhD in Agricultural Economics and a Master's Degree in Food Marketing. He currently teaches Agricultural Economics and Management of Agri-food industries at the Faculty of Agriculture, University of Extremadura. His main research lines focus on the study of competitiveness of agri-food industries, farm management and consumer preferences for food. He has been the leader of several competitive international and Spanish research projects dealing with his research interests. He is the author of more than 50 research papers, published in some of the top international scientific journals. He has also published several books and book chapters with Spanish and international publishers. He has been a visiting professor in different universities of Italy, México and the USA, and has been invited to deliver conferences in different scientific and professional events in Spain, Portugal, Italy and México.

Odysseas Moschidis is an Associate Professor at the Department of Business Administration of University of Macedonia, Greece, Applied Statistics with emphasis in Business Administration. He graduated with a degree in Mathematics from Aristotle University of Thessaloniki. His postgraduate studies are in the field of Applied Statistics and Applied Mathematics and he holds a Doctor's degree in 'Contribution to the study of scales evaluation with the method of multidimensional data analysis' from the Department of Applied Informatics of the University of Macedonia, Greece. He held the positions of General Manager and President from 2004 to2008, at the General Hospital of Thessaloniki "Agios Dimitrios". He has participated in several research projects working on the areas of evaluating business data and market research and multidimensional statistical analysis. His research interests include the areas of scales evaluation, multidimensional statistical analysis and evaluation of business data and market research. He has authored chapters and numerous publications in international journals and conferences.

Suphan Nasir is Professor at Department of Business Administration, İstanbul University. She is chairman of production management and marketing subdicipline and coordinator of business administration e-learning program. She worked as chief editor at International Journal of E-Entrepreneurship and Innovation 2010-2014, and she is also associate editor of Yükseköğretim Dergisi/Journal of Higher Education since 2010. She also undertaken important roles such as organization committee member and advisory board member of international conferences. Her articles were published in reputable academic journals and she has written chapters for well-known books. Her latest books entitled Modern Entrepreneurship and E-Business Innovations and CRM Strategies in the Digital Era were published in USA in

2013 and 2015, respectively. She has also presented her papers at major international conferences and attended some international conference as keynote speaker. CRM as well as mobile marketing, consumer behavior, innovation management and entrepreneurship are the scope of her interest area.

V. Aslihan Nasir recevied her Bachelor's degree in Economics and MBA degree in International Management from Istanbul University. In 2004, she earned a Ph.D. degree in Marketing from Bogazici University. In 2005, she joined the Department of Management Information Systems at Boğaziçi University and she is currently Professor of Marketing at the same department. She is the author of several articles and research reports that are published in prominent journals and conference proceedings. Her research and publications focus on consumer behaviour, consumer food consumption, digital marketing, and CRM.

Amir Noor has completed his Bachelor from Department of Computer Science, London Metropolitan University, 166-220 Holloway Rd, Holloway, London N7 8DB, United Kingdom. He is currently associated with software development farm.

Shahrina Binti Md Nordin has an academic experience in the fields of language and communication. She obtained her B.HSc in Language and Literature from International Islamic university Malaysia. Her Master's degree was in Education from Universiti Teknologi Malaysia where her project was on technical communication in the engineering industry. Her PhD thesis was on second language writing in technical communication in Universiti Sains Malaysia. She is currently Head of Centre of Social Innovation (CoSI) and Director of Project Management Office, Universiti Teknologi PETRONAS (UTP). She has taught undergraduate and postgraduate programs in the area of Academic Writing, Technical and Professional Writing, Professional Communication Skills, Research Methodology, Business Communication, and Corporate Communication. Her current research areas are in communications for innovation diffusion, CSR communications for community empowerment and organizational communications. She has published many papers, two books and has received a number of awards including Most Commendable Paper by Emerald. She is currently an advocate of Social Innovation that draws upon the strength of multidisciplinary effort in impacting the communities.

Panos M. Pardalos serves as Distinguished Professor of Industrial and Systems Engineering at the University of Florida. Additionally, he is the Paul and Heidi Brown Preeminent Professor in Industrial & Systems Engineering. He is also an affiliated faculty member of the Computer and Information Science Department, the Hellenic Studies Center, and the Biomedical Engineering Program. He is also the Director of the Center for Applied Optimization. Dr. Pardalos is a world leading expert in global and combinatorial optimization. His recent research interests include network design problems, optimization in telecommunications, e-commerce, data mining, biomedical applications, and massive computing.

Giovanni Quaranta received his PhD in Agricultural Economics and Policy from the University of Naples Federico II, in 1994. From 1994 the University of Basilicata, Italy, has hired him as permanent researcher till 2002, and as Associate Professor from then on. Currently he teaches Natural Resources Economics and Ecological Economics. His is also the chief of Business Economics and Natural and Cultural Resources Economics by the same University. He is author of numerous papers published by international journals, mainly focusing on Natural resources management and policies.

Rosanna Salvia received her PhD in Agro Food and Natural Resources Economics from the University of Naples Parthenope, in 2001. From 2002 the University of Basilicata, Italy, has hired her as appointed researcher, where currently she teaches Economic Development and Natural Resources and International Trade. She is author of numerous papers published by international journals, mainly focusing on socio-economic aspects of land degradation and rural development.

Georgios Tsaples is a PhD student in the Business Administration Department at the University of Macedonia., Greece working on Multicriteria Decision Analysis. He received his MSc in Engineering and Policy Analysis from Delft University of Technology, where he was introduced to simulation and decision analysis. He has worked as a research fellow at the Center for Cyber Intelligence and Information Security of the University of Sapienza in Rome, where he was part of the System Dynamics Group and worked on two European projects related to infrastructure security (CRISADMIN project HOME/2011/CIPS/AG/4000002116 and ATTACS project HOME/2013/CIPS/AG/4000005083). In parallel he continues his research on topics such as: population dynamics, urban development and methodological techniques (exploratory modelling and analysis, integration of decision-making techniques, etc.).

Dimitrios C. Tselempis is post-doctoral research associate at the Department of Agricultural Technology–Agricultural Economics, Alexander Technological Educational Institute of Thessaloniki. He has received his doctoral degree in Agricultural Economics from Aristotle University of Thessaloniki after studying the effect of public policy on the implementation of quality certification systems in the fruits and vegetables supply chain. He has worked as a quality and phytosanitary auditor for more than 15 years and has participated in several research projects, in collaboration with several Universities and Institutes.

Loukas K. Tsironis is an Assistant of Operations Management and a member of the Business Excellence Laboratory (BEL) at the Department of Business Administration of the University of Macedonia. He received his Ph.D. (2001) from Technical University of Crete, Department of Production Engineering & Management. His research interests extended in the Operations and Supply Chain Management, Total Quality Management, Business Process Modeling and Management, on which he recently published several articles in journals and referred conferences.

Denis Veliu has a Ph.D. on Applied Mathematics in Finance from "La Sapienza" University of Rome. Now, he is Head of Department of Software Engineering of Canadian Institute of Technology, Tirana, Albania.

George Vlontzos is an Agricultural Economics Assistant Professor at the University of Thessaly, Greece. His research interests focus on operational and environmental efficiency of agriculture.

Geoffrey Wall is Professor of Geography and Environmental Management at University of Waterloo, Canada. He is interested in the implications of tourism of different types for destination areas with different characteristics and the planning and management implications of such information. Much of his research has been conducted in Asia, especially China, including Taiwan, and Indonesia. He has written a number of books on tourism, most recently: Planning for Indigenous Tourism by Ashgate Publishing Limited in 2006. In 2011, he was given the Award for Scholarly Distinction in Geography of the Canadian Association of Geographers. He is an Honorary Professor of Nanjing University and Dalian University of Technology and holds a Friendship Awards from Hainan, Xinjiang Uygur Autonomous Region and the Peoples Republic of China. He is a founder member and past President of the International Academy for the Study of Tourism.

Sarina Yusuf received her MSc. Communication Technology certificate from Universiti Putra Malaysia in 2015. She is currently pursuing her PhD programme in Youth Studies, focusing on youth's online aggressive behaviour studies at the Institute of Social Sciences (IPSAS) of the university. She has conducted several research studies and authored some researcher articles and book chapters on inmate rehabilitation programmes, cyberbullying, children's online safety, social media and ICT for children development.

Index

S

T

U

V

W

www.ingramcontent.com/pod-product-compliance
Lightning Source LLC
LaVergne TN
LVHW060406300926
849162LV00006B/1355